PLANT BREEDING AND GENETICS

Plant Breeding and Genetics

Edited by **Raimund Scheinberg**

ISBN: 9781681179438
Library of Congress Control Number: 2017949081

SCITUS Academics LLC,
616, Corporate Way, Suite 2, 4766,
Valley Cottage, NY 10989

www.scitusacademics.com

Printed in United States of America.

Preface

The evolution from hunting and gathering to farming happened about 10,000 years ago, independently and diffusely in several places in the world. Plant breeding has been part of agriculture since its beginning. Plant breeding is a critical tool in the fight for food security and responsible environmental stewardship in the 21st century. For more than one hundred years, Plant Breeding and Genetics has been widely recognized for developing novel breeding methodologies and discovering economically important genes and varieties. In fact, it is difficult to say whether agriculture influenced plant breeding or vice-versa. Most probably, both have evolved together towards enhancing the quality and yield of cultivated crops. From the beginning of agriculture until today, plant breeding has undergone many changes but even more changes are likely to occur in the future. During the last 50 years plant breeding has entered a molecular era based on molecular tools to analyze DNA, RNA and proteins and associate such molecular results with plant phenotype. Breeding better cultivars has become a highly efficient way to improve plant production for yield, quality and reduced input. Still plant breeders, scientists as well as society have ample interest in widespread public understanding of the use of new as well as old technologies for improvement of our cultivated plants. This is not least to avoid future communication problems with the general public like experienced with genetically modified plants during recent years.

This volume 'Plant Breeding and Genetics' aims to present some of the recent advances of 21st century plant breeding, exemplifying novel views, approaches, research efforts, achievements, challenges and perspectives in breeding of some crop species. The book chapters have presented the latest advances and comprehensive information on selected topics that will enhance the reader's knowledge of contemporary plant breeding. It also provides some updated discussions on current advances, challenges, and future perspectives of plant genome studies and applications. The book should prove useful to students, researchers, and experts in the area of conservation biology, genetic diversity, and molecular biology.

Table of Contents

CHAPTER 1

Recent Advances in Fruit Species Transformation

Hülya Akdemir[1], Jorge Gago[2], Pedro Pablo Gallego[2] and Yelda[a] - Ozden Çiftçi[1]

[1] *Gebze Institute of Technology, Department of Molecular Biology and Genetics,Plant Biotechnology Laboratory, Kocaeli, Turkey*

[2] *Applied Plant and Soil Biology, Faculty of Biology,University of Vigo, Vigo, Spain*

1. INTRODUCTION

Rapid increase of human population together with global climate variability resulted in increased demand of plant based food and energy sources (Varshney et al., 2011). Fruits and nuts have essential role to enhance quality of humankind life since a diet based on cereal grains, root and tuber crops, and legumes is generally lacked a wide range of products such as fiber, vitamin, provitamins or other micronutrients and compounds exist in fruit and nut species (Heslop-Harrison, 2005). According to last FAOSTAT statistics, totally about 594.5 million t fruit crops (except melons) were produced in the world in 2009 (http://faostat.fao.org). Because an increase demand exists in global food production, many economically important fruit crops production need to be improved, however, conventional breeding is still limited due to genetic restrictions (high heterozygosity and polyploidy), long juvenile periods, self-incompatibility, resources restricted to parental genome and exposed to sexual combination (Akhond & Machray, 2009; Malnoy et al., 2010; Petri et al., 2011). Thus, there is an urgent need for the biotechnology-assisted crop improvement, which ultimately aimed to obtain novel plant traits (Petri & Burgos, 2005).

Plant genetic engineering has opened new avenues to modify crops, and provided new solutions to solve specific needs (Rao et al., 2009). Contrary to conventional plant breeding, this technology can integrate foreign DNA into different plant cells to produce transgenic plants with new desirable traits (Chilton et al., 1977; Newell, 2000). These biotechnological approaches are a great option to improve fruit genotypes with significant commercial properties such as increased biotic (resistance to disease of virus, fungi, pests and bacteria) (Ghorbel et al., 2001; Fagoaga et al., 2001; Fagoaga et al., 2006; Fagoaga et al., 2007) or abiotic (temperature, salinity, light, drought) stress tolerances (Fu et al., 2011); nutrition; yield and quality (delayed fruit ripening and longer shelf life) and to use as bioreactor to produce proteins, edible vaccines and biodegradable plastics (Khandelwal et al., 2011).

Currently, public concerns and reduced market acceptance of transgenic crops have promoted the development of alternative marker free system technology as a research priority, to avoid the use of genes without any purpose after the transformation protocol as selectable and reporter marker genes. Typically, it is employed for the selection strategy that confers resistance to antibiotics and to herbicides (Miki & McHugh, 2004; Manimaran et al., 2011). A large proportion of European consumers considered genetically modified crops as highly potential risks for human health and the environment. European laws are restrictive and do not allow the deliberate release of plant modified organism (Directive 2001/18/EEC of the European Parliament and the Council of the European Union). Under these premises, great efforts have also been realized to develop alternative marker free technologies in fruit species. Recently, it was demonstrated in apple and in plum, that transgenic plants without marker genes can be recovered and confirmed its stability by molecular analysis (Malnoy et al., 2010; Petri et al., 2011). In 2011, for first time it was described authentically "cisgenic" plants in apple cv. Gala (Schouten et al., 2006a,b; Vanblaere et al., 2011).

Efficient regeneration systems for the generation of transgenic tissues still appear as an important bottleneck for most of the species and cultivars. In the literature, different protocols were described to transform fruit cells using various DNA delivery techniques, however the attempts generally focused on transformation via *Agrobacterium* or microprojectile bombardment. In this chapter, a detailed application of these techniques in fruit transformation is summarized together with usage of proper marker and selection systems and *in vitro* culture techniques for regeneration of the transgenic plants.

2. TECHNIQUES USED TO TRANSFORM FRUIT SPECIES

Improvement of the plant characteristics by transfer of selected genes into fruit plant cells is possible mainly through two principal methods: *Agrobacterium*-mediated transformation and microprojectile bombardment (also called "biolistic" or "bioballistic"). Soil-borne Gram negative bacteria of the genus *Agrobacterium* infect a wound surface of the plants via a plasmid called Ti-plasmid containing three genetically important elements; *Agrobacterium* chromosomal virulence genes (*chv*), T-DNA (transfer DNA) and Ti plasmid virulence genes (*vir*) that constitute the T-DNA transfer machinery. Since Ti plasmid encodes mechanisms of integration of T-DNA into the host genome, it is used as a vector to transform plants.

Since direct gene transfer procedures involve intact cells and tissues as targets, in some species breaching of the cell wall is needed in order to enable entrance of DNA to cell (Petolino, 2002). This is accomplished by making some degree of cell injury or totally enzymatic degradation of the cell wall. Advantages of microprojectile bombardment can be summarized as i) transfer of multiple DNA fragments and plasmids with co-bombardment, ii) unnecessity pathogen (such as *Agrobacterium*) infection and usage of specialized vectors for DNA transfer (Veluthambi et al., 2003). Although microprojectile bombardment eliminates species-dependent and complex interaction between bacterium and host genome, stable integration is lower in this technique in comparison to *Agrobacterium*-mediated transformation (Christou, 1992). Moreover, the existence of truncated and rearranged transgene DNA can also lead transgene silencing in the transgenic plants (Pawlowski & Somers, 1996; Klein & Jones, 1999; Paszkowski & Witham, 2001). On the other hand, other important requirement for this technique is that the explants or target cells have to be physically available for the bombardment (Hensel et al., 2011). Also, it was described that transgenic explants regenerated can be chimeric (Sanford et al., 1990). Nevertheless, application of both of the techniques for the transfer of foreign DNA results in "transient" or "stable" expression of the DNA fragment. In the following sections, recent advances in genetic transformation of fruit species via *Agrobacterium*-mediated and direct gene transfer techniques are presented.

2.1. *Agrobacterium*-Mediated Gene Transfer

2.1.1. A complex relationship

In the decade of the 80´s, the first reports were published related to the introduction of foreign DNA in plant genome thanks to the Ti plasmid of *Agrobacterium tumefaciens* (De Block et al., 1984; Horsch et al., 1985). After more than 25 years, *Agrobacterium*-mediated gene transfer is still the most used method for fruit species transformation including apple, almond, banana, orange, grapevine, melon etc. (Table 1; Rao et al., 2009).

Plant transformation by using *Agrobacterium* has some advantages since the technique is relatively simple; transfer and integration of foreign DNA sequences with defined ends (left and right borders of T-DNA) is precise; stable transformation is high; transgene silencing is typically low and long T-DNA sequences (>150 kb) can be transferred (Veluthambi et al., 2003). However, it is still far from to be a routine transformation application in plants because of its host-range restrictions (Gelvin, 1998).

The initial drawback of *Agrobacterium*-mediated transformation method is the host-range restrictions. However, the bacterium and the target tissue can be manipulated to overcome this obstacle (Trick & Finer, 1997). These authors proposed a new approach to facilitate *Agrobacterium* penetration into plant tissues, the sonication assisted *Agrobacterium*-mediated transformation (SAAT) method. This method consists the use of ultrasounds to produce cavitations on and below the plant surfaces and into the membrane cells, wounding the tissues to enhance *Agrobacterium* infection (Trick & Finer, 1997, 1998).. Also, this method can be combined with vacuum infiltration to promote bacteria agglutination around the tissues to increase the *Agrobacterium* infection as it was demonstrated in kidney bean (Liu et al., 2005) and in woody plants as *Eucalyptus* (Villar et al., 1999; Gallego et al., 2002; Gallego et al., 2009). Today, application of these modifications solely or in combination with other approaches has made it possible to transfer foreign DNA via *Agrobacterium* even to monocots (Hiei et al., 1994; Ishida et al., 1996; Hensel et al., 2011) which were initially transformed with direct gene transfer methods since *Agrobacterium* is not a natural host. Following its first successful usage in soybean and Ohio buckeye (Trick & Finer, 1997), SAAT was also applied recently to fruit species including orange (Oliveira et al., 2009); banana (Subramanyam et al., 2011) and grapevine (Gago et al., 2011). In the last paper the developed efficient methodology that combined SAAT with vacuum infiltration allowed to obtain reporter gene expression in different newly formed organs such as stems, petioles and leaves. Expression was related to vascular tissues due to the *EgCCR* promoter of *Eucalyptus gunnii* and demonstrated that its activity is conserved and fully functional in grapevine as it was shown by *uidA* (GUS) and *gfp* reporter marker genes. Transgenic grapevine lines were verified by Southern blot analysis in five randomly chosen transgenic lines showing simple integration patterns in four lines with different band length indicating independent transformation events into the grapevine genome. We also applied the optimized protocol to pistachio nodes to reveal out if this method of transformation and vascular-specific promoter of eucalyptus, also works in this species. Histological observations of GFP activity presence in vascular bundles and leaves (Fig. 1) together with PCR amplification of 858 bp fragment of *nptII* and 326 bp of *uidA* (Fig. 2) genes confirmed not only gene integration but also showed that SAAT in combination with vacuum infiltration and vascular specific promoter could also be used for pistachio transformation. With PCR amplification, four out of five GFP+ putative transgenic shoots showed the amplified bands of *nptII* and *uid*A genes (Fig 2).

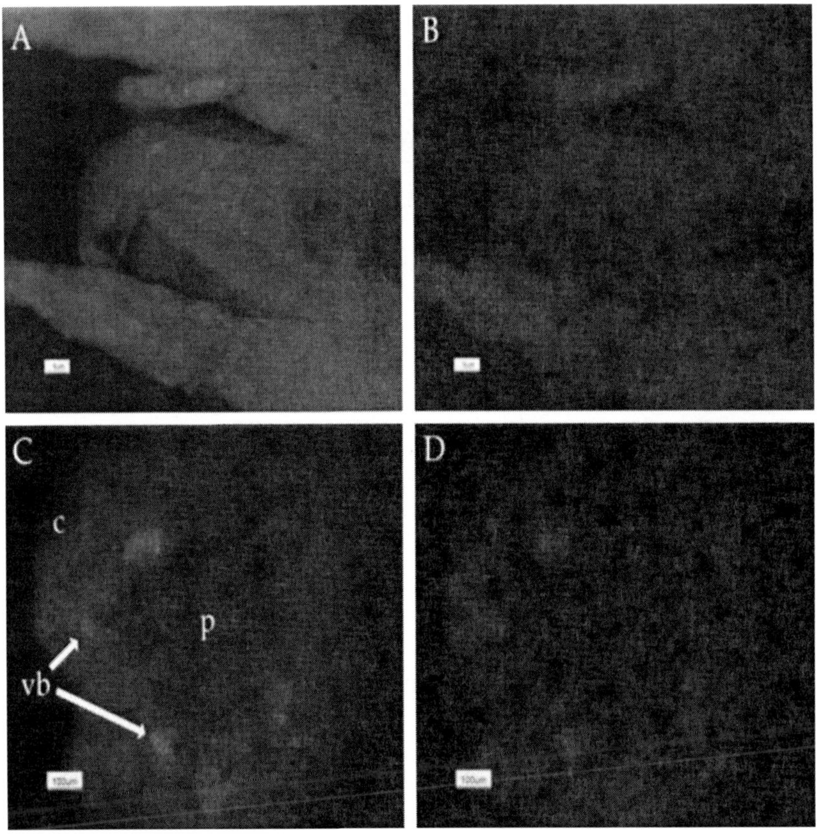

FIGURE 1. Expression of GFP in pistachio transformed with *EgCCR-GFP-GUS* construct. Fluorescent images of different tissues and organs were taken 6 months after *Agrobacterium*-mediated transformation. GFP fluorescence in shoot apex (A-B, bars represent 5μm) and transverse section of the transgenic microshoot (C-D, bars represents 100μm) were carried out using a 480/40 nm exciter filters, and two-barrier filter >510 nm (wide range) and 535/550 nm (specific filter for GFP fluorescence). (Abbreviations: vascular bundles (vb), pith, p; cortex, c).

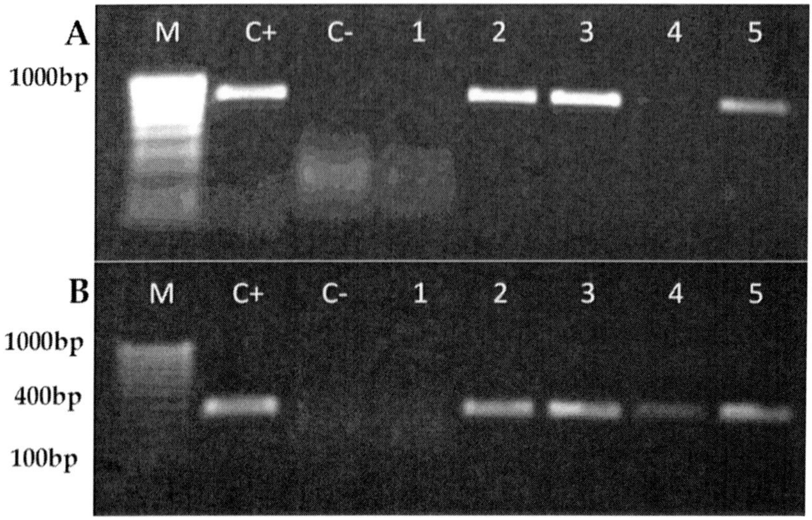

FIGURE 2. Analysis of putative transgenic pistachio plants by PCR amplification using primers designed for 858 bp fragment of the *nptII* gene (A) and for a fragment of 326 bp of *uid*A gene (B). (M: DNA 1000 bp ladder, C+: positive control, C-: untransformed plant, T1-T5 putative transgenic shoots lines).

Agrobacterium-mediated transformation is highly genotype dependent for many plants (Pérez-Piñeiro, 2012) but also for fruit species. Different reports described that some cultivars were completely found to be as highly recalcitrant for transformation process whereas others are completely successful (Galun & Breiman, 1998; Petri & Burgos, 2005). This problem is widely described in different fruit species as apricot, grapevine and others (López-Pérez et al., 2008; Wang, 2011). López-Pérez and collaborators (2009) described that grapevine cultivars "Crimson Seedless" and "Sugraone" obtained different transformation efficiencies depending on the optical densities tested. Transformation of hypocotyls obtained from germination of mature seeds and nodal explants of apricot cultivars Dorada, Moniquí, Helena, Canino, Rojo Pasión and Lorna resulted in different transformation efficiencies (Wang, 2011). Some authors pointed out that one of the main goals of plant genetic engineering must be the development of genotype-independent transformation procedures, however due to this highly complex plant-pathogen interaction it will be very difficult to achieve this with the currently available technologies (Petri & Burgos, 2005).

TABLE 1. Some important reports on genetic transformation of fruit species via A. tumefaciens or A. rhizogenes.

Species	Aim	Plasmid	Transgenes	References
Apple				
Malus x domestica	Method optimisation	pBIN6	*nptII, nos*	James et al., 1989
M. x domestica	Investigation of early events in transformation	pDM96.0501	*sgfp, gusA,nptII*	Maximova et al., 1998
M. x domestica	Investigation of influence of rolA gene on shoot growth	pMRK10	*rolA, nptII*	Holefors et al., 1998
M. x domestica	Scab resistance	p35S-ThEn42, pBIN19ESR	*ech42, nptII*	Bolar et al., 2000
M. x domestica	Resistance to fireblight	pLDB15	*attE, nptII, gusA*	Ko et al., 2000
M. x domestica	Improve rooting ability	pCMB-B	*rolB, nptII, gusA*	Zhu et al., 2001
M. x domestica	Scab resistance	pBIN(Endo+Nag)	*ech42, nag70, nptII*	Faize et al., 2003
M. x domestica	Self-fertility	pGPTV-KAN	*S_3RNase, nptII*	Broothaerts et al., 2004
M. x domestica	Method optimisation Enhance rooting	pCMB-B	*rolB, nptII, gusA*	Radchuk & Korkhovoy, 2005
M. x domestica	Method optimisation	pNOV2819	*pmi, nptII, gusA*	Degenhardt et al., 2006
M. x domestica	Investigation of function of ARRO-1 in adventitious rooting	pK7GWIWG2 (II)	*ARRO-1, nptII*	Smolka et al., 2009
M. x domestica	Stability of scab resistance	pMOG402.hth.gus.intron	*Hth, nptII, gusA*	Krens et al., 2011
M. x domestica	Development of selection system	pCAMBIAVr-ERE-GUS	*VrERE, gusA*	Chevreau et al., 2011
M. x domestica	Transformation without selectable marker gene	pPin2Att.35SGUSint+.n pPin2MpNPR1.GUS−.n ptII		Malnoy et al., 2010
Almond				
Prunus dulcis	Method optimisation	pBI121mgfp-5-ER pNOV2819	*nptII pmi*	Ramesh et al., 2006
Avocado				
Persea americana Mill.	Method optimisation	pMON9749, pTiT37-SE	*nptII, gusA*	Cruz-Hernandez et al., 1998

Banana				
Musa spp.	Method optimisation (Agro + SAAT+ Vacuum infiltration)	pCAMBIA1301	*hptII, gusA*	Subramanyam et al., 2011
Musa spp.	Resistance to Fusarium wilt	pBI121-PFLP	*pflp, nptII*	Yip et al., 2011
Blueberry				
Vaccinium spp.	Method optimisation	p35SGUS-int	*gusA*	Cao et al., 1998

Species	Aim	Plasmid	Transgenes	References
Blueberry				
V. corymbosum L.	Method optimisation	pBISN1	*nptII, gusA*	Song & Sink, 2004
Grapevine				
Vitis vinifera	Method optimisation	Nr	*gusA, nptII*	Nakano et al., 1994; Gago et al., 2011
V. rootstocks	Resistance to viruses and crown gall	pBIN19 pGA482G	*mutant virE2, nptII GLRaV-3cp*	Xue et al., 1999
V. vinifera	Resistance to fungal pathogens	pBI121	*nptII rice chitinase gene*	Yamamoto et al., 2000
V. vinifera	Fungal resistance	pGJ42	*chitinase, rip, nptII*	Bornhoff et al., 2005
V. vinifera	Method optimisation	pGA643	*nptII, GFLVCP*	Maghuly et al., 2006
V. vinifera	Method optimisation	Nr	*egfp, nptII*	Dutt et al., 2007
V. vinifera	Resistance to powdery mildew	pGL2	*ricechitinase gene, hgt*	Nirala et al., 2010
V. vinifera	Method optimisation	pBin19-sgfp	*nptII, sgfp*	Pérez-López et al., 2008
V. vinifera	Method optimisation	pSGN	*nptII, egfp*	Li et al., 2006
V. vinifera	Method optimisation	pCAMBIA2301	*nptII, gusA*	Wang et al., 2005
Grapefruit				
Citrus paradisi	Resistance to Citrus tristeza virus	pGA482GG	*CP, RdRp, gusA, nptII*	Febres et al., 2003
C. paradisi	Resistance to Citrus tristeza virus	pGA482GG	*CP, gusA, nptII*	Febres et al., 2008
Kiwifruit				
Actinidia spp.	Hairy root induction	A722,C58, ICMP8302, ICMP8326, ID1576, LBA 4404, A4T	*gusA, nptII*	Atkinson et al., 1990
Actinidia spp.	Method optimisation	pLAN411, pLAN421	*gusA, nptII*	Uematsu et al., 1991
A.deliciosa	Improved rooting	pBIN19	*nptII, rol A,B,C*	Rugini et al., 1991
A. eriantha	Method optimisation	pART27-10	*gusA, nptII*	Wang et al., 2006
A. deliciosa	Manipulation of plant architecture	pBI121	*ipt*	Honda et al., 2011

Species	Aim	Plasmid	Transgenes	References
Mango				
Magnifera indica L.	Method optimisation	pTiT37-SE::pMON9749	*nptII, gusA*	Mathews et al., 1992
M.indica L.	Methodoptimisation	pGV3850::1103	*nptII*	Mathews et al., 1993
M.indica L.	Mediate ethylene biosynthesis	pBI121	*nptII, gusA antisense ACC oxidase, antisense ACC synthase*	Cruz Hernandez et al., 1997
M. indica L.	Rooting enhancement	Nr	*rol B*	Chavarri et al., 2010
Melon				
Cucumis melo	Resistance to ZYMV, TEV, PVY	FLCP core AS	*nptII, ZYMV coatpr.*	Fang & Grumet, 1993
C. melo	Salt resistance	pRS655	*nptII, gusA, hal1*	Bordas et al., 1997
C. melo	Resistance to ZYMV	pBI-ZCP3'UTR	*nptII, ZYMV coatpr.*	Wu et al., 2009
Nectarberry				
Rubus arcticus	Method optimisation	pFAJ3001	*gusA*	Kokko & Kärenlampi, 1998
Orange				
Citrus sinensis	Method optimisation (Agro + SAAT+ vacuum infiltration)	pGA482GG	*gusA, nptII*	Oliveira et al., 2009
C. sinensis	Research on expression of Mt-GFP	pBI. mgfp4.coxIV	*Mt-gfp*	Xu et al., 2011
C. sinensis	Influence of methylation on gene expression	pBIN.mgfp5-ER	*gfp, nptII*	Fan et al., 2011
C. sinensis	Modification of gibberellin levels	pBinJIT-CcGA20ox1-sense pBinJITCcGA20ox1-antisense	*nptII, CcGA20ox1 nptII, CcGA20ox1*	Fagoaga et al., 2007
C. sinensis	Resistance to fungi	pBI121.*P23*	*nptII, PR-5*	Fagoaga et al., 2001
C. aurantifolia	Resistance to virus	pBin19-sgfp	*nptII, sgfp, p23*	Fagoaga et al., 2006
Poncirus trifoliate	Enhanced salt tolerance	pBin438	*nptII, AhBADH*	Fu et al., 2011
Papaya				
Carica papaya	Resistance to PRSV	pRPTW	*PRSV replicase gene, neo*	Chen et al., 2001

Species	Aim	Plasmid	Transgenes	References
Pear				
Pyrus communis	Alter growth habit	pGA-GUSGF	*rolC, gusA, nptII*	Bell et al., 1999
P. communis	Method optimisation	pPZP pME504	*gusA, nptII*	Yancheva et al., 2006
P. communis	Method optimisation	PBISPG	*nptII, gusA*	Sun et al., 2011
Peanut				
Arachis hypogaea	Production of edible vaccines for *Helicobacter pylori*	pBI121.Oleosin-UreB	*ureB, nptII*	Yang et al., 2011
A. hypogaea	Improvement of salt and drought resistance	pGNFA-(pAHC17)	*AtNHX1*	Asif et al., 2011
A. hypogaea	Production of vaccines for Peste des petits ruminants (PPR)	pBI121	*Hn*	Khandelwal et al., 2011
Plum				
Prunus armeniaca	Method optimisation	pBIN19-sgfp	*nptII, gfp*	Petri et al., 2004
P. armeniaca	Method optimisation	pBIN19-sgfp, p35SGUSint	*nptII, gfp/nptII, gusA*	Petri et al., 2008
P. domestica	Transformation of marker free plants	pCAMBIAgfp94(35S) / pGA482GGi ihpRNAE10'	*nptII, gfp, gusA, ppv-cp*	Petri et al., 2011
P. domestica	New selection system with hygromycin	pC1381, pC1301, pC2301	*gusA, hpt, nptII*	Tian et al., 2009
P. domestica	Control of PPV infection	pGA482GG	*nptII, gusA, PRVcp*	Scorza et al., 1995
P. salicina	Method optimisation	pCAMBIA2202	*nptII, gfp*	Urtubia et al., 2008
Pomegranate				
Punica granatum	Method optimisation	pBIN19-sgfp	*nptII, gfp*	Terakami et al., 2007
Strawberry				
Fragaria spp.	Method optimisation	pBI121	*nptII, gusA*	Barcelo et al., 1998
Fragaria x ananassa Duch.	Modulation of fruit softening	pBI121	antisense of *endo-β-1,4-glucanase*	Lee & Kim, 2011
White mulberry				
Morus alba	Method optimisation	pBI121	*nptII, gusA*	Agarwal & Kanwar, 2007

Some abbreviations: AtNHX1: a vacuolar type Na^+/H^+ antiporter gene; gfp: green fluorescent protein coding gene; hal1: yeast salt tolerance gene; hpt: hygromycin phosphotransferase coding gene; ipt: isopentyl transferase gene; neo: neomycin phosphorate transferase coding gene; nos, nopaline synthase coding gene; nptII, neomycin phosphtransferase II coding gene; pmi: phosphomannose isomerase coding gene; ppv: Plum pox virus; prsv: papaya ringspot virus; pvy: potato virus Y; tev: tobacco etch virus; gusA (uidA): β-glucuronidase coding gene; UreB: antigen gene; zymv: zucchini yellow mosaic virus.

2.2. Direct Gene Transfer

Direct gene transfer techniques include microprojectile bombardment, microinjection, electroporation, and usage of whiskers. Among them, microprojectile bombardment is an alternative technique of Agrobacterium-mediated transformation since its physical nature overcomes biological barriers and enables naked DNA delivery directly into host genome or alternatively into mitochondria and chloroplasts. In this technique, plasmid or linearized DNA-coated metal microparticles (gold or tungsten) at high velocity is bombarded to intact cells or tissues (Sanford et al. 1987; Klein et al. 1987; Sanford, 1988). Furthermore, biological projectiles such as bacteria (i.e., E. coli, Agrobacterium), yeast and phage associated with tungsten can also be used in microprojectile bombardment (Bidney, 1999; Kikkert et al. 1999).

Microprojectile bombardment was developed in the 1980s for transformation of plants which were recalcitrant to Agrobacterium–mediated transformation (Paszkowski et al., 1984) such as agronomically important cereals. Following the development of the first particle delivery system (Sanford et al. 1987; Sanford 1988), different effective devices such as PDS-1000/He, Biolistic® particle delivery system; particle inflow gun; electrical discharge particle acceleration; ACCELL™ technology and microtargeting bombardment device were also evolved to improve transformation capacity. Among them, PDS-1000/He, Biolistic® particle delivery system (BIO-RAD), which is a modified version of Sanford's system, is the most used system for biolistic transformation due to its efficient and relatively simple application and acquisition of reproducible results between laboratories (Taylor & Fauquet, 2002). Particle inflow gun can be an alternative to other biolistic systems due to its very low cost and it was used successfully in banana transformation (Becker et al., 2000). Electrical discharge particle acceleration, ACCELL™ technology uses high voltage electrical discharge into a droplet water to generate shock waves and project microprojectiles to different cell layers of target tissues (McCabe & Christou, 1993). Microtargeting bombardment device was designed for shoot meristem transformations (Sautter, 1993) but it is not widely used for plant transformation. All of the microprojectile bombardment systems are not depend on any plant cell type but target cells which will be bombarded need to be physically accessible (Hensel et al., 2011).

Particle bombardment were carried out not only to optimize plant transformation but also to transfer gene constructs encoding for various antimicrobial peptides or proteins for fungal resistance against to Fusarium oxysporum f. sp. cubense and Mycospaerella fijiensis or preharvest and postharvest diseases Verticillium theobromae or Trachysphaera fructigen (i.e., Remy et al., 2000; Sagi et al., 1998; Tripathi, 2003), virus (i.e., Fitch et al., 1992; Tennant et al., 1994; Gonsalves et al., 1994; Scorza et al., 1996), pest (i.e., Serres et al., 1992) and herbicide tolerance (i.e., Zeldin et al., 2002). This technique has been applied to transformation of various fruit species including banana, cranberries, citrus, grapevine, melon, papaya and peanut (Table 2).

TABLE 2. Some important reports on genetic transformation of fruit species via direct gene transfer.

Species	Aim	Transfer system	Plasmid	Transgenes	References
Apple					
Malus x domestica	Method optimisation	PEG-mediated	pKR10	*Gfp*	Maddumage et al., 2002
Banana					
Musa spp.	Method optimisation	Particle bombardment	pUbi-BtintORF1 pBT6.3-Ubi-NPT pUbi-BTutORF5 pBT6.3-Ubi-NPT pUGR73 pDHkan	*nptII, gusA, BBTV*	Becker et al., 2000
Musa spp.	Tolerance to Sigatoka leaf spot	Particle bombardment	pYC39	*ThEn-42, StSy, Cu, Zn-SOD*	Vishnevetsky et al., 2011
Musa spp.	Resistance to virus	Particle bombardment	pAB6, pAHC17,pH1	*gusA, bar,ubi, BBTV-G-cp*	Ismail et al., 2011
Cranberry					
Vaccinim macrocarpon	Method optimisation & Pest control	Particle bombardment	pTvBTGUS	*nptII, gusA, Bt*	Serres et al., 1992
V. macrocarpon	Herbicide resistance	Particle bombardment	pUC19	*bar, aphII*	Zeldin et al., 2002
Grapevine					
Vitis vinifera	Method optimisation	Biolistic	pBI426	*nptII, gusA*	Hebert et al., 1993
V. vinifera	Method optimisation	Particle bombardment & Agro	pGA482GG	*nptII, gusA TomRSV-CP*	Scorza et al., 1996
V. vinifera	Method optimisation	Biolistic	pSAN237	*nptII, magainin,PGL*	Vidal et al., 2003
V. vinifera	Comparison of minimal cassette with standard circular plasmids	Biolistic	pSAN168, pSAN237	*Magainin, nptII*	Vidal et al., 2006
Kiwifruit					
Actinidia spp.	Method optimisation	PEG 4000	pDW2	*Cat*	Oliveira et al., 1991
Actinidia spp.	Method optimisation	Electroporation	pBI121, pTi35SGUS	*gusA, nptII*	Oliveira et. al., 1994
A. deliciosa	Method optimisation	PEG 4000	p35SGUS	*gusA*	Raquel & Oliveira, 1996

Species	Aim	Transfer system	Plasmid	Transgenes	References
Melon					
Cucumis melo	Protection against	Particle bombardment Infection & Agro	pGA4822GG/ CP	*nptII, gusA, CMV-WLCP*	Gonsalves et al., 1994
Papaya					
Carica papaya	PRV resistance	Particle bombardment	pGA482GG	*PRV, nptII*	Fitch et al., 1992
C. papaya	PRV resistance	Particle bombardment	pGA482GG	*nptII, gusA, cpPRVHA*	Tennant et al., 1994
C. papaya	Control of PRSV	Particle bombardment	pGA482GG	*cpPRSV-pHA5, nptII, gusA*	Cai et al., 1999
C. papaya	Method optimisation	Particle bombardment	pCAMBIA1303 pML202	*hpt, nptII, mgfp5'*	Zhu et al., 2004
C. papaya	Use of PMI/Man	Particle bombardment	pNOV3610	*Pmi*	Zhu et al., 2005

A successful protocol was studied very recently in banana cv. Williams apical meristems with microprojectile bombardment of a new construct pRHA2 plasmid containing bar and coat protein of banana bunchy top nanovirus (BBTV-cp) genes that encoded the viral coat protein by using Biolistic™ PDS-1000/He system, 650 psi helium pressures and 5 µg DNA/shot for acquisition of virus resistance (Ismail et al., 2011). After bombardment, 62% of apical meristems were survived on the selective medium and 80% of explants produced shoots in the following first subculture and all shoots were rooted (Ismail et al., 2011). In addition to those disease-based studies, others were also carried out in order to develop efficient transformation protocols via biolistic transformation (Sagi et al., 1995; Becker et al., 2000). Among them, Sagi and co-workers (1995) reported the transformation of embryogenic cell suspensions of cooking banana 'Bluggoe' (ABB genome) and plantain 'Three Hand Planty' (AAB genome) via particle bombardment. Then, Cavendish banana cv. Grand Nain embryogenic suspension cells were co-bombarded with the plasmid containing nptII selectable marker gene under the control of BBTV promoter or the cauliflower mozaic virus (CaMV) 35S promoter, the β-glucuronidase (gusA) reporter gene and BBTV genes under the control of the maize polyubiquitin promoter by using particle inflow gun and stably integration was obtained in all of the tested transformed plants (Becker et al., 2000). Very recently, microprojectile bombardment was also applied to induce tolerance to Sigatoka leaf spot caused by Mycosphaerella fijiensis in banana by transferring endochitinase gene of ThEn-42 from Trichoderma harzianum together with the grape stilbene synthase gene (StSy) under the control of 35 S promoter and the inducible PR-10 promoter, respectively (Vishenevetsky et al., 2011). Moreover, in order to improve scavenging of free radicals generated during fungal attack, the superoxide dismutase gene

(Cu, Zn-SOD) of tomato was also included to this gene cassette under the control of ubiquitin promoter. Both PCR and Southern blot analysis confirmed the stable integration of the transgenes and 4-year field trial showed that several transgenic banana lines had improved tolerance not only to Sigatoka but also other fungus such as Botrytis cinerea. Gene transfer via microprojectile bombardment was also carried out in American cranberry (Vaccinium macrocarpon) firstly to increase productivity by transferring Bacillus thuringiensis subsp. Kurstaki crystal protein gene (Bt) for pest resistance (Serres et al., 1992), and latter on, by bar gene to confer tolerance to the phosphinothricin-based herbicide glufosinade (Zeldin et al., 2002). Although preliminary bioassays for efficiency of the Bt gene against an important lepidopteran demonstrated no consistently effective control in former, stable transmission and expression of herbicide tolerance was observed in both inbred and outcrossed progeny of cranberry trans clone in latter.

In tangelo (Citrus reticulata Blanco × C. paradisi Macf.) cv. Page embryogenic suspension cells were bombarded with tungsten coated plasmid containing gusA and nptII genes (Yao et al., 1996). Following to bombardment, 600 transient and 15 stable transformants were obtained and integration of the interest genes confirmed by PCR and Southern blot analyses. A large of kanamycin-resistant embryogenic calli showed also GUS activity. In another study, Kayim and associates (1996) bombarded tungsten-coated plasmid (pBI221.2) containing the gusA gene into lemon cv. Kütdiken nucellar cells by biolistic device and expression of the gusA gene was histochemically confirmed.

Feronia limonia L. is important fruit tree because of its edible fruits. It is suitable for cultivation in semi-arid tropics and also can be used for reforestation and wasteland reclamation projects (Sing et al., 1992; Purohit et al., 2007). Feronia limonia L. hypocotyl segments were also bombarded with tungsten-coated plasmid pBI121 having gusA reporter gene driven by CaMV35S promoter and nptII as a selective marker under control of nos promoter using Biolistic™ PDS-1000/He particle delivery system at different rupture disc pressures (1100 and 1350 psi) and target distances (6 and 9 cm) (Purohit et al., 2007). This study revealed that 1100 psi/6 cm and 1350 psi/9 cm were the optimal bombardment condition with supplying a maximum 90% of GUS transient expression.

In grapevine, the initial transformation studies via microprojectile bombardment were performed for method optimization with transferring nptII and gus genes as selective and reporter marker genes, respectively (Hebert et al., 1993; Kikkert et al., 1996; Scorza et al., 1996). Later, Vidal and co-workers (2003) studied the efficiency of biolistic cotransformation in grapevine for multiple gene transfer of nptII and antimicrobial genes (magainin and peptidyl-glycine-leucine). The stable transformation was confirmed by gus gene expression, followed by PCR and Southern blot analyses of nptII and antimicrobial genes showed. Three years later, same research group (Vidal et al., 2006) reported the efficient biolistic transformation of grapevine by using minimal gene cassettes, which are linear DNA fragments lacking the vector backbone sequence.

Papaya is economically important and preferred another fruit species because of its nutritional and medicinal properties grown in tropical and subtropical regions (Tripathi et al., 2011). Papaya ringspot virus (PRSV) is major limiting factor in papaya production in Hawaii (Gonsalves, 1998; Fuchs & Gonsalves, 2007). First PRSV resistant papaya plants (cv. SunUp) were obtained by PDS/1000-He particle bombardment device of cv. Sunset with the transformation vector pGA482GG/cpPRV4 containing the prsv coat protein (CP) gene (Fitch et al., 1992). The PRSV resistant papaya has been commercialized, reached to end user and improved papaya is now under production in Hawaii (Tripathi et al., 2008). This study was followed by other reports mainly on improvement of PRSV tolerance in papaya via microprojectile bombardment-based transformation (Tennant et al., 1994; Cai et al., 1999; Guzman-Gonzalez et al., 2006). The deployment of transgenic papayas has showed that virus CP protein supplies durable and stable resistance to homologous strains of PRSV (Fermin et al., 2010). Moreover, no ecological influence of transgenic papayas on adjacent non-transgenic papaya trees, microbial flora and beneficial insects was evident (Sakuanrungsirikul et al., 2005). However, political and social factors have negatively affected the technology in Thailand (Davidson, 2008).

Although there are various wild peanut species having disease resistance traits, hybridization between wild and cultivars is difficult due to self-incompatibility, low frequency of hybrid seed production and linkage drag (Stalker & Simpson, 1995) and because of that genetic transformation is a practical tool to improve disease resistant cultivars. Singsit and associates (1997) transformed peanut somatic embryos with gold-coated plasmid constructs containing both Bacillus thuringiensis cryIA(c) and hph genes driven by CaMV35S promoters by PDS 1000 He biolistic device for resistance lepidopteran insect larvae of lesser cornstalk borer. The embryogenic cell lines showed hygromycin resistance and integration of hph and Bt genes were confirmed by PCR and/or Southern blot analyses in regenerated plants and a progeny. 18% CryIA(c) protein of total soluble protein was detected by ELISA immunoassay in the hygromycin resistant plants. Production of peanut stripe virus (PStV) resistant peanut is another attempt for biotechnologists since the virus negatively affects seed quality and yield in Asia and China (Higgins et al., 1999). Somatic embryos of peanut cv. Gajah and cv. NC-7 were transformed by co-bombardment of hph gene and one of two forms of the PStV coat protein genes and both of the transgenic plants showed high level resistance to the homologous virus isolate (Higgins et al., 2004). Transfer of anti-apoptotic genes originated from mammals, nematods or virus into plants is another approach for enhancement of plant resistance against to biotic and abiotic stresses (Chu et al., 2008a). With this aim, peanut cv. Georgia Green embryogenic callus was bombarded with anti-apoptotic Bcl-xL gene by microprojectile bombardment. Although Bcl-xL protein was detected in four transgenic lines, just one transgenic line (25-4-2a-19) had stable protein expression and showed tolerance to 5μM paraquat (commercial herbicide) (Chu et al., 2008a). Around 0.6% of total population in USA is affected of IgE-mediated allergic reaction following to peanut consumption (Sicherer et al., 2003). To produce hypoallergenic peanut, peanut cv. Georgia Green embryogenic cultures were also transformed via microprojectile bombardment and silenced peanut allergens (Ara h 2 and Ara h 6) by RNA interference. Expression of these allergens was not decreased effectively but, binding of IgE to the two allergens, significantly declined (Chu et al., 2008b).

Apart from microprojectile bombardment, electroporation (Oliviera et al., 1994) and PEG-mediated transformation were also carried out in apple (Maddumage et al., 2002) and kiwifruit (Raquel & Oliveira, 1996) in order to optimize transformation protocol by transferring gusA, gfp and/or nptII.

3. MARKERS AND SELECTION OF TRANSFORMANTS

3.1. Reporter Genes

Reporter genes or non-selectable marker genes are commonly used components of the plasmid constructs allowing the verification of transformation and the detection of the putative transformed cells. In many fruit transformation studies, histochemical analyses of transformed cells are visualized by using β-glucuronidase (GUS) expression as a reporter gene (Jefferson, 1987; Table 1). This enzyme is encoded by E. coli uidA (gusA) gene and histochemical localization of the gene expression is detected in subcellular levels (Daniell et al., 1991). High levels of GUS is not toxic for plant and the enzyme is very stable in cells, however, the assay is destructive to plants (Miki & McHugh, 2004). gusA generally co-transformed with other selective marker genes to enable the selection of transformants. The gene gfp encodes for the protein green fluorescent protein (GFP) (Chalfie et al., 1994). This is one of the mostly used reporter marker gene in fruit transformation protocols for monitoring transformed cells in vivo and in real time just by application of UV-light for the excitation of the fluorescent protein. GFP has not any cytotoxic effect on transformed plant cells (Stewart, 2001; Manimaran et al., 2011). In vivo detection may permit the manual selection of transformed tissues with focusing in the areas where the signal is more brightly. Fusion of GFP with other proteins of interest provides precise visualizing of intracellular localization and transport in transformed plant (Miki & McHugh, 2004; Manimaran et al., 2011). In some fruit species, it is reported that chlorophyll red autofluorescence can mask GFP expression making the detection really difficult or even impossible in species as apricot, peach and plum (Billinton & Knight, 2001; Padilla et al., 2006; Petri et al., 2008; Petri et al., 2011). However, it was described as an efficient reporter gene in other woody fruit plants, such as citrus (Ghorbel et al., 1999) and peach (Pérez-Clemente et al., 2004). These contrary results confirm the highly variability of the reporter gfp gene which is described by Hraška and co-workers (2008). Other reporter gene, luciferase (luc) (Gould & Subramani, 1988) also let the monitorization of the transgene putative cells in living tissues, however, it is not so widely employed as the gfp (van Leeuwen et al., 2000; Miki & McHugh, 2004).

3.2. Selection Systems, A Critical Step

Selection of transformed regenerants is a critical step in any transformation procedure (Burgos & Petri, 2005). Selection systems can be classified as positive or negative, and conditional or non-conditional. Positive selection systems are those that promote the growth of transformed cells and tissues, by the contrary, negative selection systems are those that promote the death of the transgenic cells. Both systems can be conditioned by an external substrate to perform their activity. Currently, negative selection systems are used in combination with positive selection systems to eliminate transformed cells with incorrect molecular programmed excision of the T-DNA (Schaart et al., 2004; Vamblaere et al., 2011). Typically, in positive conditional selection systems the selectable marker gene encodes for an enzyme conferring resistance to some specific toxic substrate that enable the growth of the transformed cell, tissues and inhibiting or killing non transformed tissues (more information in the comprehensive review of Miki & McHugh, 2004). In the literature approximately 50 selection marker genes are described for genetic plant transformation, however, just only three genes of positive conditional selection system (nptII and hpt, resistance to the antibiotics kanamycin and hygromycin, respectively, and bar gene encoding resistance to herbicide phosphinothricin) are commonly employed in more than 90% of the papers (Miki & McHugh, 2004). These three selectable genes are also the most used ones to transform fruit species as it can be seen in Table 1 and Table 2. Escherichia coli nptII gene (also known neo) encoded protein (neomycin phosphotransferase, NPTII) inactives aminoglycoside antibiotics such as kanamycin, neomycin, geneticin (G418), and paramomycin that inhibit protein translation in the transformed cells (Padilla & Burgos, 2010). Hygromycin B is another aminoglycoside antibiotic that inhibits protein synthesis with a broad spectrum activity against prokaryotes and eukaryotes and especially it is very toxic in plants. Escherichia coli hpt (aphIV, hph) gene codes for the hygromycin phosphotransferase to detoxify hygromycin B by phosphorylation via an ATP-dependent phosphorylation of a 7”-hydroxyl group and it is generally used as another selection marker gene when nptII was not effective in plant transformation studies (Twyman et al., 2002; Miki & McHugh, 2004).

Similar to antibiotics, herbicides have different specific target sites in plants. The resistance can be achieved by various mechanisms such as usage of natural isozyme or generation of enzyme mutagenesis or detoxification of the herbicides by metabolic processes. Phosphinothricin (PPT; ammonium glufosinate) is an active component of commercial herbicides formulations and analogous to glutamate, the substrate of glutamate synthase. In plants, this enzyme catalyzes the conversion of glutamate to glutamine by removing ammonia assimilation from the cell. Inhibition of the enzyme results in ammonia accumulation and disruption of chloroplast and finally cell death due to photosynthesis inhibition (Lindsey, 1992; OECD, 1999). In plant transformation studies, as herbicide resistance selection marker gene, pat from S. viridochromogenes (Wohlleben et al., 1988) and bar gene from S. hygroscopicus (bialophos resistance; Thompson et al., 1987) encoding the enzyme phosphinothricin N-acetyltransferase (PAT) are extensively used for resistance to PPT. PAT converts PPT to a non-herbicidal acetylated form by transferring the acetyl group from acetyl CoA to the free amino group of PPT (Miki & McHugh, 2004).

Currently, an alternative to these highly employed “toxic” approaches conditional positive selection markers based on the promotion of a metabolic advantage to transformed cells are used. Some authors mentioned that this kind of selection can improve considerably the selection of the transformants, since others such as antibiotics generally cause considerable necrosis (produced by the death of non-transformed cells) that often inhibits regeneration from adjacent tissues (Petri & Burgos, 2005). Previously, results obtained with this approach demonstrated higher yields than when the toxic selective agents were employed, and seems to be broadly applicable to crop plants (Miki & McHugh, 2004). Some of the most widely used are the AtTPS1/glucose (Leyman et al., 2006); galT/galactose (Joersbo et al., 2003); xylose isomerase (Haldrup et al., 1998); D-aminoacid/dao1 (Alonso et al., 1998) and the pmi/mannose (Joersbo et al., 1998). Probably, one of the most used one in fruit species is the gene pmi that encodes the enzyme phosphomannose isomerase (EC 5.3.1.8) that catalyzes the reversible interconversion of mannose 6-phosphate and fructose 6-phosphate. This enzyme is present in bacteria as E. coli and also, in humans, however it is not present in plants, as exception of soybean and other legumes. Using a media with mannose as the unique carbon source, only transformed cells can grow and develop. Glycolysis is inhibited due to the accumulation of mannose-6-phosphate converted from mannose by hexokinase with preventing cell growth and development in non-transformed cells (Miki & McHugh, 2004). Sensitivity to the toxic effect of mannose-6-phosphate is different between species, and can be avoided by combining with other sugars such as sucrose, maltose and fructose (Joersbo et al., 1999). Diverse fruit trees were selected with this system, alone or in combination with sucrose, i.e., 12 g/L mannose and 5 g/L sucrose in orange (Ballester et al., 2008); 30 g/L mannose without any sugar more in papaya (Zhu et al., 2005); 2,5 g/L mannose and 5 g/L sucrose in almond (Ramesh et al., 2006) or 1-10 g/L mannose and 5-30 g/L in apple (Degenhardt et al., 2006). In Citrus sinensis, the best results were obtained when 13 g/L mannose as unique source of carbon was added into the selection media. Mannose combined with other sugars promoted reduction in transformation efficiencies and escapes (Boscariol et al., 2003). Apricot cv. Helena and Canino required the lower combination of mannose with sucrose (1,25 g/L mannose and 20 g/L sucrose) in comparison with other woody fruit trees to obtain the most effective selection procedure. Moreover, safety assessments were revealed that there is no any adverse effect of the enzyme on mammalian allergenicity and toxicity (Reed et al., 2001).

Other selective strategies were developed as positive non-conditional systems, or in other words, using selectable marker genes that "promote" plant regeneration. Currently, there is more information about the genetic and biochemical control of organogenesis than embryogenesis for plant regeneration. Because of this, commonly genes related with cytokinins synthesis are employed for shoot organogenesis. More efforts are required to discover molecular mechanisms of embryogenesis to use these strategies in species highly dependent on embryogenesis regeneration to develop transgenic plants. Genes as cki1 or the most employed isopentenyl transferase ipt gene encoding the enzyme IPT, catalyze the synthesis of isopentyl-adenosine-5-monophosphate, which is the first step in cytokinin biosynthesis (Miki & McHugh, 2004). This gene modify the endogenous balance between cytokinins and auxins, stimulating cell division and differentiation of the cells that promote an altered morphology, development and physiology of transgenic plants (Sundar & Sakthivel, 2008). Some authors observed that the ipt gene improved transformation efficiency in apricot leaf explants in comparison with the selection through nptII (López-Noguera et al., 2009).

3.3. A Differential Transgene Expression: Constitutive Versus Specific Promoters

Currently, an important debate is carrying out about the risks of the "unpredictable" behavior and recombinogenic potential of constitutive promoters (Gittins et al., 2003) and to avoid the public concerns about the risks of ubiquitous transgene expression in crops.

Commonly, most of the fruit species have been transformed with plasmidic constructions harbouring constitutive or ubiquitous promoters, as the Cauliflower Mosaic virus 35S (CaMV35S). In this sense, different authors described that constitutive expression may be harmful for the host plant, causing sterility, retarded development, abnormal morphology, yield penalty, altered grain composition or transgene silencing (Cai et al., 2007 and references therein) and its expression level is dependent on the cell type, the developmental stage and on the perception of environmental triggers (Hensel et al., 2011). Moreover, under constitutive promoters reporter and selectable marker, and genes of interest are expressed continuously in all tissues without any temporal control. In this sense, specific-promoters appear as an alternative approach to avoid the undesirable side effects of constitutive promoters and to target transgene expression in a spatial or temporal specific way (Gago et al., 2011; Hensel et al., 2011).

Recently, vascular specific promoter EgCCR from Eucaliptus gunnii was checked in pistachio in this study as mentioned above as well as other fruit species such as kiwifruit and grapevine (Paradela et al., 2006; Gago et al., 2011) and results demonstrated that this promoter is conserved and fully functional in these species. Vascular promoters can drive resistance to biotic or abiotic stresses related with vascular tissues. Specific promoters could be useful to synchronize transgene activity spatially and/or temporally to control with more accuracy the pathogenic process (Gago et al., 2011).

3.4. Alternative Transformation Systems: Transgenics Without Marker Genes

A highly desirable approach to promote public acceptation for future commercialization of transgenic plants and products is focused on the elimination of marker genes from transformed plants or the direct production of marker-free transgenics (Kraus, 2010). These newly and promising approaches are highly dependent on previously established highly efficient regeneration protocols that may be based on organogenesis or embryogenesis (Petri et al., 2011). There are various technologies such as homologous recombination, co-transformation, site-specific recombination (Cre/loxP site specific recombination system, R/RS system, FLP/FRT system etc) or marker elimination by transposons to remove selective marker genes (Hao et al., 2011; Manimaran et al., 2011). However, there are still few marker-free fruit species transformation protocols.

Strawberry leaf explants were transformed with site-specific recombinase for the precise elimination of undesired DNA sequences and a bifunctional selectable marker gene used for the initial positive selection of transgenic tissue and subsequent negative selection for fully marker-free plants (Schaart et al., 2004).

MAT (multi-auto-transformation) (Ebinuma et al., 1997) combined with the Agrobacterium oncogene ipt gene, for positive selection with the recombinase system R/RS for removal of marker genes acting as "molecular scissors" after transformation were used as alternative approach in citrus plants (Ballester et al., 2007; 2008). Also, in apricot (López-Noguera et al., 2009) a similar strategy was used. Regeneration of apricot transgenic shoots was significantly improved to non-transformed plants (regenerated in non-selective media). Moreover, it was significantly higher in comparison with previous published data using resistance to kanamycin mediated by nptII gene. The lack of ipt differential phenotype promoted difficulties to assess the excision of the marker genes, that require periodic assays. Complete excision of marker genes ranged from 5 to 12 months, however, only 41% of the regenerated transgenic shoots R-mediated recombination occurs correctly. In Citrus sp., it was also reported that anomalous excision of marker genes promoting failures in the expression of the reporter genes (Ballester et al., 2007, 2008).

Apple (Malnoy et al., 2010) and pineapple sweet orange (Ballester et al., 2010) transformation using "clean" binary vector including only the transgene of interest were carried out to create marker-free transformants. Very recently, melon (C. melo L. cv Hetao) was transformed with a marker-free and vector-free antisense 1-aminocyclopropane-1-carboxylic acid oxidase construct via the pollen-tube pathway and transgenic lines are choosen by PCR without using any selectable marker agent (Hao et al., 2011).

In plum (Prunus domestica), transformation was carried out without reporter or selectable marker genes using a high-throughput transformation system (Petri et al., 2011). Previously, authors checked the efficiency of the regeneration of transformed shoots using conventional constructs harbouring reporter marker such as gusA and gfp, and nptII gene. Transformation efficiency varied from 5.7-17.7%. Using a marker free construct, the intron-hairpin-RNA (ihpRNA) harbouring the Plum Pox Virus coat protein (ppv-cp) gene, these authors regenerated five transgenic lines confirmed by Southern blot. It is important to take into account that this kind of free marker strategy is widely dependent on highly yields in regeneration systems.

3.5. Cisgenesis, The P-Dna Technology And Multigene Transformation

Other relevant advance in fruit species transformation was the proposal made by Schouten and coworkers (2006), the "cisgenesis". This term means the use of recombinant DNA technology to introduce genes from crossable donors plants, isolated from within the existing genome or sexually compatible relative species for centuries therefore, unlikely to alter the gene pool of the recipient species. Cisgenesis includes all the genetic events of the T-DNA as introns, flanking regions, promoters, and terminators (Vanblaere et al., 2011).

This methodology proposes to transfer the own plant DNAs, the P-DNAs. The use of this technology requires the construction of whole plant derived vector from the target species. Within the target species genome, it must be a DNA fragment with two T-DNA border-like sequences oriented as direct repeats ideally about 1-2 kb apart with suitable restriction sites for cloning of a desirable gene.

In the last years, different works were considered to step towards introducing regulatory elements and genes of interest from crossable donor plants, however with some foreign elements as marker genes in species as melon and apples (Benjamin et al., 2009; Joshi, 2010; Szankowski et al., 2009). Up to 2011 there is no any report of real "cisgenesis" plantlets, in agreement with Schouten et al. (2006) definition of the topic. In 2011, Vanblaere and coworkers developed apple cv. Gala cisgenic plants by expressing the apple scab resistance gene HcrVf2 encoding resistance to apple scab. Marker-free system was employed for the development of three cisgenic lines containing one insert of the P-DNA after removing by recombination with using chemical induction. These lines were not observed different from non-transformed cv. Gala plants.

Cisgenic plants are essentially the same as the traditionally bred varieties, and they might be easier to commercialise than the "problematic" transgenic plants (Schouten et al., 2006; Rommens et al., 2007). Critical opinions to these proposals also were clearly exposed, the uncontrolled P-DNA integration into the plant target genome can cause mutations or affect to the expression of other native genes, altering the behaviour of that cisgenic plants in an unpredictable manner (Schubert & Willims, 2006; Akhond & Machray, 2009). Recently, interesting approaches are being proposed for genome editing using ZFNs (Zinc finger nucleases) that can promote induction of double-strand breaks at specific genomic sites and promote the replacement of native DNA with foreign T-DNA (Weinthal et al., 2010).

The multigen transfer (MGT) methodology consist in introducing more than one gene at once. Commonly, most of the transgenic plants are generated by introducing just one single gene of interest, but now MGT are being developed to obtain more ambitious phenotypes as the complete import of metabolic pathways, whole protein complex and the development of transgenic fruit species with various new traits simultaneously transferred (Naqvi et al., 2009). In this sense, this technology would be highly desirable for commercial fruit species cultivars to obtain new traits related with large fruit size, high coloration of the fruit epidermis, flesh firmness and virus resistance (Petri et al., 2011) at the same time without the need of several rounds of introgressive backcrossing.

4. IN VITRO CULTURE TECHNIQUES FOR THE RECOVERY OF TRANSGENIC PLANTS

Plants are complex, diverse organisms and have adapted evolutionarily to almost every ecological niche on the planet. Development of successful transformation protocol depends on a reliable and highly efficient regeneration system. Explant types are highly variable since it depends on the selected organogenetic process optimized for each species. Commonly, the genetic transformation protocols of fruit species employed explants such as ovules, anthers, seedlings, zygotic and somatic embryos, cotyledons, epicotyles, hypocotyles, leaf pieces, roots, meristems (Fagoaga et al., 2007; Lopez- Perez et al., 2008; Petri et al., 2008; Husaini, 2010; Malnoy et al., 2010; Bosselut et al., 2011; Petri et al., 2011; Gago et al., 2011). Typically, it is recommended that those tissues have high and active cell division to enhance the regeneration of the transgenic lines (Mante et al., 1991; Schuerman & Dandekar, 1993; Wang, 2011). Ideally, fruit species transformation must be done with somatic tissues such as leaves and roots to transform varieties already well known and accepted in the market by the consumers. Recently, some authors also proposed the possibility of the use of transgenic seedlings to develop new fruit varieties through subsequent cross-breeding. These transgenic seedlings can add new traits impossible to obtain in the species genome-pool (Petri et al., 2011).

Organogenesis was the strategy selected in different species to develop most of the known and efficient regeneration protocols for fruit species, concretely for fruit trees (Petri et al., 2011). Almond (Costa et al., 2006); apple (Smolka et al. 2009; Lau & Korban, 2010; Vanblaere et al., 2011);

banana (Subramanyam et al., 2011); fig (Yancheva et al., 2005); kiwifruit (Tian et al., 2011); peach (Padilla et al., 2006); strawberry (Mercado et al., 2010); peanut (Asif et al., 2011); watermelon (Huang et al., 2011) and pear (Sun et al., 2011) are some examples of transformed cultivars for some fruit species that the transformed tissues were regenerated via organogenesis. Since organogenesis protocols are developed for many different fruit species, it is easier to adapt the regeneration system into genetic transformation methods (Frary & Eck, 2005). However, some risks also are assumed in using this regeneration system. Generally, it is considered that the origin of the new adventitious shoots is based on the involvement of few cells (George et al., 2008), enhancing the risks of chimera development.

Somatic embryogenesis that leads the formation of an embryo from somatic cells is another procedure to regenerate fruit transformants such as banana (Vishnevetsky et al., 2011); papaya (Zhu et al., 2001); grapevine (Nirala et al., 2010) and mango (Chavarri et al., 2010). Regeneration from transformed embryos can be achieved via direct germination or shoot organogenesis and the method is useful for large-scale and rapid propagation of transformants. In grapevine most of the approaches are being performed by using embryogenic cultures from different tissues such as zygotic embryos, leaves, ovaries and anther filaments to provide cells amenable to gene insertion and regeneration (Mezzetti et al., 2002; Dutt et al., 2007; López-Noguera et al., 2009). However, these techniques are highly genotype dependent and for many cultivars they have been difficult to obtain successful results (Dutt et al., 2007). Moreover, it is considered that anther filaments, as commonly employed in grapevine for embryogenic calli, are laborious, cultivar-dependent, depend on availability of immature flowers and may affect strongly the phenotype of the regenerated plantlets (Mezzetti et al., 2002). However, it is really interesting to take into account that regeneration from somatic embryos and secondary somatic embryos are currently assumed that they are derivatives of single cell origin.

In the decade of the 90´s some unsuccessful efforts were reported to transform meristems from micropropagated shoot tips due to high explant mortality and uncontrolled Agrobacterium overgrowth after coculture stages (Ye et al., 1994; Druart et al., 1998; Scorza et al., 1995). Mezzetti and co-workers (2002) described in grapevine the development of meristematic bulk tissues, a highly aggregate of meristematic cells produced after three months in increased concentrations of BA (N^6-benzyladenine) and the removal of the apical meristem. After 90 days, under the previous conditions, these highly regenerative tissues produced easily adventitious shoots and can be transformed by Agrobacterium, being able to regenerate several transgenic lines. Other interesting approach was the genetic transformation of shoot apical meristems, previously subjected to a dark growth stage after wounding for transformation. Authors reported that 1%

of shoot tips produced stable transgenic lines after weeks (Dutt et al., 2007). Ismail and co-workers (2011) transformed successfully banana apical meristems via microprojectile bombardment and regenerated 80.3% percent of the transformed meristematic tissues.

4.1. The Chimeric Question: Are My Transgenic Plants Genetically Uniform?

This is one of the most exciting questions that plant biotechnology researchers ask to themselves after all the long extensive, intensive and difficult labour needed to transform most of the fruit species. Some of the transformed regenerants can be chimeras, a mix of transformed and non transformed cells in the tissues, in other words, non genetically uniform organisms (Hanke et al., 2007). Recently, Petri and collaborators (2011) described that most of the known and efficient regeneration methods for fruit trees are based on organogenesis, where new adventitious shoot formation is originated from a determined number of cells. So, it comes hard to detect non chimeric and stable transgenic lines without using a selectable marker gene. Very recently, different authors using marker free technology as alternative systems or with genetically programmed marker excision reported the appearance of chimeric transformants in apple, strawberry, lime, citrus or plum (Domínguez et al., 2004; Schaart et al., 2004; Ballester et al., 2007; Malnoy et al., 2011; Petri et al., 2011).

Strawberry is highly sensitive to kanamycin selection, and it was described that selection of transgenic regenerants in these sensitive tissues can be associated with chimeric shoots containing transgenic and non-transgenic sections (Husaini, 2010). It was observed that increasing antibiotic concentration gradually avoid chimerisms in strawberry (Mathews et al., 1998; Husaini et al., 2010). Even under this strictly methodology some authors pointed out the inactivation events on the selection agent must be performed through the transformed cells, so, non transformed cells can develop and grown (Petri & Burgos, 2005; Wang, 2011). A useful methodology was also proposed for the quick and low-cost identification of chimeras by Faize and collaborators (2010) in tobacco and in apricot based in quantitative real-time PCR even in early developmental stages, and also let to monitor their dissociation.

5. FUTURE PERSPECTIVES AND CONCLUDING REMARKS

Currently, most of the fruit genetic transformation protocols integrated the new genes randomly and in unpredictable copy numbers influencing negatively its expression. Also public concerns and reduced market acceptance of transgenic crops have promoted the development of alternative marker free technologies in fruit species. For those reasons development of protocols to obtain transgenic fruits without marker genes and the use of the own plant DNA resources, such as "cisgenic" fruit plants, are the big challenges. ZFNs have also been succesfuly used to drive the replacement of native DNA sequences with foreign DNA molecules and to mediate the integration of the targeted transgene into native genome sequences.

Most of the fruit transgenic plants are generated by introducing just one single new character (gene of interest), however, some authors proposed that multigene transfer technology (MGT) needs to be developed to obtain new traits related at the same time. The combination of multiple traits can be a highly interesting approach as it could be applied to achieve resistance to several biotic or abiotic stresses and traits related to fruit quality such as large fruit size, high coloration of the fruit epidermis, increase flesh firmness to improve ripening without the need of several rounds of introgressive backcrossing.

The future of fruit genetic transformation is required of genotype-independent protocols, accuracy molecular tools to drive the T-DNA insertion and its expression, and efficiency and highly-yield selection and regeneration in vitro culture methodology. But Agrobacterium-mediated transformation procedure is a high non linear complex biological process, and its complexity can be understood with the composition of many different and interacting elements governed by non-deterministic rules and influenced by external factors. In this sense, the emergent technology dedicated to meta-analysis can be really useful to increase our understanding of fruit genetic transformation, making possible to identify relationships among several factors and extracting useful information generating understable and reusable knowledge (Gago et al., 2011; Gallego et al., 2011; Perez-Pineiro et al., 2012) Under these perspectives, modeling any fruit transformation procedure (Agrobacterium-mediated, biolistics, electroporation etc.) including the genetic engineering, in vitro plant tissue culture and regeneration stages will be improved for the next years.

6. ACKNOWLEDGEMENTS

The authors wish to thank to Dr. Mariana Landín and to Dr. María José Clemente for critical reading of this work. This work was supported by Regional Government of Xunta de Galicia: exp.2007/097 and PGIDIT02BTF30102PR. This study was also partially supported by # TBAG-209T030 from TUBITAK—The Scientific and Technical Research Council of Turkey. PG thanks to Minister of Education of Spain for funding the sabbatical year at Faculty of Science, University of Utrecht, Netherlands.

REFERENCES

1. S. Agarwal, K. Kanwar, 2007 Comparison of genetic transformation in Morus alba L. via different regeneration systems. Plant Cell Rep 26, 177185
2. M. A. Y. Akhond, G. C. Machray, 2009 Biotech crops: technologies, achievements and prospects. Euphytica 166, 4759 .
3. J. Alonso, J. L. Barredo, B. Diez, E. Mellado, F. Salto, J. L. Garcia, E. Cortes, 1998 D-Amino-acid oxidase gene from Rhodotorula gracilis (Rhodosporidium toruloides) ATCC 26217. Microbiology, 144, 10951101
4. Asif, Y. Zafar, J. Iqbal, Iqbal, U. Rashid, G. M. Ali, A. Arif, F. Nazir, 2011 Enhanced expression of AtNHX1, in transgenic groundnut (Arachis hypogaea L.) Improves Salt and Drought Tolerence. Molecular Biotechnology 49, 250256 .
5. R. G. Atkinson, C. J. Candy, R. C. Gardner, 1990 Agrobacterium infection of five New Zealand fruit crops. Zealand of Crop and Horticultural Science, 18, 153156
6. A. Ballester, M. Cervera, L. Peña, 2007 Efficient production of transgenic Citrus plants using isopentenyl transferase positive selection and removal of the marker gene by site-specific recombination. Plant Cell Reports, 26, 3945
7. A. Ballester, M. Cervera, L. Peña, 2008 Evaluation of selection strategies alternative to nptII in genetic transformation of citrus. Plant Cell Reports 27, 10051015
8. A. Ballester, M. Cervera, L. Pena, 2010 Selectable marker-free transgenic orange plants recovered under non-selective conditions and through PCR analysis of all regenerants. Plant Cell Tissue and Organ Culture, 102, 329336
9. M. Barceló, I. El -Mansouri, J. A. Mercado, Quesada, F. P. Alfaro, 1998 Regeneration and transformation via Agrobacterium tumefaciens of the strawberry cultivar Chandler. Plant Cell Tissue and Organ Culture, 54, 2936
10. D. Becker, B. Dugdale, M. K. Smith, R. M. J. Harding, J. L. Dale, 2000 Genetic transformation of Cavendish banana (Musa spp. AAA group) cv. 'Grand Nain' via microprojectile bombardment. Plant Cell Reports, 19, 229234
11. R. L. Bell, R. Scorza, C. Srinivasan, K. Webb, 1999 Transformation of 'Beurre Bosc' pear with the rolC Gene. J. Amer. Soc. Hort. Sci. 124(6), 570574
12. I. Benjamin, D. Kenigsbuch, M. Galperin, J. A. Abrameto, Y. Cohen, 2009 Cisgenic melons over expressing glyoxylate-aminotransferase are resistant to downy mildew. European Journal of Plant Pathology, 125, 355365
13. BIORAD 2002 http://www.bio-rad.com.
14. D. Bidney, 1999 Plant transformation method using Agrobacterium species adhered to microprojectiles. United States Patent, 5, 932, 782
15. N. Billinton, A. W. Knight, 2001 Seeing the wood through the trees: a review of techniques for distinguishing green fluorescent protein from endogenous autofluorescence. Analytical Biochemistry, 291, 175197

16. J. P. Bolar, J. L. Norelli, K. W. Wong, C. K. Hayes, G. E. Harman, H. S. Aldwinckle, 2000 Expression of endochitinase from Trichoderma harzianum in transgenic apple increases resistance to apple scab and reduces vigor. Phytopathology, 90, 7277
17. M. Bordas, C. Montesinos, M. Dabauza, A. Salvador, L. A. Roig, R. Serrano, V. Moren, 1997 Transfer of the yeast salt tolerance gene HAL1 to Cucumis melo L. cultivars and in vitro evaluation of salt Tolerance. Transgenic Research, 6, 4150
18. B. A. Bornhoff, M. Harst, E. Zyprian, R. Töpfer, 2005 Transgenic plants of Vitis vinifera cv. Seyval blanc. Plant Cell Reports, 24, 433438 .
19. R. L. Boscariol, W. A. B. Almeida, M. T. V. C. Derbyshire, F. A. A. Mourâo-Filho, B. M. J. Mendes, 2003 The use of the PMI/mannose selection system to recover transgenic sweet orange plants (Citrus sinensis L. Osbeck). Plant Cell Reports, 22, 122128
20. N. Bosselut, CV. Ghelder, M. Claverie, R. Voisin, JP. Onesto, MN. Rosso, D. Esmenjaud, , 2011 Agrobacterium rhizogenes-mediated transformation of Prunus as an alternative for gene functional analysis in hairy-roots and composite plants. Plant Cell Reports, 30(7), pp. 1313-1326
21. W. Broothaerts, J. Keulemans, I. Van Nerum, 2004 Self-fertile apple resulting from S-RNase gene silencing. Plant Cell Reports, 22, 497501
22. M. Cai, J. Wei, L. Xianghua, X. Caiguo, W. Shiping, 2007 A rice promoter containing both novel positive and negative cis-elements for regulation of green tissue specific gene expression in transgenic plants, Plant Biotechnology Journal, 5, 664674
23. W. Q. Cai, C. Gonsalves, P. Tennant, G. Fermin, M. Souza, N. Sarindu, F. J. Jan, H. Y. Zhu, D. Gonsalves, 1999 A protocol for efficient transformation and regeneration of Carica papaya L. In Vitro Cellular and Developmental Biology-Plant, 35(1), 61-69
24. X. Cao, Q. Liu, L. J. Rowland, F. A. Hammerschlag, 1998 GUS expression in blueberry (Vaccinium spp.): factors influencing Agrobacterium-mediated gene transfer efficiency. Plant Cell Reports, 18, 266270
25. M. Chalfie, Y. Tu, G. Euskirchen, W. Ward, D. Prasher, 1994 Green fluorescent protein as a marker for gene expression. Science, 263 802805
26. M. Chavarri, A. V. García, A. Y. Zambrano, Z. Gutiérrez, J. R. Demey, 2010 Insertion of Agrobacterium rhizogenes rolB gene in Mango. Interciencia, 35 (7), 521525 .
27. G. Chen, C. M. Ye, J. C. Huang, M. Yu, B. J. Li, 2001 Cloning of the papaya ringspot virus (PRSV) replicase gene and generation of PRSV-resistant papayas through the introduction of the PRSV replicase gen. Plant Cell Reports, 20, 272277
28. E. Chevreau, F. Dupuis, J. P. Taglioni, S. Sourice, R. Cournol, C. Deswartes, A. Bersegeay, J. Descombin, M. Siegwart, K. Loridon, 2011 Effect of ectopic expression of the eutypine detoxifying gene Vr-ERE in transgenic apple plants. Plant Cell Tissue and Organ Culture, 106, 161168
29. Chilton, M. H. Drummond, D. J. Merio, D. Sciaky, A. L. Montoya, M. P. Gordon, E. W. Nester, 1977 Stable incorporation of plasmid DNA into higher plant cells: the molecular basis of crown gall tumorigenesis. Cell, 11, 263271
30. P. Christou, 1992 Genetic transformation of crop plants using microprojectile bombardment. Plant Journal, 2, 275281
31. Y. Chu, P. Faustinelli, M. L. Ramos, M. Hajduch, S. Stevenson, J. J. Thelen, S. J. Maleki, H. Cheng, P. Ozias-Akins, 2008b Reduction of IgE binding and nonpromotion of Aspergillus flavus fungal growth by simultaneously silencing Ara h 2 and Ara h 6 in peanut. J. Agric. Food Chem. 56, 1122511233
32. Y. Chu, X. Y. Deng, P. Faustinelli, P. Ozias-Akins, 2008a Bcl-xL transformed peanut (Arachis hypogaea L.) exhibits paraquat tolerance. Plant Cell Reports, 27, 8592
33. Costa, C. M. Miguel, Oliveira, 2006 An improved selection strategy and the use of acetosyringone in shoot induction medium increase almond transformation efficiency by 100-fold. Plant Cell, Tissue and Organ Culture, 85, 205209
34. A. Cruz-Hernández, Litz. R. E. Witjaksono, Lim. M. Gomez, 1998 Agrobacterium tumefaciens-mediated transformation of embryogenic avocado cultures and regeneration of somatic embryos. Plant Cell Reports, 17, 497503
35. A. Cruz-Hernandez, Lim. Gomez, , R. E. Litz, (1997, (1997).Transformation of mango somatic embryos. Acta Horticulturae, 455 292298

36. H. Daniell, M. Krishnan, B. F. Mc Fadden, 1991 Transient expression of β-glucuronidase in different cellular compartments following biolistic delivery of foreign DNA into wheat leaves and calli. Plant Cell Reports, 9, 615619
37. S. N. Davidson, 2008 Forbidden Fruit: Transgenic Papaya in Thailand. Plant Physiology, 147, 487493
38. M. De Block, L. Herrera-Estrella, M. Van Montagu, J. Schell, P. Zambriski, 1984 Expression of foreign genes in regenerated plants and their progeny. EMBO Journal, 3, 16811689
39. J. Degenhardt, A. Poppe, J. Montag, I. Szankowski, 2006 The use of the phosphomannose-isomerase/mannose selection system to recover transgenic apple plants. Plant Cell Reports, 25, 11491156
40. A. Dominguez, J. Guerri, M. Cambra, L. Navarro, P. Moreno, L. Pena, 2000 Efficient production of transgenic citrus plants expressing the coat protein gene of citrus tristeza virus. Plant Cell Reports, 19, 427433
41. P. Druart, F. Delporte, M. Brazda, C. Ugarte-Ballon, Câmara. da, A. Machado, da. Laimer, Machado. M. Câmara, Jacquemin, & Watillon, B. 1998 Genetic transformation of cherry trees. Acta Horticulturae, 468, 7176
42. M. Dutt, Z. T. Li, S. A. Dhekney, D. J. Gray, 2007 Transgenic plants from shoot apical meristems of Vitis vinifera Thompson Seedless via Agrobacterium-mediated transformation. Plant Cell Reports, 26, 21012110
43. H. Ebinuma, K. Sugita, E. Matsunaga, S. Endo, K. Yamada, 1997 Selection of marker-free transgenic plants using the isopentyl transferase gene. Proceedings of National Academy of Science USA, 94, 21172121
44. C. Fagoaga, F. R. Tadeo, D. J. Iglesias, L. Huerta, I. Lliso, A. M. Vidal, M. Talon, L. Navarro, J. L. Garcia-Martinez, L. Peña, 2007 Engineering of gibberellin levels in citrus by sense and antisense overexpression of a GA 20-oxidase gene modifies plant architecture. Journal of Experimental Botany, 58(6), 14071420
45. C. Fagoaga, C. López, Mendoza. A. Hermoso de, P. Moreno, L. Navarro, R. Flores, L. Peña, 2006 Post-transcriptional gene silencing of the p23 silencing supperssor of Citrus tristeza virus confers resistance to the virus in transgenic Mexican lime. Plant Molecular Biology, 60, 153165
46. C. Fagoaga, I. Rodrigo, V. Conejero, C. Hinarejos, J. J. Tuset, J. Arnau, J. A. Pina, L. Navarro, L. Peña, 2001 Increased tolerance to Phytophthora citrophthora in transgenic orange plants constitutively expressing a tomato pathogenesis related PR-5. Molecular Breeding, 7, 175185
47. M. Faize, M. Malnoy, F. Dupuis, M. Chevalier, L. Parisi, E. Chevreau, 2003 Chitinases of Trichoderma atroviride induce scab resistance and some metabolic changes in two cultivars of apple. Phytopathology, 93, 14961504
48. M. Faize, L. Faize, L. Burgos, 2010 Using quantitative real-time PCR to detect chimeras in transgenic tobacco and apricot and to monitor their dissociation. BMC Biotechnology, 10, 53
49. J. Fan, X. Liu, S. X. Xu, Q. Xu, W. W. Guo, 2011 T-DNA direct repeat and 35S promoter methylation affect transgene expression but do not cause silencing in transgenic sweet orange. Plant Cell Tissuse and Organ Culture, 107, 225232
50. G. Fang, R. Grumet, 1993 Genetic engineerring of potyvirus resistance using constructs derived from the zucchini yellow mosaic virus coat protein gene. Molecular Plant-Microbe Interactions 6, 358367
51. V. J. Febres, R. F. Lee, G. A. Moore, 2008 Transgenic Resistance to Citrus Tristeza Virus in Grapefruit. Plant Cell Reports, 27(1), 93104
52. V. J. Febres, C. L. Niblett, R. F. Lee, G. A. Moore, 2003 Characterization of grapefruit plants (Citrus paradisi Macf.) transformed with Citrus Tristeza Closterovirus genes. Plant Cell Reports, 21, 421428
53. Fitch, R. M. Manshardt, D. Gonsalves, J. L. Slightom, J. C. Sanford, 1992 Virus resistant papaya derived from tissues bombarded with the coat protein gene of papaya ringspot virus. Bio/Technology, 10, 14661472

54. A. Frary, J. V. Eck, 2005 Organogenesis From Transformed Tomato Explants. In: Methods in molecular biology, 286 Transgenic plants: methods and protocols. L., Pena; NJ., Totowa (Eds.), 141-151, Humana Press Inc.
55. X. Fu, E. U. Khan, S. S. Hu, Q. J. Fan, J. H. Liu, 2011 Overexpression of the betaine aldehyde dehydrogenase gene from Atriplex hortensis enhances salt tolerance in the transgenic trifoliate orange (Poncirus trifoliate L. Raf). Environmental and Experimental Botany (in press)
56. M. Fuchs, G. Gonsalves, 2007 Safety of virus-resistant transgenic plants two decades after their introduction: lesson from realistic field risk assessment studies. Annual Review of Phytopathology, 47, 173202
57. J. Gago, 2009 Biotechnology of Vitis vinifera L.: Modellization through Artificial Intelligence. Doctoral Thesis. University of Vigo (Spain) (in Spanish).
58. J. Gago, J. Grima-Pettenati, P. P. Gallego, 2011 Vascular-specific expression of GUS and GFP reporter genes in transgenic grapevine (Vitis vinifera L. cv. Albariño) conferred by the EgCCR promoter of Eucalyptus gunnii. Plant Physiology and Biochemistry, 49, 413419
59. P. P. Gallego, R. Rodriguez, F. de la Torre, B. Villar, 2002 Genetic transformation of Eucalyptus globulus. In: Sustainable Forestry Wood Products and Biotechnology. S. Espinel, Y. Barredo, & E. Ritter (Eds.), 163170 , DFA-AFA Press, Vitoria-Gasteiz, Spain
60. P. P. Gallego, M. Landín, J. Gago, 2011 Artiticial neural networks technology to model and predict plant biology process. In: Artificial Neural Networks- Methodological Advances and Biomedical Applications. K. Suzuki (Ed.), 197216 , Intech Open Access Publisher: Croatia
61. P. P. Gallego, R. Rodríguez, F. de la Torre, B. Villar, 2009 Procedimiento para transformar material vegetal procedente de árboles adultos. Patent ES2299285 A1), Universidade de Vigo, Spain.
62. E. Galun, A. Breiman, 1998 Transgenic Plants. Imperial College Press, London, UK.
63. S. B. Gelvin, 1998 The introduction and expression of transgenes in plants. Current Opinion in Biotechnology, 9, 227232
64. E. F. George, Hall, G. J. De Klerk, 2008 Adventitious regeneration. In: Plant Propagation by Tissue Culture, 3rd edn. E.F. George, M.A. Hall & G-J, De Klerk. (Eds.), 355401 , Dordrecht, Netherlands: Springer
65. R. Ghorbel, J. Juarez, L. Navarro, L. Peña, 1999 Green fluorescent protein as a screenable marker to increase the efficiency of generating transgenic woody fruit plants. Theoretical and Applied Genetics, 99, 350358
66. R. Ghorbel, C. López, C. Fagoaga, P. Moreno, L. Navarro, R. Flores, L. Peña, 2001 Transgenic Citrus plants expressing the citrus tristeza virus p23 protein exhibit viral-like symptoms. Molecular Plant Pathology 2, 2736
67. J. R. Gittins, T. K. Pellny, S. Biricolti, E. R. Hiles, A. J. Passey, D. J. James, 2003 Transgene expression in the vegetative tissues of apple driven by the vascular-specific rolC and CoYMV promoters. Transgenic Research, 12, 391402
68. C. Gonsalves, B. Xue, M. Yepes, M. Fuchs, K. Ling, S. Namba, 1994 Transferring cucumber mosaic virus-white leaf strain coat protein gene into Cucumis melo L. and evaluating transgenic plants for protection against infections, Journal of American Society for Horticultural Science, 119, 345355 .
69. D. Gonsalves, 1998 Control of papaya ringspot virus in papaya: a case study. Annual Review of Phytopathology, 36, 415437
70. S. J. Gould, S. Subramani, 1988 Firefly luciferase as a tool in molecular and cell biology. Analytical Biochemistry, 175, 513
71. S. Guzmán-González, P. Valádez-Ramírez, R. E. Robles-Berber, L. Silva-Rosales, J. L. Cabrera-Ponce, 2006 Biolistic genetic transformation of Carica papaya L. using HeliosTM gene gun. HortScience, 41, 10531056
72. A. Haldrup, S. G. Petersen, F. T. Okkels, 1998 Positive selection: a plant selection principle based on xylose isomerase, an enzyme used in the food industry. Plant Cell Reports, 18, 7681

73. M. V. Hanke, M. Reidel, S. Reim, H. Flachowsky, 2007 Analysis of tissue uniformity in transgenic apple plants. Acta Horticulturae, 738, 301306
74. J. Hao, Y. Niu, B. Yang, F. Gao, L. Zhang, J. Wang, A. Hasi, 2011 Transformation of a marker-free and vector-free antisense ACC oxidase gene cassette into melon via the pollen-tube pathway. Biotechnology Letters, 33, 5561
75. D. Hebert, J. R. Kikkert, F. D. Smith, B. I. Reisch, 1993 Optimization of biolistic transformation of embryogenic grape cell suspensions. Plant Cell Reports, 12, 585589
76. G. Hensel, A. Himmelbach, W. Chen, D. K. Douchkov, J. Kumlehn, 2011 Transgene expression systems in the Triticeae cereals. Journal of Plant Physiology, 168, 3044
77. J. S. Heslop-Harrison, 2005 Introduction. In: Biotechnology of Fruit and Nut Crops (Ed. RE Litz). CAB International, London, UK. p. xix.
78. Y. Hiei, S. Ohta, T. Komari, T. Kumashiro, 1994 Efficient transformation of rice (Oryza sativa L.) mediated by Agrobacterium and sequence analysis of the boundaries of the T-DNA. The Plant Journal, 6, 271282
79. C. M. Higgins, R. G. Dietzgen, M. Akin, H. Sudarsono, K. Chen, Z. Xu, 1999 Biological and molecular variability of peanut stripe potyvirus. Curr Topics Virol, 1, 126
80. C. M. Higgins, R. M. Hall, N. Mitter, A. Cruickshank, R. G. Dietzgen, 2004 Peanut stripe potyvirus resistance in peanut (Arachis hypogaea L.) plants carrying viral coat protein gene sequences Transgenic Research, 13, 5967
81. A. Holefors, Z. T. Xue, M. Welander, 1998 Transformation of the apple rootstock M26 with the rolA gene and its influence on growth. Plant Science, 136, 6978
82. C. Honda, S. Kusaba, T. Nishijima, T. Moriguchi, 2011 Transformation of kiwifruit using ipt gene alters tree architecture. Plant Cell Tissuse and Organ Culture 107, 4550
83. R. B. Horsch, J. E. Fry, N. L. Hoffman, D. Eichholtz, D. Rogers, R. T. Fraley, 1985 A simple and general method for transferring genes into plants. Science, 227, 12291231
84. M. Hraška, V. Heřmanová, S. Rakouský, V. Ĉurn, 2008 Sample topography and position within plant body influence the detection of the intensity of green fluorescent protein fluorescence in the leaves of transgenic tobacco plants. Plant Cell Reports, 27, 6777
85. Y. C. Huang, C. H. Chiang, C. M. Li, T. A. Yu, 2011 Transgenic watermelon lines expressing the nucleocapsid gene of Watermelon silver mottle virus and the role of thiamine in reducing hyperhydricity in regenerated shoots. Plant Cell Tissuse and Organ Culture 106, 2129
86. A. M. Husaini, 2010 Pre- and post-agroinfection strategies for efficient leaf disk transformation and regeneration of transgenic strawberry plants. Plant Cell Reports, 29, 97110
87. Y. Ishida, H. Satio, S. Ohta, Y. Hiei, T. Komari, T. Kumashiro, 1996 High efficiency transformation of maize (Zea mays L.) mediated by Agrobacterium tumefaciens. Nature Biotechnology, 14, 745750
88. R. M. Ismail, F. M. El -Domyati, Wagih, Sadik, A. Z. E. Abdelsalam, 2011 Construction of banana bunchy top nanovirus-DNA-3 encoding the coat protein gene and its introducing into banana plants cv. Williams. Journal of Genetic Engineering and Biotechnology, 9 (1), 3541
89. James, A. J. Passey, D. J. Barbara, M. Bevan, 1989 Genetic transformation of apple (Malus pumila Mill) using a disarmed Ti-binary vector. Plant Cell Reports, 7, 658661
90. R. A. Jefferson, 1987 Assaying chimeric genes in plants: the gus gene fusion system. Plant Molecular Biology Reporter, 5, 387405
91. M. Joersbo, I. Donaldson, J. Kreiberg, Petersen. S. Guldager, J. Brunstedt, F. T. Okkels, 1998 Analysis of mannose selection used for transformation of sugar beet. Molecular Breeding, 4, 111117
92. M. Joersbo, K. Joregensen, J. Brunstedt, 2003 A selection system for transgenic plants based on galactose as selective agent and a UDP-glucose: galactose-1-phosphate uridyltransferase gene as selective gene. Molecular Breeding, 11, 315323

93. M. Joersbo, S. G. Petersen, F. T. Okkels, 1999 Parameters interacting with mannose selection employed for the production of transgenic sugar beet. Physiologia Plantarum, 105, 109115
94. S. G. Joshi, 2010 Towards durable resistance to apple scab using cisgenes. Wageningen, The Netherlands: Wageningen University, PhD Thesis
95. M. Kayim, N. M. Koc, M. Tor, 1996 Gene Transfer into citrus (Citrus limon L.) nucellar cells by particle bombardment and expression of gus activity. Turkish Journal of Agriculture and Forestry, 20, 349352
96. A. Khandelwal, G. J. Renukaradhya, M. Rajasekhar, G. L. Sita, Shaila, 2011 Immune responses to hemagglutinin-neuraminidase protein of peste des petits ruminants virus expressed in transgenic peanut plants in sheep. Veterinary Immunology and Immunopathology, 140, 291296
97. J. R. Kikkert, D. Hebert-Soule, P. G. Wallace, Striem, B. I. Reisch, 1996 Transgenic plantlets of 'Chancellor' grapevine (Vitis sp.) from biolistic transformation of embryogenic cell suspensions. Plant Cell Reports, 15, 311316
98. J. R. Kikkert, G. A. Humiston, M. K. Roy, J. C. Sanford, 1999 Biological projectiles (phage, yeast, bacteria) for genetic transformation of plants. In Vitro Cellular & Developmental Biology-Plant, 35, 4350
99. T. M. Klein, E. D. Wolf, R. Wu, et al. 1987 High velocity microprojectiles for delivering nucleic acids into living cells. Nature, 327, 7073
100. T. M. Klein, T. J. Jones, 1999 Methods of genetic transformation: the gene gun. In: Molecular Improvement of Cereal Crops: Advances in Cellular and Molecular Biology of Plants, 5 I.K. Vasil, (Ed.), 21-42, Dordrecht, The Netherlands: Kluwer Academic Publishers
101. K. Ko, J. L. Norelli, J. P. Reynoird, E. Boresjza-Wysocka, S. K. Brown, H. S. Aldwinckle, 2000 Effect of untranslated leader sequence of AMV RNA 4 and signal peptide of pathogenesis-related protein 1b on attacin gene expression, and resistance to fire blight in transgenic apple. Biotechnology Letters, 22, 373381
102. H. I. Kokko, S. O. Kärenlampi, 1998 Transformation of arctic bramble (Rubus arcticus L.) by Agrobacterium tumefaciens. Plant Cell Reports, 17, 822826
103. J. Kraus, 2010 Concepts of Marker Genes for Plants. In: Genetic Modification of Plants. F., Kempken, C., Jung (Eds.), 39-60, Biotechnology in Agriculture and Forestry, Volume 64, 1, Springer-Verlag, Berlin-Heidelberg
104. F. A. Krens, J. G. Schaart, R. Groenwold, A. E. J. Walraven, T. Hesselink, J. T. N. M. Thissen, 2011 Performance and long-term stability of the barley hordothionin gene in multiple transgenic apple lines. Transgenic Research, 20, 11131123
105. J. M. Lau, S. S. Korban, 2010 Transgenic apple expressing an antigenic protein of the human respiratory syncytial virus. Journal of Plant Physiology, 167, 920927
106. Y. K. Lee, I. J. Kim, 2011 Modulation of fruit softening by antisense suppression of endo-b-1,4-glucanase in strawberry. Molecular Breeding, 27, 375383
107. B. Leyman, N. Avonce, M. Ramon, P. Van Dijck, G. Iturriaga, J. M. Thevelein, 2006 Trehalose-6-phosphate synthase as an intrinsic selection marker for plant transformation. Journal of Biotechnology, 121, 309317
108. Z. T. Li, S. Dhekney, M. Dutt, M. Van Aman, J. Tattersall, K. T. Kelley, D. J. Gray, 2006 Optimizing Agrobacterium-mediated transformation of grapevine. In Vitro Cellular and Developmental Biology-Plant, 42, 220227
109. K. Lindsey, 1992 Genetic manipulation of crop plants. Journal of Biotechnology 26, 128
110. Z. Liu, B. J. Park, A. Kanno, T. Kameya, 2005 The novel use of a combination of sonication and vacuum infiltration in Agrobacterium-mediated transformation of kidney bean (Phaseolus vulgaris L.) with leagene. Molecular Breeding, 16, 189197
111. S. López-Noguera, C. Petri, L. Burgos, 2009 Combining a regeneration-promoting ipt gene and site-specific recombination allows a more efficient apricot transformation and the elimination of marker genes. Plant Cell Reports, 28, 17811790
112. A. J. Lopez-Perez, L. Velasco, M. Pazos-Navarro, M. Dabauza, 2008 Development of highly efficient genetic transformation protocols for table grape Sugraone and

Crimson Seedless at low agrobacterium density. Plant Cell Tissue and Organ Culture, 94, 189199

113. R. Maddumage, R. M. W. Fung, H. Ding, J. L. Simons, A. C. Allan, 2002 Efficient transformation of suspension culturederived apple protoplasts. Plant Cell Tissue and Organ Culture, 70, 7782

114. F. Maghuly, S. Leopold, A. Machado, Fernández. E. Borroto, Khan, G. Gambino, I. Gribanno, A. Schartl, M. Laimer, 2006 Molecular characterization of grapevine plants transformed with GFLV resistance genes: II. Plant Cell Reports, 25, 546553

115. M. Malnoy, Boresjza, J. L. Norelli, Flaishman, D. Gidoni, H. S. Aldwinckle, 2010 Genetic transformation of apple (Malus x domestica) without use of a selectable marker gene. Tree Genetics and Genome, 6, 423433

116. P. Manimaran, G. Ramkumar, K. Sakthivel, R. M. Sundaram, Madhav, S. M. Balachandran, 2011 Suitability of non-lethal marker and marker-free systems for development of transgenic crop plants: Present status and future prospects. Biotechnology Advances, 29, 703714 .

117. S. Mante, P. H. Morgens, R. Scorza, J. M. Cordts, A. M. Callahan, 1991 Agrobacterium-mediated transformation of plum (Prunus domestica L) hypocotyls slices and regeneration of transgenic plants. Bio-Technology, 9, 853857

118. H. Mathews, V. Dewey, W. Wagner, R. K. Bestwick, 1998 Molecular and cellular evidence of chimaeric tissues in primary transgenics and elimination of chimaerism through improved selection protocols. Transgenic Research, 7, 123129

119. H. Mathews, R. E. Litz, D. H. Wilde, S. Merkel, H. Y. Wetzstein, 1992 Stable integration and expression of β-glucuronidase and NPT II genes in mango somatic embryos. In Vitro Cellullar and Developmental Biology- Plant, 28, 172178

120. H. Mathews, R. E. Litz, H. D. Wilde, H. Y. Wetzstein, 1993 Genetic transformation of mango. Acta Horticulturae, 341, 9397

121. S. N. Maximova, A. M. Dandekar, Guiltinan, 1998 Investigation of Agrobacterium-mediated transformation of apple using gren fluorescent protein: high transient expression and low stable transformation suggest that factors other than T-DNA transfer are rate-limiting. Plant Molecular Biology, 37, 549559

122. D. Mc Cabe, P. Christou, 1993 Direct DNA transfer using electrical discharge particle acceleration (ACCELL™ technology). Plant Cell Tissue and Organ Culture, 33, 227236

123. J. A. Mercado, L. Trainotti, L. Jiménez-Bermúdez, N. Santiago-Doménech, S. Posé, R. Donolli, M. Barceló, G. Casadoro, F. Pliego-Alfaro, Quesada, 2010 Evaluation of the role of the endo-β-(1,4)-glucanase gene FaEG3 in strawberry fruit softening. Postharvest Biology and Technology, 55, 814

124. B. Mezzetti, T. Pandolfini, O. Navacchi, L. Landi, 2002 Genetic transformation of Vitis vinifera via organogenesis. BMC Biotechnology, 2, 1827

125. B. Miki, S. Mc Hugh, 2004 Selectable marker genes in transgenic plants: applications, alternatives and biosafety. Journal of Biotechnology, 107, 193232

126. M. Nakano, Y. Hoshino, M. Mii, 1994 Regeneration of transgenic plants of grapevine (Vitis vinifera L.) via Agrobacterium rhizogenes- mediated transformation of embryogenic calli. Journal of Experimental Botany, 45, 649656

127. S. Naqvi, G. Farré, G. Sanahuja, T. Capell, C. Zhu, P. Christou, 2009 When more is better: multigene engineering in plants. Trends in Plant Science, 15, 4856

128. Newell, 2000 Plant transformation technology; developments and applications. Molecular Biotechnology, 16, 5365

129. N. K. Nirala, D. K. Das, P. S. Srivastava, S. K. Sopory, K. C. Upadhyaya, 2010 Expression of a rice chitinase gene enhances antifungal potential in transgenic grapevine (Vitis vinifera L.). Vitis, 49, 181187

130. OECD. 1999 Concensus document on general information concerning the genes and their enzymes that confer tolerance to phosphinothricin herbicide. Series on Harmonization of Regulatory Oversight in Biotechnology, 11

131. M. L. P. Oliveira, V. J. Febres, M. G. C. Costa, G. A. Moore, W. C. Otoni, 2009 High-efficiency Agrobacterium-mediated transformation of citrus via sonication and vacuum infiltration. Plant Cell Reports, 28, 387395

132. Oliveira, J. G. Borrosa, M. Martins, Pais, 1994 In Y. P. S. Bajaj (Ed.), Biotechnology in Agriculture and Forestry, 29 189210 . Berlin: Springer

133. Oliveira, J. Barroso, Pais, 1991 Direct gene transfer into kiwifruit protoplasts: analysis of transient expression of the CAT gene using TLC autoradiography and a GC-MS based method. Plant Molecular Biology, 17, 235242

134. I. M. G. Padilla, L. Burgos, 2010 Aminoglycoside antibiotics: structure, functions and effects on in vitro plant culture and genetic transformation protocols. Plant Cell Reports, 29, 12031213

135. I. M. G. Padilla, A. Golis, A. Gentile, C. Damiano, R. Scorza, 2006 Evaluation of transformation in peach Prunus persica explants using green fluorescent protein (GFP) and beta-glucuronidase (GUS) reporter genes. Plant Cell Tissue and Organ Culture, 84, 309314

136. M. R. Paradela, P. De Paz, P. P. Gallego, 2006 Comparation of two kiwifruit transformation methods, Proceedings of 11th International Association for Plant Tissue Culture & Biotechnology Congress. Biotechnology and Sustainable Agriculture 2006 and Beyond, 76 P-1121), Beijing China, September 27-30, 2006

137. J. Paszkowski, R. D. Shillito, M. Saul, V. Mandak, T. Hohn, B. Hohn, et al. 1984 Direct gene transfer to plants. EMBO Journal, 3, 27172722

138. J. Paszkowski, S. A. Witham, 2001 Gene silencing and methylation processes. Current Opinion in Plant Biology, 4, 123129

139. W. P. Pawlowski, D. A. Somers, 1996 Transgene inheritance in plants genetically engineered by microprojectile bombardment. Molecular Biotechnology, 6, 1730

140. R. M. Pérez-Clemente, A. Pérez-Sanjuán, L. García-Férriz, J. P. Beltrán, L. A. Cañas, 2004 Transgenic peach plants (Prunus persica L.) produced by genetic transformation of embryo sections using the green fluorescent protein (GFP) as an in vivo marker. Molecular Breeding, 14, 419427

141. P. Pérez-Piñeiro, J. Gago, P. Landín, P. P. Gallego, 2012 Agrobacterium-mediated transformation of wheat: general overview and new approaches to model and identify the key factors involved. In: Transgenic Plants. Y., Ozden Çiftçi (Ed.), XX_XX, Intech Open Access Publisher: Croatia (in press)

142. J. Petolino, 2002 Direct DNA delivery into intact cells and tissues. In: Transgenic Plants and Crops. G. Khachatourians, A. McHughen, R. Scorza, W-K. Nip &Y. Hui, (Eds.), 137141 , Mercel Dekker Inc.; New York

143. C. Petri, L. Burgos, 2005 Transformation of fruit trees. Useful breeding tool or continued future prospect? Transgenic Research, 14, 1526

144. C. Petri, N. Alburquerque, S. García-Castillo, J. Egea, L. Burgos, 2004 Factors affecting gene transfer efficiency to apricot leaves during early Agrobacterium-mediated transformation steps. Journal of Horticultural Science and Biotechnology, 79, 704712

145. C. Petri, J. M. Hily, C. Vann, C. Dardick, R. Scorza, 2011 A high-throughput transformation system allows the regeneration of marker-free plum plants (Prunus domestica). Annals of Applied Biology, 159, 302315

146. C. Petri, H. Wang, N. Alburquerque, M. Faize, L. Burgos, 2008 Agrobacterium-mediated transformation of apricot (Prunus armeniaca L.) leaf explants. Plant Cell Reports, 27, 13171324

147. S. D. Purohit, S. Raghuvanshi, A. K. Tyagi, 2007 Biolistic-mediated DNA delivery and transient expression of GUS in hypocotyls of Feronia limonia L.- A fruit tree. Indian Journal of Biotechnology, 504507

148. V. V. Radchuk, V. I. Korkhovoy, 2005 The rolB gene promotes rooting in vitro and increases fresh root weight in vivo of transformed apple scion cultivar 'Florina'. Plant Cell Tissue and Organ Culture, 81, 203212

149. S. Ramesh, B. N. Kaiser, T. K. Franks, G. G. Collins, M. Sedgley, 2006 Improved methods in Agrobacterium-mediated transformation of almond using positive (mannose/pmi) or negative (kanamycin resistance) selection-based protocols. Plant Cell Reports, 25, 821828

150. A. Q. Rao, A. Bakhsh, S. Kiani, K. Shahzad, Shahid, T. Husnain, S. Riazuddin, 2009 The myth of plant transformation. Biotechnology Advances, 27, 753763
151. M. H. Raquel, Oliveira, 1996 Kiwifruit leaf protoplasts competent for plant regeneration and direct DNA transfer. Plant Science, 121, 107114
152. J. Reed, L. Privalle, M. L. Powell, M. Meghji, J. Dawson, E. Dunder, J. Suttie, A. Wenck, K. Launis, C. Kramer, Y. F. Chang, G. Hansen, M. Wright, 2001 Phosphomannose isomerase: an efficient selectable marker for plant transformation In Vitro Cellular and Developmental Biology-Plant, 37, 127132
153. S. Remy, A. Buyens, B. P. A. Cammue, R. Swennen, L. Sagi, 2000 Production of transgenic banana plants expressing antifungal proteins. International Symposium on Banana in the Subtropics. Acta Horticulturae, 490, 219277
154. C. M. Rommens, Haring, K. Swords, H. V. Davies, W. R. Belknap, 2007 The intragenic approach as a new extension to traditional plant breeding. Trends in Plant Science, 12, 397403
155. E. Rugini, A. Pellegrineschi, M. Mecuccini, D. Mariotti, 1991 Increase of rooting ability in the woody species kiwi (Actinidia deliciosa A. Chev.) by transformation with Agrobacterium rhizogenes rol genes. Plant Cell Reports, 6, 291295
156. L. Sagi, B. Panis, S. Remy, H. Schoofs, K. Smet, R. Swennen, B. P. A. Gammue, 1995 Genetic transformation of banana and plantain (Musa spp.) via particle bombardment. Biotechnology, 13, 481485
157. L. Sagi, S. Remy, R. Swennen, 1998 Genetic transformation for the improvement of bananas- a critical assessment. In: INIBAP Annual Report. 3336 . Montpellier: FRA.
158. S. Sakuanrungsirikul, N. Sarindu, V. Prasartsee, S. Chaikiatiyos, R. Siriyan, M. Sriwatanakul, P. Lekananon, C. Kitprasert, P. Boonsong, P. Kosiyachinda, et al. 2005 Update on the development of virus-resistant papaya: virus-resistant transgenic papaya for people in rural communities of Thailand. Food Nutr Bull, 26, 422426
159. J. C. Sanford, T. M. Klein, E. D. Wolf, N. Allen, 1987 Delivery of substances into cells and tissues using a particle bombardment process. Journal of Paniculate Science and Technology, 6, 559563
160. Sanford JC. 1988 The Biolistic Process. Trends in Biotechnology 6, 299302
161. J. C. Sanford, 1990 Biolistic plant transformation. Physiologia Plantarum 79, 206209
162. C. Sautter, 1993 Development of a microtargeting device for particle bombardment of plant meristems. Plant Cell Tissue and Organ Culture, 33, 251257
163. J. Schaart, G., F. A. Krens, K. T. Pelgrom, O. Mendes, G. J. Rouwendal, 2004 Effective production of marker-free transgenic strawberry plants using inducible site-specific recombination and a bifunctional selectable marker gene. Plant Biotechnology Journal, 2, 233240
164. H. J. Schouten, F. A. Krens, E. Jacobsen, 2006a Do cisgenic plants warrant less stringent oversight? Nature Biotechnology, 24, 753
165. H. J. Schouten, F. A. Krens, E. Jacobsen, 2006b Cisgenic plants are similar to traditionally bred plants. EMBO Reports, 7, 750753
166. D. Schubert, D. Willims, (2006).'Cisgenic' as a product designation. Nature Biotechnology, 24 13271329
167. P. L. Schuerman, A. M. Dandekar, 1993 Transformation of temperate woody crops-Progress and Potentials. Scientia Horticulturae, 55, 101124
168. R. Scorza, J. M. Cordts, D. J. Gray, D. Gonsalves, R. I. Emershad, D. W. Ramming, 1996 Producing transgenic 'Thompson Seedless' grape (Vitis vinifera L.) plants. Journal of American Society for Horticultural Science, 121, 616619
169. R. Scorza, L. Levy, V. Damsteegt, L. M. Yepes, J. Cordts, A. Hadidi, J. Slightom, D. Gonsalves, 1995 Transformation of plum with the papaya ringspot virus coat protein gene and reaction of transgenic plants to plum pox virus. Journal of American Society for Horticultural Science, 120, 943952
170. R. Serres, E. Stang, D. Mc Cabe, D. Russell, D. Mahr, B. Mc Cown, 1992 Gene transfer using electric discharge particle bombardment and recovery of transformed cranberry plants. Journal of American Society for Horticultural Science, 117, 174180

171. S. H. Sicherer, A. Munoz-Furlong, H. A. Sampson, 2003 Prevalence of peanut and tree nut allergy in the United States determined by means of a random digit dial telephone survey: a 5-year followup study. Journal of Allergy and Clinical Immunology, 112, 12031207
172. S. Singh, 1992 Fruit crops of wasteland. Scientific Publishers, Jodhpur, India.
173. C. Singsit, Adang, R. Lynch, W. F. Anderson, A. Wang, G. Cardineau, P. Ozias-Akins, 1997 Expression of a Bacillus thuringiensis cryIA(c) gene in transgenic peanut plants and its efficacy against lesser cornstalk borer. Transgenic Research, 6, 169176
174. A. Smolka, M. Welander, P. Olsson, A. Holefors, L. H. Zhu, 2009 Involvement of the ARRO-1 gene in adventitious root formation in apple. Plant Science, 177, 710715
175. G. Q. Song, K. C. Sink, 2004 Agrobacterium tumefaciens-mediated transformation of blueberry (Vaccinium corymbosum L.). Plant Cell Reports, 23, 475484
176. H. T. Stalker, Simpson, 1995 Genetic resources in Arachis. In: Advances in Peanut Science, Pattee HE., & Stalker HT, 1453 . American Peanut Research and Educational Society, Stillwater, OK, USA
177. C. N. Stewart, 2001 The utility of green fluorescent protein in transgenic plants. Plant Cell Reports, 20, 376382
178. K. Subramanyam, K. Subramanyam, K. V. Sailaja, M. Srinivasulu, K. Lakshmidevi, 2011 Highly efficient Agrobacterium-mediated transformation of banana cv. Rasthali (AAB) via sonication and vacuum infiltration. Plant Cell Reports, 30, 425436
179. Q. Sun, Y. Zhao, H. Sun, R. W. Hammond, R. E. Davis, L. Xin, 2011 High-efficiency and stable genetic transformation of pear (Pyrus communis L.) leaf segments and regeneration of transgenic plants. Acta Physiologiae Plantarum, 33, 383390
180. I. K. Sundar, N. Sakthivel, 2008 Advances in selectable marker genes for plant transformation. Journal of Plant Physiology, 165, 16981716
181. I. Szankowski, S. Waidmann, J. Degenhardt, A. Patocchi, R. Paris, E. Silfverberg-Dilworth, G. Broggini, C. Gessler, 2009 Highly scab-resistant transgenic apple lines achieved by introgression of HcrVf2 controlled by different native promoter lengths. Tree Genetics and Genomes, 5, 349358
182. N. J. Taylor, C. M. Fauquet, 2002 Microparticle Bombardment as a Tool in Plant Science and Agricultural Biotechnology. DNA and Cell Biology, 21, 963977
183. P. F. Tennant, C. Gonsalves, K. S. Ling, Fitch, R. Manshardt, L. J. Slightom, D. Gonsalves, 1994 Differential protection against papaya ringspot virus isolates in coat protein gene transgenic papaya and classically cross-protected papaya. Phytopathology, 84, 13591366
184. S. Terakami, N. Matsuda, T. Yamamoto, S. Sugaya, H. Gemma, J. Soejima, 2007 Agrobacterium-mediated transformation of the dwarf pomegranate (Punica granatum L. var. nana). Plant Cell Reports, 26, 12431251
185. C. J. Thompson, N. R. Movva, R. Tizard, R. Crameri, J. E. Davies, M. Lauwereys, J. Botterman, 1987 Characterization of the herbicide-resistance gene bar from Streptomyces hygroscopicus. EMBO Journal, 6, 25192523
186. N. Tian, J. Wang, Z. Q. Xu, 2011 Overexpression of Na+/H+ antiporter gene AtNHX1 from Arabidopsis thaliana improves the salt tolerance of kiwifruit (Actinidia deliciosa). South African Journal of Botany, 77, 160169
187. L. Tian, F. A. Canli, X. Wang, Sibbald, 2009 2009. Genetic transformation of Prunus domestica L. using the hpt gene coding for hygromycin resistance as the selectable marker. Scientia Horticulturae, 119, 339343
188. H. N. Trick, J. J. Finer, 1998 Sonication-assisted Agrobacterium-mediated transformation of soybean Glycine max L. Merrill embryogenic suspension culture tissue. Plant Cell Reports, 17, 482488
189. H. N. Trick, J. J. Finer, 1997 SAAT: sonication-assisted Agrobacterium mediated transformation. Transgenic Research, 6, 329336
190. S. Tripathi, J. Y. Suzuki, J. B. Carr, G. T. Mc Quate, S. A. Ferreira, R. M. Manshardt, K. Y. Pitz, Wall, D. Gonsalves, 2011 Nutritional composition of Rainbow papaya, the first

commercialized transgenic fruit crop. Journal of Food Composition and Analysis, 24, 140147

191. L. Tripathi, 2003 Genetic engineering for improvement of Musa production in Africa. African Journal of Biotechnology, 2, 503508
192. S. Tripathi, J. Y. Suzuki, S. A. Ferreira, D. Gonsalves, 2008 Papaya ringspot virus-P: characteristics, pathogenicity, sequence variability and control. Molecular Plant Pathology, 9, 269280
193. R. M. Twyman, E. Stöger, A. Kohli, T. Capell, P. Christou, 2002 Selectable and screenable markers for rice transformation. Molecular Methods of Plant Analysis, 22, 117
194. C. Uematsu, M. Murase, H. Ichikawa, J. Imamura, 1991 Agrobacterium-mediated transformation and regeneration of kiwi fruit. Plant Cell Reports, 10, 286290
195. C. Urtubia, J. Devia, A. Castro, P. Zamora, C. Aguirre, E. Tapia, P. Barba, P. Dell'Orto, M. R. Moynihan, C. Petri, R. Scorza, H. Prieto, 2008 Agrobacterium-mediated genetic transformation of Prunus salicina. Plant Cell Reports, 27, 13331340
196. W. van Leeuwen, M. J. M. Hagendoorn, T. Ruttink, R. van Poecke, L. H. W. van der Plas, A. R. van der Krol, 2000 The use of the luciferase reporter system for in planta gene expression studies. Plant Molecular Biology Reports, 18, 143a143t
197. T. Vanblaere, I. Szankowski, J. Schaart, H. Schouten, H. Flachowsky, G. A. L. Broggini, C. Gessler, 2011 The development of a cisgenic apple plant. Journal of Biotechnology, 154, 304311
198. R. K. Varshney, C. Kailash, K. C. Bansal, P. K. Aggarwal, S. K. Datta, P. Q. Craufurd, 2011 Agricultural biotechnology for crop improvement in a variable climate: hope or hype? Trends in Plant Science, 16, 363371
199. K. Veluthambi, A. K. Gupta, A. Sharma, 2003 The current status of plant transformation technologies. Current Science, 84, 368380
200. J. R. Vidal, J. R. Kikkert, P. G. Wallace, B. I. Reisch, 2003 High-efficiency biolistic co-transformation and regeneration of 'Chardonnay' (Vitis vinifera L.) containing npt-II and antimicrobial peptide genes. Plant Cell Reports, 22, 252260
201. J. R. Vidal, J. R. Kikkert, Malnoy, P. G. Wallace, J. Barnard, B. I. Reisch, 2006 Evaluation of transgenic Chardonnay (Vitis vinifera) containing magainin genes for resistance to crown gall and powdery mildew. Transgenic Research, 15, 6982
202. B. Villar, J. J. Oller, C. Teulieres, A. M. Boudet, P. P. Gallego, 1999 In planta transformation of adult clones of Eucalyptus globulus sp using an hypervirulent Agrobacterium tumefaciens strain. In: Application of Biotechnology to Forest Genetic. Espinel, S. & Ritter, E. (Eds.), 2225 , DFA-AFA Press, Vitoria-Gasteiz, Spain.
203. J. Vishnevetsky, T. L. White Jr, A. J. Palmateer, M. Flaishman, Y. Cohen, Y. Elad, M. Velcheva, U. Hanania, N. Sahar, O. Dgani, A. Perl, 2011 Improved tolerance toward fungal diseases in transgenic Cavendish banana (Musa spp. AAA group) cv. Grand Nain. Transgenic Research, 20, 6172
204. T. Wang, Y. Ran, R. G. Atkinson, A. P. Gleave, D. Cohen, 2006 Transformation of Actinidia eriantha: A potential species for functional genomics studies in Actinidia. Plant Cell Reports, 25, 425431
205. H. Wang, 2011 Development of a genetic transformation protocol genotype-independent in apricot. Doctoral Thesis. CEBAS-CSIC, University of Murcia (Spain) (in Spanish).
206. Q. Wang, P. Li, U. Hanania, N. Sahar, M. Mawassi, R. Gafny, I. Sela, E. Tanne, A. Perl, 2005 Improvement of Agrobacterium-mediated transformation efficiency and transgenic plant regeneration of Vitis vinifera L. by optimizing selection regimes and utilizing cryopreserved cell suspensions. Plant Science, 168, 565571
207. D. Weinthal, A. Tovkach, V. Zeevi, T. Tzfira, 2010 Genome editing in plant cells by zinc finger nucleases. Trends in Plant Science, 15, 308321
208. W. Wohlleben, W. Arnold, I. Broer, D. Hilleman, E. Strauch, A. Puhler, 1988 Nucleotide sequence of the phosphinothricin N-acetyltransferase gene from Streptomyces viridochromogenes Tü494 and its expression in Nicotiana tabacum. Gene,70, 2537

209. H. W. Wu, T. A. Yu, J. A. J. Raja, H. C. Wang, S. D. Yeh, 2009 Generation of transgenic oriental melon resistant to Zucchini yellow mosaic virus by an improved cotyledon-cutting method. Plant Cell Reports, 28, 10531064

210. S. X. Xu, X. D. Cai, B. Tan, W. W. Guo, 2011 Comparison of expression of three different sub-cellular targeted GFPs in transgenic Valencia sweet orange by confocal laser scanning microscopy. Plant Cell Tissue and Organ Culture, 104, 199207

211. B. Xue, K. S. Ling, C. L. Reid, S. Krastanova, M. Sekiya, E. A. Momol, S. Sule, J. Mozsar, D. Gonsalves, T. J. Burr, 1999 Transformation of five grape rootstocks with plant virus genes and a virE2 gene from Agrobacterium tumefaciens. In Vitro Cellular and Developmental Biology- Plant, 35, 226231

212. T. Yamamoto, H. Iketani, H. Ieki, Y. Nishizawa, K. Notsuka, T. Hibi, T. Hayashi, Matsuta, 2000ishizawa, Y.; Notsuka, K.; Hibi T, Hayashi T, Matsuta N (2000) Transgenic grapevine plants expressing a rice chitinase with enhanced resistance to fungal pathogens. Plant Cell Reports, 19, 639646

213. S. D. Yancheva, L. A. Shlizerman, S. Golubowicz, Z. Yabloviz, A. Perl, U. Hanania, Flaishman, 2006 The use of green fluorescent protein (GFP) improves Agrobacterium-mediated transformation of 'Spadona' pear (Pyrus communis L.). Plant Cell Reports, 25 183189

214. S. D. Yancheva, S. Golubowicz, Z. Yablowicz, A. Perl, Flaishman, 2005 Efficient Agrobacterium-mediated transformation and recovery of transgenic fig (Ficus carica L.) plants. Plant Science, 168, 14331441

215. C. Yang, S. Chen, G. Duan, 2011 Transgenic peanut (Arachis hypogaea L.) expressing the urease subunit B gene of Helicobacter pylori. Curr Microbiol, 63, 387391

216. J. L. Yao, J. H. Wu, A. P. Gleave, B. A. M. Morris, 1996 Transformation of citrus embryogenic cells using particle bombardment and production of transgenic embryos. Plant Science, 113, 175183

217. X. J. Ye, S. K. Brown, R. Scorza, J. M. Cordts, J. C. Sanford, 1994 Genetic transformation of peach tissues by particle bombardment. Journal of the American Society of Horticultural Science, 119, 367373

218. M. K. Yip, S. W. Lee, K. C. Su, Y. H. Lin, T. Y. Chen, T. Y. Feng, 2011 An easy and efficient protocol in the production of pflp transgenic banana against Fusarium wilt. Plant Biotechnology Reports, 5, 245254

219. E.. Zeldin, T. P. Jury, R. A. Serres, B. H. Mc Cown, 2002 Tolerance to the herbicide glufosinate in transgenic cranberry (Vaccinium macrocarpon Ait.) and enhancement of tolerance in progeny. Journal of American Society for Horticultural Science, 127, 502507

220. Y. J. Zhu, R. Agbayani, H. Mc Cafferty, H. H. Albert, P. H. Moore, 2005 Effective selection of transgenic papaya plants with the PMI/Man selection system. Plant Cell Reports, 24, 426432

221. Y. J. Zhu, R. Agbayani, P. H. Moore, 2004 Green fluorescent protein as a visual selection marker for papaya (Carica papaya L.) transformation Plant Cell Reports, 22, 660667

222. L. H. Zhu, A. Holefors, A. Ahlman, Z. T. Xue, M. Welander, 2001 Transformation of the apple rootstock M.9/29 with the rolB gene and its influence on rooting and growth. Plant Science, 160, 433439

CHAPTER 2

Genetic Dissection of Blackleg Resistance Loci in Rapeseed (*Brassica napus* L.)

Harsh Raman[1], Rosy Raman[1] and Nick Larkan[2]

[1] *Graham Centre for Agricultural Innovation (an alliance between NSW Department of Primary Industries and Charles Sturt University), Wagga Wagga Agricultural Institute, Wagga Wagga, Australia*

[2] *Saskatoon Research Centre, Agriculture and Agri-Food Canada, Saskatoon, Canada*

1. INTRODUCTION

Blackleg disease caused by the heterothallic ascomycete fungus *Leptosphaeria maculans* (Desm.) Ces. et de Not. (anamorph: *Phoma lingam* Tode ex Fr.), is the major disease of Brassica crops such as turnip rape *(Brassica rapa* L. syn. *B. campestris;* 2n = 2x = 20, genome AA), cabbage (*B. oleracea* L.; 2n = 2x = 18, genome CC), rapeseed (syn. canola or oilseed rape *B. napus* L.; 2n = 4x = 38, genome AACC), and *B. juncea* L. (Indian or brown mustard; 2n = 4x = 36, genome AABB) grown in temperate regions of the world. It was recorded for the first time on stems of red cabbage [1]. *B. napus* originated as a result of natural interspecific hybridization and genome doubling between the monogenomic diploid species, *B. rapa* and *B. oleracea,* in southern Europe approximately 10,000–100,000 years ago [2, 3]. However, it was selected and grown as an oilseed crop only 300-500 years ago [4, 5]. *B. napus* originally evolved as a spring or semi-winter type under the Mediterranean climates, and spread rapidly from southern to northern Europe after the development of winter *B. napus* varieties [6]. Both spring and winter types are affected by blackleg disease, particularly in Australia, Europe and North America. Currently *B. napus* is the world's third most important oilseed crop, grown on an area of over 23 million hectares and produce almost 53.3 million tonnes annually [7]. Increase in *B. napus* production has been attributed to the development and release of high yielding superior varieties including hybrids having traits such as high oil content, improved protein quality and herbicide resistance for better crop management.

Among the bacterial, fungal, viral and phytoplasmic-like diseases, blackleg is the most important global disease of *B. napus* crops and causes annual yield losses of more than $900 million in Europe, North America and Australia [8-10]. *L. maculans* has an ability to kill plants even at the seedling stage, infecting cotyledons, leaves, stems, roots and pods. Under epiphytotic conditions, this disease can cause yield losses of up to 90 per cent [11 - 13]. Therefore, control of blackleg disease has been one of the major objectives of many *B. napus* breeding programs.

2. SYMPTOMS

Blackleg disease causes two distinct symptoms; leaf lesions and stem canker. Outbreak of the fungus is characterised by dirty-whitish spots on leaves with small dark fruiting bodies (pycnidia). Black lesions are generally also seen on the leaves and deep brown lesions with a dark margin can be seen on the base of stem [11]. In severe epidemic conditions fungus girdles the stem at the crown, leading to lodging of the plant and possible severance of the stem. Typical lesions of blackleg can also occur on pods. Pod infection may leads to premature pod shatter and seed infection.

3. BIOLOGY OF THE PATHOGEN AND EPIDEMIOLOGY OF THE *L. MACULANS*

The pathogen can infect several crucifers, including cruciferous weeds. Up to 28 crucifer species have been reported as hosts [14]. During infection, the pathogen grows systemically down towards the tap root of the plant, producing severe disease symptoms at the adult plant stage characterised by stem cankers. *L. maculans* reproduces both asexually and sexually on host species and can complete several disease cycles during a single growing season. The fungus survives as mycelium, pycnidia and pseudothecia on crop residues, mainly on stubble [15, 16] subsisting from one season to the next. Sexual mating occurs on crop residues, resulting in the production of ascospores which can travel up to 8 km [17]. High humidity and moderate temperatures during vegetative growth promote disease development [18].

In Australia and most parts of Europe, *L. maculans* infection generally occurs during the seedling stage from infected seed and wind-dispersed ascospores (sexual spores), released from pseudothecia. In western Canada and Poland, asexual pycnidospores are the primary source of inoculum [19], dispersed largely by rain-splash. Under high humidity conditions, ascospores and pycnidiospores adhere to cotyledons or young leaves and germinate to produce hyphae which penetrate through stomata and wounds [9, 20, 21] and grow into sub-stomatal cavities without forming appressoria [22]. After entering into substomatal cavities, the fungus grows between the epidermis and palisade layer and then into intercellular spaces in the mesophyll of lamina. The fungus then reaches the vascular strands and grows within the plant asymptomatically, until eventually invading and killing cells of the stem cortex and causing the stem canker symptom [22-24]. Variability for virulence in *L. maculans* for the first time was reported in 1927 [25]. Australian populations of *L. maculans* have a high level of genetic variability as compared to European and North American isolates [26], along with a high diversity of avirulence genes [27]. Molecular analyses of populations of *L. maculans* have shown high gene flow within and between populations. Isolates of *L. maculans* are usually classified either on the basis of their aggressiveness or into pathogenicity groups [28].

4. MANAGEMENT OF THE *L. MACULANS*

Various practices such as crop rotation, stubble management, time of sowing, seed dressing and foliar application of fungicide, and deployment of genetic resistance have been employed to control this disease and subsequently reduce yield losses [9, 29]. Deployment of host resistance has been used as the most cost-effective and environmentally sound measure for disease control in various crops including in rapeseed. This strategy has been extensively used to manage blackleg disease especially in Australia, Canada, France, and Germany.

5. EVALUATION OF GERMPLASM FOR *L. MACULANS* RESISTANCE

An efficient and reliable method for phenotyping resistance to *L. maculans* is required for germplasm evaluation and predictive breeding including molecular mapping and gene cloning research. Various criteria are used to assess disease severity, such as severity of cotyledon or stem canker lesions, which rely principally on scales or estimates of the percent of diseased leaf tissue at either seedling (intact and detached leaf) or at adult plant stages. Symptom expression can vary with the environmental conditions, test locations (glasshouse, environment chamber and field conditions), and the method of inoculations (cotyledon, leaf and stem).

Resistance of *B. napus* germplasm to *L. maculans* is tested on the basis of disease reaction under glasshouse and/or field conditions. Cotyledon inoculations, performed under controlled conditions in either a growth chamber or glasshouse, allow for large scale and efficient screening of germplasm. Various environmental conditions such as temperature, light intensity and humidity can be reliably controlled, expediting the development of suitable resistant cultivars [30] as selections can be performed at early stages of plant development. This method also overcomes some of the uncertainties inherent in field testing with its dependence upon growing environment and further reduce the genotype by environment (G x E) interactions. Growth conditions are typically maintained with at 18°C to and 22°C. For uniform infection, a spore suspension is used to inoculate wounded cotyledons of 7 to 15 day-old seedlings [31-33]. Alternatively, seedlings can be sprayed with a spore suspension at up to the third leaf stage and kept at 100 % humidity for 48-72 hr. Spore suspensions of *L. maculans* are generally raised from single-spore isolate cultures grown on different media such as V8-agar, malt-agar and rapeseed leaf extract-agar [21, 22, 34]. Published studies used spore concentrations in the range from 4×10^6 to 1×10^8 spores per ml [31 - 33].

Doubled haploid (DH) populations were screened for resistance to *L. maculans* in the glasshouse at three plant growth stages: cotyledon, true leaf and adult plant, as well as under field conditions and reported a high correlation ($r \geq 0.82$) for disease severity between glasshouse and field grown lines [33]. Similar observations were also made by McNabb et al [35]. High correlation coefficient values suggest that the resistance to *L. maculans* can be evaluated at all three stages [33]. However among three stages, cotyledon stage was the most promising as inoculum-droplets can be kept at the inoculation site as compared to true leaves.

Assessment of adult plants for resistance to *L. maculans* populations under field conditions is considered very important for the selection of resistant germplasm by the rapeseed breeders. Inoculum is provided by either spreading infected stubble in a disease nursery or spraying plants with fungal spore suspension. Two measures; disease severity and disease incidence are commonly used for evaluating resistance to *L. maculans*. However, disease severity is much more difficult to estimate than disease incidence, due to the G x E interactions and unreliable and inconsistent estimation of canker lesions, even within the same genotype, particularly when infection is not uniform. The use of increased sample size (25 to 50 plants/genotype) and reliable and congenial growing conditions for the disease development will allow better estimation of canker lesions.

Assessment of blackleg resistance under field conditions is usually performed by exposing the plants to a mixed population of *L. maculans* races, which can make the detection of race-specific *R*-genes difficult. No relationship between the degree of cotyledon-lesion development at the seedling stage and crown canker development in mature plants was observed in the intercross population derived from Maluka/Niklas [36]. This study concluded the limited value of the cotyledon test in screening for adult plant blackleg resistance. Similarly a lack of correlation between cotyledon (seedling) resistance and stem (adult plant) resistance in *B. napus* and B genome sources has also been reported [37]. Recently, a poor correlation between seedling and field reactions was reported in the DH from Skipton/Ag-Spectrum which could have been due to the prevalence of different pathotypes under field conditions as contrary to cotyledon test, where often a specific isolate is used for phenotyping [32]. In order to mimic field conditions and increase reliability of disease development, an ascospore shower test [38] has been used for germplasm evaluation and varietal release in Australia. In this test, stubble with mature pseudothecia is sprayed with distilled water until run-off, producing 'ascospore shower'. The infected plants can then be assessed for resistance at both the cotyledon and adult plant stages. This method has shown a high correlation with canker lesions scored under field conditions [39].

6. NATURAL GENETIC VARIATION FOR RESISTANCE TO *L. MACULANS*

The introgression of blackleg resistance (*R*) genes into *B. napus* germplasm for blackleg disease management is one of the major objectives of breeding programs aiming to release cultivars in disease-prone areas. Genetic variation for resistance to *L. maculans* exists within *B. napus* germplasm [39, 40, 41]. Some other Brassica species such as *B. rapa*, *B. juncea*, *B. nigra* (black mustard; 2n = 2x = 16, genome BB) and *B. carinata* (Abyssinian or Ethiopian mustard; 2n = 34, genome BBCC), as well as other crucifers such as *Sinapis arvensis* have been reported to carry resistance [42-53]. Some of these sources were utilised in transferring resistance into *B. napus* breeding lines and cultivars. A continuous variation for blackleg resistance in a world-wide collection of *B. rapa* genotypes was reported [54]. None of genotypes were completely susceptible or completely resistant to either *L. maculans* pathotypes used. However, some *B.*

rapa accessions that were either highly resistant or completely susceptible were identified (Raman et al., unpublished) in a set of differential cultivars currently being used in Australia [39].

It has been reported that all B genome Brassica species; *B. nigra, B. carinata* and *B. juncea* carry complete resistance to *L. maculans* which remains effective throughout the life of the plant [40], however susceptible *B. juncea* cultivars have also been identified [55] demonstrating that complete resistance is not a feature of all B genomes. Some B genome resistance genes have been introgressed into *B. napus* lines. [47, 56-59]. Earlier studies have shown that C genome species of the *Brassica* are susceptible to blackleg [50, 53, 60]. However, a recent study [61] evaluated three accessions of *B. oleracea* var. *virids,* collected from the USDA germplasm collection and found that the accession NSL6146 was moderately resistant to *L. maculans.*

Genetic resources for adult plant resistance are very limited and most of them are derived from the French cultivar Jet Neuf [62]. Efforts are currently being made to identify both qualitative and quantitative resistance in the Australian Brassica Germplasm Improvement Programs.

7. INHERITANCE OF RESISTANCE TO *L. MACULANS*

Genetic inheritance studies revealed that resistance to *L. maculans* is complex. Resistance is either described as qualitative (also referred as monogenic/seedling/race-specific resistance/vertical resistance) or quantitative (also referred as polygenic/adult plant/race non-specific resistance/horizontal resistance) in *Brassica*.

7.1. Qualitative Resistance

Monogenic inheritance was reported in several spring and winter cultivars of *B. napus* such as Cresor, Maluka, Dunkeld, Maluka, Skipton, and Major [32, 63-67]. Eighteen major genes for resistance to *L. maculans*; *Rlm1* to *Rlm11*, *RlmS*, *LepR1* to *LepR4*, *BLMR1* and *BLMR2*, have been identified in Brassica species; *B. rapa*, *B. napus*, *B. juncea* and *B. nigra* [31, 32, 40, 45, 68-73].

Six of them, *Rlm1*, *Rlm2*, *Rlm3*, *Rlm4*, *Rlm7* and *Rlm9* were identified in *B. napus*, all of them except *Rlm2* were clustered genetically on chromosome A07 [74]. *Rlm2* was mapped on chromosome A10 [45]. The *Rlm5* and *Rlm6* were identified in *B. juncea*; *Rlm8* and *Rlm11* in *B. rapa*, and *Rlm10* was identified in *B. nigra*. Four resistance genes; *LepR1, LepR2, LepR3,* and *LepR4* were introgressed into *B. napus* from *B. rapa* subsp. *sylvestris* (Table 1).

TABLE 1. Molecular mapping of qualitative genes for resistance to *Leptosphaeria maculans* in *Brassica.** BC: Backcross population, DH: Doubled haploid population. # loci not mapped with molecular markers to date.

Species	Locus	*Population	Phenotyping stage	Marker type	Mapping strategy	chromosome	Linked markers/interval	Reference
B. napus	*Rlm1*	Maxol/S006 (140 DH)	Cotyledon inoculation	RAPD	Bulked segregant analysis	A7	T04 680 (14cM)	74
		Quinta/Score (110 F_2)	Cotyledon inoculation	RAPD	Bulked segregant analysis	A7	C02 1375/O15.1360 (19cM)	57, 74
		Maxol/Westar-10 (96 DH)	Cotyledon inoculation and stem canker	SSR, DArT	Whole genome mapping	A7	Xna12a-02a/Xra2-a05b	82
		Columbus/Westar-10	Cotyledon inoculation and stem canker	SSR	A7 chromosome specific mapping	A7	Xol12-e03a/Xna12-a02a	82
B. napus	*Rlm2*	Glacier/Score (110 F_2)	Cotyledon inoculation	RAPD	Bulked segregant analysis	A7	M08.1200, M08.600, P02.700	57, 74
		Glacier/Yudal (BC189)	Cotyledon inoculation	RAPD	Bulked segregant analysis	A7	M08 1200, M08 600, P02.700	74
		Darmor/Samourai (133 DH)	Cotyledon, Field	RAPD	Bulked segregant analysis	A7	M08.1200 (10cM),	74
B. napus	*Rlm3*	Maxol/S006 (140DH)	Cotyledon inoculation	RAPD	Bulked segregant analysis	A7	Q12 750 (7cM)	74
B. napus	*Rlm4*	Quinta/Score (110F2)	Cotyledon inoculation	RAPD	Bulked segregant analysis	A7	C02 1375 (3.6 cM)/ O15.1360 (˙33 cM)	74
		Skipton/Ag-Spectrum	Cotyledon and Stem canker	SSR	Whole genome mapping		BRMS075 (˙0.7 cM)	32
B. juncea	*Rlm6*	Recombinant lines (*B. napus-B. juncea*)	Cotyledon and field test	RAPD/ RFLP	Bulked segregant analysis	B8	OPG02.800, OPT01	47
B. napus	*Rlm7*	2311.1/Darmor (221 F_2)	Cotyledon	RAPD	Bulked segregant analysis	A7	T12 650 (4cM)	74, 85
B. napus	*Rlm9*	Darmor-bzh/Yudal (132 DH)	Cotyledon	RAPD	Bulked segregant analysis	A7	T12.650/C02 1375	74
B. nigra	*Rlm10*	Addition lines (Darmor/Junius)	Cotyledon test	Isozyme RAPD	Whole genome mapping	B4	OPA11.1200, OPC19.3300	83, 84

B. rapa ssp. *sylvestris*	*LepR1*	6270/Springfield (DHP95)	Cotyledon inoculation and field resistance	RFLP	Whole genome analysis	A2 (N2)	pR4b, pO85h, pW180b, pN181a, pW207a	31
B. rapa ssp. *sylvestris*	*LepR2*	6279/3027 (DHP96)	Cotyledon inoculation and field resistance	RFLP	Whole genome analysis	A10 (N10)	pN21b, pR34b, pN53b	31
B. napus	*LepR3*	Surpass400/Westar (N-o-1)-BC	Cotyledon inoculation	SSR	A1 and A10 chromosome specific mapping		sR12281a (2.2 cM) sN2428Rb (0.7 cM)	69
		Topas (DH16516)/ Surpass400	Cotyledon inoculaton	SSR, SCAR	A10 chromosome specific mapping	A10 (N10)	Ind10-12	79
B. rapa ssp. *sylvestris*	*LepR4*	16S/PAS12//16S (BC$_3$S$_2$)	Cotyledon inoculation Disease nursery (field)	SSR	A genome specific marker analysis	A6	sN2189b (8.8cM) sR9571a (8.3 cM)	77
B. napus	*BLMR1*	Surpass400/Westar (1513 F$_3$BC$_2$)	Cotyledon inoculation	SRAP, SNP	Selective genotyping	A10 (N10)	80E24a (0.1 cM)	70
B. napus	*BLMR2*	Surpass400/Westar (1513 F$_3$BC$_2$)	Cotyledon inoculation	SRAP, SNP	Selective genotyping	A10 (N10)	R278 (1.2 cM)	70
B. napus	*LmFr1*	Cresor (resistant)/ Westar (susceptible)	Field/artificial inoculation	RFLP	Whole genome analysis	Linkage group 6 (A7)	cDNA011/cDNA110	64
B. napus	*LEM1*	Major (resistant)/Stellar (susceptible)	Cotyledon Stem inoculation	RFLP	Whole genome analysis	A7	TG5D9b/WG5A1A	65
B. napus	*cLmR1*	Shiralee/90-3046 (153 DH lines)	Cotyledon incoulation	RFLP, RAPD	Bulked segregant analysis	A7	RAPD654 (~4.8cM)	66
B. napus	*cLmR1*	Shiralee/PSA12 (BC1 lines)	Cotyledon incoulation	RFLP, EST, SCAR	Bulked segregant analysis	A7	est126M9a/est149M9d	140
B. napus	*cLmR1*	DH12075/PSA12 (BC1 lines)	Cotyledon incoulation	RFLP, EST, SCAR	Bulked segregant analysis	A7	est126M9a/est149M9d	140
B. napus	*cLmR1*	Maluka/90-3046 (34 DH lines)	Cotyledon incoulation	RFLP, RAPD	Bulked segregant analysis	A7	RAPD654 (~4.8cM)	66
B. napus	*cRLM (cRLMm)*	Maluka/Westar	Cotyledon, Adult	RFLP, AFLP, RAPD	Bulked segregant analysis	A7	22-25	67
B. napus	*cRLM (cRLMrb)*	RB87-62/Westar	Cotyledon, adult plant	RFLP, AFLP, RAPD	Bulked segregant analysis	A7	22-25	67
B. napus	*cRLM (cRLMc)*	Cresor/Westar	Cotyledon, adult plant	RFLP, AFLP, RAPD	Bulked segregant analysis	A7	22-25	67
B. napus	*Rlm.wwai-A1*	Skipton/Ag-Spectrum (DH)	Cotyledon incoulation	SSR	Whole genome analysis	A1	Xpbcessrna16-Xbrms017b	32
B. napus	*ORlm.wwai-A10a*	Skipton/Ag-Spectrum (DH)	Cotyledon inoculation	SSR	Whole genome analysis	A10	Xcb10079d-Xcb10079c	32

B. napus	*Rpg$_3$Dun*	Westar/Dunkeld (F_2)	Cotyledon inoculation	SRAP	Bulked segregant analysis	A7	$NA_{12}A_{02}$-200/$NA_{12}A_{02}$-190, $BG_{20}SA_{12}$-480/$BG_{20}SA_{12}$-475/ BN204	81
B. juncea	*#Rlm5*	150-2-1, 151-2-1, Aurea, Picra	Cotyledon inoculation	-	-	-	-	71
B. rapa	*#Rlm8*	156-2-1	Cotyledon inoculation	-	-	-	-	71
B. rapa	*#Rlm11*	02-159-4-1	Cotyledon inoculation	-	-	-	-	72
B. napus	*#RlmS*	Surpass400	Cotyledon inoculation	-	-	-	-	73
B. juncea	*LMJR1*	AC Vulcan/UM3132 (F_2)	Cotyledon test	RFLP, SSR	Whole genome mapping	J13 (B3)	PN199RV (22.1 cM), sBb31143F (8.7 cM)	43
B. juncea	*LMJR2*	AC Vulcan/UM3132 (F2)	Cotyledon test	RFLP, SSR	Whole genome mapping	J18 (B8)	PN120cRI, sB1534	43
B. juncea	*rjlm2*	B genome introgression lines	cotyledon	RAPD, RGA & SCAR	B genome-specific	not defined	B5-1520, C5-1000, RGALm	80

Recently, two genes *BLMR1* and *BLMR2* in Surpass 400; an Australian cultivar developed from an interspecific cross between wild *B. rapa* subsp. *sylvestris* (resistant) from Sicily and *B. oleracea* subsp. *alboglabra* were identified [70, 76]. However, *LepR1* to *LepR4* genes are thought not be related with *Rlm* genes on the basis of their map locations, except for *Rlm2* and *LepR3*, which are phenotypically different [31, 69, 77]. It appears that loci *LepR3*, *BLMR1* and *BLMR2* localised on chromosome A10 control resistance to *L. maculans* in Surpass 400. However, Van de Wouw et al. [73] demonstrated that two independently segregating *L. maculans* avirulence (Avr) genes, *AvrLm1* corresponding to *Rlm1* (on chromosome A7) and *AvrLmS*, are responsible for inducing resistance in this cultivar. Subsequently, Larkan *et al.* [78] investigated the interaction of *AvLm1* and *AvLmS* isolates with *B. napus* populations segregating for the resistance genes *Rlm1* (from the French cultivar Quinta) and *LepR3* (from Surpass 400). This study reported that (i) *AvrLm1* interacts in a gene-for-gene manner with both *Rlm1* and *LepR3*, (ii) *AvrLmS* is not responsible for triggering the *LepR3* mediated defence response, (iii) Surpass 400 does not contain *Rlm1,* and (iv) *Rlm1* and *LepR3* may be the same genes located in two distinct loci or may have evolved as two functional genes. Recently, *LepR3* has become the first functional *B. napus* resistance gene to be cloned and was shown to encode a receptor-like protein. Additionally, *LepR3*-transgenic *B. napus* and *AvrLm1*-transgenic *L. maculans* were used to demonstrate that *AvrLm1* conveys avirulence to *LepR3*. The shared genomic location of *LepR3* and *BLMR1* also suggested that these were the same gene [79]. Several other genes such as *LmR1, ClmR1,LmFr1, cRLMm, cRLMrb, aRLMrb,* and *LEM1* have also been identified using uncharacterised isolates, which are thought to be allelic to known *R*-genes [45, 68, 74]. Qualitative resistance conferred by single major genes is usually dominant and expressed at the seedling growth stage. Qualitative *R*-genes explain majority of phenotypic variation for blackleg resistance at adult plant stage [32, 74]. However, digenic mode of inheritance has also been reported in *B. napus and B. juncea* populations [40, 80].

7.2. Quantitative Resistance

Quantitative inheritance for field resistance has been reported in segregating populations derived from *B. napus, B. juncea* and their hybrid derivatives [30, 32, 65, 80, 86]. Some of the QTLs identified are given in Table 2. Quantitative genetic analysis revealed that significant non-additive genetic variance for all measures of disease severity indicated the presence of strong dominance/epistasis at loci controlling blackleg resistance [36]. In the literature, the term 'QTL' as a quantitative locus has been used even when a large percent of genotypic variation is explained by the major locus. In classical genetics, QTL refers to genes that have, low heritability, non-Mendelian and quantitative accumulative effects. The majority of genetic analyses have utilised doubled-haploid (DH) populations, which are not suitable to infer modes of inheritance. Advanced intercross populations are required to interpret such phenomena, as used in [74].

8. GENE-FOR GENE INTERACTIONS

Host resistance genes (*R*-genes) interact in paired combination with pathogen avirulence (*Avr*) genes to condition resistance [89]. Two types of interactions may occur; compatible and incompatible. Compatible interaction occurs when there is an absence of an effective host defence response, due to a lack of a resistance allele in host (*r*) or an allele for virulence (*avr*) at the corresponding pathogen locus. An incompatible interaction occurs when there is no disease development due to the presence of both an effective host resistance allele (*R*) with an allele for *Avr* at the corresponding pathogen locus [90]. Biochemically, gene-for-gene interactions have been interpreted as the interaction of a race-specific pathogen elicitors with either cultivar-specific plant receptors or alternatively with a cultivar-specific signal transduction compounds [91].

TABLE 2. Significant QTLs associated with blackleg resistance (scored as Internal infection due to canker development at adult plant stage) identified from mapping populations, * QTL with consistent effect, # range of LOD and R^2 varied with method of regression analysis (simple and composite interval mapping).* refers to predicted markers from supplementary figures ESM7-10 shown in Kaur et al [81]

Mapping Population	Stubble, Location	Flanking markers	Chromosome	LOD# score	%Genetic variance (R^2)#	Additive effect	Reference
Av-Sapphire/ Westar10	*B. napus*, Lake Bolac, Australia	E34M15_S190/ E35M53_S416	A1	2.5-5.6	14-16	Not known (-)	86*
		E34M15_S218/ E35M53_S350	A2	1.3-3.8	4-26	-	
	Dahlen, Australia	CB10443_W258_ S269	C1	0.8-2.9	3-8	-	
	Lake Bolac	E36M47_W197/ E34M62_W127	LG1	1.0-3.6	4-10	-	
Caiman$_3$/Westar	*B. napus*, Lake Bolac	BRMS056/ E34M50_W140	A1	2.9-3.5	20-22.7	-	86*
	Dahlan	E35M53_C455/ E34M15_W271	A10	1.4-3.0	5-34	-	
		Not shown	C5	4.4-5.6	19-23	-	
Camberra$_4$/Westar	Lake Bolac	E36M55_C306/ E33M57_C306	A5	0.5-2.6	1.5-33	-	86*
	Dahlen	Na12D10_w203	A1/C1	2.9-5.1	17-18	-	
	Lake Bolac	E36M62_W414/0 1ju1fE07_cl_3b	A10	0.3-2.7	2-31	-	
		Not shown	C7-2	2.7-3.7	13-24	-	
		E33M59_W107/ E33M53_C75	LG2	2.1-2.8	14-28	-	
Darmor/Samourai	*B. napus*, Le Rheu	*BN483*	A1	2.3	6.7	Samourai	88
		BN239	A2	2.3-3.02	8.1-14.6	Darmor	
		BN182.1	A6	1.9-2.8	6.2-10.0	Samourai	
		At17	A10	2.7	11.0	Samourai	
		BN204	C2	2.0-2.2	8.0-8.4	Darmor	

		Vers6.9	C8	1.9	-	Samourai	
Darmor/Yudal	*B. napus*, Le Rheu, France	*OPE02.1200*	A2	2.4-5.5	3.8-8.5	Darmor	87
		OPW08.1620	A4	3.3	4.8	Darmor	Delourme et al, 2008; comm. pers.)
		OPW05.750-Bzh	A6	5-12.2	7.2-20	Darmor	
		Bras023	A7	4.5	6.9	Darmor	
		CB10026b	A8	7.2	13.0	Darmor	
		OPW15.1470	A9	3.3	4.8	Darmor	
		Fad8	C2	5.5-6.6	8.3-13.3	Yudal	
		OPD08.1310	C4	4.7-9.5	6.7-15.2	Darmor	
		OPH06.CD1	C8	4.2	6.2	Darmor	
Rainbow/Av-Sapphire	Lake Bolac	E33M57_R105	A9-2	3.7	13	-	
Skipton/Ag-Spectrum	*B. napus* Mixed stubble, Wagga, Australia	*Xbras123/Xem1-bg11-237*	A2	7.0	11.5	Ag-Spectrum	32
		Xbrms319-Xbrms176	A9	2.9	5.0	Skipton	
		Xcb10172-BnFLC10	A10b	2.2	6.2	Skipton	
		Xbrms287a-Xcb10034	C1	4.2	11.5	Ag-Spectrum	
		Xol10-c10/Xna12-c03	C2a	6.8	16.6	Skipton	
		Xpbcessrna13/Xol13-d02a	C3	4.2	24.5	Skipton	
		Xem1-bg23-89/Xol12-e03	C6	6.1	14.5	Ag-Spectrum	
	B. napus, ATR Beacon stubble, Wagga, Australia	*Xol12-f11/Xpbcessrbr21*	A1a	6.1	26.1	Ag-Spectrum	

Differential interaction between specific *R*-genes in the host (Brassica) and corresponding *Avr* genes of the pathogen *(L. maculans)* was first studied at the seedling stage using a cotyledon inoculation test in *B. napus* [92] and subsequently verified [57]. Qualitative and quantitative resistance differ with respect to host-pathogen interaction, as the latter does not appear to (but not proven) follow the gene-for gene hypothesis, being more effective against diverse pathogen populations (non-race specific). While quantitative resistance normally provides partial resistance to the pathogen and it is less likely to be rapidly overcome by shifting pathogen populations.

At least ten *Avr* genes have been identified in *L. maculans*, many of which map to two gene clusters; *AvrLm1-AvrLm2-AvrLm6* and *AvrLm3-AvrLm4-AvrLm7-AvrLepR1* ([71, 72, 86, 87]. Four of the *Avr* genes; *AvrLm1, AvrLm6, AvrLm4-7* and *AvrLm11* have been cloned. It has shown that although *AvrLm1* and *AvrLm6* are physically clustered together in the *L.*

maculans genome, they are not allelic forms of a single gene [85, 96]. However, *AvrLm4* and *AvrLm7* are allelic variants of a single *Avr* gene that corresponds to the two resistance genes; *Rlm4* and *Rlm7* [71, 85]. It has also been demonstrated that *AvrLm1* interacts with two distinct resistance loci; *Rlm1* and *LepR3*, though these loci are located on different chromosomes (A7 and A10, respectively) [78]. The cloning and characterisation of additional Brassica *R*-genes and *L. maculans* Avr genes will lead to a better understanding of how these functional redundancies developed. In the recent years, understanding of *L. maculans/Brassica* interactions has increased our ability to deploy appropriate *R*-genes in new cultivars and manage blackleg disease with the increased knowledge of the distribution of *Avr* alleles in *L. maculans* populations [27, 94, 98]. Currently, it seems that the genes involved in race-specific resistance and polygenic non-specific resistance are distinct. A better understanding of the mechanisms underlying quantitative resistance would help our understanding of the relationships between quantitative and major resistance genes [99].

9. ALIEN GENE INTROGRESSION FOR BLACKLEG RESISTANCE

Deployment of *R*-genes has been used as the most cost-effective and environmentally sound measure for disease control in various crops since a century ago when first *R*-genes were identified [100]. Conventional plant breeding methodologies have played an important role in gene introgression for disease resistance, especially in easily-crossed genetic backgrounds. As a result several cultivars rated for resistance to *L. maculans* now dominate commercial cultivation worldwide. There has been a continuous threat of 'breakdown' of resistance, especially when a resistant cultivar is grown extensively on large acreages over long period of time. For example, 'breakdown' of resistance in cultivar Surpass 400 occurred within three years of its release [101, 102] due to the evolution and spread of more virulent strain of *L. maculans*. 'Breakdown' of resistance implies that the resistance has not changed rather the pathogen population has shifted/been selected for virulence. The effectiveness of *Rlm1* in France was also greatly reduced from 1997 to 2000 following wide deployment of *Rlm1* varieties, effectively selecting for enrichment of the virulent *avrLm1* allele in *L. maculans* populations [34]. Interestingly, a similar enrichment for the virulent *avrLm1* allele was documented after the 'breakdown' of *LepR3* resistance in Australia [103]. Due to the threat of current resistance being rendered ineffective by shifting *L. maculans* populations, new effective sources of resistance are constantly in demand. In order to enlarge genetic variation for resistance to *L. maculans*, interspecific and intergeneric donor sources have been utilised. This has been achieved by conventional sexual crossing [44, 52, 75, 104] or via laboratory tools such as somatic hybridization [105], and embryo culture. Roy [52] crossed *B. juncea* and *B. napus* to introgress genes for blackleg resistance but none of the interspecific hybrids achieved the same level of *B. juncea* resistance as the donor parent. Wide hybrids (interspecific, intergeneric or intertribal) have also been produced either by sexual crossing followed by embryo culture or by somatic hybridisation as a result of protoplast fusion to transfer genes for blackleg resistance [106, 107]. Previous studies have reported hybrids between *B. napus* and *Arabidopsis thaliana*, belonging to different tribes; the Brassiceae and Sisymbrieae, respectively [108]. These hybrids were further utilized for identifying genetic regions associated with blackleg resistance [49]. Two regions localised on chromosome 3 of *A. thaliana* were shown to be linked with resistance to *L. maculans*.

Crouch *et al.* [75] transferred genes for resistance to *L. maculans* derived from *B. rapa* subsp. *sylvestris* into *B. napus*, using a resynthesised amphidiploid, as a result of hybridisation between *B. rapa* subsp. *sylvestris* and *B. oleracea* subsp. *alboglabra*. As a result, several cultivars derived from the re-synthesized *B. napus* lines were released for commercial cultivation in Australia such as Surpass 400, Surpass 404CL, Surpass 501TT, Surpass 603CL, Hyola 43, and Hyola 60. The *R*-genes *LepR1*, *LepR2* and *LepR4* have also been introgressed into *B. napus* via conventional interspecific crosses [75, 109]. Introgression of genes for resistance to *L. maculans* from *Sinapis arvensis*, *Coincya momensis* and *B. juncea* into *B. napus* was attempted [110]. Hybrid derivatives of *B. napus* and *S. arevensis*, and *B. napus* and *C. momensis* showed a high levels of resistance at the seedling (cotyledon) and/or adult plant stages. The offspring from asymmetric hybrids between *B. napus* and *B. nigra, B. juncea* and *B. carinata* were analysed for the presence of B genome markers and resistance to *L. maculans* [111]. This study revealed that resistance is conserved in one triplicate region in the B genome. Often, the majority of wide-hybrid derivatives exhibit unwanted traits and low frequencies of recombination between the different species which complicate the development of *B. napus* cultivars resistant to *L. maculans* by traditional breeding [43, 47]. Linkage drag due to suboptimal/undesired genes can be eliminated using the application of high density genome-wide molecular markers such as SNPs [112]. However, Rouxel and Balesdent [93] cautioned that before important breeding efforts are devoted to introgression of resistance genes from distant species into Brassica, there is a need thoroughly to evaluate their genetic control, putative redundancy and potential durability in the field.

Using transgenic technology, *R*-genes from other organisms can also be transferred irrespective of natural barriers to crossing. However, it is possible that transferred genes may not always contribute novel resistance specificities to the transgenic crop. Although several approaches have been used to induce host resistance in plants [113, 114] no major breakthrough has been made for an efficient management of blackleg disease. For example, Hennin *et al.* [115] demonstrated the expression of *Cf9* gene, which confers *Avr9*-dependant resistance to *Cladosporium fulvum* in tomato, along with co-expression of *Avr9* produced increased resistance to *L. maculans* in transgenic *B. napus* plants. Manipulation of plant defense responses is resource-expensive [116] and may be deleterious to the plant. Plants need to be selected for both appropriate expression of beneficial defense responses and avoidance of unnecessary ones [117], making artificially-induced constitutive expression of these responses an impractical solution to engineering resistance.

10. DURABILITY OF RESISTANCE TO *L. MACULANS*

Durable disease resistance can be achieved by utilisation of one or more single dominant *R*-genes [118]. However, the effectiveness of the specific *R*-genes depends on the *L. maculans* population structure, i.e. on the frequency of the corresponding *Avr* allele, which is known to differ according to regions/countries [27, 94] and the rapid evolution of virulent pathotypes. For example, the mean number of virulence alleles per isolates was reported to be higher in Australia (5.11 virulence alleles) than in Europe (4.33) and Canada (3.46) [27]. It has been suggested that there is a fitness cost associated with pathogen evolution from avirulence to virulence to overcome host resistance [38, 119].

Previous research has shown that different qualitative gene sources for resistance vary in providing effective durable resistance over period of time. For example, Light *et al.* [120] reported that the adult plant survival of French winter lines such as Doublol (*Rlm1*), Capitol (*Rlm1, Rlm3*), Columbus*1 (*Rlm1, Rlm3*), Carolus (*Rlm1, Rlm2, Rlm3*) and Rlm_EX (*Rlm7*) was higher than the Australian cultivar, AV-Sapphire and concluded that French winter canola cultivars have effective resistance under Australian conditions.

Single resistance genes do not always provide a durable resistance as has been shown in a field experiment using the *Jlm1/Rlm6* gene introgressed into *B. napus* from *B. juncea* [121]. Several incidences on the breakdown/ineffectiveness of race-specific resistance genes in Surpass 400 ((*LepR3, RlmS*)), in Vivol and Capitol (*Rlm1*), and *Rlm6* genes in *Brassica* have been reported in literature particularly when they were grown extensively [34, 94, 122]. As a consequence, breeders have to develop new cultivars and replace 'old' cultivars in order to change pathogen specificity of *R-gene* even without the knowledge of comprehensive distribution of *Avr* genes. The latter is now feasible and being used in order to monitor the pathogen population [123]. In order to avoid selection pressure against a particular *Avr* gene in the pathogen population, pyramiding of several host *R*-genes and deployment of quantitative resistance is being practiced in several crops such as in wheat, and barley. However, this strategy has not resulted in greater durability of resistance [124, 125]. In contrast, a recent study [121] demonstrated that a major *R*-gene (*Rlm6*) is more durable when expressed in a genetic background that also has quantitative resistance, indicating the need to identify and combine both qualitative and quantitative loci for blackleg resistance. Although the proposed strategy may be useful for blackleg disease management in areas where 'less' disease pressure and low variability with *L. maculans* populations exists, in Australia polygenic resistance derived from the French cultivar Jet Neuf [87], was reported to become less effective over time [37]. Additionally, several Australian cultivars which are reported to harbour both qualitative and quantitative loci for blackleg resistance are susceptible to natural populations of *L. maculans* Delourme et al [99]. It is difficult to know whether this evolution results from a change in virulence, or in aggressiveness in the pathogen populations since these polygenic-resistance cultivars may also carry specific *R*-genes [99]. In order to keep the frequency of isolates virulent towards any race–specific gene under a 'threshold' level, an integrated approach based upon best farm practices such as crop rotation, stubble management, application of fungicides and deployment of resistance genes including rotation of race-specific genes [126] needs to be implemented for sustainable canola production, especially in areas where *L. maculans* populations are highly diverse and rapidly evolving.

11. MOLECULAR DISSECTION OF QUALITATIVE AND QUANTITATIVE RESISTANCE LOCI

Molecular markers have been applied to identify loci associated with resistance to *L. maculans*, which relies on the availability of sequence variation among parental genotypes of mapping populations and diversity panels. Several genotyping methods based upon DNA hybridisation such as Restriction Fragment Length Polymorphism (RFLP) and Diversity Arrays Technology (DArT); PCR-based techniques such as Randomly Amplified Polymorphic DNA (RAPD), Simple Sequence Repeats (SSR) and Amplified

Fragment Length Polymorphism (AFLP); Sequence-related amplified polymorphism (SRAP); and sequence-based analysis such as Single Nucleotide Polymorphism (SNP); Restriction site-associated DNA, (RAD) and Genotyping by Sequencing (GBS) have been developed for molecular analyses [81, 127 - 136]. RFLP, RAPD, SSR, SRAP and AFLP markers have all been used to map loci for resistance to *L. maculans* (Table 1). New marker technologies such as DArT, 60K SNP Infinium array, RAD and GBS are currently being developed and applied for mapping of blackleg resistance loci. These high-throughput approaches are expected to complement or replace low-throughput marker assays that were used previously to facilitate genetic and physical map-based cloning of resistance genes.

Loci for resistance to *L. maculans* have been mapped using linkage/QTL mapping and association mapping approaches [32, 74, 82, 86, 137] using structured (F_2, doubled-haploid (DH) and backcross) and unstructured (diversity sets/breeding lines) populations (Table 1). Bulked Segregant Analysis approach, used for the first time [138], is particularly useful when a limited number of traits are to be mapped and resources (money and time) required for extensive genotyping are limited [81]. Whole-genome analysis has been used to locate both qualitative and quantitative loci associated with resistance to *L. maculans* [32, 86]. Generally, it requires the framework linkage map of all 19 chromosomes (linkage groups) for linkage (QTL) analysis.

11.1. Qualitative Resistance

The majority of genes for resistance to *L. maculans* have been genetically mapped with molecular markers (Table 1) on chromosomes A1, A2, A6, A7, A10, B3, B4 and B8 in *Brassica species: B. rapa, B. napus, B juncea* and *B. nigra* [31, 32, 45, 68-70, 99]. None of the race-specific genes have been mapped on the C genome yet. Previous linkage mapping studies revealed that at least five resistance genes (*Rlm1, Rlm3, Rlm4, Rlm7* and *Rlm9)* are localised in a cluster within a 35 cM genomic region on chromosome A7 [32, 45, 64-68, 74, 82]. This genomic region showed extensive inter- and intra-genomic duplications, as well as intra-chromosomal tandem duplications [140]. Whether some of these *R*-genes are allelic remains unknown. For example, it was concluded that at least four resistance genes *Rlm3, Rlm4, Rlm7,* and *Rlm9* could correspond to a cluster of tightly linked genes, to a unique gene with different alleles, or to a combination of these two hypotheses. However, *Rlm1* has been shown to be linked with *Rlm3* but is not allelic [74].

A major gene named *LmFr1* controlling adult plant resistance to blackleg was tagged in the DH population from French cultivar Cresor (resistant to *L. maculans*) and Westar (susceptible to *L. maculans*) with RFLP markers [64].

Similar study [65] mapped loci for blackleg resistance in a DH population from Major/Stellar and found that genetic control of resistance vary with inoculation techniques. A major gene designated as *LEM1* was mapped to linkage group 6 based on qualitative/quantitative scores of the interaction phenotype on inoculated cotyledons with a single ascospore-derived PG2 isolate, PHW1245. However, four other putative QTL for resistance were also identified on linkage groups LG8, LG17 and pair 4. This study further showed that none of the QTL that were associated with resistance at the seedling and stem stage had a significant effect in conferring resistance in the field. This may be attributed due to use of different pathogen population (PHW 1245 in cotyledon and stem experiments and natural *L. maculans* population in field experiment). The *Rpg3Dun* gene was mapped in an F_2 population from Westar/Dunkeld and identified a suite of SCAR markers that showed cosegregation with resistance to *L. maculans* [81]. Recently, the whole genome average interval mapping approach was applied to localise both qualitative and quantitative trait loci controlling blackleg resistance [32] in a DH population derived from the Australian *B. napus* vernalisation responsive cultivars, Skipton and Ag-Spectrum. Marker regression analyses revealed that at least fourteen genomic regions were associated with blackleg resistance, explaining 19.5% to 88.9% of genotypic variation. A major qualitative locus, designated *RlmSkipton (Rlm4),* was mapped on chromosome A7, within 0.8 cM of the SSR marker BRMS075 (Table 1).

Genomic regions of chromosome A10 harbours *Rlm2*, which has been shown to be the most common *R*-gene in winter *B. napus* varieties, such as Samourai, Eurol, Bristol, Symbol, Andol, Kintol, Akamar, Colvert, Synergy, and Tapidor, [41]. Chromosome A10 also harbours *LepR2*, *LepR3* and *BLMR2* genes derived from *B. rapa* subsp. *sylvestris* sources [31, 69, 70]. *LepR1* and *LepR2* were mapped on chromosomes A2 and A10, respectively with RFLP markers [31]. Genetic analysis revealed that both genes confer resistance independently and therefore are additive. *LepR1* was a dominant nuclear gene while *LepR2* was an incompletely dominant gene. This study further showed that *LepR1* generally conferred a higher level of resistance than *LepR2*. Both genes exhibited race-specific interactions with pathogen isolates.

The blackleg resistance gene *Rlm6* has been identified on B genome chromosome 8 [47]. *Rlm6* has been successfully introgressed to *B. napus* (AACC) from *B. juncea* (AABB) [47, 141] and provides excellent resistance to *L. maculans* isolates [58], though this gene has not yet been deployed in commercial cultivars [47, 58].

11.2. Quantitative Resistance

The genetic basis of quantitative resistance has been investigated only in limited *B. napus* cultivars such as in Darmor; a derivative of Jet Neuf [119, 137, 142]. However, a number of DH populations have recently been utilised for identification of loci for quantitative resistance under field conditions [32, 86, 143], and are currently being validated (Raman *et al.*, unpublished, Larkan *et al.*, unpublished). Thirteen quantitative trait loci (QTL) on 10 linkage groups associated with quantitative field resistance to *L. maculans* were identified in a DH population from Darmor-*bzh*/Yudal [87]. Their detection was dependent upon phenotypic method used; seven QTL for mean disease index and six QTL for per cent survival (percentage of lost plants due to canker) and were also dependant on growing environment (year of evaluation). However, only four of the QTL were stable across experiments. These QTL accounted from 23% to 57% of the genotypic variation (Table 2). The unexplained variation was described due to non-detected additive QTL, G x E interaction and incomplete map coverage. This study further showed that resistance to *L. maculans* is influenced with growth habit. For example, one QTL, located close to a dwarf gene (*bzh*), was detected with a very strong effect, masking the detection of other QTL. This study further showed that these dwarfing genes also affect other traits such as earliness, and glucoinsolate content.

In order to validate the stability of QTL for field resistance to *L. maculans*, QTL were mapped and characterised in $F_{2:3}$ population from Darmor (resistant)/Samourai (susceptible) revealing only four QTL on LG3, LG11 and DY5 and DS6 that were consistent in Darmor/Yudal and Darmor/Samourai populations [143]. This study found that the genetic background and inoculum pressure are the major factors of the QTL instability and therefore suggested that QTL mapping must be carried out separately for each population. The genomic regions carrying the most consistent resistance QTL in Darmor do not correspond to the two regions on N7 (A7) and N10 (A10) identified as carrying race specific resistance genes to *L. maculans* [74]. The position of *Rlm2* on N10 (A10) corresponds to a QTL identified for adult plant resistance in the Darmor/Samourai DH population [88]. The cultivar Samourai carries both the resistance allele at this QTL and *Rlm2*. Since it has been reported that no French isolates of *L. maculans* carry *AvrLm2* [34], two hypotheses were proposed to explain this co-location; either the *Rlm2* gene has a residual effect at the adult plant stage, similar to that suggested in other pathosystems, or genes linked to *Rlm2* are responsible for part of variation for resistance at this QTL [99].

QTL for blackleg resistance were identified in four mapping populations derived from the crosses Caiman/Westar_{10}, Camberra/Westar_{10}, ^{AV}Sapphire/Westar_{10} and Rainbow/^{AV}Sapphire [86]. Multiple QTLs were identified accounting for 13–33% of phenotypic variance. A recent study [32] identified seven significant QTL associated with blackleg resistance, scored on the basis of internal disease score, on chromosomes A2, A9, A10, C1, C2, C3 and C6 in a DH population derived from Skipton/Ag-Spectrum. The genotypic variation explained by the individual QTL ranged from 5% to 24.5%. Both parents contributed the alleles for blackleg resistance. This study showed poor correlation between canker lesion scores over the two years (2009, 2010). Some of the genomic regions for blackleg resistance may be the same as reported previously that have been identified using both classical QTL and association mapping approaches [31, 69, 87, 137, 144, 145]. The conservation of QTL between Australian and French studies is interesting and suggests the non-specificity of these QTL, irrespective of the environment, genetic background and G x E interactions [32]. However, it is possible that some of the original donor gene sources in French and Australian parental lines used for mapping resistance genes may be the same.

The majority of mapping populations used to map blackleg resistance genes in *B. napus* so far have been comparatively small (Table 1). The development of a high density map utilising larger populations, comprising several hundred to thousands lines, will allow for the precise mapping of resistance loci. Stability of QTL resistance needs to be tested in different environments. Although QTL mapping studies provide comprehensive information on the nature of inheritance, location, magnitude and allelic effects of QTL, much of the information tends to be 'population' specific. In biparental (structured) populations, generally two alleles at each locus are sampled and therefore trait-marker association may not be highly relevant to diverse genetic backgrounds. The validation of trait-marker association is necessary before their use for routine marker-assisted breeding (MAS). Association mapping can be utilised for investigating linkage disequilibrium close to loci of interest in a diverse germplasm [145-149] and therefore offers an alternative to linkage and QTL mapping. This approach has been applied in determining and confirming the markers located within the QTL associated with resistance to *L. maculans* previously identified in Darmor and established their usefulness in MAS [137]. A diverse set of an oilseed rape collection, comprised of 128 lines showing a large spectrum of responses to infection by *L. maculans*, was characterised using 72 SSR and other markers. At least 61 marker alleles were found to be associated with resistance to stem canker. Some of these markers were associated with previously identified QTL, which confirms their usefulness in MAS. Markers located in regions not harbouring previously identified QTL were also associated with resistance, suggesting that new QTL or allelic variants are present in the collection [137]. Genome-wide association based on 1513 markers enabled identification and validation of genomic loci associated with blackleg resistance. This study detected significant marker - race-specific blackleg resistance associations (P<0.01) at the seedling and adult plant stages. Loci for resistance were located on chromosomes A1, A2, A3, A5, A6, A7, A10, C1, and C2. Both studies suggested that association mapping is an efficient approach for identifying novel loci/alleles associated with blackleg resistance in diverse germplasm [137, 142]. Superior molecular marker allele(s) associated with resistance to *L. maculans* may be captured by canola breeding programs. Molecular markers associated with seedling and stem canker resistance will help identify accessions carrying desirable alleles and facilitate QTL introgression to develop elite germplasm having new gene/allele combinations for blackleg resistance [32].

12. HOST *R*-GENE CLONING AND CANDIDATE GENE ANALYSIS

At least 20 *R*-genes and several allele variants and haplotypes of cloned *R*-genes have been identified in plants [151-158]. Molecular analyses revealed that these genes belong to large multiple gene families, which encode nucleotide binding site- leucine–rich repeats (NBS-LRRs), serine-threonine-kinases, leucine zipper and protein kinase domains, and toll/interleukin-1 receptor domains [159-164]. These genes are often clustered in many plant species including crops such as rice, maize and soybean and transduce the hypersensitive response to defend against pathogen attack [164-167]. At least 30 CC-NBS-LRR and TIR-NBS-LRR non-redundant genes have been identified in *B. rapa* [167]. Two major gene clusters for resistance to *L. maculans* exist on chromosomes A7 [74] and A10 [31, 69, 70], along with other genes dispersed on different chromosomes. It is possible that some of these *R*-genes may represent to multiple copies of the same functional gene. A recent study has shown that at least eight functional copies of *FLOWERING TIME LOCUS C* (*FLC*) exist within *B. napus* [6] which may modulate flowering time and other functions in different cultivars [168].

In *B. napus*, only few studies aimed at characterizing the genes underlying the resistance to *L. maculans* have been attempted. The recent cloning of the first functional *B. napus* resistance gene *LepR3* revealed a receptor-like protein responsible for conferring resistance to *AvrLm1 L. maculans* isolates [79]. Resistance genes effective against *L. maculans* have also been cloned in *A. thaliana* [169-171], which encode Toll interleukin-1 receptor-nucleotide binding (TIR-NB) or TIR-NB-LRR class proteins. Based on the synteny between *B. napus* and *A. thaliana*, it was deduced that several *B. napus* resistance genes are localised in a region of A7 (N7) that corresponds to the chromosome segment on *Arabidopsis* chromosome 1 which harbours *RLM1*Col [139, 167]. However, a recent report detailing the gene responses to *L. maculans* infections suggests very different responses in *B. napus* and *A. thaliana* [172]. Both salicylic acid and ethylene signaling was triggered in *B. napus*, possibly due to the hemibiotrophic nature of the infection. This stands in contrast to the JA signaling observed in *A. thaliana*, suggesting *L. maculans* is acting as a necrotroph during infection of susceptible *A. thaliana* lines. Since many *R*-genes are conserved and share sequence similarity, degenerated primers based on conserved motifs of *R*-genes have also been used to localise potential resistance gene loci in Brassica species such as *B. oleracea* (on chromosomes C1 (O1), C4 (O4), C8 (O8) and C9 (O9) and *B. napus* on linkage groups LG1a, LG1b, LG2, LG5, LG8, LG12, LG13, LG14, LG15 and LG18 [173, 174]. However their association with loci controlling resistance to *L. maculans* have not yet been established/validated.

In order to clone genes controlling blackleg resistance in *B. napus* population, high resolution mapping of *LmR1* and *ClmR1* loci was performed using 2500 backcross lines from two crosses between PSA12 and Shiralee, and PSA12 and Cresor, respectively [140], and reported that both resistance loci are located in a highly duplicated genomic region on chromosome A7. This region contained several genes encoding protein kinases or LRR domains. It is reported that the SCAR marker (BN204) that showed cosegregation with *RpgDun* locus for resistance to *L. maculans* is derived from a region showing 92% amino acid identity with the defense-related gene serine threonine 20 (ste-20) protein kinase of *Arabidopsis thaliana* [81]. A proteomic approach has also been utilised to understand gene expression in response to *L. maculans* infection [176]. However, candidacy of any of these genes has not yet been reported.

Recently an alternative approach for identifying candidate *R*-genes has been employed based on genomics [177]. Next-generation massively parallel sequencing platforms such as the Roche 454 genome sequencer FLX instrument, the Illumina Genome Analyser (HiSeq), and the ABI SOLiD System have revolutionized genome sequencing by providing high throughput and cost-effective high coverage sequencing [179-182] and has enabled much quicker identification of candidate genes [178]. Molecular markers associated with *RlmSkipton* (*Rlm4*) locus in the DH population from Skipton/Ag-Spectrum were aligned with the complete genome sequence *B. rapa* as reported in [32]. Eighteen candidate genes, designated as *BLR1-18* with disease resistance characteristics, several of which were clustered around a region syntenic to *Rlm4*. Among candidates, *BLR2* and *BLR11* were the promising candidates for *Rlm4*-mediated resistance [178]. High resolution mapping and gene sequencing of different sources of *L. maculans* resistance will allow for a better understanding of the structural organisation and function of *R*-genes.

Recently, the reference genome of *B. rapa* has been published [182] and genomes of *B. oleracea*, *B. nigra* and *B. napus* are expected to be published in coming years. Re-sequencing of whole genomes of known blackleg-resistant genotypes will allow identification of genetic variation between individuals, which can provide molecular genetic markers and insights into gene function [183]. Sequencing of different *R*-genes and understanding their function will also enable us to manipulate resistance to *L. maculans*, as genes with different specificities can be created.

13. PREDICTIVE BREEDING FOR RESISTANCE TO *L. MACULANS* USING MOLECULAR MARKERS

Success of new disease resistance genes relies heavily on the successful transfer of target genomic regions from donor sources and the development of rigorous selection methods. Molecular markers have been used to improve the effectiveness and efficiency of selection strategies in predictive breeding in several agricultural crops. However, the development of molecular markers in *B. napus* and their application in breeding is a challenging exercise due to the large genome size, amphidipliod (4X) nature, open-pollination and lower research funding as compared to other key crops such as wheat, barley, maize and soybean. The *B. napus* genome is highly complex and homologous recombination plays a major role in chromosome rearrangements such as duplications and reciprocal translocations. These arrangements further add to the complexity of molecular analysis and interpretation. *B. napus* chromosomes C6 and A7, which harbours *Rlm1*, *Rlm3*, *Rlm4*, *Rlm7* and *Rlm9* genes for resistance, produced a reciprocal translocation in some cultivars such as in Westar, Marnoo, Monty and Maluka [185, 186] which makes analysis of resistance genes difficult [142].

In most of the breeding programs, selection for blackleg is conducted once a year during the growing season, hampering selection efficiency. Several studies suggest a significant correlation between cotyledon test and canker lesion scores. Therefore, cotyledon tests can be used for selection for resistance to *L. maculans*. However, in many developed countries, it is costly and laborious to perform, particularly as compared to molecular marker analysis, when several tests need to be carried to screen large populations. Furthermore, analysis of different blackleg resistance genes in a canola breeding program using a differential set of *L. maculans* isolates at various stages of the breeding cycle is a very slow process [39]. Interpretation of *R*-gene content using a differential set of control *B. napus* varieties, especially of Australian origin, is a challenging exercise, as majority of cultivars used are heterozygous and/or heterogeneous [32, 41]. In addition, phenotypic tests are dependent upon the growing environment (microclimate conditions and other factors such as powdery and downy mildew), which can complicate scoring of inoculated seedlings. Molecular markers generally out-perform conventional seedling assays, in both efficiency and reliability. It is also possible to identify haplotypes using molecular markers and then validate trait-marker associations, in conjunction with comprehensive phenotyping and conventional allelism tests.

The published literature suggests that little effort has been made to evaluate the allelic relationship among the known genes from different sources, to test stability of majority of QTL or qualitative genes identified over diverse growing environments, or to test their usefulness in achieving long term durable control of the disease. Table 1 also suggests that majority of markers are not very closely linked (<1cM) with resistance loci. Diagnostic or perfect markers for resistance genes are required for routine MAS and will assist allele enrichment strategies in breeding programs, although this is not always possible, even if the complete gene is cloned and characterised for its functionality [187]. The linkage between molecular markers and *Xbn204* flanking the *RlmSkipton* locus was verified in an F_2population derived from Skipton/Ag-Spectrum [32]. The results showed that SSR markers linked to *RlmSkipton* are suitable for enrichment of favourable alleles for blackleg resistance in breeding programs. A separate study [82] validated the map location of *Rlm1* in the DH population derived from Maxol/Westar with SSR and DArT markers. Previously, *Rlm1* and *Rlm3* genes were mapped on chromosomes A7 in the Maxol (resistant to blackleg)/S006 (susceptible to blackleg) utilising RAPD markers and with single spore isolates with known *Avr* genotypes in the *B. napus* European cultivars, Columbus and Maxol [41, 71, 74]. RAPD markers are not amenable for high throughput marker analysis, as they are assayed on low-throughput agarose or polyacrylamide gel systems. Validation of a large array of genes for blackleg resistance in diverse segregating populations representing *B. napus* germplasm is a challenging exercise. However, an association mapping approach can be employed to test trait-marker associations in a large set of germplasm as demonstrated recently [137, 142].

14. CONCLUSIONS

It is now clear that major resistance genes will be overcome in time, as has been seen in many crop plants. Therefore, there is constant need to identify new sources of both qualitative and quantitative resistance loci and to properly utilise the resources available to us so that resistance can be deployed long term. Recent advances in molecular marker systems, such as the development of highly-parallel systems for genotyping and sequencing, have created new opportunities and strategies to select for qualitative and quantitative traits, including resistance to *L. maculans*. Strategies for deploying resistance in breeding programs will vary with individual breeding programs; monitoring introgression of specific loci, using whole-genome marker scans (genomic selection) or identifying individual plants that may offer the greatest opportunity for genetic gain. This is now becoming reality as several genome-wide signals associated with blackleg resistance have been identified (but need to be validated) and alleles at these loci can be selected efficiently and at a cheaper rate with new marker technologies. Development and validation of tightly-linked molecular markers amenable to high throughput marker screening with both qualitative and quantitative resistance and cost effective systems will enable the increased adoption in *B. napus* breeding programs. In addition to genetic resistance, deployment of agronomic practices such as use of rotation and stubble management will remain key management tools for reducing pathogen inoculum for subsequent crops.

15. ACKNOWLEDGEMENTS

Authors are thankful to Dr Regine Delourme, INRA, Le Rheu, Cedex France for providing critical comments and QTL information for quantitative resistance in the Darmor/Yudal population.

REFERENCES

1. Tode HI. *Fungi Mecklenburgenses Selecti*. Fasciculus. 1791;II(51, Plate XVI, Fig 126).
2. Olsson G. Species Crosses within the Genus *Brassica*. I: Artificial *Brassica napus*. Hereditas. 1960;46:351-96.
3. U N. Genomic Analysis in Brassica with Special Reference to the Experimental For-mation of *B. napus* and Peculiar Mode of Fertilisation. Jpn J Bot. 1935;7:389-452.
4. Gómez-Campo C, Prakash S. Origin and Domestication. In: C.Gómez-Campo (Ed.), Biology of Brassica Coenospecies. Elsevier, Netherlands, 33-58. 1999.
5. Baranyk P, Fábry A. History of the Rapeseed (*Brassica Napus* L.) Growing and Breed-ing from Middle Age Europe to Canberra. Proceedings 10th International Rapeseed Congress, Canberra, Australia (http://regionalorgau/au/gcirc/4/374htm). 1999.
6. Zou X, Suppanz I, Raman H, Hou J, Wang J, Long Y, et al. Comparative Analysis of *FLC* Homologues in Brassicaceae Provides Insight into Their Role in the Evolution of Oilseed Rape. PLoS One 2012 ; 7(9): e45751.
7. FAO: Food and Agriculture Organisation (2012) http://faostat.fao.org/.
8. Fitt B, Brun H, Barbetti M, Rimmer S. World-Wide Importance of Phoma Stem Can-ker (*Leptosphaeria maculans* and *L. biglobosa*) on Oilseed Rape (*Brassica napus*). Europe- an Journal of Plant Pathology. 2006;114(1):3-15.

9. West JS, Kharbanda PD, Barbetti MJ, Fitt BDL. Epidemiology and Management of *Leptosphaeria maculans* (Phoma Stem Canker) on Oilseed Rape in Australia, Canada and Europe. Plant Pathol. 2001;50(1):10-27.
10. Howlett BJ. Current Knowledge of the Interaction between *Brassica napus* and *Leptosphaeria maculans*. Canadian Journal of Plant Pathology. 2004;26(3):245-52.
11. Marcroft S, Bluett C. Blackleg of Canola. Agriculture Notes, State of Victoria, Department of Primary Industries, May 2008, Ag1352, ISSN 1329-8062. 2008.
12. Henderson MP. The Black-Leg Disease of Cabbage Caused by *Phoma lingam* (Tode) Desmaz. Phytopathology. 1918;8:379-431.
13. Sosnowski MR, Scott ES, Ramsey MD. Infection of Australian canola cultivars (*Brassica napus*) by *Leptosphaeria maculans* is influenced by cultivar and environmental conditions. Australasian Plant Pathology. 2004;33(3):401-11
14. Petrie GA. Variability in *Leptosphaeria maculans* (Desm.) Ces et de Not., the Cause of Blackleg of Rape. PhD Thesis, University of Saskatchwen, Canada. 1969.
15. Hall R. Epidemiology of Blackleg of Oilseed Rape. Canadian Journal of Plant Pathology 1992;14:46-55.
16. Alabouvette C, Brunin B. Recherches Sur La Maladie Du Colza Due a *Leptospharia maculans* (Desm.) Ces. Et De Not. 1. Role Des Restes De Culture Dans La Conservation Et La Dissemination Du Parasite. Ann Phytopathol 2(3): 463-475. 1970.
17. Bokor A, Barbetti MJ, Brown AGP, MacNish GC, Poole ML, Wood P. Blackleg - Major Hazard to the Rapeseed Industry. Journal of Agriculture, Western Australia. 1975;16:7-10.
18. Ghanbarnia K, Fernando WGD, Crow G. Developing Rainfall- and Temperature-Based Models to Describe Infection of Canola under Field Conditions Caused by Pycnidiospores of *Leptosphaeria maculans*. Phytopathology 2009;99(7):879-86.
19. Ghanbarnia K, Fernando WGD, Crow G. Comparison of Disease Severity and Incidence at Different Growth Stages of Naturally Infected Canola Plants under Field Conditions by Pycnidiospores of *Phoma lingam* as a Main Source of Inoculum. Canadian Journal of Plant Pathology. 2011;33(3):355-63.
20. Chen CY, Howlett BJ. Rapid Necrosis of Guard Cells Is Associated with the Arrest of Fungal Growth in Leaves of Indian Mustard (*Brassica juncea*) Inoculated with Avirulent Isolates of *Leptosphaeria maculans*. Physiological and Molecular Plant Pathology. 1996;48(2):73-81.
21. Hua L, Sivasithamparam K, Barbetti MJ, Kuo J. Germination and Invasion by Ascospores and Pycnidiospores of *Leptosphaeria maculans* on Spring-Type *Brassica napus* Canola Varieties with Varying Susceptibility to Blackleg. Journal of General Plant Pathology 2004;70(5):261.
22. Hammond KE, Lewis BG, Musa TM. A Systemic Pathway in the Infection of Oilseeed Rape Plants by *Leptosphaeria maculans*. Plant Pathology. 1985;34:557-65.
23. Travadon R, Marquer B, Ribule A, Sache I, Masson JP, Brun H, et al. Systemic Growth of *Leptosphaeria maculans* from Cotyledons to Hypocotyls in Oilseed Rape: Influence of Number of Infection Sites, Competitive Growth and Host Polygenic Resistance. Plant Pathology. 2009;58:461-9.
24. Sprague SJ, Watt M, Kirkegaard JA, Howlett BJ. Pathways of Infection of *Brassica napus* Roots by *Leptosphaeria maculans*. New Phytologyst. 2007;176: 211-222.
25. Cunningham GH. Dry-Rot of Swedes and Turnips: Its Causes and Control. Wellington, New Zealand: New Zealand Department of Agriculture: Bulletin 133. 1927.
26. Kutcher HR, van den Berg CGJ, Rimmer SR. Variation in Pathogenicity of *Leptosphaeria maculans* on *Brassica Spp.* Based on Cotyledon and Stem Reactions. Can J Plant Pathol 1993;15:253-8.
27. Balesdent MH, Barbetti MJ, Li H, Sivasithamparam K, Gout L, Rouxel T. Analysis of *Leptosphaeria maculans* Race Structure in a Worldwide Collection of Isolates. Phytopa- thology. 2005 Sep;95(9):1061-71.

28. Koch E, Song K, Osborn TC, Williams PH. Relationship between Pathogenicity Based on Restriction Fragment Length Polymorphism in *Leptosphaeria maculans*. Molecular Plant-Microbe Interactions. 1991;4:341-9.
29. Aubertot J, West J, Bousset-Vaslin L, Salam M, Barbetti M, Diggle A. Improved Resistance Management for Durable Disease Control: A Case Study of Phoma Stem Canker of Oilseed Rape (*Brassica napus*). European Journal of Plant Pathology. 2006;114(1):91-106.
30. Cargeeg LA, Thurling N. Contribution of Host-Pathogen Interactions to the Expression of the Blackleg Disease of Spring Rape (*Brassica napus* L.) Caused by *Leptosphae- ria maculans* (Desm.) Ces. et de Not.. Euphytica. 1980;29:465-76.
31. Yu F, Lydiate DJ, Rimmer SR. Identification of Two Novel Genes for Blackleg Resistance in *Brassica napus.* Theoretical and Applied Genetics. 2005;110:969-79.
32. Raman R, Taylor B, Marcroft S, Stiller J, Eckermann P, Coombes N, et al. Molecular Mapping of Qualitative and Quantitative Loci for Resistance to *Leptosphaeria maculans*; Causing Blackleg Disease in Canola (*Brassica napus* L.). Theoretical and Applied Genetics. 2012;125(2):405-18.
33. Bansal VK, Kharbanda PD, Stringam GR, Thiagarajah MR, Tewari J. A Comparison of Greenhouse and Field Screening Methods for Blackleg Resistance in Doubled Haploid Lines of *Brassica napus.* Plant Disease 1994;78:276-81.
34. Rouxel T, Penaud A, Pinochet X, Brun H, Gout L, Delourme R, et al. A 10-Year Survey of Populations of *Leptosphaeria maculans* in France Indicates a Rapid Adaptation Towards the *Rlm1* Resistance Gene of Oilseed Rape. European Journal of Plant Pathology. 2003;109(8):871-81.
35. McNabb WM, Van Den Berg CGJ, Rimmer SR. Comparison of Inoculation Methods for Selection of Plant Resistant to *Leptosphaeria Maculans* in *Brassica napus.* Can J Plant Sci 1993;73:1199-207.
36. Pang ECK, Halloran GM. Genetics of Virulence in *Leptosphaeria maculans* (Desm.) Ces. et De Not., the Cause of Blackleg in Rapeseed (*Brassica napus* L.). Theoretical and Applied Genetics. 1996;93(3):301-6.
37. Salisbury P, Ballinger DJ, Wratten N, Plummer K, Howlett BJ. Blackleg Disease on Oilseed Brassica in Australia: A Review. Australian Journal of Experimental Agricultural. 1995;35:665-72.
38. Huang YJ, Li ZQ, Evans N, Rouxel T, Fitt BDL, Balesdent MH. Fitness Cost Associat- ed with Loss of the *Avrlm4* Avirulence Function in *Leptosphaeria maculans* (Phoma Stem Canker of Oilseed Rape). European Journal of Plant Pathology. 2006;114(1): 77-89.
39. Marcroft SJ, Elliott VL, Cozijnsen AJ, Salisbury PA, Howlett BJ, Van de Wouw AP. Identifying Resistance Genes to *Leptosphaeria maculans* in Australian *Brassica napus* Cultivars Based on Reactions to Isolates with Known Avirulence Genotypes. Crop and Pasture Science. 2012;63(4):338-50.
40. Rimmer SR, van den Berg CGJ. Resistance to Oilseed Brassica Spp. To Blackleg Caused by *Leptosphaeria maculans.* Can J Plant Pathol. 1992;14: 56-66.
41. Rouxel T, Willner E, Coudard L, Balesdent M-H. Screening and Identification of Resistance to *Leptosphaeria maculans* (Stem Canker) in *Brassica napus* Accessions. Euphy- tica. 2003;133(2):219-31.
42. Marcroft SJ, Purwantara A, Salisbury P, Potter TD, Wratter N, Khangura R, et al. Reaction of a Range of *Brassica Species* under Australian Conditions to the Fungus, *Leptosphaeria maculans,* the Causal Agent of Blackleg. Australian Journal of Experimental Agriculture. 2002;42(5):587-94.
43. Christianson JA, Rimmer SR, Good AG, Lydiate DJ. Mapping Genes for Resistance to *Leptosphaeria maculans* in *Brassica juncea*. Genome. 2006;49(1):30-41.
44. Roy NN. Interspecific Transfer of *Brassica juncea*-Type High Blackleg Resistance to *Brassica napus*. Euphytica. 1984;33: 295-303.
45. Delourme R, Chevre AM, Brun H, Rouxel T, Balesdent MH, Dias JS, et al. Major Gene and Polygenic Resistance to *Leptosphaeria maculans* in Oilseed Rape (*Brassica napus*). European Journal of Plant Pathology. 2006;114(1):41-52.

46. Pang ECK, Halloran GM. The Genetics of Blackleg [*Leptosphaeria maculans* (Desm.) Ces. et de Not.] Resistance in Rapeseed (*Brassica napus* L.). II. Seedling and Adult-Plant Resistance as Quantitative Traits. Theoretical and Applied Genetics. 1996;93:941-9.
47. Chèvre AM, Barret P, Eber F, Dupuy P, Brun H, Tanguy X, et al. Selection of Stable *Brassica napus-B. juncea* Recombinant Lines Resistant to Blackleg (*Leptosphaeria macu- lans*). 1. Identification of Molecular Markers, Chromosomal and Genomic Origin of the Introgression. Theoretical and Applied Genetics. 1997;95(7):1104-11.
48. Snowden RJ, Winter H, Diestel A, Sacristan MD. Development and Characterisation of *Brassica napus-Sinapus arvensis* Addition Lines Exhibiting Resistance to *Leptosphae- ria maculans*. Theoretical and Applied Genetics. 2000;101:1008-14.
49. Bohman S, Wang M, Dixelius C. *Arabidopsis thaliana*-Derived Resistance against *Leptosphaeria maculans* in a *Brassica napus* Genomic Background. Theoretical and Applied Genetics. 2002;105(4):498-504.
50. Sjödin C, Glimelius K. Screening for Resistance to Blackleg *Phoma lingam* (Tode Ex Fr.) Desm. Within Brassicaceae. Journal of Phytopathology. 1988;123(4):322-32.
51. Leflon M, Brun H, Eber F, Delourme R, Lucas MO, Vallee P, et al. Detection, Introgression and Localization of Genes Conferring Specific Resistance to *Leptosphaeria maculans* from *Brassica rapa* into *B. napus*. Theoretical and Applied Genetics. 2007 Nov;115(7):897-906.
52. Roy NN. A Study on Disease Variation in the Populations of an Interspecific Cross of *Brassica juncea* L. x *B. napus* L. Euphytica 27:145-149. 1978
53. Mithen RF, Lewis BG, Heaney RK, Fenwick GR. Resistance of Leaves of Brassica Species to *Leptosphaeria maculans*. Trans Br Mycol Soc 88:525-531. 1987.
54. Mitchell-Olds T, James RV, Palmer MJ, Williams PH. Genetics of *Brassica rapa* (Syn. *Campestris*). 2. Multiple Disease Resistance to Three Fungal Pathogens: *Peronospora parasitica*, *Albugo candida* and *Leptosphaeria maculans*. Heredity. 1995;75(4):362-9.
55. Keri M, van den Berg CJG, McVetty PBE, Rimmer SR. Inheritance of Resistance to *Leptosphaeria maculans* in *Brassica juncea*. Phytopathology. 1997;87:594-8.
56. Brun H, Levivier S, Somda I, Ruer D, Renard M, Chevre AM. A Field Method for Evaluating the Potential Durability of New Resistance Sources: Application to the *Leptosphaeria maculans-Brassica napus* Pathosystem. Phytopathology. 2000;90(9):961-6.
57. Ansan-Melayah D, Balesdent MH, Delourme R, Pilet ML, Tanguy X, Renard M, et al. Genes for Race-Specific Resistance against Blackleg Disease in *Brassica napus* L. Plant Breeding. 1998;117(4):373-8.
58. Plieske J, Struss D, Röbbelen G. Inheritance of Resistance Derived from the B-Genome of Brassica against *Phoma lingam* in Rapeseed and the Development of Molecu- lar Markers. Theoretical and Applied Genetics. 1998;97(5):929-36.
59. Barret P, Guérif J, Reynoird JP, Delourme R, Eber F, Renard M, et al. Selection of Stable *Brassica napus-Brassica juncea* Recombinant Lines Resistant to Blackleg (*Leptosphaeria maculans*). 2. A 'To and Fro' Strategy to Localise and Characterise Interspecific Introgressions on the *B. napus* Genome. Theoretical and Applied Genetics. 1998;96(8): 1097-103.
60. Monteiro AA, Williams PH. The Exploration of Genetic Resources of Portuguese Cabbage and Kale for Resistance to Several Brassica Diseases. Euphytica. 989; 41:215-225. 1.
61. Ananga AO, Cebert E, Soliman K, Kantety R, Pacumbaba RP, Konan K. Rapd Mark- ers Associated with Resistance to Blackleg Disease in Brassica Species. African Jour- nal of Biotechnology. 2006;5(22):2041-8.
62. Wittern I. Wittern I (1984) Untersuchungen Zur Erfassung Der Resistenz Von Winter- raps (*Brassica napus* L. Var. *oleifera* Metzger) Gegenüber *Phoma lingam* (Tode Ex Fr.) Desm. Und Zu Der Durch Den Erreger Verursachten

Wurzelhals- Und Stengel- fäule. PhD Thesis, Universität Göttingen, Göttingen, Germany 1984.

63. Stringam GR, Bansal VK, Thiagarajah MR, Tewari JP. Genetic Analysis of Blackleg (*Leptosphaeria maculans*) Resistance in *Brassica napus* L. Using the Doubled Haploid Method. XIII Eucarpia Congress, July 6-11, Angers, France. 1992.
64. Dion Y, Gugel RK, Rakow GFW, Séguin-Swartz G, Landry BS. RFLP Mapping of Resistance to the Blackleg Disease [Causal Agent, *Leptosphaeria maculans* (Desm.) Ces. Et De Not.] in Canola (*Brassica napus* L.). Theoretical and Applied Genetics. 1995;91(8): 1190-1194.
65. Ferreira ME, Rimmer SR, Williams PH, Osborn TC. Mapping Loci Controlling *Brassica napus* Resistance to *Leptosphaeria maculans* under Different Screening Conditions. Phytopathology. 1995; 85:213-217.
66. Mayerhofer R, Bansal VK, Thiagarajah MR, Stringam GR, Good AG. Molecular Mapping of Resistance to *Leptosphaeria maculans* in Australian Cultivars of *Brassica napus*. Genome. 1997;40:294-301.
67. Rimmer SR, Borhan MH, Zhu B, Somers D. Mapping Resistance Genes in *Brassica napus* to *Leptosphaeria maculans*. Proceeding 10th International Rapeseed Congress, Can- berra, Australia. 1999. http://www.regional.org.au/au/gcirc/3/47.htm
68. Rimmer SR. Resistance Genes to *Leptosphaeria maculans* in *Brassica napus*. Can J Plant Pathol-Rev Can Phytopathol. 2006;28:S288-S297.
69. Yu F, Lydiate DJ, Rimmer SR. Identification and Mapping of a Third Blackleg Resistance Locus in *Brassica napus* Derived from *B. rapa* Subsp. *sylvestris*. Genome. 2008;51(1):64-72.
70. Long Y, Wang Z, Sun Z, Fernando DW, McVetty PB, Li G. Identification of Two Blackleg Resistance Genes and Fine Mapping of One of These Two Genes in a *Brassica napus* Canola Cultivar 'Surpass 400'. Theoretical and Applied Genetics. 2011;122(6): 1223-31.
71. Balesdent MH, Attard A, Kuhn ML, Rouxel T. New Avirulence Genes in the Phytopathogenic Fungus *Leptosphaeria maculans*. Phytopathology. 2002;92(10):1122-33.
72. Balesdent MH, Fudal I, Ollivier B, Bally P, Grandaubert J, Eber F, et al. The dispensable chromosome of *Leptosphaeria maculans* shelters an effector gene conferring avirulence towards *Brassica rapa*. New Phytol. 2013;198(3):887-98.
73. Van de Wouw AP, Marcroft SJ, Barbetti MJ, Hua L, Salisbury PA, Gout L, et al. Dual Control of Avirulence in *Leptosphaeria maculans* Towards a *Brassica napus* Cultivar with '*Sylvestris*-Derived' Resistance Suggests Involvement of Two Resistance Genes. Plant Pathology. 2008;58(2):305-13
74. Delourme R, Pilet-Nayel ML, Archipiano M, Horvais R, Tanguy X, Rouxel T, et al. A Cluster of Major Specific Resistance Genes to *Leptosphaeria maculans* in *Brassica napus*. Phytopathology. 2004;94(6):578-83.
75. Crouch JH, Lewis BG, Mithen RF. The Effect of a Genome Substitution on the Resistance of *Brassica napus* to Infection by *Leptosphaeria maculans*. Plant Breeding.1994; 112(4):265-278.
76. Buzza G, Easton A. A New Source of Blackleg Resistance from *Brassica sylvestris*. In GCIRC Technical Meeting Poznan, Poland Bulletin No18. 2002.
77. Yu F, Lydiate DJ, Hahn K, Kuzmisz Kuzmicz S, Hammond C, Rimmer SR. Identification and Mapping of a Novel Blackleg Resistance Locus *LepR4* in the Progenies from *Brassica napus* x *B. rapa* Subsp. *sylvestris*. Proc. 12th International Rapeseed Conference, Wuhan, China, March 2007.
78. Larkan NJ, Kuzmicz S, Yu F, Lydiate D, Genetic Evidence for the Recognition of the *Leptosphaeria maculans* Avirulence Gene *AvrLm1* by Two *Brassica napus* Resistance Genes: *Rlm1* and *LepR3*. Proceedings 17th Crucifer-Genetics Workshop; 2010; Saska- toon, Canada.
79. Larkan N.J., D.J. Lydiate, I.A.P. Parkin, M.N. Nelson, D.J. Epp, W.A. Cowling, et al. The *Brassica napus* blackleg resistance gene *LepR3* encodes a receptor-like protein

triggered by the *Leptosphaeria maculans* effector *AvrLM1* New Phytologist. 2013, 197:595-605

80. Saal B, Struss D, RGA- and RAPD-derived SCAR markers for a Brassica B-genome introgression conferring resistance to blackleg in oilseed rape. Theoretical and Applied Genetics, 2005; 111: 281-290
81. Dusabenyagasani M, Fernando WGD. Development of a Scar Marker to Track Canola Resistance against Blackleg Caused by *Leptosphaeria maculans* Pathogenicity Group 3. Plant Disease. 2008;92(6):903-8.
82. Raman R, Taylor B, Lindbeck K, Coombes N, Barbulescu D, Salisbury P, et al. Molecular mapping and validation of *Rlm1* genes for resistance to *Leptosphaeria maculans* in canola (*Brassica napus* L). Crop & Pasture Science. 2012;63:1007–1017.
83. Delourme R, Piel N, Horvais R, Pouilly N, Domin C, Vallée P, et al. Molecular and Phenotypic Characterization of near Isogenic Lines at QTL for Quantitative Resistance to *Leptosphaeria maculans* in Oilseed Rape (*Brassica napus* L.). Theoretical and Applied Genetics. 2008(117):1055-67
84. Eber F., Lourgant K., Brun H., Lode M., Huteau V., Coriton O., Alix K., Balesdent M., Chevre A.M. Analysis of *Brassica nigra* Chromosomes Allows Identification of a New Effective *Leptosphaeria maculans* resistance Gene Introgressed in *Brassica napus*. Proceeding of the 13th International rapeseed congress, Prague 5–9 June 2011.
85. Parlange F, Daverdin G, Fudal I, Kuhn M-L, Balesdent M-H, ne, et al. *Leptosphaeria maculans* Avirulence Gene *Avrlm4-7* Confers a Dual Recognition Specificity by the *Rlm4* and *Rlm7* Resistance Genes of Oilseed Rape, and Circumvents *Rlm4*-Mediated Recognition through a Single Amino Acid Change. Molecular Microbiology. 2009;71:851-63.
86. Kaur S, Cogan NOI, Ye G, Baillie RC, Hand ML, Ling AE, et al. Genetic Map Construction and QTL Mapping of Resistance to Blackleg (*Leptosphaeria maculans*) Disease in Australian Canola (*Brassica napus* L.) Cultivars. Theoretical and Applied Genetics. 2009 Dec;120(1):71-83.
87. Pilet ML, Delourme R, Foisset N, Renard M. Identification of Loci Contributing to Quantitative Field Resistance to Blackleg Disease, Causal Agent *Leptosphaeria maculans* (Desm.) Ces. Et De Not., in Winter Rapeseed (*Brassica napus* L.). Theoretical and Applied Genetics. 1998;96:23-30.
88. Delourme R, Piel N, Horvais R, Pouilly N, Domin C, Vallée P, et al. Molecular and Phenotypic Characterization of near Isogenic Lines at QTL for Quantitative Resistance to *Leptosphaeria maculans* in Oilseed Rape (*Brassica napus* L.). Theoretical and Ap- plied Genetics. 2008;117(7):1055-67.
89. Flor HH. The Current Status of the Gene-for-Gene Concept. Annu Rev Phytopathol. 1971;9:275-96.
90. Keen NT. Gene-for-Gene Complementarity in Plant-Pathogen Interactions. Annu Rev Genet. 1990;24:447-63
91. Flor HH. Inheritance of Pathogenicity in *Melampsora lini*. Phytopathology 1942;32:653-69.
92. Williams PH, Delwiche PA, editors. Screening for Resistance to Blackleg of Crucifers in the Seedling Stage. Proceedings Eucarpia Conference, Breeding of Cruciferous Crops, Wageningen, Netherlands: 164-170; 1979.
93. Rouxel T, Balesdent MH. The Stem Canker (Blackleg) Fungus, *Leptosphaeria maculans*, Enters the Genomic Era. Mol Plant Pathol. 2005;6(3):225-41.
94. Balesdent MH, Louvard K, Pinochet X, Rouxel T. A Large-Scale Survey of Races of *Leptosphaeria maculans* Occurring on Oilseed Rape in France. European Journal of Plant Pathology. 2006;114(1):53-65.
95. Ghanbarnia K, Lydiate D, Rimmer SR, Li G, Kutcher HR, Larkan N, et al. Genetic mapping of the *Leptosphaeria maculans* avirulence gene corresponding to the *LepR1* resistance gene of *Brassica napus*. Theoretical and Applied Genetics. 2012;124(3):505-13.

96. Gout L, Fudal I, Kuhn ML, Blaise F, Eckert M, Cattolico L, et al. Lost in the Middle of Nowhere: The *Avrlm1* Avirulence Gene of the Dothideomycete *Leptosphaeria maculans*. Mol Microbiol 2006;60:67-80.
97. Fudal I, Ross S, Gout L, Blaise F, Kuhn ML, Eckert MR, et al. Heterochromatin-Like Regions as Ecological Niches for Avirulence Genes in the *Leptosphaeria maculans* Genome: Map-Based Cloning of *Avrlm6*. Mol Plant Microbe Interact 2007;20:459-70.
98. Stachowiak A, Olechnowicz J, Jedryczka M, Rouxel T, Balesdent MH, Happstadius I, et al. Frequency of Avirulence Alleles in Field Populations of *Leptosphaeria maculans* in Europe. European Journal of Plant Pathology. 2006;114(1):67-75.
99. Delourme R, Barbetti MJ, Snowdon R, Zhao J, Maria J. Manzanares-Dauleux. Genetics and Genomics of Disease Resistance Edwards D. Batley J, Parkin IAP, Kole C (Eds), USA: Science Publishers, CRC Press; 2011.
100. Jackson AO, Taylor CB. Plant-Microbe Interaction:Life and Death at the Interface. The Plant Cell. 1996; 8:1651-68.
101. Sprague S, Balesdent M-H, Brun H, Hayden H, Marcroft S, Pinochet X, et al. Major Gene Resistance in *Brassica napus* (Oilseed Rape) Is Overcome by Changes in Virulence of Populations of *Leptosphaeria maculans* in France and Australia. European Journal of Plant Pathology. 2006;114(1):33-40.
102. Li H, Sivasithamparam K, Barbetti MJ. Breakdown of a *Brassica rapa* Subsp. *sylvestris* Single Dominant Blackleg Resistance Gene in *B. napus* Rapeseed by *Leptosphaeria maculans* Field Isolates in Australia. Plant Disease. 2003;87(6):752.
103. Van de Wouw AP, Cozijnsen AJ, Hane JK, Brunner PC, McDonald BA, Oliver RP, et al. Evolution of linked avirulence effectors in *Leptosphaeria maculans* is affected by genomic environment and exposure to resistance genes in host plants.. PLoS Pathogens 2010;6: e1001180.
104. Howell PM, Sharpe AG, Lydiate DJ. Homoeologous Loci Control the Accumulation of Seed Glucosinolates in Oilseed Rape (*Brassica napus*). Genome. 2003;46:454-60.
105. Sjödin C, Glimelius K. Transfer of Resistance against *Phoma lingam* to *Brassica napus* L. By Asymmetric Somatic Hybridisation Combined with Toxin Selection. Theoretical and Applied Genetics. 1989;78:513-20.
106. Waara S, Glimelius K. The Potential of Somatic Hybridization in Crop Breeding Euphytica. 1995;85:217-33.
107. Gerdemann-Kncörck M, Sacristan MD, Braatz C, Schieder O. Utilization of Asymmetric Somatic Hybridization for the Transfer of Disease Resistance from *Brassica ni- gra* to *Brassica napus*. Plant Breeding. 1994;113(2):106-13.
108. Bauer-Weston B, Keller WA, Webb J. Production and Characterization of Asymmetric Somatic Hybrids between *Arabidopsis thaliana* and *Brassica napus*. Theoretical and Applied Genetics 1993;86:150-158
109. Yu F, Lydiate DJ, Gugel RK, Sharpe AG, Rimmer SR. Introgression of *Brassica rapa* Subsp. *sylvestris* Blackleg Resistance into *B. napus*. Molecular Breeding 2012;30:1495-1506.
110. Winter H, Gaertig S, Diestel A, Sacristan MD. Blackleg Resistance of Different Origin Transferred into *Brassica napus*. Proceeding 10th International Rapeseed Congress, Canberra, Australia. 1999. http://www.regional.org.au/au/gcirc/4/593.htm
111. Dixelius C, Wahlberg S. Resistance to *Leptosphaeria maculans* Is Conserved in a Specific Region of the Brassica B Genome. Theoretical and Applied Genetics. 1999;99(1): 368-72.
112. Michael TP, Alba R. The Tomato Genome Fleshed Out. Nature Biotechnology. 2012;30(8): 765-7.
113. Strittmatter G, Janssens J, Opsomer C, Botterman J. Inhibition of Fungal Disease Development in Plants by Engineering Controlled Cell Death. Nature Biotechnology. 1995;13(10):1085-9.
114. Keller H, Pamboukdjian N, Ponchet M, Poupet A, Delon R, Verrier J-L, et al. Pathogen-Induced Elicitin Production in Transgenic Tobacco Generates a Hypersensitive Response and Nonspecific Disease Resistance. The Plant Cell. 1999;11:223-35.

115. Hennin C, Hofte M, Diederichsen E. Functional Expression of *Cf9* and *Avr9* Genes in *Brassica napus* Induces Enhanced Resistance to *Leptosphaeria maculans*. Molecular Plant-Microbe Interactions. 2001;14(9):1075-85.
116. Bolton MD. Primary Metabolism and Plant Defense-Fuel for the Fire. Mol Plant-Microbe Interact. 2009;22:487-97.
117. Katagiri F, Tsuda K. Understanding the Plant Immune System. Molecular Plant-Microbe Interactions. 2010;23(12): 1531-1536.
118. Johnson R, Law CN. Cytogenetic Studies of the Resistance of the Wheat Variety Bersée to *Puccinia Striformis*. Cereal Rusts Bulletin 1973;1:38-43.
119. Vera Cruz C, Bai J, On˜a I, Leung H, Nelson R, Mew T-W, et al. Predicting Durability of a Disease Resistance Gene Based on an Assessment of the Fitness Loss and Epidemiological Consequences of Avirulence Gene Mutation. Proceedings of the National Academy of Sciences USA 2000;97: 13500-13505.
120. Light KA, Gororo NN, Salisbury PA. Usefulness of Winter Canola (*Brassica napus*) Race-Specific Resistance Genes against Blackleg (Causal Agent *Leptosphaeria maculans*) in Southern Australian Growing Conditions. Crop and Pasture Science. 2011;62 162-8.
121. Brun H, Chevre AM, Fitt BDL, Powers S, Besnard AL, Ermel M, et al. Quantitative Resistance Increases the Durability of Qualitative Resistance to *Leptosphaeria maculans* in *Brassica napus*. New Phytol. 2010;185(1):285-99.
122. Sprague SJ, Marcroft SJ, Hayden HL, Howlett BJ. Major Gene Resistance to Blackleg in *Brassica napus* Overcome within Three Years of Commercial Production in Southeastern Australia. Plant Disease. 2006;90(2):190-8.
123. Van de Wouw AP, Stonard JF, Howlett BJ, West JS, Fitt BDL, Atkins SD. Determining Frequencies of Avirulent Alleles in Airborne *Leptosphaeria maculans* Inoculum Using Quantitative PCR. Plant Pathol. 2010;59(5):809-18.
124. Kolmer JA, Dyck PL, Roelfs AP. An Appraisal of Stem and Leaf Rust Resistance in North American Hard Red Spring Wheats and the Probability of Multiple Mutations to Virulence in Populations of Cereal Rust Fungi. Phytopathology 1991;81:237-9.
125. Crute R, Pink D. Genetics and Utilization of Pathogen Resistance in Plants. Plant Cell. 1996;8:1747-55.
126. Marcroft S, van De Wouw A, Salisbury P, Potter T, Howlett BJ. Rotation of Canola (*Brassica napus*) Cultivars with Differential Complements of Blackleg Resistance Genes Decreases Disease Severity. Plant Pathol. 2012;
127. Raman H, Raman R, Kilian A, Detering F, Long Y, Edwards D, et al. A Consensus Map of Rapeseed (*Brassica napus* L) Based on Diversity Array Technology Markers: Applications in Genetic Dissection of Qualitative and Quantitative Traits. BMC Genomics 2013; 14:277.
128. Li G, Quiros CF. Sequence-Related Amplified Polymorphism (SRAP), a New Marker System Based on a Simple PCR Reaction: Its Application to Mapping and Gene Tagging in Brassica. Theoretical and Applied Genetics. 2001;103(2):455-61.
129. Sun Z, Wang Z, Tu J, Zhang J, Yu F, McVetty P, et al. An Ultradense Genetic Recombination Map for *Brassica napus* Consisting of 13551 SRAP Markers. Theoretical and Applied Genetics. 2007;114(8):1305-17.
130. Durstewitz G, Polley A, Plieske J, Luerssen H, Graner EM, Wieseke R, et al. SNP Discovery by Amplicon Sequencing and Multiplex SNP Genotyping in the Allopolyploid Species *Brassica napus*. Genome. 2010;53(11):948-56.
131. Trick M, Long Y, Meng J, Bancroft I. Single Nucleotide Polymorphism (SNP) Discovery in the Polyploid *Brassica napus* Using Solexa Transcriptome Sequencing. Plant Bi- otechnology Journal. 2009;7:334-46.
132. Suwabe K, Iketani H, Nunome T, Kage T, Hirai M. Isolation and Characterization of Microsatellites in *Brassica rapa* L. Theoretical and Applied Genetics 2002;104:1092-8.
133. Suwabe K, Tsukazaki H, Iketani H, Hatakeyama K, Kondo M, Fujimura M, et al. Simple Sequence Repeat-Based Comparative Genomics between *Brassica*

rapa and *Arabi- dopsis thaliana*: The Genetic Origin of Clubroot Resistance. Genetics. 2006;173(1):309-19.
134. Lombard V, Delourme R. A Consensus Linkage Map for Rapeseed (*Brassica napus* L.): Construction and Integration of Three Individual Maps from DH Populations. Theoretical and Applied Genetics. 2001;103(4):491-507.
135. Baird NA, Etter PD, Atwood TS, Currey MC, Shiver AL, Lewis ZA, et al. Rapid Snp Discovery and Genetic Mapping Using Sequenced RAD Markers. PLoS ONE. 2008;3(10):e3376.
136. Miller MR, Dunham JP, Amores A, Cresko WA, Johnson GA. Rapid and Cost-Effec- tive Polymorphism Identification and Genotyping Using Restriction Site Associated DNA (RAD) Markers. Genome Research. 2007;17:240-8.
137. Jestin C, Lodé M, Vallée P, Domin C, Falentin C, Horvais R, et al. Association Mapping of Quantitative Resistance for *Leptosphaeria maculans* in Oilseed Rape (*Brassica napus* L.). Mol Breed. 2011;27:271-87.
138. Michelmore RW, Paran I, Kesseli RV. Identification of Markers Linked to Disease-Resistance Genes by Bulked Segregant Analysis: A Rapid Method to Detect Markers in Specific Genomic Regions by Using Segregating Populations (Random Amplified Polymorphic DNA/Restriction Fragment Length Polymorphism). Proc Natl Acad Sci USA. 1991;88:9828-32,.
139. Raman H, Milgate A. Molecular Breeding for Resistance to Septoria Tritici Blotch. Cereal Res Commun. 2012; 40:451-466
140. Mayerhofer R, Wilde K, Mayerhofer M, Lydiate D, Bansal VK, Good AG, et al. Complexities of Chromosome Landing in a Highly Duplicated Genome: Toward Map-Based Cloning of a Gene Controlling Blackleg Resistance in *Brassica napus*. Genetics. 2005;171(4):1977-88.
141. Chèvre AM, Brun H, Eber F, Letanneur JC, Vallee P, Ermel M, Glais I, Li H, Sivasithamparam K, Barbetti MJ. Stabilization of Resistance to *Leptosphaeria maculans* in *Brassica napus - B. juncea* Recombinant Lines and Its Introgression into Spring-Type *Brassica napus*. Plant Disease. 2008;92(8):1208-14.
142. Raman H, Raman R, Taylor B, Lindbeck K, Coombes N, Eckermann P, et al. Blackleg Resistance in Rapeseed: Phenotypic Screen, Molecular Markers, and Genome-Wide Linkage and Association Mapping. Proceedings of the 17th Australian Research Assembly on Brassicas, 15-17 August, 2011, Wagga Wagga, pp 61-64.
143. Pilet ML, Duplan G, Archipiano M, Barret P, Baron C, Horvais R, et al. Stability of QTL for Field Resistance to Blackleg across Two Genetic Backgrounds in Oilseed Rape. Crop Sci 2001;41:197-205.
144. Piquemal J, Cinquin E, Couton F, Rondeau C, Seignoret E, Doucet I, et al. Construction of an Oilseed Rape (*Brassica napus* L.) Genetic Map with SSR Markers. Theoretical and Applied Genetics. 2005;111(8):1514-23.
145. Breseghello F, Sorrells ME. Association Analysis as a Strategy for Improvement of Quantitative Traits in Plants. Crop Sci. 2006;46:1323-30.
146. Breseghello F, Sorrells ME. Association Mapping of Kernel Size and Milling Quality in Wheat (*Triticum aestivum* L.) Cultivars. Genetics. 2006;172(2):1165-77.
147. Buckler ES, Holland JB, Bradbury PJ, Acharya CB, Brown PJ, Browne C, et al. The Genetic Architecture of Maize Flowering Time. Science. 2009;325(5941):714-8.
148. Flint-Garcia SA, Thornsberry JM, Buckler ES. Structure of Linkage Disequilibrium in Plants. Annu Rev Plant Biol 2003 54:357-74
149. Hirschhorn JN, Daly MJ. Genome-Wide Association Studies for Common Diseases and Complex Traits. Nat Rev Genet. 2005;6:95-108.
150. Raman H, Stodart B, Ryan P, Delhaize E, Emberi L, Raman R, et al. Genome Wide Association Analyses of Common Wheat (*Triticum aestivum* L) Germplasm Identifies Multiple Loci for Aluminium Resistance. Genome. 2010;53(11):957-66.
151. Cloutier S, McCallum BD, Loutre C, Banks TW, Wicker T, Feuillet C, et al. Leaf Rust Resistance Gene *Lr1*, Isolated from Bread Wheat (*Triticum aestivum* L.) Is a Member of the Large *Psr567* Gene Family. Plant Mol Biol. 2007;65:93-106.

152. Feuillet C, Travella S, Stein N, Albar L, Nublat A, Keller B. Map-Based Isolation of the Leaf Rust Disease Resistance Gene *Lr10* from the Hexaploid Wheat (*Triticum aestivum* L.) Genome. Proc Natl Acad Sci U S A. 2003;100(25):15253-8.
153. Krattinger SG, Lagudah ES, Spielmeyer W, Singh RP, Huerta-Espino J, McFadden H, et al. A Putative ABC Transporter Confers Durable Resistance to Multiple Fungal Pathogens in Wheat. Science. 2009;323:1360-3.
154. Michelmore RW, Meyers BC. Clusters of Resistance Genes in Plants Evolve by Divergent Selection and a Birth-and-Death Process. Genome Res 1998;8:1113-30.
155. Srichumpa P, Brunner, S., Keller, B. & Yahiaoui, N. Allelic Series of Four Powdery Mildew Resistance Genes at the *Pm3* Locus in Hexaploid Bread Wheat. Plant Physiology 139, 885-895. 2005.
156. Martin G, Brommonschenkel S, Chunwongse J, Frary A, Ganal M, Spivey R, et al. Map-Based Cloning of a Protein Kinase Gene Conferring Disease Resistance in Tomato. Science. 1993;262(5138):1432-6.
157. Fu D, Uauy C, Distelfeld A, Blechl A, Epstein L, Chen X, et al. A Kinase-Start Gene Confers Temperature-Dependent Resistance to Wheat Stripe Rust. Science. 2009;323(5919):1357-60.
158. Baumgarten A, Cannon S, Spangler R, May G. Genome-Level Evolution of Resistance Genes in *Arabidopsis thaliana*. Genetics. 2003;165 309-19.
159. Persson M, Staal J, Oid S, Dixelius C. Layers of Defense Responses to *Leptosphaeria Maculans* Below the *Rlm1* - and Camelexin - Dependent Resistances. New Phytologist. 182, 470-482. 2009.
160. Dangl JL, Jones JDG. Plant Pathogens and Integrated Defence Responses to Infection. Nature 2001;411:826-33.
161. Dixon MS, Hatzixanthis K, Jones DA, Harrison K, Jones JDG. The Tomato *Cf-5* Disease Resistance Gene and Six Homologs Show Pronounced Allelic Variation in Leucine-Rich Repeat Copy Number. Plant Cell 10: 1915-1926. 1998.
162. Dixon MS, Jones D, Keddie JS, Thomas CM, Harrison K, Jones JD. The Tomato *Cf-2* Disease Resistance Locus Comprises Two Functional Genes Encoding Leucine-Rich Repeat Proteins. Cell 84: 451-459. 1996.
163. Song WY, Pi LY, Wang GL, Gardner J, Holsten T, Ronald PC. Evolution of the Rice *Xa21* Disease Resistance Gene Family. Plant Cell 9: 1279–1287. 1997.
164. Ellis J, Jones D. Structure and Function of Proteins Controlling Strain-Specific Pathogen Resistance in Plants. Current Opinion in Plant Biology. 1998;1:288-93.
165. Meyer JDF, Silva DCG, Yang C, Pedley KF, Zhang C, Mortel Mvd, et al. Identification and Analyses of Candidate Genes for *Rpp4*-Mediated Resistance to Asian Soybean Rust in Soybean. Plant Physiology. 2009;150:1 295-307
166. Hulbert SH, Bennetzen JL. Recombination at the *Rp1* Locus of Maize. Mol Gen Genet 226: 377-382. 1991.
167. Mun JH, Yu HJ, Park S, Park BS. Genome-Wide Identification of NBS-Encoding Resistance Genes in *Brassica rapa*. Molecular and General Genetics. 2009;282:617-31.
168. Deng W, Ying H, Helliwell CA, Taylor JM, Peacock WJ, Dennis ES. Flowering Locus C (*FLC*) Regulates Development Pathways Throughout the Life Cycle of *Arabidopsis*. Proceedings of the National Academy of Sciences. 2011;108 (16):6680-5
169. Staal J, Kaliff M, Bohman S, Dixelius C. Transgressive Segregation Reveals Two Arabidopsis TIR-NB-LRR Resistance Genes Effective against *Leptosphaeria maculans*, Causal Agent of Blackleg Disease. Plant J. 2006;46(2):218-30.
170. Staal J, Dixelius C. *Rlm3*, a Potential Adaptor between Specific *TIR*-NB-LRR Receptors and DZC Proteins. Communicative & Integrative Biology 2008;1(1):59-61.
171. Staal J, Kaliff M, Dewaele E, Persson M, Dixelius C. *Rlm3*, a TIR Domain Encoding Gene Involved in Broad-Range Immunity of Arabidopsis to Necrotrophic Fungal Pathogens. Plant Journal 2008;55(2):188-200.
172. Šašek V, Nováková M, Jindřichová B, Bóka K, Valentová O, BurketováLenka. Recognition of Avirulence gene *AvrLm1* from hemibiotrophic ascomycete *Leptosphaeria*

maculans triggers salicylic acid and ethylene signaling in *Brassica napus*. Molecular Plant-Microbe Interactions 25:9, 1238-1250. 2012.

173. Vincente JG, King GJ. Characterisation of Disease Resistance Gene-Like Sequences in *Brassica oleracea* L. Theoretical and Applied Genetics. 2001;102:555-63.

174. Fourmann M, Charlot F, Froger N, Delourme R, Brunel D. Expression, Mapping, and Genetic Variability of *Brassica napus* Disease Resistance Gene Analogues. Genome. 2001;44(6):1083.

175. Sharma N, N. Hotte, M.H. Rahman, M. Mahammadi, M.K. Deyholos, N.N.V. Kav. Towards Identifying Brassica Proteins Involved in Mediating Resistance to *Leptosphaeria maculans*: A Proteomics-Based Approach. Proteomics. 2008;8:3516-35.

176. Marra R, Li H, Barbetti MJ, Sivasithamparam K, Vinale F, Cavallo P, et al. Proteomic Analysis of the Interaction between *Brassica napus* Cv. Surpass 400 and Virulent or Avirulent Isolates of *Leptosphaeria maculans*. Journal of Plant Pathology. 2010;92:89-101.

177. Tollenaere R, Hayward A, Dalton-Morgan J, Campbell E, Lee JRM, Lorenc M, et al. Identification and Characterization of Candidate *Rlm4* Blackleg Resistance Genes in *Brassica napus* Using Next-Generation Sequencing. Plant Biotechnology Journal. 2012;10(6):709-15.

178. Wheeler DA, Srinivasan M, Egholm M, Shen Y, Chen L, McGuire A, et al. The Complete Genome of an Individual by Massively Parallel DNA Sequencing. Nature. 2008;452(7189):872-6.

179. Metzker ML. Sequencing Technologies- the Next Generation. Nat Rev Genet. 2010;11(1):31-46.

180. Ju YS, Kim J-I, Kim S, Hong D, Park H, Shin J-Y, et al. Extensive Genomic and Transcriptional Diversity Identified through Massively Parallel DNA and Rna Sequencing of Eighteen Korean Individuals. Nat Genet. 2011;43(8):745-52.

181. Consortium TPGS. Genome Sequence and Analysis of the Tuber Crop Potato. Nature. 2011;475(7355):189-95.

182. Wang X, Wang H, Wang J, Sun R, Jian Wu, Shengyi Liu, et al. The Genome of the Mesopolyploid Crop Species *Brassica rapa*. Nature Genetics. 2011;43:1035-9.

183. Imelfort M, Edwards D. Next Generation Sequencing of Plant Genomes. Briefings in Bioinformatics. 2009;10:609-18.

184. Osborn TC, Butrulle DV, Sharpe AG, Pickering KJ, Parkin IAP, Parker JS, et al. Detection and Effects of a Homeologous Reciprocal Transposition in *Brassica napus*. Genetics. 2003;165(3):1569-77.

185. Kelly AL. The Genetic Basis of Petal Number and Pod Orientation in Oilseed Rape (*Brassica napus*). PhD Thesis, University of New Castle, UK. 1996.

186. Wang J, Kaur S, Cogan N, Dobrowolski M, Salisbury P, Burton W, et al. Assessment of Genetic Diversity in Australian Canola (*Brassica napus* L.) Cultivars Using SSR Markers. Crop & Pasture Sci. 2009;60:1193-201.

187. Raman H, Ryan PR, Raman R, Stodart BJ, Zhang K, Martin P, et al. Analysis of *TALMT1* Traces the Transmission of Aluminum Resistance in Cultivated Common Wheat (*Triticum aestivum* L.). Theoretical and Applied Genetics. 2008;116:343–54.

CHAPTER 3

Plant Breeding for Harmony between Modern Agriculture Production and the Environment

João Carlos da Silva Dias

Instituto Superior de Agronomia, University of Lisbon, Lisbon, Portugal

ABSTRACT

The world population is estimated to be 9.2 billion in 2050. To sufficiently feed these people, the total food production will have to increase 60% - 70%. Climate models predict that warmer temperatures and increases in the frequency and duration of drought during the present century will have negative impact on agricultural productivity. These new global challenges require a more complex integrated agricultural and breeding agenda that focuses on livelihood improvement coupled with agro-ecosystem resilience, eco-efficiency and sustainability rather than just on crop productivity gains. Intensifying sustainability agro-ecosystems by producing more food with lower inputs, adapting agriculture to climate change, conserving agro-biodiversity through its use, and making markets to work for the small farmers are needed to address the main issues of our time. Plant breeding has played a vital role in the successful development of modern agriculture. Development of new cultivars will be required while reducing the impact of agriculture on the environment and maintaining sufficient production. Conventional plant breeding will remain the backbone of crop improvement strategies. Genetic engineering has the potential to address some of the most challenging biotic constraints faced by farmers, which are not easily addressed through conventional plant breeding alone. Protective measures and laws, especially patenting, must be moderated to eliminate coverage so broad that it stifles innovation. They must be made less restrictive to encourage research and free flow of materials and information. Small farmers have an important role in conserving and using crop biodiversity. Public sector breeding must remain vigorous, especially in areas where the private sector does not function. This will often require benevolent public/private partnerships as well as government support. Active and positive connections between the private and public breeding sectors and large-scale gene banks are required to avoid a possible conflict involving breeders' rights, gene preservation and erosion. Plant breeding can be a powerful tool to bring "harmony" between agriculture and the environment, but partnerships and cooperation are needed to make this a reality.

Keywords: Breeding, Modern Agriculture, Climate Changes, Biodiversity, Environment, Ecosystems, Transgenic Crops, Small Farmers

1. INTRODUCTION

There are now almost 7.25 billion human beings inhabiting this planet, and it has been projected that world population growth may exceed 70 million annually over the next 40 years. The world population will be approximately 9.2 billion in 2050, when the concentration of carbon dioxide and ozone will be 550 ppm and 60 ppm, respectively and the climate will be warmer by 2°C [1] . At that time it is expected that approximately 90% of this global population will reside in Asia, Africa, and Latin American countries [2] [3] . Currently, about 1 billion human beings suffer from hunger; 3 billion malnourished people suffer one or more micronutrient deficiencies (especially vitamin A, iodine and iron) and live with less than 2 US dollars per day; and anthropogenic climate change continues to affect food output and quality [4] [5] . By 2050, to sufficiently feed all these people, the total food production will have to increase 60% to 70% to meet a net demand of 1 billion tonnes of cereal for food and to feed, and 200 million tonnes of meat [6] -[8] , depending on assumptions of population growth, income growth and dietary changes. This projected increase of global crop demand is partly due to a growing global population, but a larger driver is increasing global affluence and associated changes in diet due to higher incomes [4] [8] . As global incomes increase, diets typically shift from those comprised of mostly grains, to diets that contain more vegetables and fruits and a greater proportion of meat, dairy products, and eggs [4] [8] -[10] .

This shift from plant-based diets to more intensive demand for animal products is termed as the "Livestock Revolution" [11] and it is estimated that approximately 40% of the world's population will undergo this revolution to more animal consumption by the year 2050 [8] . In order to meet these demands, global livestock production systems are shifting from using mostly waste products, crop residues, and marginal lands to more industrial systems which require less land and use of higher value feed crops [11] [12] . In developing countries with high rates of increasing animal product demands, a greater proportion of cereals are being directed towards animals [13] . Increasing demand for meat and dairy is also of importance to the global environment because their production requires more land and other resources than plant-based foods [14] -[16] . In fact, livestock production is the single largest anthropogenic use of land. According to a 2011 analysis, 75% of all agricultural land (including crop and pasture land) is dedicated to animal production [17] . Livestock production is also responsible for other environmental impacts. Livestock production is estimated to be responsible for 18% of total greenhouse gas emissions [18] , and animal products generally have a much higher water footprint than plant-based foods [19] .

In addition to growing meat and dairy demands, affluent nations are also directing a growing proportion of high-value feedstock to bio-fuel production. A great majority of bio-fuel feedstock are human-edible, especially from maize in the United States and sugarcane in Brazil. In 2010 global bio-fuel production represented 2.7% of global fuel for road transportation (at 107 billion liters produced), which is more than a 450% increase from the year 2000 [20] . To produce these fuels the United States and Brazil combined dedicated over 460 million tonnes of maize and sugarcane, respectively, to bio-fuel production in 2010, which is 6% of global crop production [21] .

In 2008, the world's arable land amounted to 1386 M ha, out of a total 4883 M ha land used for agriculture [22] . Each year, arable and agricultural land is lost due to deforestation, overgrazing, agricultural activities, gathering and overexploitation for fuel-wood, urbanization and industrialization. The most direct negative impact of agriculture on biodiversity is due to the considerable loss of natural habitats, which is caused by the conversion of natural ecosystems into agricultural land. The arable land is limited. Increases in arable land can only be done by deforestation. Agricultural production should be increased without further deforestation. This requires innovation and better technologies, as well as substantial investment, to increase yields on existing agricultural land.

Climate models predict that warmer temperatures and increases in the frequency and duration of drought during the twenty-first century will have negative impact on agricultural productivity [23] -[28] . For example, maize production in Africa could be at risk of significant yield losses as researchers predict that each degree-day that the crop spends above 30°C reduces yields by 1% if the plants receive sufficient water [27] . These predictions are similar to those reported for maize yield in the United States [29] . Lobell et al. [27] further showed that maize yields in Africa decreased by 1.7% for each degree-day the crop spent at temperatures of over 30°C under drought. Wheat production in Russia decreased by almost one-third in 2010, largely due to the summer heat wave. Similarly, wheat production declined significantly in China and India in 2010, largely due to drought and sudden rise in temperature respectively, thereby causing forced maturity [30] . Warming at +2°C is predicted to reduce yield losses by 50% in Australia and India [31] [32] . Likewise, the global maize and wheat production, as a result of warming temperatures during the period of 1980 to 2008, declined by 3.8% and 5.5%, respectively [28] . So climate change poses a serious threat to species fitness [33] [34] , and to agro-ecosystems essential to food production [35] .

Climatic variation and change are already influencing the distribution and virulence of crop pest and diseases, but the interactions between the crops, pests and pathogens are complex and poorly understood in the context of climate change [36] . We will need to integrate plant biology into the current paradigm with respect to climate change to succeed in defeating emerging pests and pathogens posing a new threat to agriculture due to climate change [37] -[39] .

In this context we can ask: can we feed and clothe the growing world population while simultaneously preserving or improving ecosystems and the natural environment?

History shows that modern agriculture has the potential to feed the world population but also to be worst and even catastrophically with the natural environment. Some examples are deforestation, overgrazing and erosion, in many parts of the world, which contributed to the outright collapse of ecosystems. One classical example is Madagascar's central highland plateau that has become virtually totally barren (about ten percent of the country), as a result of slash-and-burn deforestation, an element of shifting cultivation practiced by many natives. Intensification of production systems have also led to reduction in crop and livestock biodiversity, and increased genetic vulnerability and erosion. In contrast, the "Green Revolution", which began providing high-yielding crop cultivars and high-input management techniques to developing countries in the 1960s, has prevented mass starvation and improved living standards throughout the world [40] . Dwarfing, photoperiod insensitive genes and host plant resistance genes to pathogens and pests were bred for various crops during the "Green Revolution" [41] . Production in agriculture was increased in many nations of Asia and Latin America by the "Green Revolution". Crop yields in the developing world would have been at least 20% less and food prices about 19% higher than they were in 2000 without the innovations of the "Green Revolution" [42] . Calorie consumption would have dropped by about 5% and the number of malnourished children would have increasing by at least 2%; i.e., the "Green Revolution" helped to improve the health status of 32 to 42 million pre-school children. Since the beginning of the "Green Revolution" in 1960, land devoted to crops increased some 10%, land under irrigation has doubled, pesticide use by agriculture has tripled, fertilizer use is up 23-fold, pesticide use is up by a factor of 53. Nowadays, forty per cent of crop production comes from the 16% of agricultural land that is irrigated. Irrigated lands account for a substantial portion of increased yields obtained during the "Green Revolution". However, the global rate of increase in the irrigated area is declining, per capita irrigated area has declined by 5% since 1978, and new dam construction may allow only a 10% increase in water for irrigation over the next 30 years [43] . The enhancement of yield achieved in the "Green Revolution" (29% in food supplies per capita since 1960) may have been associated with an increased level of greenhouse gas emissions associated with higher fertilizer production and application, but, overall, its net effect has been calculated to have reduced CO_2 emission by some 161 gigatons of carbon (GtC) over the period 1961-2005 [44] , implying that gains in crop productivity can make a positive contribution to reducing greenhouse gas emissions.

Developing sustainable agriculture in environmentally sensitive systems is the great challenge of the coming decades. More food, animal feed, fiber, fuel, and forest products must be produced with less available land, water, and nutrients, to meet basic human needs and improve the sustainability of production [45] . In addition, pressure from an increasing global human population will necessitate more efficient and diversified land use. High crop yields of main staples will still be needed for freeing land to cash or more nutritious crops such as fruits and vegetables, as well as to prevent biodiversity losses and protect the environment by avoiding the use of today's forests, woodlands, pastures, rangelands and mountain sides for agriculture. Furthermore, recent research shows that investment in crop yield gains compares favorably with other commonly proposed climate change mitigation strategies, and should be a priority target to reduce greenhouse gas emissions.

Identifying the most appropriate technologies and practices to achieve these objectives are critical. This requires the building of a knowledge base to support such tasks. Fortunately, the World already possesses the know-how and research capability required to achieve the main objective to transform agriculture into a fully natural resource-based system which manages eco-efficiently its surrounding environment and associated biodiversity. Agro-ecological approaches are known to improve farming system productivity, reduce pollution through sound methods of nutrient and pest management, maintain biodiversity reserves, and enhance habitat quality through careful management of soil, water, and natural vegetation. Infrastructure (particularly roads and irrigation), incentives, institutions and other innovations beyond agriculture are also needed to implement such a global change of rural landscapes in the short- to mid-terms. The agenda for a new "Green Revolution" needs to consider new approaches to promote innovations in plant science, agricultural and management practices and benefits to farmers and consumers.

Modern production agriculture in the developed world is highly industrialized. There is considerable discussion about the inadequacy of the dominant model of agricultural intensification and growth, which relies on increased use of capital inputs, such as fertilizer and pesticides [43] . Technology and purchased inputs, e.g. fertilizer, pesticides and water are required to maintain high levels of production, and use of these inputs continues to increase in the developing world. Despite the critical need for agricultural production and continued improvements in management practices, current systems are still not in "harmony" with the environment because they can create many problems for ecosystems and human communities. The generation of unacceptable levels of environmental damage and problems of economic feasibility are cited as key problems with this model of industrial agriculture [45] [46] . Specific external costs of industrial agriculture which should be improved include soil deterioration, erosion, declining surface water and groundwater quality, limited recycling of nutrients, excessive use of off-farm fertilizers and pesticides, diminished biodiversity within the agricultural system (both in terms of the variety of crops sown and coexisting species), lapses in food safety, and the loss of rural employment. By developing new field crops, and trees that meet societal needs, plant breeding plays a distinctive and crucial role in addressing these challenges, which must be dealt with immediately to develop sustainable agronomic systems for the future.

In this article two general ways are described in which plant breeders can engage in environmental issues: 1) by breeding plants that are better adapted to environment and environmental stresses, producing more with less and where productivity can be maintained in the face of increasingly variable weather patterns and sub-optimal conditions, as well as pest and disease pressures; and 2) by breeding plants that can alter and "improve" environments, as breeding alternative crops and crops for new uses or breeding for local adaptation and sustainable solutions. It is also presented the impacts of transgenic crops, commonly referred to as genetically modified crops on biodiversity.

2. PLANT BREEDING, AGRICULTURE AND ENVIRONMENT

2.1. Introduction

Farming and plant breeding have been closely associated since the early days when crops were first domesticated. The domestication of staple crops, for example, rice and soybean in eastern Asia; wheat in the Middle East; sorghum in Africa; and maize, beans, and potatoes in the Americas [47] , began independently, in multiple locales, 5000 - 12,000 years ago [47] . For thousands of years, these crops were grown and morphologically altered by farmers, who selected the most desirable and adaptable cultivars to plant in the next growing season. Without understanding the science behind it, early farmers saved the seed from the best portion of their crop each season. Over the years, they selected the traits which they liked best, transforming and domesticating the crops they grew.

After the discoveries of Darwin and Mendel, scientific knowledge was applied to plant breeding in the late 1800s [40] . Commercial hybridization of crop species began in the United States in the middle of the 1920s with sweet corn and followed by onions in the 1940s [4] . With the implementation of hybrid crop breeding, yield per unit land area rapidly increased in the United States [48] and since that time, public and private breeding companies have been placing more and more emphasis on the development of hybrids, and many species have been bred as hybrid cultivars for the marketplace. Besides heterosis, hybrids also allow breeders to combine the best traits and multiple disease and stress resistances. Furthermore, if the parents are homozygous, the hybrids will be uniform, an increasingly important trait in commercial market production. The creation of hybrid cultivars requires homozygous inbred parental lines, which provide a natural protection of plant breeders' rights without legal recourse and ensure a market for seed companies.

In the 1970's breeders' rights protection has been provided through International Union for the Protection of New Varieties of Plants (UPOV), which coordinates an international common legal regime for plant variety protection. Protection was granted for those who develop or discover cultivars that are new, distinct, uniform, and stable [49] . Cultivars may be either sexually or asexually propagated. Coverage for herbaceous species is 20 years. Protective ownership was extended by UPOV in 1991 to include essentially derived cultivars [49] . At the same time, the farmer's exemption (which permitted farmers to save seed for their own use) was restricted; giving member states the option to allow farmers to save seed. Additionally, in Europe after 1998 and the United States after 2001, plant breeding companies can take advantages of patent laws to protect not only the cultivar itself but all of the plant's parts (pollen, seeds), the progeny of the cultivar, the genes or genetic sequences involved, and the method by which the cultivar was developed [50] . The seed can only be used for research that does not include development of a commercial product i.e., another cultivar, unless licensed by the older patent. The patents are considered the ultimate protective device allowing neither a farmer's exemption nor a breeder's exemption (that permitted the protected cultivar to be used by others in further breeding to create new cultivars) [51] . The use of patents for transgenic crops introduces additional problems according to the IAASTD report [46] developed with the contribution from 400 scientists around the world, and adopted by 58 governments. In developing countries, especially instruments such as patents may boost up costs and restrict experimentation by individual farmers whereas potentially undermining local practices for securing food and economic sustainability. Thus, there is particular concern regarding present intellectual property rights instruments, which may inhibit seed-saving, exchange, sale, and access to proprietary materials of vital importance to the independent research community, specifically in view of the need for analyses and long term experimentation on climate change impacts [49] [50] .

Research and development (hereafter R & D) for improved seed development is expensive. Such product protection has presented a business incentive to corporations to invest in the seed industry, which supported an enormous increase in private R & D leading to strong competition in the marketplace between the major seed companies. The majority of current crop cultivars sold nowadays are proprietary products developed by private R & D. A significant consequence of this increase in R & D has been a reduction in public breeding programs. As a result, the cost for R & D to develop new crop cultivars is shifting from the publicly supported research programs to the customers of the major seed companies [4] [52] .

One of the main factors to determine success in plant breeding is crop biodiversity and genetic capacity. Access to genetic variation, biodiversity, is required to achieve crop cultivar improvement. No practical breeding program can succeed without large numbers of lines (genotypes) to evaluate, select, recombine and inbreed (fix genetically). This effort must be organized in order for valid conclusions to be reached and decisions to be made. Scientists, breeders, support people and facilities, budgets, and good management are requirements to assure success in the seed business. Science must be state-of-the-art to maximize success in a competitive business environment. The continued need for fundamental breeding research is critical to support development of new technology and expansion of the knowledge base which supports cultivar development, competition among proprietary cultivar results in owner-companies striving to do the best possible research to develop their own products and to compete on genetic and physiological quality of crop seed in the marketplace. Reasonable profit margins are essential to pay back the R & D costs to the owner and to fund future research on developing even better crop cultivars to stay competitive. There is considerable genetic variation within the numerous crop species, which can be exploited in the development of superior proprietary cultivars. The consequences of this dynamic situation will mean relatively short-lived cultivars replaced by either the owner of the cultivar or a competitor seed company. This intense competition means constantly improved and more sophisticated cultivars. Seed companies are in the business of manipulating genes to improve plant cultivar performance for a profit. The success of the research is judged by the success of the product in making a reasonable profit. The research must improve economic performance starting with the seed production costs and including the farmer-ship- per/processor and the end user. If any link in this sequence of events is weak or broken, the new cultivar will likely fail [4] [53] .

Modern plant breeding is the science of improving plants to achieve farmer needs and better fit production environments, but it is a long-term proposition. Each released cultivar represents a culmination of a decade or more of work, from initial crosses through final testing. The rate of improvement is a function of the amount of heritable genetic variation present in a population, the time it takes to complete a breeding cycle (from seed production through selection to seed production again), which can range from multiple generations per year (e.g. maize on field sites in both hemispheres) to decades (some trees require 8 years of growth before flowering). In hybrid crops, several years (multiple breeding cycles) are necessary to develop inbred lines that must then be tested in hybrid combinations. Many years of testing under various environmental conditions must be conducted to ensure that the new cultivar (inbred, hybrid, or population) will perform well for the farmer, consumer, or end-user before any substantial additional investment is made to increase production and distribution of the cultivar.

Biotechnology is a new and potentially powerful tool that has been added by all the major seed corporations to their crop breeding research programs, and is part of an ongoing public research for developing genetic engineered crop projects. It can augment and/or accelerate conventional cultivar development programs through time saved, better products, and more genetic uniformity, or achieve results not possible by conventional breeding [54] . Genetic engineering provides innovative methods for modern plant breeding to adapt crops to agricultural systems facing new challenges brought by the changing climate. New breeding methods, relying on genetic engineering, can accelerate the pace to improve crops, or be more precise in transferring desired genes into plant germplasm. Some limited target traits already available in transgenic cultivars include those adapting agriculture to climate change and reducing their emissions of greenhouse gases.

Plant breeding may benefit from recent advances in genotyping and precise phenotyping, and by increasing the available agro-biodiversity through the use of genomics-led approaches. Today marker-assisted breeding is applied to a broad range of crops and could facilitate domesticating entirely new crops. Marker-assisted selection is particularly important for improving complex, quantitatively inherited traits that alter yield, and for speeding up the breeding process [55] . Crop genomics has also been improving in the last decade and today there are faster and cheaper systems being increasingly used in gene banks, genetic research and plant breeding, e.g. for studying interactions between loci and alleles such as heterosis, epistasis and pleiotropy, or analyzing genetic pathways. Advances in crop genomics are providing useful data and information for identifying DNA markers, which can be further used for both germplasm characterization and marker-assisted breeding. Genomics-assisted breeding approaches along with bioinformatics capacity and metabolomics resources are becoming essential components of crop improvement programs worldwide [49] [56] .

Progress in crop genome sequencing, high resolution genetic mapping and precise phenotyping will accelerate the discovery of functional alleles and allelic variation associated with traits of interest for plant breeding. Genome sequencing and annotation include an increasing range of species such as banana/plantain, cassava, citrus, grape, model legumes, maize, potato, rice, sorghum, sugarcane, soybean, among other species. Perhaps, one day further research on the genome of a plant species from a drought-prone environment may assist in breeding more hardy and water efficient related crops due to gene synteny.

Transgenic breeding involves the introduction of foreign DNA. The use of transgenic crops remains controversial worldwide after more than 1.5 decades of introducing them into the agro-ecosystems using specific frameworks to regulate their release and commercialization. While conventional plant breeding utilizing non- transgenic approaches will remain the backbone of crop improvement strategies, transgenic crop cultivars should not be excluded as products capable of contributing to development goals. Breeding of transgenic crops so far seems to have been responsible and regulatory agencies have proceeded with caution in releasing transgenic crops. Available commercial transgenic crops and products are at least as safe in terms of food safety as those ensuing from conventional plant breeding [54] [57] -[59] . Farm level profitability will ultimately determine whether farmers adopt and retain new transgenic crop technology, which may also depend on much more than technical performance particularly in the developing world; e.g. national research capacity, environmental and food safety regulations, intellectual property rights and agricultural input markets.

Decisions, policies and procedures about monitoring transgenic crops should be science-based, and this approach requires education. There will be continuing assessment on the need for, and type of monitoring as new (and unique) products are developed and released to agro-ecosystems. For example, when deploying transgenic crops with host plant resistance to an insect pest (e.g. expressing Cry insecticidal proteins derived from Bacillus thurigiensis) numerous experiments are conducted to determine effective insect resistance management strategies for farmers. Collecting baseline data is indeed essential for effective monitoring and guiding research on transgenic crops. Environmental, food and feed safety aspects should also be investigated before releasing trans- genic crops. Another research area should focus on the unintentional spread of transgenic traits into conventionally-bred crop or landrace gene pools of the same species, particularly in the centers of crop diversity or origin. Transgene flow raises a new set of ecological and economic issues for scientists and policymakers to consider for transgene containment. Appropriate measurements should also be taken when transgenic and conventional crops of the same species coexist in the same locations if some farmers wish to grow crops for transgenic-free markets. The global spread of transgenic crops also has significant implications for organizations involved in germplasm conservation and genetic enhancement. Protocols, which are most likely based on polymerase chain reaction (PCR) markers for detecting specific recombinant DNA sequences in bulked samples collected from sentinel plots, are therefore needed to monitor unintentional transgene flow in gene banks and breeding plots. Using plant-derived genes to introduce useful traits and plant-derived promoters, will also overcome some concerns about the development of genetically engineered crops. In this regard, cisgenesis addresses some negative views regarding the use of genes from non-crossable species for breeding crops. Cisgenesis involves only genes from the plant itself or from a crossable close relative, and these genes could also be transferred by conventional breeding methods. Crop wild relatives are therefore a valuable source of traits for cisgenesis. The impact of transgenic crops on biodiversity is discussed in chapter 3.

Plant breeders need to understand the various valuation strategies very early in the breeding process if they are to direct long-term selection toward reducing agriculture's negative environmental impacts and achieving greater sustainability while maintaining productivity. Regardless of method, breeding objectives can be broadened to include traits which reduce the environmental footprint of traditional production systems (e.g. nutrient and water use efficiencies that reduce off-farm inputs), to adapt crops to new climates, to host plant resistance to tackle old and emerging pathogen epidemics, or new cultivars for new production systems (e.g. perennial polycultures that mimic the biodiversity of natural systems), albeit with some reduction in rate of gain for the traditional agronomic traits of interest. Interdisciplinary crop improvement strategies accounting for ecological, socio-economic and stakeholder considerations will help identify traits leading to plant cultivars using fewer inputs, less land, and less energy, thereby resulting in a more sustainable agricultural ecosystem.

The impact of breeding on crop production is dependent upon the complex relationships involving the farmers, the cultivars available to them, and the developers of those cultivars. Farmers consist of commercial producers with varying size land holdings ranging from moderately small farms to very large ones, and subsistence farmers with small farms often on marginal lands. The subsistence farmers are usually poor. Several types of cultivars are available. The least sophisticated in terms of methods of development are landraces, also known as local cultivars. Modern cultivars consist of development by crossing and selection alone, those developed by crossing and selection with specific important improvements are often obtained from crosses with wild species or by transgenic methods, and F_1 hybrids between desirable inbred lines. The developers of landraces are usually farmers themselves, and are obtained by repeated simple selection procedures of generation after generation. Improved cultivars and hybrids are created either by public sector breeders or seed companies.

Nearly 70% of the world's farmers, from 570 million world exploitations, are small/subsistence and poor farmers. They feed 1.5 billion of the world's population. So they are also a key for biodiversity and for improving the sustainability. For these farmers improved cultivars, hybrids or transgenic seeds tend to be riskier than landraces, since the higher costs associated with seeds and production impose a greater income risk. The lack of capital available denies them the opportunity to invest in production inputs. Small farmers may have lower production costs with landraces because they achieve adequate yields with fewer inputs. In addition, profits from improved hybrid or transgenic cultivars tend to be more variable. Yields are often higher but market prices tend to be inconsistent. For example in India states of Andhra Pradesh and Maharashtra, farmers have been promised higher yields and lower pesticide costs when using Bt cotton, thus they acquired loans to afford the costly seeds (Monsanto has control over 95% of the Indian Bt cotton seed market and this near monopoly has resulted in great increased prices). When, in many cases, the farmers found the yields failed to meet their expected result, the consequences were usually very serious and many farmers died by committing suicide over the past 15 years, perhaps due to this reason. This situation of using Bt cotton seeds was explained by the absence of irrigation systems combined with specialization in high-cost crops, and played low market prices. Without collateral help these farmers are usually unable to secure a loan from a bank or money lender [53] [60] . Rates are often unmanageably high for those able to get a loan, with strict penalties for late payments. Similarly, a lack of education, resources, skill training and support prevent these farmers from using improved cultivars and then to generate a stable income from their production. In addition, governments do not usually regulate the price of crops or even provide market information. Improving market information systems for crops and facilitating farmers' access to credit are then essential components for a strategy to enable poor farmers to grow improved cultivars. A major obstacle to success in crop production using improved cultivars is the shortage of affordable credit. Desperate for cash, subsistence farmers are forced to sell their crops immediately after the harvest to middlemen or their creditors at unfavorable prices. Low cost quality seeds are essential for these poor farmers to improve their life [60] .

2.2. Breeding to Adapt Plants to the Environment

2.2.1. Producing More with Less

In the coming decades we will need to produce more with less. Fresh water suitable for irrigation is expected to become increasingly scarce and the costs of fertilizer and other agricultural inputs will increase as fossil-fuel costs rise. Nevertheless, continuing gains in production per hectare must be realized to offset the loss of premium agricultural lands (e.g. from urbanization and industrialization), while supplying a growing population. By developing resource efficient plants, plant breeders can continue to improve the sustainability of agricultural ecosystems. Plants requiring fewer off-farm input applications (specifically water, pesticides, nitrogen, phosphorus, and other nutrients) decrease the cost of production, lower fossil energy use, and reduce contamination of water systems, which help to improve public health and stabilize rural economies [61] [62] .

Although modern plant breeding efforts initially focused on improving uptake of inputs, recent efficiency gains have been made in physiologically increasing yield and biomass production without further increasing inputs.

Many crops already have genetic variation in nutrient use efficiency, utilization, and uptake [63] -[65] and plant breeding will further improve these traits. Intensive agro-ecosystems, central to food security and reducing rural poverty, should emphasize improvements in system productivity, host plant resistance and enhance use- efficiency of inputs such as water and fertilizers.

Water use-efficiency and water productivity are therefore being sought by agricultural researchers worldwide to address water scarcity in drought-prone environments across the world. Under water scarcity, yields, of some crops, are a function of the amount of water used by the crop, how efficiently the crop uses this water for biomass-growth (i.e., water use efficiency or above-ground biomass/water use), and the harvest index; i.e., the proportion of grain yield to above-ground biomass. Water use efficiency is the ratio of total dry matter accumulation to evapotranspiration and other water losses; i.e., water entering and being lost from the system which is not transpired through the plant. An increase in transpiration efficiency or reduction in soil evaporation will increase water use efficiency. More recently, water productivity was re-defined at the crop level as the ratio of biomass with economic value produced (for example grain yield of cereals) compared to the amount of water transpired. This water productivity has been labeled as "productive" because transpiration is the only water flow in a field actually passing through the crop. Both water use efficiency and water productivity may be improved through plant breeding, as biomass accumulation and harvest index. Farooq et al. [66] discuss the advances in transgenic breeding for drought-prone environments. In their review, they noted the testing of 10 transgenic rice events [unique DNA recombination taking place in one plant cell and thereafter to be used for generating entire transgenic plant(s)] under water scarcity. It seems the transgenic expression of some stress-regulated genes leads to increased water use efficiency.

Agriculture contributes significantly to greenhouse gas emissions. Nitrous oxide and dioxide are potent greenhouse gases released by manure or nitrogen (N) fertilizer, particularly in intensive cropping systems. Nitrous oxide (N_2O), which is a potent greenhouse gas susceptible to denitrification, is generated through use of manure or N fertilizer. In many intensive cropping systems common N fertilizer practices lead to high fluxes of N_2O and nitric oxide (NO). Several groups of heterotrophic bacteria use NO_3 as a source of energy by converting it to the gaseous forms N_2, NO, and NO_2 (nitrous dioxide). N_2O is therefore often unavailable for crop uptake or utilization. Proper amounts and timing of N applications can reduce NO_2 emissions (50% less) in intensive irrigated agro-ecosystems without significantly affecting crop yields. An optical, hand held sensor which calculates the normalized differential vegetative index, thereby assessing yield potential as plants grow, can reduce unneeded N fertilizer inputs, saving farmers' money and protecting the environment by reducing trace gas emissions. Genetic enhancement of crops shows great potential for reducing N_2O emissions from soils into the atmosphere. Some plants possess the capacity to modify nitrification in situ because they produce chemicals which inhibit nitrification in soil. This release of chemical compounds from plant roots suppressing soil nitrification has been called biological nitrification inhibition, which seems to vary widely among and within species, and appears to be a widespread phenomenon in some tropical pasture grasses, e.g. Brachiaria humidicola. Nitrification inhibition also enhances agro-ecosystem fertility in a sustainable way especially under high nitrate leaching and denitrification fluxes, which may account for the ecological advantage of African grasses over indigenous grasses of other tropical pastures. Biological nitrification inhibition may be an interesting target trait of crop genetic engineering for mitigating climate change.

Almost one-fifth of global methane emissions are from enteric fermentation in ruminant animals. Apart from various rumen manipulation and emission control strategies, genetic engineering is a promising tool to reduce these emissions. The amount of methane produced varies substantially across individual animals of the same ruminant species. Efforts are ongoing to develop low methane-emitting ruminants without impacting reproductive capacity and wool and meat quality. A recent study by Shi et al. [67] , to understand why some sheep produce less methane than others, deployed high-throughput DNA sequencing and specialized analysis techniques to explore the contents of the rumens of sheep. The study showed that the microbiota present in sheep rumen was solely responsible for the differences among high and low methane emitting sheep. It was further observed that the expression levels of genes involved in methane production varied more substantially across sheep, suggesting differential gene regulation. There is an exciting prospect that low-methane traits can be slowly introduced into sheep.

Crops are bred for N use efficiency (hereafter NUE) because this trait is a key factor for reducing N fertilizer pollution, improving yields in N limited environments, and reducing fertilizer costs. The use of plant species or genotypes of same species efficient in absorption and utilization of N is an important strategy in improving NUE in sustainable agricultural systems. Whole plant physiology, quantitative genetics, and forward and reverse genetic approaches are providing a better understanding of the physiological and molecular controls of N assimilation in crops under varying environments. Crops are being bred for NUE because this trait will be a key factor for reducing N fertilizer pollution as well as for improving yields in N limiting environments. There are various genetic engineering activities for improving NUE in crops [64] [68] . The gene Alanine aminotransferase from barley, which catalyzes a reversible transamination reaction in the N assimilation pathway, seems to be a promising candidate for accomplishing this plant breeding target. Transgenic plants over-expressing this enzyme can increase N uptake especially at early stages of growth. This gene technology was licensed to a private biotechnology company, which was founded in the last decade with the aim of promoting sustainable agriculture by running a profitable company [69] . A patent issued a few years ago gave this company the rights to use the NUE gene technology in major cereals (wheat, rice, maize, sorghum, and barley), as well as, in sugarcane. Today they are testing the technology with rice in China, and researching further with rice and wheat in India, and planning to assess its value for maize and rice in sub-Saharan Africa through private-public partnerships.

Keeping N in ammonium form will affect how N remains available for crop uptake and will improve N recovery, thus reducing losses of N to streams, groundwater and the atmosphere. There are genes in tropical grasses such as B. humidicola and in the wheat wild relative Leymus racemosus that inhibit or reduce soil nitrification by releasing inhibitory compounds from roots and suppressing Nitrosomonas bacteria [70] . Their value for genetic engineering crops for reducing nitrification needs to be further investigated.

2.2.2. Adapting to Global Climate Change and for Abiotic and Biotic Stress Tolerance

Extreme weather events are expected to increase in both number and severity in coming years [71] . Climate change impacts agro-ecosystems through changes over the long-term in key variables affecting plant growth (e.g. rising temperatures) and through increasing the variability (frequency and intensity) of weather conditions (rainfall, drought, waterlogging and elevated temperatures). These changes affect both crop productivity and quality. In addition to physically destroying crops, climate change has altered host-pathogen relationships and resulted in increased disease incidence, in insect-pest borne stress in crop plants, and in invasive pests which feed and damage them.

There are two ways to adapt crops to new environments: developing new crops (long-term endeavor starting with domestication) and introducing target traits into existing crops through plant breeding, which includes genetic engineering. However, the job of crop improvement is becoming increasingly difficult. Cultivars which are not only high yielding but are also efficient in use of inputs are needed, tailored to ever more stringent market demands, able to maintain stability under increasing climate variability, and potentially contribute to climate mitigation. These multi-trait demands for new cultivars provide significant challenges for crop breeders, and standard selection approaches struggle under such complexity. To maintain productivity in the face of increased climatic variability, both the population and the plant cultivars will need to be continually developed to withstand "new" climate extremes and the stresses which these will entail [72] .

Many breeding programs are already developing plants which tolerate extreme weather conditions, including drought, heat, and frost [73] [74] . Plant breeders are also beginning to address expected changes due to increased climate variability, by increasing genetic diversity sources and by adjusting selection and testing procedures [75] .

More frequent weather extremes will likely affect the existing ranges of not only agronomic cultivars but also local native plant species [76] . Because some genetic variation useful for climate change adaptation will be found only in wild plant relatives of cultivated crops, preserving genetic biodiversity is essential in order for breeders to select plants that will be well-suited for future environmental conditions [77] .

Global climate change notwithstanding, additional stress tolerances in crop species are needed to maintain productivity and survival. In the near future, tolerance to various soil conditions including acidic, aluminum-rich soils (particularly in the tropics) and saline soils (especially those resulting from irrigation), will be increasingly important for production on marginal agricultural lands or as the salt content of irrigated lands increases [78] . Bhatnagar-Mathur et al. [79] suggested that genetic engineering could accelerate plant breeding to adapt crops to stressful environments. They further underline that engineering the regulatory machinery involving transcription factors (TF; a protein binding specific DNA sequences and thereby governing the flow of genetic information from DNA to messenger RNA) provides the means to control the expression of many stress-responsive genes. There are various target traits for adapting crops, through genetic engineering, to high CO_2 and high O_3 environments of the changing climate [80] . Ortiz [81] , Jewell et al. [82] , and Dwivedi et al. [83] [84] provide the most recent overviews on research advances in genetic engineering for improved adaptation to drought, salinity or extreme temperatures in crops. The most cited include TF, and genes involved in: 1) signal sensing, perception, and transduction; 2) stress-responsive mechanisms for adaptation; and 3) abscisic acid biosynthesis for enhanced adaptation to drought. Transporter, detoxifying and signal transduction genes as well as TF are cited for tolerance to salinity. Genes related to reactive oxygen species, membrane and chaperoning modifications, late abundance embryogenesis proteins, osmoprotectants/compatible solutes and TF are pursued in crop genetic engineering for temperature extremes.

Transgenic crops provide the means to adapt crops to climate change, particularly in terms of drought and salinity. Duration and intensity of drought has increased in recent years, consistent with expected changes of the hydrologic cycle under global warming. Drought dramatically reduces crop yields. Genetic engineering may be one of the biotechnology tools for developing crop cultivars with enhanced adaptation to drought [85] . It should be seen as complementary to conventional plant breeding rather than as an alternative to it. The function of a TF such as the Dehydration-Responsive Element Binding (DREB) gene in water stress-responsive gene expression has been extensively investigated [86] . The main research goal was to gain a deep understanding of TF in developing transgenic crops targeting drought-prone environments [87] . For example, the DREB1A gene was placed under the control of a stress-inducible promoter from the rd29A gene and inserted via biolistic transformation into wheat bread [88] . Plants expressing this transgene demonstrated significant adaptation to water stress when compared to controls under experimental greenhouse conditions as manifested by a 10-day delay in wilting when water was held. Saint Pierre et al. [89] indicated, however, that these transgenic lines did not generally out-yield the controls under water deficit in confined field trials. Nonetheless, they were able to identify wheat lines combining acceptable or high yield under enough irrigation which also showed stable performance across the water deficit treatments used in their experiments; i.e., severe stress, stress starting at anthesis, and terminal stress.

Soils affected by salinity are found in more than 100 countries, and about 1/5 of irrigated agriculture is adversely affected by soil salinity. Therefore, breeding salt-tolerant crops should be a priority because salinity will most likely increase under climate change. Mumms [90] lists some candidate genes for salinity tolerance, indicating the putative functions of these genes in the specific tissues in which they may operate. Genes involved in tolerance to salinity in plants, limit the rate of salt uptake from the soil and the transport of salt throughout the plant, adjust the ionic and osmotic balance of cells in roots and shoots, and regulate leaf development and the onset of plant senescence. The most promising genes for the genetic engineering of salinity tolerance in crops, as noted by Chinnusamy et al. [91] , are those related to ion transporters and their regulators, as well as the C-re- peat-binding factor. The recent genome sequencing of Thellungiella salsuginea, a close relative of Arabidopsis thriving in salty soils, will provide more resources and evidence about the nature of defense mechanisms constituting the genetic basis underlying salt tolerance in plants [92] .

In the quest for breeding transgenic rice and tomato, advances showing salt tolerance have occurred. Plett et al. [93] were able to show an improved salinity tolerance in rice by targeting changes in mineral transport. They initially observed that cell type-specific expression of AtHKT1 (a sodium transporter) improved sodium (Na^+) exclusion and salinity tolerance in

Arabidopsis. Further research explored the GAL4-GFP enhancer trap (transgenic construction inserted in a chromosome and used for identifying tissue-specific enhancers in the genome) to drive expression of AtHKT1 in the root cortex in transgenic rice plants. The transgenic rice plants had a higher fresh weight under salinity stress due to a lower concentration of Na^+ in the shoots. They also noted that root-to- shoot transport of $^{22}Na^+$ decreased and was correlated with an up-regulation of OsHKT1, the native transporter responsible for Na^+ retrieval from the transpiration stream. Moghaieb et al. [94] bred transgenic tomato plants producing ectoine (a common compatible solute in bacteria living in high salt concentrations). Ectoine synthesis was promoted in the roots of transgenic tomato plants under saline conditions, which led to increased concentration of photosynthesis in improving water uptake. Likewise, the photosynthetic rate of ectoine-transgenic tomato plants increased through enhancing cell membrane stability in oxidative conditions under salt stress.

Transgenic crops can also contribute to climate change mitigation efforts by reducing input use intensity [95] . The integration of genetic engineering with conventional plant breeding, within an interdisciplinary approach, will likely accelerate the development and adoption of crop cultivars with enhanced adaptation to climate change related stresses [96] . Global warming will reduce yields in many crops about 6% and 5% average yield loss per 1°C in C_3 and C_4 crops, respectively, whose optimum temperature ranges are 15°C - 20°C and 25°C - 30°C [97] . The extent of yield loss depends on crop, cultivar, planting date, agronomy and growing area. For instance, an increase of 1°C in the night time maximum temperature translates into a 10% decrease in grain yield of rice, whereas a rise of 1°C above 25°C shortens the reproductive phase and the grain-filling duration in wheat by at least 5%, thereby reducing grain yield proportionally. Heat stress will exacerbate climate change impacts in the tropics, while it may put agriculture at risk in high latitudes where heat-sensitive cultivars are grown today. Hence, new cultivars must be bred to address heat stress. Ainsworth and Ort [98] suggested giving priority to traits improving photosynthesis for adapting to heat stress. However, plants have various mechanisms to cope with high temperatures, e.g. by maintaining membrane stability, or by ion transporters, proteins, osmoprotectants, antioxidants, and other factors involved in signaling cascades and transcriptional control [99] [100] . Furthermore, Gao et al. [101] noted that bZIP28 gene (a gene encoding a membrane-tethered TF) up-regulated in response to heat in Arabidopsis. Some of these genes can be used in crop genetic engineering to enhance plant adaptation to heat stress. For example, some stress-associated genes such as ROB5, a stress inducible gene isolated from bromegrass, enhanced performance of transgenic canola and potato at high temperatures [102] . Likewise, Katiyar-Agarwal et al. [103] introduced hsp101 gene (a heat shock protein gene from Arabidopsis) in basmati rice.

This transgenic rice had a better growth in the recovery phase after suffering heat stress.

Globalization has, among other consequences, led to the rapid spread of plant disease and invasive pests. Being immobile, plants are unable to escape pathogens causing plant disease and pests which feed and damage them. Plant disease is mainly caused by fungi, bacteria, viruses, and nematodes. Approximately 70,000 species of pests exist in the world, but of these, only 10% are considered serious [104] . Synthetic pesticides have been applied to crops since 1945 and have been highly successful in reducing crop losses to some pest insects, plant pathogens, weeds and in increasing crop yields [104] . One estimate suggests that without pesticides, crop losses to pests might increase by 30%. Despite pesticide use, insects, pathogens and weeds continue to exact a heavy toll on world crop production, approaching 40% [104] [105] . Pre-harvest losses are globally estimated at 15% for insect pests, 13% for damage by pathogens, and about 12% for weeds [104] . Developing resistant cultivars reduces the need for expensive and environmentally damaging pesticides to be applied. For example, a recent outbreak of Xanthomonas campestris pv. musacearum led to the devastating Xanthomonas wilt of banana in the Great Lakes Region of Africa, thereby threatening the food security and income of millions of East and Central African people who depend on this crop. Transgenic banana plants with the hypersensitivity response-assisting protein (Hrap) gene from sweet pepper did not show any infection symptoms after artificial inoculation of potted plants with Xanthomonas wilt in the screen house [106] . Selected transgenic banana plants with putative host plant resistance to Xanthomonas wilt are ongoing confined field-testing in East Africa, where elevated temperatures, due to the changing climate, will likely favor banana production.

Weather influences how pathogens and pests affect and interact with crops and their host plant resistance, and thus climate change can also have wide-ranging impacts on pests and diseases [84] . Late blight, which is caused by Phytophthora infestans, ranks as the most damaging potato pest. Late blight accounts for 20% of potato harvest failures worldwide, translating into 14 million tonnes valued at 7.6 billion US dollars. Global warming will increase late blight spread, e.g. expanding its range above 3000 meters in the Andes [107] . Chemical control may lead to more aggressive strains of the pathogen and chemical control is often regarded as being environmentally damaging. Cisgenic potato cultivars with late blight resistance are becoming available and will impact growers, consumers and the environment favorably [108] . Related wild Solanum species can be a source of alleles to enhance host plant late blight resistance in potato. For example, S. bulbocastum (a wild relative with high resistance to late blight from Mexico) was used to breed the cultivar "Fortuna" using genetic engineering.

Cisgenesis allows inserting several host plant resistance genes from wild crop species in one step without linkage drag (reduction in cultivar fitness).

2.3. Breeding Plants to Improve the Environment

In general, plants are bred for their most obvious end products, including grain, fiber, sugar, biomass yield, fruit quality, or ornamental qualities. However, plants deployed across the landscape in agricultural or forestry settings affect the environment in measurable ways. Perennial crops have environmentally beneficial properties not present in annual crops, such as helping to prevent erosion in agricultural systems, providing wildlife habitat, and acting as sinks for carbon and nutrients. Traditionally, perennial crops have not been a major focus of breed- ing programs because they generally take more time and scientific knowledge to improve, and therefore, products such as new cultivars are often not produced within the timeframe of funding cycles. Current tree breeding programs are developing elms (Ulmus spp.), chestnuts (Castanea dentata), hemlocks (Tsuga spp.), and other species which are resistant to introduced diseases and insects [109] [110] . As compared with natural selection, artificial selection via plant breeding has overcome these stresses more effectively by rapidly incorporating diverse exotic genetic sources of resistance, hybridizing to include multiple, different genetic resistances into the same plant, and making use of off-season locations or artificial conditions to shorten generation cycles. A more complex example which may be feasible in the future is tree breeding for larger and improved root systems to decrease soil erosion, sequester carbon, and improve soil quality by increasing soil organic matter.

New crop cultivars developed by plant breeders must help improve soil health, reduce soil erosion, prevent nutrient and chemical runoff, and maintain biodiversity. The goal to breed projects for forages, which include several species, is to produce a high yield of leaf and stem biomass, as opposed to grain, for ruminant animals. In the tropics many forages are perennial, providing year-round erosion control, improving water infiltration as compared with that, from annual cropping systems, and in some cases, sequestering carbon. The forage breeding program at the University of Georgia (UG) has developed cultivars in several species and has been proactive in developing agreements with private-sector commercial partners to oversee seed production and marketing of new cultivars. Among the cultivars developed at UG is "Jesup MaxQ" tall fescue, a cultivar carrying a non-toxic endophytic fungus that was both highly persistent under grazing and greatly improved animal weight gain and feed efficiency over standard cultivars. In addition, this program developed the first true dual purpose, grazing and hay, alfalfa cultivar "Alfagraze", followed by several further improved alfalfa cultivars like "Buldog 805" which persist through summer under cattle grazing [111] .

Cover crops are annual species planted in rotation with crops to specifically improve soil conditions and to control weeds, soil-borne diseases, and pests [112] -[114] . Continuous cover crops can reduce on-farm erosion, nutrient leaching, and grain losses due to pest attacks and build soil organic matter as well as improve the water balance, leading to higher yields [115] [116] . For instance, Kaumbutho and Kienzle [117] showed that maize yield increased from 1.2 to 1.8 - 2.0 t/ha in Kenya with the use of a mucuna (velvet bean) cover crop using case studies conducted from 2004 to 2007; and Pretty and Hine [118] found that farmers who adopted mucuna cover cropping benefited from higher yields of maize with less labor input for weeding (maize following mucuna yields 3 - 4 t/ha without application of nitrogen fertilizer, similar to yields normally obtained with recommended levels of fertilization of 130 Kg·N/ha).

Many current perennial and cover crop cultivars are essentially wild species bred from germplasm collections and developed to increase success in managed agro-ecosystems. As compared with non-native vegetation, plant species native to a particular region are generally thought to survive on less water, use fewer nutrients, require minimal pesticide applications, and be non-invasive; however, counter examples for both native and non-native species are plentiful [119] . As potentially valuable species are identified, breeding to improve them for traits of consumer importance will be needed to broaden available biodiversity in cultivated landscapes. With a changing climate, species considered critical to the landscape may require human-assisted hybridization with distant relatives to better ensure survival from threats posed by novel pests or diseases.

Alternative crops are also being bred for new uses, such as removing toxic chemicals and excess nutrients and improving degraded soils, including mine spoils [120] . Phytoremediation is a biotechnology to clean the contaminated sites of toxic elements (e.g. Cd, Cu, Zn, As, Se, Fe) via plant breeding, plant extracting, and plant volatilizing [121] . Phytoremediation of land contaminated with inorganic and/or organic pollutants has attracted much attention and research over the last decade [122] -[124] . Among the various approaches comprising phytoremediation, phytoextraction of metals and metalloids is probably the most challenging task. Where soils are impacted by industrial or mining activities, the degree of pollution is usually severe, making phytoextraction unfeasible within a reasonable time frame because of the high quantity of the pollutants present in the soil. Simple mass-balance calculations show that phytoextraction is potentially feasible only in low or moderately contaminated soils. For more heavily contaminated soils, phytostabilization with tolerant plants may be used to stabilize the contaminated sites and reduce the risk of erosion and leaching of pollutants to water bodies. Hyper-accumulation of metals or metalloids is important for the phytoextraction strategy [125] [126] . The last few years have seen a steady expansion in the list of hyper-accumulator species, which could be valuable plant resources for phytoremediation. A word of caution however, as the hyper-accumulation ability of some reported "hyper-accumulators" have yet to be confirmed in studies using field contaminated soils. Chaney et al. [125] discussed situations where phytoextraction may be applicable, such as paddy soils contaminated with moderate levels of cadmium (Cd) giving rise to Cd concentrations in rice grain exceeding the safe limit. An Indica-Japonica hybrid cultivar of rice was found to be an effective Cd phytoextractor, removing 7% - 14% of soil Cd; this had the effect of decreasing subsequent Cd accumulation in soybean seeds in a pot study by 24% - 46% [127] . The efficient translocation of Cd from roots to shoots appears to be the main reason for the efficient Cd accumulation in this rice cultivar, although the underlying mechanisms remain unclear. A major QTL responsible for the root to shoot translocation of Cd has been identified in an F_2 population constructed from the parental lines of rice differing in Cd accumulation in shoots by 13-fold [128] . While the trait for high Cd accumulation in shoots is useful for phytoextraction, the opposite is true for the strategy of breeding crops low in Cd accumulation for the benefit of food safety [129] . Many of the previous studies on phytoextraction have focused on hyper-accumulators. Small-scale field trials have shown that an ecotype of the Zn/Cd hyper-accumulator Thlaspi caerulescens from southern France was able to phytoextract Cd efficiently through the different seasons with good growth of biomass [130] [131] . This ecotype possesses a high-affinity Cd uptake system which is not suppressed by Zn [132] . The Chinese brake fern Pteris vittata has a strong ability to hyper-accumulate arsenic (As) and shows promising potential for phytoextraction of and from contaminated soils under field conditions [133] [134] , but the plant thrives only in the humid tropic/subtropical climates. Phytoextraction using high biomass plants such as willow (Salix sp.) and poplar (Populus sp.) has also been proposed [135] . Some Salix species are good accumulators of Cd and Zn, and up to 20% of soil Cd was removed by cropping three S. asmithiana in a lysimeter study [136] . A large proportion of the metals are stored in leaves, so plants either have to be harvested before leaves fall or the fallen leaves are to be collected. Such biomass plants may be grown on contaminated areas not suitable for food production, allowing gradual phytoextraction of metals while the biomass may be used to generate energy by pyrolysis [137] .

New perennial crops and tree species [e.g. switchgrass (Panicum virgatum), poplar (Populus spp.), Miscanthus, Arundo, etc.] are being developed as improved cellulosic feedstock for biofuels that will have a higher yield and energy content than was previously available [138] [139] . Cellulosic biofuels provide one approach for mitigating the impacts of global warming associated with fossil-fuel combustion, but concerns over appropriate implementation and environmental impacts remain [140] . Frontier approaches should be applied to study the possible advantages of perennial biofuel crops more photosynthetically productive, entail lower input costs, and improve soil nutrient input and retention. Through alliances with the bio-energy industry, research should also adapt industrial processes to biomass sources and sources to promising processes. Biofuels should form therefore part of a global, cross-cutting agenda of agricultural research, involving partners in the farming and energy sectors.

Simply developing more productive feedstock does not necessarily lead to enhanced environmental health. Without crop rotation, further monocultures of grain maize or increased palm oil production could have net negative environmental effects in the long term, but such efforts may be a necessary transition to facilitate infrastructure development for cellulosic feedstock. There is a concern however, that energy crops might inadvertently compete for land currently allocated for food crop production, and thereby raise food prices, must be considered carefully. Breeding alternative crops needs to be undertaken in close consultation with agronomists, economists, ecologists, and the commercial sector or industry, to ensure that new cultivars have the proper traits making them both profitable and sustainable.

The increasing demand, particularly by the industrialized world, for biofuels should take into account the agro-ecosystems and its biodiversity to ensure their healthy management. The agricultural systems required for producing biofuels need to be sustainable for an efficient use of biomass, and partitioning it among energy, feed, food and CO_2 fixation demands. They should be more eco-efficient for using existing farmland or marginal (dry, waterlogged, saline) tracts. Although some advocate that bio-energy can play a role for mitigating climate change by reducing greenhouse emissions, appropriate life-cycle analysis will be needed on a case by case basis to determine the use of land resources and estimate net carbon emissions of each suggested renewable energy technology. The agenda for plant breeding may include increasing plant grain and biomass productivity, optimizing the chemical and physical attributes of biofuel sources, and improving specific traits in first- and second-generation biofuel crops, within a framework of sustainable agriculture.

A major goal of harmonizing agriculture with the environment is to "tailor" crops to individual landscapes. Plant breeding has always maximized production by selecting for adaptation in the target environments of interest, using local environmental forces for plant selection [75] . By selecting breeding germplasm growing under local environmental conditions, individual cultivars can be optimized for small regional areas of production that fit prevailing environmental and weather patterns. Likewise, plants could be tailored to provide specific ecosystem services to local environments, to address local needs. One cost-effective way to achieve this is through participatory plant breeding, which involves local farmers in the breeding process.

Alternative crop rotations, planting densities, and tillage systems may make production more environmentally benign but will require altering breeding targets and an understanding that systems biology is complex and rarely has simple solutions. For example, no-tillage systems used for soil conservation can lead to colder soils in spring and change the prevalence and onset of various soilborne diseases, thus requiring the addition of specific disease resistances in the breeding objectives [141] . Breeders must select from conditions prevailing under new management practices to ensure cultivars will be optimally productive.

3. IMPACT OF TRANSGENIC CROPS ON CROP BIODIVERSITY AND AGRO-BIODIVERSITY

The potential impact of transgenic crops on biodiversity has been a topic of interest both in general as well as specifically in the context of the Convention on Biological Diversity. In a recent review, Carpenter [142] took a biodiversity lens to the substantial body of literature existing on the potential impacts of genetically modified crops on the environment, considering the impacts at the crop, farm, and landscape scales. Overall, the review finds that currently commercialized transgenic crops have reduced the impacts of agriculture on biodiversity, through enhanced adoption of conservation tillage practices, reduction of insecticide use, and use of more environmentally benign herbicides. Increasing yields also alleviate pressure to convert additional land into agricultural use.

Respecting crop biodiversity is widely accepted, that greater varietal and species diversity would enable agricultural systems to maintain productivity over a wide range of conditions. With the introduction of transgenic crops, concern has been raised that crop genetic biodiversity will decrease since breeding programs will concentrate on a smaller number of high value cultivars. Three studies (two in the United States on cotton and soybean and one in India on cotton) have analyzed the impact on the introduction of transgenic crops within-crop genetic biodiversity [142] . Studies in the United States of genetic diversity on cotton and soybean both concluded that the introduction of transgenic cultivars was found to have little or no impact on biodiversity. In contrast, the introduction of Bt cotton in India initially resulted in a reduction of on-farm varietal biodiversity due to the introduction of technology in only a small number of cultivars, which has since been offset by more Bt cultivars becoming available over time [142] . Carpenter [142] concluded that from a broader perspective, transgenic crops may actually increase crop biodiversity by enhancing underutilized alternative crops, making them more suitable for widespread domestication.

Respecting agro-biodiversity plants have a major influence on soil communities of micro- and other organisms fundamental to many functions of soil systems, such as nitrogen cycling, decomposition of wastes, and mobilization of nutrients. The potential impact of Bt crops on soil organisms is well studied. A comprehensive review of the available literature, by Icoz and Stotzky [143] , on the effects of Bt crops on soil ecosystems included the results of 70 scientific articles. The review found that, in general, few or no toxic effects of Cry proteins on woodlice, collembolans, mites, earthworms, nematodes, protozoa, and the activity of various enzymes in soil have been reported. Although some effects ranging from no effect to minor and significant effects, of Bt plants on microbial communities in soil have been reported, they were mostly the result of differences in geography, temperature, plant cultivar, and soil type and, in general, were transient and not related to the presence of the Cry proteins. Studies published since the Icoz and Stotzky review [143] have reached similar conclusions, including novel studies on snails.

In a study conducted in the northeastern part of the United States, Hoheisel and Fleischer [144] investigated the seasonal dynamics of coccinellids and their food (aphids and pollen) in a vegetable farm system containing plantings of Bt sweet corn, Bt potato, and transgenic insect-resistant squash [145] . The results indicated the transgenic vegetable crops provided conservation of coccinellids and resulted in a 25% reduction in insecticide use. In a similar study with these same crops, Leslie et al. [145] compared the soil surface-dwelling communities of Coleoptera and Formicidae in the transgenic crops and their isolines and found no differences in species richness and species composition but found that the transgenic vegetables required fewer insecticide applications. Such results are clear that genetically modified technology can be introduced within vegetable integrated pest management (IPM) systems and transgenic vegetables can offer novel and effective ways of controlling insects and the pathogens they transmit [54] [59] .

Another main concern of transgenic crops is the unintentional spread of transgenic traits into weedy species [146] . There are examples of transgene escape and some evidence for selective advantage of herbicide resistance picked up by weeds [147] [148] . The risk of herbicide resistant genes from a transgenic crop cultivar being transferred to weed relatives has been demonstrated in field crops such as canola/oilseed rape [149] [150] and sugar beet [151] . Rose et al. [152] has demonstrated that a "transgenic mitigation strategy" may impart a negative genetic load to hybrids ensuing from crosses between the weed field mustard and oilseed rape crop. The transgenic mitigation measure was a fitness-mitigating dwarfing gene beneficial for crops but deleterious for weeds (i.e., the hybrid weed is dwarfed due to this mitigation gene and is therefore outgrown by its non-trans- genic counterparts). This finding challenges the view that a transgenic plant might always endow a wild relative with a so-called fitness gene, making it harder and giving it the potential to become a "super-weed". Furthermore, Palaudelmàs et al. [153] found that transgenic maize volunteers had low plant vigor, rarely had cobs, and produced pollen which cross-fertilized neighbor plants only at low levels (0.16% in the worst-case scenario and was below the Regulation EC 1830/2003 establishing the adventitious threshold of 0.9% for coexistence). None- theless, transgene flow raises a new set of ecological and economic issues for scientists and policy makers to consider for transgene containment.

Crop production practices also have significant effects on the composition of weed communities. Changes in the types of weeds that are important locally are termed weed shifts. Such shifts are particularly relevant for managing weeds in herbicide tolerant crop systems in which tillage practices and herbicide use both play major roles in shaping the weed community. There are reports in literature of fourteen weed species or groups of closely related species that have increased in abundance in glyphosate resistant crops [142] . At the same time in six states of the United States, in a survey of transgenic maize, soybean, and cotton growers, between 36% and 70% of growers indicated that weed pressure had declined after implementing rotations using glyphosate resistant crops. The use of herbicides can also result in changes to weed communities through the development of herbicide tolerant weed populations. Globally, glyphosate resistant weeds have been confirmed for 21 weeds in 15 countries [142] . Most of these cases have been reported where glyphosate resistant crops are commonly grown. The development of weeds resistant to glyphosate will most likely require modification to weed control programs where practices, in addition to applying glyphosate, are needed to control the resistant populations.

The introduction of herbicide tolerant transgenic crops has been associated with the increased adoption of conservation tillage practices, which decreases run-off, increases water infiltration and reduces erosion. Trends in the adoption of conservation tillage have been studied in the United States and Argentina, the largest growers of herbicide tolerant transgenic crops. While conservation tillage was already being adopted by some growers prior to the introduction of genetically modified herbicide tolerant crops in both countries, studies have shown a positive two-way causal relationship between the adoption of conservation tillage and the adoption of genetically modified herbicide tolerant crops.

As stated, the most direct negative impact of agriculture on biodiversity is due to the considerable loss of natural habitats, which is caused by the conversion of natural ecosystems into agricultural land. Increases in crop yields allow less land to be dedicated to agriculture than it would otherwise be necessary. According to Carpenter [142] a large and growing body of literature has shown that the adoption of transgenic crops has increased yields, particularly in developing countries. A review of the results of global farmer surveys made by Carpenter [154] , found the average yield increases for developing countries range from 16% for insect-resistant corn to 30% for insect-resistant cotton, with an 85% yield increase observed in a single study on herbicide-tolerant corn. On average, developed-country farmers report yield increases ranging from no change for herbicide-tolerant cotton to a 7% increase for herbicide-tolerant soybean and insect-resistant cotton. Brookes et al. [155] have estimated the benefit of these yield improvements by reducing conversion of land into agricultural use. They estimate that 2.64 million hectares of land would probably be brought into grain and oilseed production if biotechnology traits were no longer used.

Storer et al. [156] stated that the most direct landscape level effects of growing Bt crops would be expected for target pest species for which the crop is a primary food source and is mobile across the landscape. Area-wide pest suppression not only reduces losses to adopters of technology, but may also benefit non-adopters and growers of other crops by reducing crop losses and/or the need to use pest control measures such as insecticides [142] .

Several studies have investigated the impact on the introduction of Bt corn and cotton on regional outbreaks of pest populations, reporting evidence of regional pest suppression in Bt corn and cotton in various areas of the United States and in Bt cotton growing regions of China [142] [157] -[161] . The effects of transgenic crops on above-ground non-target invertebrates have been the subject of a large number of laboratory and field studies. By the end of 2008, over 360 original research papers had been published on non-target effects of Bt crops [161] . A comprehensive review of the literature by Naranjo [161] included 135 laboratory-based studies on nine Bt crops from 17 countries and 63 field-based studies on five Bt crops from 13 countries, which were analyzed using meta-analysis techniques. In general, laboratory studies identified greater levels of hazard than field studies, at least partially explained by differences in organisms studied, and frequently higher protein exposure in lab studies compared to exposure levels in the field [161] [162] . Field studies demonstrated few harmful non-

4. CONSERVATION AND USE OF BIODIVERSITY: OPPORTUNITIES FOR COOPERATION AND NEW PARTNERSHIPS

Plant genetic resources for food and agriculture are the quintessential global public good. No nation is self-re- liant. A viable market for their conservation and trade does not exist. The conservation of plant genetic resources is a prerequisite for addressing climate change, as well as water and energy constraints, which will grow in importance in the next decades. The Svalbard Global Seed Vault is an International Treaty which establishes a multilateral System to facilitate access and benefit sharing of plant genetic resources. The Treaty has an insurance policy and provides legal framework for a cooperative and global approach to manage this essential resource. The Svalbard Global Seed Vault has a mechanism for ensuring the permanent conservation of unique crop biodiversity, the Global Crop Diversity Trust, which is structured as an endowment fund [169] .

Plant breeding is vital to protect the yield gains made to date, and to further increase the genetic yield potential of all crops. As a result of the Green Revolution, global productivity of the main food staples steadily rose since the 1960s. Such achievements ensued from crop genetic enhancement partnerships. They are models illustrating partnering for exchange, evaluation, release and use of plant genetic resources worldwide. These partnerships include national agricultural research institutes and international agricultural research centers. For many decades the global wheat yield increased due to an effective International Wheat Improvement Network (IWIN) officially founded as an international organization in 1966 [170] . This wheat network deployed cutting-edge science alongside practical multi disciplinary applications, resulting in the development of genetically enhanced wheat germplasm, which has improved food security and the livelihoods of farmers in the developing world [171] . The spring wheat germplasm bred in Mexico under the leadership of Nobel Peace Laureate Norman Borlaug was further used for launching the Green Revolution in India, Pakistan and Turkey [172] . The network was broadened during the 1970s to include Brazil, China and other major developing country wheat producers. It resulted in wheat cultivars with broader host plant resistance (especially to rusts), better adaptation to marginal environments, and tolerance to acid soils. Nowadays IWIN, an international "alliance", operates field evaluation trials in more than 250 locations, in roughly 100 countries it tests improved breeding lines of wheat in different environments. The number of wheat cultivars released annually in the developing world doubled to more than 100 cultivars by early 1990s due to this networking and the strengthening of national capacity [173] . The widespread adoption of newly bred wheat cultivars, especially in South Asia and Latin America, due to yield increases, led to 50% average annual rates of investment returns [174] . The urban poor also benefited significantly because grain harvest increases drove wheat prices down. Every year, nursery sets and trials are sent to various researchers worldwide, who share their data from these trials to catalogue and analyze. The returned data are used to identify parents for subsequent crosses and to incorporate new genetic variability into advanced wheat lines that are consequently able to cope with the dynamics of abiotic and biotic stresses affecting wheat farming systems. The full pedigree and selection histories are known and phenotypic data cover yield, agronomic, patho- logical and quality data [171] .

The International Network for Genetic Evaluation of Rice (INGER) is one more example of world cooperation. It was established in 1975 as a consortium of national agricultural research systems of rice-growing countries and Centers of today's CGIAR Consortium. INGER was initially founded as an International Rice Testing Program, but soon became an integral component of world national rice breeding program. INGER partners can share rice breeding lines. Every year partners provide about 1000 genetically diverse breeding lines, which have been grown in about 600 experiment stations from 80 countries. This network facilitated the release of 667 cultivars worldwide, which translated into 1.5 billion US dollars of economic benefits. It was estimated that ending INGER could lead to a reduction of 20 rice cultivars per year and to an economic loss of 1.9 billion US dollars [175] . Further analysis by Jackson and Huggan [176] has shown how genetic conservation of landraces can lead to significant gains in rice breeding.

Two other examples of cooperation and partnership are the Latin American Maize Project (LAMP) and the Germplasm Enhancement of Maize (GEM). The LAMP was established as a partnership between Latin America and the United States to assess national germplasm and facilitate the exchange of maize genetic resources across the American continent [177] . The United States Department of Agriculture, the participating national agricultural research systems and a multinational seed corporation provided the funding. The aim of LAMP was to obtain information about the performance of maize germplasm and to share it with plant breeders for developing genetically enhanced open pollinated and hybrid cultivars. The maize germplasm was tested for agronomic characteristics from sea level to 3300 m, and from 41°N to 34°S across 32 locations in the first stage and in 64 locations (two per region) in the second stage. These locations were clustered according to five homologous areas: lowland tropics, temperate and three altitudes.

There were a total five LAMP breeding stages [177] . In the first stage, 14,847 maize accessions belonging to a region were planted for evaluation in trials using a randomized complete block design with two replications of 10 m^2 plots at a single location, which was environmentally similar to that from where these landraces were originally collected. The next step included the assessment of the upper quintile (20%) of those accessions evaluated for agronomic performance in the previous stage. These accessions were planted in two locations with two replications, and the upper 5% were further selected according to their performance. These best selected accessions of each country were interchanged among regions belonging to the same homologous area in the third stage. They were tested in two locations with two replications in each region. The selected maize accessions from the same homologous area were mated with the best tested accession of the region in an isolated field within each region. In the fourth stage, combining ability tests of 268 selected maize accessions were carried out with a local tester using two replications at two locations within each region. The elite maize germplasm was integrated into breeding programs in the fifth stage, which was the last. The best cross combinations and heterotic pools were also determined by LAMP. Maize breeders obtained access to the most promising accessions identified by LAMP to widen the crop genetic base. A LAMP core subset has been made available for encouraging further use in broadening of maize genetic diversity [178] .

The GEM was set up to introgress useful genetic diversity from Latin American maize landraces and other tropical maize donor sources (lines and hybrids) into United States' maize germplasm, to broaden the genetic base of the "corn-belt" hybrids [179] [180] . GEM owes its existence to LAMP because it has used the Latin American landrace maize accessions selected by LAMP in crosses with elite temperate maize lines from the private seed companies in North America [177] . GEM used a pedigree breeding system to develop S_3 lines. The GEM breeders arranged their crosses into non-Stiff Stalk and Stiff Stalk heterotic groups [181] .

LAMP provided the first step through the sharing of information needed to select gene bank maize accessions for further germplasm enhancement. GEM completed the process by returning to genetically enhanced breeding materials derived from gene bank accessions. This improved germplasm can be further used in maize breeding in the United States and elsewhere. LAMP and GEM are very nice examples of international and national public-private partnerships in crop germplasm enhancement.

Agricultural plant breeding is a typical commodity- or species-oriented and solves problems within a species, rather than making breeding choices based on system wide needs. For example, maize breeders currently maximize the area in which maize can be grown, and maximize the amount of maize produced throughout that area. If environmental harmony is to be a key breeding objective, then a change in agricultural thinking to appropriately value whole cropping systems will be required. Achieving these goals will require collaboration among the private, public, and non-profit sectors, and with society as a whole. Programs within the private sector excel at breeding major, profitable crops, and have economies of scale to increase the efficiency of production and ultimately provide farmers with seed. As a valuable complement to commercial breeding programs, public and non-profit breeding programs may focus on developing alternative crops, breeding for small target regions, tackling long-term and high-risk problems, evaluating diverse genetic resources, and, importantly, conducting basic research on breeding methodology to enhance efficiency. Only publicly funded breeding programs, and in particular those based at universities, can provide the necessary education and training in plant breeding and in specialized fields such as ecology. Without trained students from public programs, private commercial breeding programs suffer from an erosion of intellectual capital. Conversely, without the private sector to commercialize public-sector-derived products, beneficial traits and new cultivars cannot easily and quickly be put in the hands of farmers, as has been seen in developing countries without a developed seed industry [182] . As stated, seed production is high technology and a cost intensive venture and only well organized seed companies with good scientific manpower and well equipped research facilities can afford seed production.

Although due to globalization, most breeding research and cultivar development in the world is presently conducted and funded in the private sector, mainly by huge multinational seed companies. Public breeders, cultivar development activities and research are disappearing worldwide. In general, this means there are fewer decision-making centers for breeding and cultivar development. This has also resulted in the focus on relatively few major crops produced worldwide, to the detriment of all the other cultivated crops. It is imperative that national governments and policymakers, as part of a social duty, invest in breeding research and cultivar development of traditional open-pollinated cultivars and in the minor crops. More investments in this area will mean less expensive seed for growers to choose from, and an increased preservation of crop biodiversity. To accomplish these goals new approaches may be required to crop breeding research and development by both the public and private sector. Until recently, breeding research and development which targets small-scale and poor farmers has largely been undertaken by public sector institutions and national agricultural research institutes. However, the capacity to undertake the work was mainly dependent on national or international funding and expertise. The work has been limited by the capacity of these institutions to pay for it. As a result, crop breeding advancement has varied enormously among countries and even within regions in developed and still developing countries. In the area of plant breeding, the process to produce improved cultivars is slow, and it requires long-term sustained commitment that may not fit the continuing changes in the national and international politics to fund research. The application of biotechnology promises acceleration in some aspects of plant breeding, but the adoption of more advanced technology raises the cost of research significantly at a time when investment funding has diminished. Public plant breeding remains a key component of crop breeding research systems worldwide, especially in developing countries. However, the increasing presence of private sector breeding and a decrease in national and international support makes it difficult for the public sector to continue operating in the traditional manner. Declining funding for public crop breeding coupled with the rapid increase of crop production and an urbanizing population has created a difficult situation. Public sector breeding must be strengthened. More public sector crop breeders are needed worldwide to select and to produce non-hybrid cultivars of the minor crops. Breeding of major crops and other minor crops must continue as a viable endeavor. This will benefit small farmers, and will safeguard biodiversity and food security in developing countries.

While the maintenance of vigorous public sector breeding programs in areas where private companies are not interested in providing low cost cultivars is highly desirable, an additional approach to maximize crop and agricultural research input would be the development of global programs with public-private partnerships. The public sector may support portions of crop and agricultural R & D, unattractive to the private sector, and feed improved breeding lines and systems to the private sector for exploitation in regions where the private sector is active, and nurture private sector development in regions where it is lacking. In recent years, private plant breeding programs have increased in number and size. Financial investment also increased, as well as interest in intellectual property protection. The spirit of original attempts to protect plant breeders' rights was that granting a certificate of protection should not inhibit the flow of information and products through continued research by the entire plant breeding community [49] [50] . In a classic sense, the patent is a defensive tool to prevent competitors from reaping benefits which rightfully belong to the inventor. In the modern context, it is an offensive weapon, to stifle competition, prevent further innovation by others and maximize income [49] [51] . The United States utility patent, it is a way to slow down the flow of progress in plant breeding research, unless the research is within the company holding the patent. While obviously benefiting that company, it is a big step backwards for the plant breeding community and by far, for agriculture itself. The intellectual property protection must encourage research and free flow of materials and information [49] [51] . Protection should be for the cultivar only. There should be no constraint against other breeders using that cultivar in further research, including further breeding. Another breeder should be free to use the protected cultivar in a cross, followed by further development through pedigree breeding. Another breeder should also be free to transfer genes controlling economic traits into the protected cultivar by the backcross method or by genetic transformation procedures [49] [50] .

5. CONCLUSIONS

The growing demand for food in the next decades poses major challenges to humanity. We have to safeguard both biodiversity and arable land for future agricultural food production, and we need to protect genetic biodiversity to safeguard ecosystem resilience. We must produce more food with less input, while deploying every effort to minimize risk. Agricultural sustainability is no longer an option, it is mandatory!

Plant breeding is the science of improving plants to further improve the human condition. Plant breeding has played a vital role in the successful development of modern agriculture via "new" cultivars. Plant breeders are continually improving the ability of cultivars to withstand various environmental conditions. By reducing the impact of agriculture on the environment while maintaining sufficient production will require the development of new cultivars.

Climate change is altering the availability of resources and the conditions crucial to plant performance. Plants respond to these changes through environmentally induced shift in phenotype. Understanding these responses is essential to predict and manage the effects of climate change on crop plants.

In the foreseeable future and an increase in population will need significant production. Breeding and modern agricultural technologies can increase yield on existing agricultural land. As a result, they can make a significant contribution to biodiversity conservation by limiting the need to expand agricultural land and by allowing nature to be maintained for conservation purposes and harmony between agriculture and the environment.

There is still an on-going debate among researchers and in the media on the best strategy to keep pace with global population growth and increasing food demand. One strategy favors the use of transgenic crops, while another strategy focuses on agricultural biodiversity. There are short research funds for agro-biodiversity solutions in comparison with funding for research in genetic modification of crops. Favoring biodiversity does not exclude any future biotechnological contributions, but favoring biotechnology threatens future biodiversity resources. The future breeding programs should encompass not only knowledge of existing practices but also conservation of a wide pool of genetic resources of existing crops and breeds, including their wild relatives, to provide the genes necessary to cope with changes in agricultural production. Therefore, agro-biodiversity should be a central element of future sustainable agricultural development, instead of just a source of traits which can be used in current breeding programs. The concept of sustainability rests on the principle that the present needs must be addressed without compromising the ability of future generations to meet their own needs. Sustainable agriculture is an alternative to solve future fundamental and applied issues related to food production in an ecological way.

Farmers in developing countries, especially small farmers, have problems specific to their cultural, economic and environmental conditions, such as limited purchasing power to access improved cultivars and proprietary technologies. These farmers have an important role in conserving and using crop biodiversity. The future of the world food security depends on stored crop genes as well as on farmers who use and maintain crop genetic diversity on a daily basis. In the long run, the conservation of plant genetic diversity depends not only on a small number of institutional plant breeders and seed banks, but also on the vast number of farmers who select, improve, and use crop diversity, especially in marginal farming environments. Their extensive farming systems using landraces or open-pollinated cultivars increase sustainability and less impact from stresses caused by drought, insect and diseases, due to long-term in situ selection of these crops cultivated as opposed to the fertilizer, herbicide, and pesticide demands in an intensive crop based system with improved, hybrid, or transgenic cultivars. That is why we should also be alerted and particularly alarmed by the current trend to exclusively use improved, hybrid, and transgenic crop cultivars. Farmers do not just save seeds; they are plant breeders who constantly adapt their crops to specific farming conditions and needs. This genetic biodiversity is the key to maintain and improve the world's food security, and agriculture sustainability.

The introduction of genetically modified technology has been hailed as a gene revolution similar to the "Green Revolution" of the 1960s. The "Green Revolution" had an explicit strategy for technology development and diffusion, targeting farmers in developing countries, in which improved germplasm was made freely available as a public good, a particular success in Asia. In contrast to the "Green Revolution", the push for genetically modified crops is based largely on private agricultural research, with cultivars provided to farmers on market terms. To date efforts on genetically modified crops have been focused on crops considered to be profitable enough by large plant breeding companies, not on solutions to problems confronted by the world's small farmers. Existing biodiversity in combination with plant breeding has much more to offer the many world's farmers and consumers, while genetically modified crops have more to offer the agro-industry and some large- scale farms, and this explains why they have received so much attention and research funding. Genetically modified crops and their creation may attract investment in agriculture, but it can also concentrate ownership of agricultural resources. In developing countries, patents may drive up costs; restrict experimentation by the public researcher or individual farmer, while undermining local practices that enhance food security and economic sustainability. There is particular concern that present intellectual property rights instruments, including genetically modified organisms, will inhibit sowing of own seeds, seed exchange, and sale.

Transgenic crops can continue to decrease pressure on biodiversity as global agricultural systems expand to feed a world population expected to continue to increase for the next 30 to 40 years. Due to higher income elasticity of demand and population growth, these pressures will be greater in developing countries. Both current and pipeline technology hold great potential in this regard. The potential of currently commercialized genetically modified crops to increase yields, decrease pesticide use, and facilitate the adoption of conservation tillage has yet to be realized, as there continue to be countries where there is a good technological fit, but they have not yet approved these technologies for commercialization. In addition to the potential benefits on expanded adoption of current technology, several pipeline technologies offer additional promises of alleviating the impacts of agriculture on biodiversity. Continued yield improvements in crops such as rice and wheat are expected with insect resistant and herbicide tolerant traits that are already commercialized in other crops. Technologies such as drought tolerance and salinity tolerance would alleviate the pressure to convert high biodiversity areas into agricultural use by enabling crop production on sub-optimal soils. Drought tolerance technology, which allows crops to withstand prolonged periods of low soil moisture, is anticipated to be commercialized within less than five years. This technology has particular relevance for areas like sub-Saharan Africa, where drought is a common occurrence and access to irrigation is limited. Salt tolerance addresses the increasing problem of saltwater encroachment on freshwater resources. Nitrogen use efficient technology is also under development, which can reduce run-off of nitrogen fertilizer into surface waters. Technology promises to decrease the use of fertilizers while maintaining yields, or increase yields achievable with reduced fertilizer rates where access to fertilizer input is limited. This technology is slated to be commercialized within the next 10 years.

One of the major arguments for genetic modified technology is that new cultivars can be developed more quickly than in traditional plant breeding, but like new cultivars derived from conventional breeding methods, transgenic cultivars require several years of field trials to ensure that the inserted traits will actually become expressed and have the desired effects in local environments. When genes coding for certain traits are transferred, typically from one plant species to another, the desired traits are not always expressed unless the environment interacts with the genes in the anticipated way triggering the desired response, which depends on the regulating sequences inserted with the gene. This means that new transgenic cultivars, developed under laboratory conditions in a controlled climate, have to be tested under field conditions, as in more traditional breeding methods, so currently there is little difference in the speed with which either method will result in the release of new cultivars.

The knowledge gained from basic plant research will underpin future crop improvements, but effective mechanisms for the rapid and effective translation of research discoveries into public good agriculture remain to be developed. Maximum benefit will be derived if robust plant breeding and crop management programs have ready access to all the modern crop biotechnological techniques, both transgenic and non-transgenic, to address food security issues. This will require additional investments in capacity building for research and development, in developing countries. Technology implementation alone is not sufficient to address such complex questions as food security. Biotechnologies will make new options available but are not a global solution. We must ensure that society will continue to benefit from the vital contribution that plant breeding offers, using both conventional and biotechnological tools. Genetic engineering has the potential to address some of the most challenging biotic constraints faced by farmers, which are not easily addressed through conventional plant breeding alone. Besides other promising traits seems to be host plant resistance to insects and pathogens. However, transgenic cultivars will have one or a few exogenous genes whereas the background genotype will still be the product of non-transgenic (or conventional) crop breeding. One should follow a pragmatic approach when deciding whether to engage in transgenic plant breeding. Biotechnology products will be successful if clear advantages and safety are demonstrated to both farmers and consumers.

João Carlos da Silva Dias There is a need of investment in research breeding and cultivar development in traditionally open-pollinated cultivars and in the minor crops. More investments in this area will mean cheaper cultivars for growers to choose from and more preservation of crop biodiversity. In recent years, private plant breeding programs have increased in number and size. Financial investment also increased, as well as interest in intellectual property protection. Protective measures, especially patenting, must be moderated to eliminate coverage so broad that it stifles innovation. The intellectual property protection laws for plants must be made less restrictive to encourage research and free flow of materials and information. Public sector breeding must remain vigorous, especially in areas where the private sector does not function. This will often require benevolent public/private partnerships as well as government support. Intellectual property rights laws for plants must be made less restrictive to encourage freer flow of materials. Active and positive connections between the private and public breeding sectors and large-scale gene banks are required to avoid a possible conflict involving breeders' rights, gene preservation and erosion. Partnerships between policy makers with public and private plant breeders will be essential to address future challenges. Many current breeding efforts remain under-funded and disorganized. There is a great need for a more focused, coordinated approach to efficiently utilize funding, share expertise, and continue progress in technologies and programs.

REFERENCES

1. Jaggard, K.W., Qi, A.M. and Ober, E.S. (2010) Possible Changes to Arable Crop Yields by 2050. Philosophical Transactions of the Royal Society B: Biological Sciences, 365, 2835-2851.
2. FAO (2012) The State of the World Population Report. By Choice, Not by Chance: Family Planning, Human Rights and Development. United Nations Population Fund, New York.
3. da Silva Dias, J.C. (2014) Guiding Strategies for Breeding Vegetable Cultivars. Agricultural Sciences, 5, 9-32.
4. Dias, J.S. and Ryder, E.J. (2011) World Vegetable Industry: Production, Breeding, Trends. Horticultural Reviews, 38, 299-356.
5. Dias, J.S. (2012) Chapter 1. Vegetable Breeding for Nutritional Quality and Health Benefits. In: Carbone, K., Ed., Cultivars: Chemical Properties, Antioxidant Activities and Health Benefits, Nova Science Publishers, Inc., Hauppauge, 1-81.
6. Tilman, D., Balzer, C., Hill, J. and Befort, B.L. (2011) Global Food Demand and the Sustainable Intensification of Agriculture. Proceedings of the National Academy of Sciences of the United States of America, 108, 20260-20264.
7. Alexandratos, N. and Bruinsma, J. (2012) World Agriculture towards 2030/2050: The 2012 Revision. Paper No. 12-03, Food and Agriculture Organization (FAO), Rome.
8. FAO (2012) How to Feed the World in 2050. FAO, Rome.
9. Delgado, C.L. (2003) Rising Consumption of Meat and Milk in Developing Countries Has Created a New Food Revolution. Journal of Nutrition, 133, 3907S-3910S.
10. Kastner, T., Rivas, M.J.I., Koch, W. and Nonhebel, S. (2012) Global Changes in Diets and the Consequences for Land Requirements for Food. Proceedings of the National Academy of Sciences of the United States of America, 109, 6868- 6872.
11. Delgado, C.L. (1999) Livestock to 2020: The Next Food Revolution. Food, Agriculture, and the Environment Discussion Paper No. 28, International Food Policy Research Institute, Washington DC.

12. The Royal Society (2009) Reaping the Benefits: Science and the Sustainable Intensification of Global Agriculture. The Royal Society Policy Document 11/09, The Royal Society, London.
13. Naylor, R., Steinfeld, H., Falcon, W., Galloway, J., Smil, V., Bradford, E., Alder, J. and Mooney, H. (2005) Losing the Links between Livestock and Land. Science, 310, 1621-1622.
14. Gerbens-Leenes, P. and Nonhebel, S. (2002) Consumption Patterns and Their Effects on Land Required for Food. Ecological Economics, 42, 185-199.
15. Wirsenius, S., Azar, C. and Berndes, G. (2010) How Much Land Is Needed for Global Food Production under Scenarios of Dietary Changes and Livestock Productivity Increases in 2030? Agricultural Systems, 103, 621-638.
16. Pimentel, D. and Pimentel, M. (2003) Sustainability of Meat-Based and Plant-Based Diets and the Environment. American Journal of Clinical Nutrition, 78, 660S-663S.
17. Foley, J.A., Ramankutty, N., Brauman, K.A., Cassidy, E.S., Gerber, J.S., Johnston, M., Mueller, N.D., O'Connell, C., Ray, D.K., West, P.C., Balzer, C., Bennett, E.M., Carpenter, S.R., Hill, J., Monfreda, C., Polasky, S., Rockström, J., Sheehan, J., Siebert, S., Tilman, D. and Zaks, D.P. (2011) Solutions for a Cultivated Planet. Nature, 478, 337-342.
18. Steinfeld, H., Gerber, P., Wassenaar, T., Castel, V., Rosales, M. and De Haan, C. (2006) Livestock's Long Shadow: Environmental Issues and Options. FAO, Rome.
19. Mekonnen, M.M. and Hoekstra, A.Y. (2012) A Global Assessment of the Water Footprint of Farm Animal Products. Ecosystems, 15, 401-415.
20. World Watch Institute (2010) Biofuel Production Up Despite Economic Downturn Vital Signs. World Watch Institute, New York.
21. Food and Agricultural Policy Research Institute (FAPRI) (2011) World Biofuels: FAPRI-ISU 2011 Agricultural Outlook. FAPRI, Ames.
22. FAO (2013) FAO Statistical Yearbook—Land Use. FAOSTAT, FAO, Rome, PA4.
23. Kucharik, C.J. and Serbin, S.P. (2008) Impact of Recent Climate Change on Wisconsin Corn and Soybean Yield Trends. Environmental Research Letters, 3, Article ID: 034003, 10 p.
24. Battisti, D.S. and Naylor, R.L. (2009) Historical Warnings of Future Food Insecurity with Unprecedented Seasonal Heat. Science, 323, 240-244.
25. Schlenker, W. and Lobell, D.B. (2010) Robust Negative Impacts of Climate Change on African Agriculture. Environmental Research Letters, 5, Article ID: 014010, 8 p.
26. Roudier, P., Sultan, B., Quirion, P. and Berg, A. (2011) The Impact of Future Climate Change on West African Crop Yields: What Does the Recent Literature Say? Global Environmental Change, 21, 1073-1083.
27. Lobell, D.B., Schlenker, W. and Costa-Roberts, J. (2011) Climate Trends and Global Crop Production since 1980. Science, 333, 616-620.
28. Lobell, D.B., Bänziger, M., Magorokosho, C. and Vivek, B. (2011) Nonlinear Heat Effects on African Maize as Evidenced by Historical Yield Trials. Nature Climate Change, 1, 42-45.
29. Schlenker, W. and Roberts, M.J. (2009) Nonlinear Temperature Effects Indicate Severe Damages to US Crop Yields under Climate Change. Proceedings of the National Academy of Sciences of the United States of America, 106, 15594- 15598.
30. Gupta, R., Gopal, R., Jat, M.L., Jat, R.K., Sidhu, H.S., Minhas, P.S. and Malik, R.K. (2010) Wheat Productivity in Indo-Gangetic Plains of India during 2010: Terminal Heat Effects and Mitigation Strategies. PACA Newsletter, 14, 1-11.
31. Asseng, S., Foster, I. and Turner, N.C. (2011) The Impact of Temperature Variability on Wheat Yields. Global Change Biology, 17, 997-1012.
32. Lobell, D.B., Sibley, A. and Ortiz-Monasterio, J.I. (2012) Extreme Heat Effects on Wheat Senescence in India. Nature Climate Change, 2, 186-189.
33. Bell, G. and Collins, S. (2008) Adaptation, Extinction and Global Change. Evolutionary Applications, 1, 3-16.
34. Kelly, A.E. and Goulden, M.L. (2008) Rapid Shifts in Plant Distribution with Recent Climate Change. Proceedings of the National Academy of Sciences of the United States of America, 105, 11823-11826.

35. Shanthi-Prabha, V., Sreekanth, N.P., Babu, P.K. and Thomas, A.P. (2011) The Trilemma of Soil Carbon Degradation, Climate Change and Food Insecurity. Disaster Risk and Vulnerability Conference 2011, The Applied Geoinformatics for Society and Environment, Germany, 107-112.
36. Gregory, P.J., Johnson, S.N., Newton, A.C. and Ingram, J.S.I. (2009) Integrating Pests and Pathogens into the Climate Change/Food Security Debate. Journal of Experimental Botany, 60, 2827-2838.
37. Patz, J.A. and Kovats, R.S. (2002) Hot Spots in Climate Change and Human Health: Present and Future Risks. Lancet, 368, 859-869.
38. Mcmichael, A., Woodruff, R.E. and Hales, S. (2006) Climate Change and Human Health: Present and Future Risks. Lancet, 367, 859-869.
39. Ziska, L.H., Epstein, P.R. and Schlesinger, W.H. (2009) Rising CO_2, Climate Change, and Public Health: Exploring the Links to Plant Biology. Environmental Health Perspectives, 117, 155-158.
40. Borlaug, N. (1983) Contributions of Conventional Plant Breeding to Food Production. Science, 219, 689-693.
41. Trethowan, R.M., Reynolds, M.P., Ortiz-Monasterio, I. and Ortiz, R. (2007) The Genetic Basis of the Green Revolution in Wheat Production. Plant Breeding Reviews, 28, 39-58.
42. Evenson, R.E. and Gollin, D. (2003) Assessing the Impact of the Green Revolution, 1960 to 2000. Science, 300, 758- 762.
43. Tilman, D., Cassman, K.G., Matson, P.A., Naylor, R. and Polasky, S. (2002) Agricultural Sustainability and Intensive Production Practices. Nature, 418, 671-677.
44. Burney, J.A., Davis, S.J. and Lobell, D.B. (2010) Greenhouse Gas Mitigation by Agricultural Intensification. Proceedings of the National Academy of Sciences of the United States of America, 107, 12052-12057.
45. Edgerton, M.D. (2009) Increasing Crop Productivity to Meet Global Needs for Feed, Food, and Fuel. Plant Physiology, 149, 7-13.
46. IAASTD (2009) Agriculture at the Crossroads. International Assessment of Agricultural Knowledge, Science and Technology for Development (IAASTD). Island Press, Washington DC.
47. Harlan, J.R. (1992) Crops and Man. American Society of Agronomy and Crop Science Society of America, Madison.
48. Pratt, R.C. (2004) An Historical Examination of the Development and Adoption of Hybrid Corn: A Case Study in Ohio. Maydica, 49, 155-172.
49. Dias, J.S. (2011) Biodiversity and Vegetable Breeding in the Light of Developments in Intellectual Property Rights. In: Grillo, O. and Verona, G., Eds., Ecosystems Biodiversity, Chapter 17, INTECH publisher, Rijeka, 389-428.
50. Dias, J.S. (2012) Impact of the Vegetable Breeding Industry and Intellectual Property Rights in Biodiversity and Food Security. In: Jones, A.M. and Hernandez, F.E., Eds., Food Security: Quality, Management, Issues and Economic Implications, Nova Science Publishers Inc., Hauppauge, 57-86.
51. Dias, J.S. (2013) Impact of Vegetable Breeding Industry and Intellectual Property Rights in Food Security. In: Nath, P., Ed., The Basics of Human Civilization-Food, Agriculture and Humanity, Vol. I. Present Scenario, Prem Nath Agricultural Science Foundation (PNASF), Bangalore & New India Publishing Agency (NIPA), New Delhi, 173-198.
52. Dias, J.S. and Ryder, E. (2012) Impact of Plant Breeding on the World Vegetable Industry. Acta Horticulturae, 935, 13-22.
53. Dias, J.S. (2010) Impact of Improved Vegetable Cultivars in Overcoming Food Insecurity. In: Nath, P. and Gaddagimath, P.B., Eds., Horticulture and Livelihood Security, Scientific Publishers, New Dehli, 303-339.
54. Dias, J.S. and Ortiz, R. (2012) Transgenic Vegetable Crops: Progress, Potentials and Prospects. Plant Breeding Reviews, 35, 151-246.
55. Dias, J.S. (1989) The Use of Molecular Markers in Selection of Vegetables. SECH, Actas de Horticultura, 3, 175-181.

56. Dias, J.S. (1991) The Use of Computers in Plant Breeding. SECH, Actas de Horticultura, 8, 367-371.
57. Dias, J.S. and Ortiz, R. (2012) Transgenic Vegetable Breeding for Nutritional Quality and Health Benefits. Food and Nutrition Sciences, 3, 1209-1219.
58. Dias, J.S. and Ortiz, R. (2013) Transgenic Vegetables for Southeast Asia. In: Holmer, R., Linwattana, G., Nath, P. and Keatinge, J.D.H., Eds., Proceedings. Regional Symposium on High Value Vegetables in Southeast Asia: Production, Supply and Demand (SEAVEG 2012), Chiang Mai, 24-26 January 2012, 361-369.
59. Dias, J.S. and Ortiz, R. (2013) Transgenic Vegetables for 21st Century Horticulture. Acta Horticulturae, 974, 15-30.
60. Dias, J.C. (2010) Impact of Improved Vegetable Cultivars in Overcoming Food Insecurity. Euphytica, 176, 125-136.
61. Tilman, D. (1999) Global Environmental Impacts of Agricultural Expansion: The Need for Sustainable and Efficient Practices. Proceedings of the National Academy of Sciences of the United States of America, 96, 5995-6000.
62. Robertson, G.P. and Swinton, S.M. (2005) Reconciling Agricultural Productivity and Environmental Integrity: A Grand Challenge for Agriculture. Frontiers in Ecology and the Environment, 3, 38-46.
63. Hirel, B., Le Gouis, J., Ney, B. and Gallais, A. (2007) The Challenge of Improving Nitrogen Use Efficiency in Crop Plants: Towards a More Central Role for Genetic Variability and Quantitative Genetics within Integrated Approaches. Journal of Experimental Botany, 58, 2369-2387.
64. Foulkes, M.J., Hawkesford, M.J., Barraclough, P.B., Holdsworth, M.J., Kerr, S., Kightley, S. and Shewry, P.R. (2009) Identifying Traits to Improve the Nitrogen Economy of Wheat: Recent Advances and Future Prospects. Field Crops Research, 114, 329-342.
65. Korkmaz, K., Ibrikci, H., Karnez, E., Buyuk, G., Ryan, J., Ulger, A.C. and Oguz, H. (2009) Phosphorus Use Efficiency of Wheat Genotypes Grown in Calcareous Soils. Journal of Plant Nutrition, 32, 2094-2106.
66. Farooq, M., Kobayashi, N., Wahid, A., Ito, O. and Basra, S.M.A. (2009) Strategies for Producing More Rice with Less Water. Advances in Agronomy, 101, 351-388.
67. Shi, W., Moon, C.D., Leahy, S.C., Kang, D., Froula, J., Kittelmann, S., Fan, C., Deutsch, S., Gagic, D., Seedorf, H., Kelly, W.J., Atua, R., Sang, C., Soni, P., Li, D., Pinares-Patiño, C.S., Mcewan, J.C., Janssen, P.H., Chen, F., Visel, A., Wang, Z., Attwood, G.T. and Rubin, E.M. (2014) Methane Yield Phenotypes Linked to Differential Gene Expression in the Sheep Rumen Microbiome. Genome Research, 24, 1517-1525.
68. Shrawat, A.K. and Good, A.G. (2008) Genetic Engineering Approaches to Improving Nitrogen Use Efficiency. Plant Research News. ISB Report, May 2008. Information Systems for Biotechnology (ISB) News Report, Blackburg http://www.isb.vt.edu /news/2008/news08.may.htm#may0801
69. Daemrich, A., Reinhardt, F. and Shelman, M. (2008) Arcadia Biosciences: Seeds of Change. Harvard Business School, Boston.
70. Subbarao, G.V., Tomohiro, B., Masahiro, K., Osamu, I., Samejima, H., Wang, H.Y., Pearse, S.J., Gopalakrishnan, S., Nakahara, K., Zakir Hossain, A.K.M., Tsujimoto, H. and Berry, W.L. (2007) Can Biological Nitrification Inhibition (BNI) Genes from Perennial Leymus racemosus (Triticeae) Combat Nitrification in Wheat Farming? Plant and Soil, 299, 55-64.
71. IPCC (Intergovernmental Panel on Climate Change) (2009) The Physical Science Basis. In: Solomon, S., Qin, D., Manning, M., Chen, Z., Marquis, M., Averyt, K.B., Tignor, M. and Miller, H.L., Eds., Contribution of Working Group I to the Fourth Assessment Report of the Intergovernmental Panel on Climate Change, Cambridge University Press, Cambridge.
72. Ortiz, R., Sayre, K.D., Govaerts, B., Gupta, R., Subbarao, G.V., Ban, T., Hodson, D., Dixon, J.M., Ortiz-Monasterio, J.I. and Reinolds, M. (2008) Climate Change: Can Wheat Beat the Heat? Agriculture, Ecosystems & Environment, 126, 46-58.
73. Araus, J., Slafer, G., Royo, C. and Serret, M.D. (2008) Breeding for Yield Potential and Stress Adaptation in Cereals. Critical Reviews in Plant Sciences, 27, 377-412.

74. Cattivelli, L., Rizza, F., Badeck, F.W., Mazzucoteli, E., Mastrangelo, A.M., Francia, E., MarÈ, C., Tondelli, A. and Stanca, A.M. (2008) Drought Tolerance Improvement in Crop Plants: An Integrated View from Breeding to Genomics. Field Crops Research, 105, 1-14.
75. Ceccarelli, S. and Grando, S. (2007) Decentralized-Participatory Plant Breeding: An Example of Demand Driven Research. Euphytica, 155, 349-360.
76. Burke, M.B., Lobell, D.B. and Guarino, L. (2009) Shifts in African Crop Climates by 2050, and the Implications for Crop Improvement and Genetic Resources Conservation. Global Environmental Change, 19, 317-325.
77. Jarvis, D.I., Brown, A.H.D., Cuong, P.H., Collado-Panduro, L., Latoumerie-Moreno, L., Gyawali, S., Tanto, T., Sawadogo, M., Mar, I., Sadiki, M., Hue, N.T., Arias-Reyes, L., Balma, D., Bajracharya, J., Castillo, F., Rijal, D., Belqadi, L., Rana, R., Saidi, S., Quedraogo, J., Zangre, R., Rhrib, K., Chavez, J.L., Schoen, D., Shapit, B., Santis, P.D., Fadda, C. and Hodgkin, T. (2008) A Global Perspective of the Richness and Evenness of Traditional Crop-Variety Diversity Maintained by Farming Communities. Proceedings of the National Academy of Sciences of the United States of America, 105, 5326-5331.
78. Witcombe, J.R., Hollington, P.A., Howarth, C.J., Reader, S. and Steele, K.A. (2008) Breeding for Abiotic Stresses for Sustainable Agriculture. Philosophical Transactions of the Royal Society B: Biological Sciences, 363, 703-716.
79. Bhatnagar-Mathur, P., Vadez, V. and Sharma, K.K. (2007) Transgenic Approaches for Abiotic Stress Tolerance in Plants: Retrospect and Prospects. Plant Cell Reports, 27, 411-424.
80. Ainsworth, E., Rogers, A. and Leakey, A.D.B. (2008) Targets for Crop Biotechnology in a Future High-CO_2 and High-O_3 World. Plant Physiology, 147, 13-19.
81. Ortiz, R. (2008) Crop Genetic Engineering under Global Climate Change. Annals of Arid Zone, 47, 343-354.
82. Jewell, M.C., Campbell, B.C. and Godwin, I.D. (2010) Transgenic Plants for Abiotic Stress Resistance. In: Kole, C., Michler, C.H., Abbott, A.G. and Hall, T.C., Eds., Transgenic Crop Plants, Springer-Verlag, Berlin-Heidelberg, 67- 132.
83. Dwivedi, S.L., Upadhyaya, H., Subudhi, P., Gehring, C., Bajic, V. and Ortiz, R. (2010) Enhancing Abiotic Stress Tolerance in Cereals through Breeding and Transgenic Interventions. Plant Breeding Reviews, 33, 31-114.
84. Dwivedi, S.L., Sahrawat, K., Upadhyaya, H. and Ortiz, R. (2013) Food, Nutrition and Agrobiodiversity under Global Climate Change. Advances in Agronomy, 120, 1-128.
85. Ruane, J., Sonnino, A., Steduto, P. and Deane, C. (2008) Coping with Water Scarcity: What Role for Biotechnologies? Land and Water Discussion Paper 7, Food and Agriculture Organization of the United Nations, Rome.
86. Sakuma, Y., Maruyama, K., Qin, F., Osakabe, Y., Shinozaki, K. and Yamaguchi-Shinozaki, K. (2006) Dual Function of an Arabidopsis Transcription Factor DREB2A in Water-Stress-Responsive and Heat-Stress-Responsive Gene Expression. Proceedings of the National Academy of Sciences of the United States of America, 103, 18822-18827.
87. Ortiz, R., Iwanaga, M., Reynolds, M.P., Wu, X. and Crouch, J.H. (2007) Overview on Crop Genetic Engineering for Drought-Prone Environments. Journal of SAT Agricultural Research, 4, 1-30.
88. Pellegrineschi, A., Reynolds, M., Pacheco, M., Brito, R.M., Almeraya, R., Yamaguchi-Shinozaki, K. and Hoisington, D. (2004) Stress-Induced Expression in Wheat of the Arabidopsis thaliana DREB1A Gene Delays Water Stress Symptoms under Greenhouse Conditions. Genome, 47, 493-500.
89. Saint Pierre, C.S., Crossa, J.L., Bonnett, D., Yamaguchi-Shinozaki, K. and Reynolds, M.P. (2012) Phenotyping Transgenic Wheat for Drought Resistance. Journal of Experimental Botany, 63, 1799-1808.
90. Mumms, R. (2005) Genes and Salt Tolerance: Bringing Them Together. New Phytologist, 167, 645-663.

91. Chinnusamy, V., Jagendorf, A. and Zhu, J.K. (2005) Understanding and Improving Salt Tolerance in Plants. Crop Science, 45, 437-448.
92. Wu, H.J., Zhang, Z., Wang, J.Y., Oh, D.H., Dassanayake, M., Liu, B., Huang, Q., Sun, H.X., Xia, R., Wu, Y., Wang, Y.N., Yang, Z., Liu, Y., Zhang, W., Zhang, H., Chu, J., Yan, C., Fang, S., Zhang, J., Wang, Y., Zhang, F., Wang, G., Yeol Lee, S., Cheeseman, J.M., Yang, B., Li, B., Min, J., Yang, L., Wang, J., Chu, C., Chen, S.Y., Bohnert, H.J., Zhu, J.K., Wang, X.J. and Xie, Q. (2012) Insights into Salt Tolerance from the Genome of Thellungiella salsuginea. Proceedings of the National Academy of Sciences of the United States of America, 109, 12219-12224.
93. Plett, D., Safwat, G., Gilliham, M., Skrumsager-Møller, I., Roy, S., Shirley, N., Jacobs, A., Johnson, A. and Tester, M. (2010) Improved Salinity Tolerance of Rice through Cell Type-Specific Expression of Athkt1;1. PLoS ONE, 5, e12571.
94. Moghaieb, R.E., Nakamura, A., Saneoka, H. and Fujita, K. (2011) Evaluation of Salt Tolerance in Ectoine-Transgenic Tomato Plants (Lycopersicon esculentum) in Terms of Photosynthesis, Osmotic Adjustment, and Carbon Partitioning. GM Crops, 2, 58-65.
95. Lybbert, T. and Sumner, D. (2011) Agricultural Technologies for Climate Change Mitigation and Adaptation in Developing Countries: Policy Options for Innovation and Technology Diffusion. ICTSD-IPC Platform on Climate Change. Agriculture and Trade Issues Brief 6. International Centre for Trade and Sustainable Development, Geneva, Switzerland.
96. Varshney, R.K., Bansal, K.C., Aggarwal, P.K., Datta, S.K. and Craufurd, P.Q. (2011) Agricultural Biotechnology for Crop Improvement in a Variable Climate: Hope or Hype? Trends in Plant Science, 16, 363-371.
97. Yamori, W., Hikosaka, K. and Way, D.A. (2013) Temperature Response of Photosynthesis in C_3, C_4, and CAM Plants: Temperature Acclimation and Temperature Adaptation. Photosynthesis Research, 119, 101-117.
98. Ainsworth, E.A. and Ort, D.R. (2010) How Do We Improve Crop Production in a Warming World? Plant Physiology, 154, 526-530.
99. Wahid, A., Gelani, S., Ashraf, M. and Foolad, M.R. (2007) Heat Tolerance in Plants: An Overview. Environmental and Experimental Botany, 61, 199-223.
100. Hasanuzzaman, M., Nahar, K., Alam, Md., Roychowdhury, R. and Fujita, M. (2013) Physiological, Biochemical, and Molecular Mechanisms of Heat Stress Tolerance in Plants. International Journal of Molecular Sciences, 14, 9643-9684.
101. Gao, H., Brandizzi, F., Benning, C. and Larkin, R.M. (2008) A Membrane-Tethered Transcription Factor Defines a Branch of the Heat Stress Response in Arabidopsis thaliana. Proceedings of the National Academy of Sciences of the United States of America, 105, 16398-16403.
102. Gusta, L. (2012) Abiotic Stresses and Agricultural Sustainability. Journal of Crop Improvement, 26, 415-427.
103. Katiyar-Agarwal, S., Agarwal, M. and Grover, A. (2003) Heat-Tolerant Basmati Rice Engineered by Over-Expression of hsp101. Plant Molecular Biology, 51, 677-686.
104. Pimentel, D. (1997) Techniques for Reducing Pesticide Use: Economic and Environmental Benefits. Wiley, New York.
105. Oerke, E.C., Dehne, H.W., Schonbeck, F. and Weber, A. (1994) Crop Production and Crop Protection: Estimated Losses in Major Food and Cash Crops. Elsevier, Amsterdam.
106. Tripathi, L., Mwaka, H., Tripathi, J.N. and Tushemereirwe, W. (2010) Expression of Sweet Pepper Hrap Gene in Banana Enhances Resistance to Xanthomonas campestris pv. Musacearum. Molecular Plant Pathology, 11, 721-731.
107. Ortiz, R., Jarvis, A., Aggarwal, P.K. and Campbell, B.M. (2014) Plant Genetic Engineering, Climate Change and Food Security. CCAFS Working Paper No. 72. CGIAR Research Program on Climate Change, Agriculture and Food Security (CCAFS), Copenhagen.
108. Haverkort, A.J., Boonekamp, P.M., Hutten, R., Jacobsen, E., Lotz, L.A.P., Kessel, G.J.T., Visser, R.F.G. and Van Der Vossen, E.A.G. (2008) Societal Costs of Late Blight in

Potato and Prospects of Durable Resistance through Cisgenic Modification. Potato Research, 51, 47-57.

109. Jacobs, D.F. (2007) Toward Development of Silvical Strategies for Forest Restoration of American Chestnut (Castanea dentata) Using Blight Resistant Hybrids. Biological Conservation, 137, 497-506.

110. Santini, A., La Porta, N., Ghelardini, L. and Mittempergher, L. (2007) Breeding against Dutch Elm Disease Adapted to the Mediterranean Climate. Euphytica, 163, 45-56.

111. Bouton, J. (2007) The Economic Benefits of Forage Improvement in the United States. Euphytica, 154, 263-270.

112. Pimentel, D., Allen, J., Beers, A., Guinand, L., Linder, R., Mclaughlin, P., Meer, B., Musonda, D., Perdue, D., Poisson, S., Siebert, S., Stoner, K., Salazar, R. and Hawkinset, A. (1987) World Agriculture and Soil Erosion. Erosion Threatens World Food Production. Bioscience, 37, 277-283.

113. Glover, J.D., Cox, C.M. and Reganold, J.P. (2007) Future Farming: A Return to Roots? Scientific American, 297, 82-89.

114. Jackson, W., Cox, S., Dehaan, L., Glover, J., Van Tassel, D. and Cox, C. (2009) The Necessity and Possibility of an Agriculture Where Nature Is the Measure. In: Bohlen, P.J. and House, G., Eds., Sustainable Agroecosystem Management, CRC Press, Boca Raton, 61-71.

115. Blanco, H. and Lal, R., Eds. (2008) Principles of Soil Conservation and Management. Springer, New York.

116. Olson, K.R., Ebelhar, S.A. and Lang, J.M. (2010) Cover Crops Effects on Crop Yields and Soil Organic Content. Soil Science, 175, 89-98.

117. Kaumbutho, P. and Kienzle, J. (2008) Conservation Agriculture as Practiced in Kenya: Two Case Studies. Food and Agriculture Organization of the United Nations, Rome.

118. Pretty, J.N. and Hine, R. (2001) Reducing Food Poverty with Sustainable Agriculture: A Summary of New Evidence. Final Report of the "SAFE-World" (The Potential of Sustainable Agriculture to Feed the World) Research Project. Centre for Environment and Society, University of Essex, Colchester.

119. Kendle, A.D. and Rose, J.E. (2000) The Aliens Have Landed! What Are the Justifications for "Native Only" Policies in Landscape Plantings? Landscape and Urban Planning, 47, 19-31.

120. Zhao, F.J. and Mcgrath, S.P. (2009) Biofortification and Phytoremediation. Current Opinion in Plant Biology, 12, 373- 380.

121. Yin, X., Yuan, L., Liu, Y. and Lin, Z. (2012) Phytoremediation and Biofortification: Two Sides of One Coin. In: Yin, X. and Yuan, L., Eds., Phytoremediation and Biofortification, Springer Briefs in Green Chemistry for Sustainable, Springer, New York, 1-6.

122. Pilon-Smits, E. (2005) Phytoremediation. Annual Review of Plant Biology, 56, 15-39.

123. Kramer, U. (2005) Phytoremediation: Novel Approaches to Cleaning up Polluted Soils. Current Opinion in Biotechnology, 16, 133-141.

124. Doty, S.L. (2008) Enhancing Phytoremediation through the Use of Transgenics and Endophytes. New Phytologist, 179, 318-333.

125. Chaney, R.L., Angle, J.S., Broadhurst, C.L., Peters, C.A., Tappero, R.V. and Parks, D.L. (2007) Improved Understanding of Hyperaccumulation Yields Commercial Phytoextraction and Phytomining Technologies. Journal of Environmental Quality, 36, 1429-1443.

126. Mcgrath, S.P. and Zhao, F.J. (2003) Phytoextraction of Metals and Metalloids from Contaminated Soils. Current Opinion in Biotechnology, 14, 277-282.

127. Murakami, M., Ae, N., Ishikawa, S., Ibaraki, T. and Ito, M. (2008) Phytoextraction by a High-Cd-Accumulating Rice: Reduction of Cd Content of Soybean Seeds. Environmental Science & Technology, 42, 6167-6172.

128. Ueno, D., Kono, I., Yokosho, K., Ando, T., Yano, M. and Ma, J.F. (2009) A Major Quantitative Trait Locus Controlling Cadmium Translocation in Rice (Oryza sativa). New Phytologist, 182, 644-653.

129. Grant, C.A., Clarke, J.M., Duguid, S. and Chaney, R.L. (2008) Selection and Breeding of Plant Cultivars to Minimize Cadmium Accumulation. Science of the Total Environment, 390, 301-310.

130. Mcgrath, S.P., Lombi, E., Gray, C.W., Caille, N., Dunham, S.J. and Zhao, F.J. (2006) Field Evaluation of Cd and Zn Phytoextraction Potential by the Hyperaccumulators Thlaspi caerulescens and Arabidopsis halleri. Environmental Pollution, 141, 115-125.

131. Maxted, A.P., Black, C.R., West, H.M., Crout, N.M.J., Mcgrath, S.P. and Young, S.D. (2007) Phytoextraction of Cadmium and Zinc from Arable Soils Amended with Sewage Sludge Using Thlaspi caerulescens: Development of a Predictive Model. Environmental Pollution, 150, 363-372.

132. Zhao, F.J., Hamon, R.E., Lombi, E., Mclaughlin, M.J. and Mcgrath, S.P. (2002) Characteristics of Cadmium Uptake in Two Contrasting Ecotypes of the Hyperaccumulator Thlaspi caerulescens. Journal of Experimental Botany, 53, 535-543.

133. Kertulis-Tartar, G.M., Ma, L.Q., Tu, C. and Chirenje, T. (2006) Phytoremediation of an Arsenic-Contaminated Site Using Pteris vittata L.: Two-Year Study. International Journal of Phytoremediation, 8, 311-322.

134. Salido, A.L., Hasty, K.L., Lim, J.M. and Butcher, D.J. (2003) Phytoremediation of Arsenic and Lead in Contaminated Soil Using Chinese Brake Ferns (Pteris vittata) and Indian Mustard (Brassica juncea). International Journal of Phytoremediation, 5, 89-103.

135. Dickinson, N.M. and Pulford, I.D. (2005) Cadmium Phytoextraction Using Short-Rotation Coppice Salix: The Evidence Trail. Environment International, 31, 609-613.

136. Wieshammer, G., Unterbrunner, R., Garcia, T.B., Zivkovic, M.F., Puschenreiter, M. and Wenzel, W.W. (2007) Phytoextraction of Cd and Zn from Agricultural Soils by Salix ssp. and Intercropping of Salix caprea and Arabidopsis halleri. Plant and Soil, 298, 255-264.

137. Lievens, C., Yperman, J., Cornelissen, T. and Carleer, R. (2008) Study of the Potential Valorisation of Heavy Metal Contaminated Biomass via Phytoremediation by Fast Pyrolysis. Part II. Characterisation of the Liquid and Gaseous Fraction as a Function of the Temperature. Fuel, 87, 1906-1916.

138. Rooney, W., Blumenthal, J., Bean, B. and Mullet, J. (2007) Designing Sorghum as a Dedicated Bioenergy Feedstock. Biofuels, Bioproducts and Biorefining, 1, 147-157.

139. Jessup, R.W. (2009) Development and Status of Dedicated Energy Crops in the United States. In Vitro Cellular & Developmental Biology-Plant, 45, 282-290.

140. Robertson, G.P., Dale, V.H., Doering, O.C., Hamburg, S.P., Melillo, J.M., Wander, M.M., Parton, W.J., Adler, P.R., Barney, J.N., Cruse, R.M., Duke, C.F., Fearnside, P.M., Follett, R.F., Gibbs, H.K., Goldemberg, J., Mladenoff, D.J., Ojima, D., Palmer, M.W., Sharpley, A., Wallace, L., Weathers, K.C., Wiens, J.A. and Wilhelm, W.W. (2008) Sustainable Biofuels Redux. Science, 322, 49-50.

141. Cook, R.J. (2006) Toward Cropping Systems that Enhance Productivity and Sustainability. Proceedings of the National Academy of Sciences of the United States of America, 103, 18389-18394.

142. Carpenter, J. (2011) Impacts of GM Crops on Biodiversity. GM Crops, 2, 7-23.

143. Icoz, I. and Stotzky, G. (2008) Fate and Effects of Insect-Resistant Bt Crops in Soil Ecosystems. Soil Biology and Biochemistry, 40, 559-586.

144. Hoheisel, G.A. and Fleischer, S.J. (2007) Coccinelids, Aphids, and Pollen in Diversified Vegetable Fields with Transgenic and Isoline Cultivars. Journal of Insect Science, 7, 1-12.

145. Leslie, T.W., Hoheisel, G.A., Biddinger, D.J., Rohr, J.R. and Fleisher, S.J. (2007) Transgenes Sustain Epigeal Insect Biodiversity in Diversified Vegetable Farm Systems. Environmental Entomology, 36, 234-244.

146. Jørgensen, R.B. and Andersen, B. (1994) Spontaneous Hybridization between Oilseed Rape (Brassica napus) and Weedy B. campestris (Brassicaceae): A Risk of

Growing Genetically Modified Oilseed Rape. American Journal of Botany, 81, 1620-1626.
147. Hansen, L.B., Siegismund, H.R. and Jørgensen, R.B. (2003) Progressive Introgression between Brassica napus (Oilseed Rape) and B. rapa. Heredity, 91, 276-283.
148. Stewart Jr., C.N., Halfhill, M.D. and Warwick, S.I. (2003) Transgene Introgression from Genetically Modified Crops to Their Wild Relatives. Nature Reviews Genetics, 4, 806-817.
149. Brown, J. and Brown, A.P. (1996) Gene Transfer between Canola (Brassica napus and B. campestris) and Related Weed Species. Annals of Applied Biology, 129, 513-522.
150. Mikkelsen, T.R., Anderson, B. and Jorgensen, R.B. (1996) The Risk of Crop Transgene Spread. Nature, 380, 31.
151. Boudry, P., Broomberg, K., Saumitou-Laprade, P., Morchen, M., Cuegen, J. and Van Dijk, H. (1994) Gene Escape in Transgenic Sugar Beet: What Can Be Learned from Molecular Studies of Weed Beet Populations? Proceedings of the 3rd International Symposium on the Biosafety, Results of Field Tests of Genetically-Modified Plants and Microorganisms, University of California, Division of Agriculture and Natural Resources, Oakland, 75-83.
152. Rose, C.W., Millwood, R.J., Moon, H.S., Rao, M.R., Halfhill, M.D., Raymer, P.L., Warwick, S.I., Al-Ahmad, H., Gressel, J. and Stewart Jr., C.N. (2009) Genetic Load and Transgenic Mitigating Genes in Transgenic Brassica rapa (Field Mustard) × Brassica napus (Oilseed Rape) Hybrid Populations. BMC Biotechnology, 9, 93.
153. Palaudelmàs, M., Peñas, G., Mele, E., Serra, J., Salvia, J., Pla, M., Nadal, A. and Messeguer, J. (2009) Effect of Volunteers on Maize Gene Flow. Transgenic Research, 18, 583-594.
154. Carpenter, J.E. (2010) Peer-Reviewed Surveys Indicate Positive Impact of Commercialized GM Crops. Nature Biotechnology, 28, 319-321.
155. Brookes, G., Yu, T.H., Tokgoz, S. and Elobeid, A. (2010) The Production and Price Impact of Biotech Corn, Canola, and Soybean Rops. AgBioForum, 13, 25-52.
156. Storer, N.P., Dively, G.P. and Herman, R.A. (2008) Landscape Effects of Insect-Resistant Genetically Modified Crops. In: Romeis, J., Shelton, A.M. and Kennedy, G.G., Eds., Integration of Insect-Resistant Genetically Modified Crops within IPM Programs, Springer, New York, 273-302.
157. Naranjo, S.E. (2005) Field Studies Assessing Arthropod Non-Target Effects of Bt Transgenic Crops: Introduction. Environmental Entomology, 34, 1178-1180.
158. Naranjo, S.E. (2005) Long-Term Assessment of the Effects of Transgenic Bt Cotton on the Abundance of Non-Target Arthropod Natural Enemies. Environmental Entomology, 34, 1193-1210.
159. Naranjo, S.E. (2005) Long-Term Assessment of the Effects of Transgenic Bt Cotton on the Function of the Natural Enemy Community. Environmental Entomology, 34, 1211-1223.
160. Wolfenbarger, L.L., Naranjo, S.E., Lundgren, J.G., Bitzer, R.J. and Watrud, L.S. (2008) Bt Crop Effects on Functional Guilds of Non-Target Arthropods: A Meta-Analysis. PLoS ONE, 3, e2118.
161. Naranjo, S.E. (2009) Impacts of Bt Crops on Non-Target Invertebrates and Insecticide Use Patterns. CAB Reviews: Perspectives in Agriculture, Veterinary Science, Nutrition and Natural Resources, 4, 1-23.
162. Duan, J.J., Lundgren, J.G., Naranjo, S. and Marvier, M. (2009) Extrapolating Non-Target Risk of Bt Crops from Laboratory to Field. Biology Letters, 6, 74-77.
163. Brookes, G. and Barfoot, P. (2012) Global Economic and Environmental Benefits of GM Crops Continue to Rise. PG Economics 2012. http://www.pgeconomics.co.uk/page/33/global-impact-2012
164. Bennet, R., Phipps, R., Strange, A. and Grey, P. (2004) Environmental and Human Health Impacts of Growing Genetically Modified Herbicide-Tolerant Sugar Beet: A Life-Cycle Assessment. Plant Biotechnology Journal, 2, 273-278.

165. National Research Council (2010) The Impact of Genetically Engineered Crops on Farm Sustainability in the United States. National Academies, Washington DC.
166. Wang, S., Just, D.R. and Pinstrup-Andersen, P. (2008) Bt-Cotton and Secondary Pests. International Journal of Biotechnology, 10, 113-120.
167. Wang, Z.J., Lin, H., Huang, J., Hu, R., Rozelle, S. and Pray, C. (2009) Bt Cotton in China: Are Secondary Insect Infestations Offsetting the Benefits in Farmer Fields? Agricultural Sciences in China, 8, 83-90.
168. Gassmann, A.J., Petzold-Maxwell, J.L., Keweshan, R. and Dunbar, M.W. (2011) Field-Evolved Resistance to Bt Maize by Western Corn Rootworm. PLoS ONE, 6, e22629.
169. Fowler, C. (2011) Conserving Diversity: The Challenge of Cooperation. Acta Horticulturae, 916, 19-24.
170. Ortiz, R., Mowbray, D., Dowswell, C. and Rajaram, S. (2007) Norman E. Borlaug: The Humanitarian Plant Scientist Who Changed the World. Plant Breeding Reviews, 28, 1-37.
171. Ortiz, R., Braun, H.J., Crossa, J., Crouch, J.H., Davenport, G., Dixon, J., Dreisigacker, S., Duveiller, E., He, Z., Huerta, J., Joshi, A.K., Kishii, M., Kosina, P., Manes, Y., Mezzalama, M., Morgounov, A., Murakami, J., Nicol, J., Ortiz-Ferrara, G., Ortiz-Monasterio, J.I., Payne, T.S., Peña, R.J., Reynolds, M.P., Sayre, K.D., Sharma, R.C., Singh, R.P., Wang, J., Warburton, M., Wu, H. and Iwanaga, M. (2008) Wheat Genetic Resources Enhancement by the International Maize and Wheat Improvement Center (CIMMYT). Genetic Resources and Crop Evolution, 55, 1095-1140.
172. Reynolds, M.P. and Borlaug, N.E. (2006) International Collaborative Wheat Improvement: Impacts and Future Prospects. Journal of Agricultural Science, 144, 3-17.
173. Lantican, M.A., Dubin, M.J. and Morris, M.L. (2005) Impacts of International Wheat Breeding Research in the Developing World, 1988-2002. Centro Internacional de Mejoramiento de Maíz y Trigo, México D.F.
174. Alston, J.M., Marra, M.C., Pardey, P.G. and Wyatt, T.J. (2000) Research Returns Redux: A Meta-Analysis of the Returns to Agricultural R&D. Australian Journal of Agricultural and Resource Economics, 44, 185-215.
175. Evenson, R.E. and Gollin, D. (1997) Genetic Resources, International Organizations, and Improvement in Rice Varieties. Economic Development and Cultural Change, 45, 471-500.
176. Jackson, M.T. and Huggan, R.D. (1993) Sharing the Diversity of Rice to Feed the World. Diversity, 9, 22-25.
177. Salhuana, W. and Pollak, L. (2006) Latin American Maize Project (LAMP) and Germplasm Enhancement of Maize (GEM) Project: Generating Useful Breeding Germplasm. Maydica, 51, 339-355.
178. Taba, S., Díaz, J., Franco, J., Crossa, J. and Eberhart, S.A. (1999) A Core Subset of LAMP from the Latin American Maize Project. CD-Rom. Centro Internacional de Mejoramiento de Maíz y Trigo, México D.F.
179. Balint-Kurti, P., Blanco, M., Milard, M., Duvick, S., Holland, J., Clements, M., Holley, R., Carson, M.L. and Goodman, M. (2006) Registration of 20 GEM Maize Breeding Germplasm Lines Adapted to the Southern USA. Crop Science, 46, 996-998.
180. Goodman, M.M. (2005) Broadening the U.S. Maize Germplasm Base. Maydica, 50, 203-214.
181. Ortiz, R., Taba, S., Chávez-Tovar, V.H., Mezzalama, M., Xu, Y., Yan, J. and Crouch, J.H. (2010) Conserving and Enhancing Maize Genetic Resources as Global Public Goods—A Perspective from CIMMYT. Crop Science, 50, 13-28.
182. Delmer, D.P. (2005) Agriculture in the Developing World: Connecting Innovations in Plant Research to Downstream Applications. Proceedings of the National Academy of Sciences of the United States of America, 102, 15739-15746.

CHAPTER 4

Molecular Markers and Marker-Assisted Breeding in Plants

Guo-Liang Jiang[1]

[1] *Plant Science Department, South Dakota State University, Brookings, USA*

1. INTRODUCTION

Molecular breeding (MB) may be defined in a broad-sense as the use of genetic manipulation performed at DNA molecular levels to improve characters of interest in plants and animals, including genetic engineering or gene manipulation, molecular marker-assisted selection, genomic selection, etc. More often, however, molecular breeding implies molecular marker-assisted breeding (MAB) and is defined as the application of molecular biotechnologies, specifically molecular markers, in combination with linkage maps and genomics, to alter and improve plant or animal traits on the basis of genotypic assays. This term is used to describe several modern breeding strategies, including marker-assisted selection (MAS), marker-assisted backcrossing (MABC), marker-assisted recurrent selection (MARS), and genome-wide selection (GWS) or genomic selection (GS) (Ribaut et al., 2010). In this article, we will address general principles and methodologies of marker-assisted breeding in plants and discuss some issues related to the procedures and applications of this methodology in practical breeding, including marker-assisted selection, marker-based backcrossing, marker-based pyramiding of multiple genes, etc., beginning with a brief introduction to molecular markers as a powerful tool for plant breeding.

2. GENETIC MARKERS IN PLANT BREEDING: CONCEPTIONS, TYPES AND APPLICATION

Genetic markers are the biological features that are determined by allelic forms of genes or genetic loci and can be transmitted from one generation to another, and thus they can be used as experimental probes or tags to keep track of an individual, a tissue, a cell, a nucleus, a chromosome or a gene. Genetic markers used in genetics and plant breeding can be classified into two categories: classical markers and DNA markers (Xu, 2010). Classical markers include morphological markers, cytological markers and biochemical markers. DNA markers have developed into many systems based on different polymorphism-detecting techniques or methods (southern blotting – nuclear acid hybridization, PCR – polymerase chain reaction, and DNA sequencing) (Collard et al., 2005), such as RFLP, AFLP, RAPD, SSR, SNP, etc.

2.1. Classical Markers

Morphological markers: Use of markers as an assisting tool to select the plants with desired traits had started in breeding long time ago. During the early history of plant breeding, the markers used mainly included visible traits, such as leaf shape, flower color, pubescence color, pod color, seed color, seed shape, hilum color, awn type and length, fruit shape, rind (exocarp) color and stripe, flesh color, stem length, etc. These morphological markers generally represent genetic polymorphisms which are easily identified and manipulated. Therefore, they are usually used in construction of linkage maps by classical two- and/or three-point tests. Some of these markers are linked with other agronomic traits and thus can be used as indirect selection criteria in practical breeding. In the green revolution, selection of semi-dwarfism in rice and wheat was one of the critical factors that contributed to the success of high-yielding cultivars. This could be considered as an example for successful use of morphological markers to modern breeding. In wheat breeding, the dwarfism governed by gene *Rht10* was introgressed into Taigu nuclear male-sterile wheat by backcrossing, and a tight linkage was generated between *Rht10* and the male-sterility gene *Ta1*. Then the dwarfism was used as the marker for identification and selection of the male-sterile plants in breeding populations (Liu, 1991). This is particularly helpful for implementation of recurrent selection in wheat. However, morphological markers available are limited, and many of these markers are not associated with important economic traits (e.g. yield and quality) and even have undesirable effects on the development and growth of plants.

Cytological markers: In cytology, the structural features of chromosomes can be shown by chromosome karyotype and bands. The banding patterns, displayed in color, width, order and position, reveal the difference in distributions of euchromatin and heterochromatin. For instance, Q bands are produced by quinacrine hydrochloride, G bands are produced by Giemsa stain, and R bands are the reversed G bands. These chromosome landmarks are used not only for characterization of normal chromosomes and detection of chromosome mutation, but also widely used in physical mapping and linkage group identification. The physical maps based on morphological and cytological markers lay a foundation for genetic linkage mapping with the aid of molecular techniques. However, direct use of cytological markers has been very limited in genetic mapping and plant breeding.

Biochemical/protein markers: Protein markers may also be categorized into molecular markers though the latter are more referred to DNA markers. Isozymes are alternative forms or structural variants of an enzyme that have different molecular weights and electrophoretic mobility but have the same catalytic activity or function. Isozymes reflect the products of different alleles rather than different genes because the difference in electrophoretic mobility is caused by point mutation as a result of amino acid substitution (Xu, 2010). Therefore, isozyme markers can be genetically mapped onto chromosomes and then used as genetic markers to map other genes. They are also used in seed purity test and occasionally in plant breeding. There are only a small number of isozymes in most crop species and some of them can be identified only with a specific strain. Therefore, the use of enzyme markers is limited.

Another example of biochemical markers used in plant breeding is high molecular weight glutenin subunit (HMW-GS) in wheat. Payne et al. (1987) discovered a correlation between the presence of certain HMW-GS and gluten strength, measured by the SDS-sedimentation volume test. On this basis, they designed a numeric scale to evaluate bread-making quality as a function of the described subunits (*Glu-1* quality score) (Payne et al., 1987; Rogers et al., 1989). Assuming the effect of the alleles to be additive, the Bread-making quality was predicted by adding the scores of the alleles present in the particular line. It was established that the allelic variation at the *Glu-D1* locus have a greater influence on bread-making quality than the variation at the others *Glu-1* loci. Subunit combination 5+10 for locus *Glu-D1* (*Glu-D1* 5+10) renders stronger dough than *Glu-D1* 2+12, largely due to the presence of an extra cysteine residue in the Dx-5 subunit compared to the Dx-2 subunit, which would promote the formation of polymers with larger size distribution. Therefore, breeders may enhance the bread-making quality in wheat by selecting subunit combination Glu-D1 5+10 instead of Glu-D1 2+12. Of course, the variation of bread-making quality among different varieties cannot be explained only by the variation in HMW-GS composition, because the low molecular weight glutinen subunit (LMW-GS) (as well as the gliadins in a smaller proportion) and their interactions with the HMW-GS also play an important role in the gluten strength and bread-making quality.

2.2. Dna Markers

DNA markers are defined as a fragment of DNA revealing mutations/variations, which can be used to detect polymorphism between different genotypes or alleles of a gene for a particular sequence of DNA in a population or gene pool. Such fragments are associated with a certain location within the genome and may be detected by means of certain molecular technology. Simply speaking, DNA marker is a small region of DNA sequence showing polymorphism (base deletion, insertion and substitution) between different individuals. There are two basic methods to detect the polymorphism: Southern blotting, a nuclear acid hybridization technique (Southern 1975), and PCR, a polymerase chain reaction technique (Mullis, 1990). Using PCR and/or molecular hybridization followed by electrophoresis (e.g. PAGE – polyacrylamide gel electrophoresis, AGE – agarose gel electrophoresis, CE – capillary electrophoresis), the variation in DNA samples or polymorphism for a specific region of DNA sequence can be identified based on the product features, such as band size and mobility. In addition to Sothern blotting and PCR, more detection systems have been also developed. For instance, several new array chip techniques use DNA hybridization combined with labeled nucleotides, and new sequencing techniques detect polymorphism by sequencing. DNA markers are also called molecular markers in many cases and play a major role in molecular breeding. Therefore, molecular markers in this article are mainly referred to as DNA markers except specific definitions are given, although isozymes and protein markers are also molecular markers. Depending on application and species involved, ideal DNA markers for efficient use in marker-assisted breeding should meet the following criteria:

- High level of polymorphism
- Even distribution across the whole genome (not clustered in certain regions)
- Co-dominance in expression (so that heterozygotes can be distinguished from homozygotes)
- Clear distinct allelic features (so that the different alleles can be easily identified)

- Single copy and no pleiotropic effect
- Low cost to use (or cost-efficient marker development and genotyping)
- Easy assay/detection and automation
- High availability (un-restricted use) and suitability to be duplicated/multiplexed (so that the data can be accumulated and shared between laboratories)
- Genome-specific in nature (especially with polyploids)
- No detrimental effect on phenotype

Since Botstein et al. (1980) first used DNA restriction fragment length polymorphism (RFLP) in human linkage mapping, substantial progress has been made in development and improvement of molecular techniques that help to easily find markers of interest on a large-scale, resulting in extensive and successful uses of DNA markers in human genetics, animal genetics and breeding, plant genetics and breeding, and germplasm characterization and management. Among the techniques that have been extensively used and are particularly promising for application to plant breeding, are the restriction fragment length polymorphism (RFLP), amplified fragment length polymorphism (AFLP), random amplified polymorphic DNA (RAPD), microsatellites or simple sequence repeat (SSR), and single nucleotide polymorphism (SNP). According to a causal similarity of SNPs with some of these marker systems and fundamental difference with several other marker systems, the molecular markers can also be classified into SNPs (due to sequence variation, e.g. RFLP) and non-SNPs (due to length variation, e.g. SSR) (Gupta et al., 2001). The marker techniques help in selection of multiple desired characters simultaneously using F_2 and back-cross populations, near isogenic lines, doubled haploids and recombinant inbred lines. In view of page limitation, only five marker systems mentioned above are briefly addressed here according to published literatures. The details about the technical methods how to develop DNA markers and the procedures how to detect in practice have been described in the recently published reviews and books in this area (Farooq and Azam, 2002a, 2002b; Gupta et al., 2001; Semagn et al., 2006a; Xu, 2010).

RFLP markers: RFLP markers are the first generation of DNA markers and one of the important tools for plant genome mapping. They are a type of Southern-Boltting-based markers. In living organisms, mutation events (deletion and insertion) may occur at restriction sites or between adjacent restriction sites in the genome. Gain or loss of restriction sites resulting from base pair changes and insertions or deletions at restriction sites within the restriction fragments may cause differences in size of restriction fragments. These variations may cause alternation or elimination of the recognition sites for restriction enzymes. As a consequence, when homologous chromosomes are subjected to restriction enzyme digestion, different restriction products are produced and can be detected by electrophoresis and DNA probing techniques.

RFLP markers are powerful tools for comparative and synteny mapping. Most RFLP markers are co-dominant and locus-specific. RFLP genotyping is highly reproducible, and the methodology is simple and no special equipment is required. By using an improved RFLP technique, i.e., cleaved amplified polymorphism sequence (CAPS), also known as PCR-RFLP, high-throughput markers can be developed from RFLP probe sequences. Very few CAPS are developed from probe sequences, which are complex to interpret. Most CAPS are developed from SNPs found in other sequences followed by PCR and detection of restriction sites. CAPS technique consists of digesting a PCR-amplified fragment and detecting the polymorphism by the presence/absence of restriction sites (Konieczny and Ausubel, 1993). Another advantage of RFLP is that the sequence used as a probe need not be known. All that a researcher needs is a genomic clone that can be used to detect the polymorphism. Very few RFLPs have been sequenced to determine what sequence variation is responsible for the polymorphism. However, it may be problematic to interpret complex RFLP allelic systems in the absence of sequence information. RFLP analysis requires large amounts of high-quality DNA, has low genotyping throughput, and is very difficult to automate. Radioactive autography involving in genotyping and physical maintenance of RFLP probes limit its use and share between laboratories. RFLP markers were predominantly used in 1980s and 1990s, but since last decade fewer direct uses of RFLP markers in genetic research and plant breeding have been reported. Most plant breeders would think that RFLP is too laborious and demands too much pure DNA to be important for plant breeding. It was and is, however, central for various types of scientific studies.

RAPD markers: RAPD is a PCR-based marker system. In this system, the total genomic DNA of an individual is amplified by PCR using a single, short (usually about ten nucleotides/bases) and random primer. The primer which binds to many different loci is used to amplify random sequences from a complex DNA template that is complementary to it (maybe including a limited number of mismatches). Amplification can take place during the PCR, if two hybridization sites are similar to one another (at least 3000 bp) and in opposite directions. The amplified fragments generated by PCR depend on the length and size of both the primer and the target genome. The PCR products (up to 3 kb) are separated by agarose gel electrophoresis and imaged by ethidium bromide (EB) staining. Polymorphisms resulted from mutations or rearrangements either at or between the primer-binding sites are visible in the electrophoresis as the presence or absence of a particular RAPD band.

RAPD predominantly provides dominant markers. This system yields high levels of polymorphism and is simple and easy to be conducted. First, neither DNA probes nor sequence information is required for the design of specific primers. Second, the procedure does not involve blotting or hybridization steps, and thus it is a quick, simple and efficient technique. Third, relatively small amounts of DNA (about 10 ng per reaction) are required and the procedure can be automated, and higher levels of polymorphism also can be detected compared with RFLP. Fourth, no marker development is required, and the primers are non-species specific and can be universal. Fifth, the RAPD products of interest can be cloned, sequenced and then converted into or used to develop other types of PCR-based markers, such as sequence characterized amplified region (SCAR), single nucleotide polymorphism (SNP), etc. However, RAPD also has some limitations/disadvantages, such as low reproducibility and incapability to detect allelic differences in heterozygotes.

AFLP markers: AFLPs are PCR-based markers, simply RFLPs visualized by selective PCR amplification of DNA restriction fragments. Technically, AFLP is based on the selective PCR amplification of restriction fragments from a total double-digest of genomic DNA under high stringency conditions, i.e., the combination of polymorphism at restriction sites and hybridization of arbitrary primers. Because of this AFLP is also called selective restriction fragment amplification (SRFA). An AFLP primer (17-21 nucleotides in length) consists of a synthetic adaptor sequence, the restriction endonuclease recognition sequence and an arbitrary, non-degenerate 'selective' sequence (1-3 nucleotides). The primers used in this technique are capable of annealing perfectly to their target sequences (the adapter and restriction sites) as well as a small number of nucleotides adjacent to the restriction sites. The first step in AFLP involves restriction digestion of genomic DNA (about 500 ng) with two restriction enzymes, a rare cutter (6-bp recognition site, *Eco*RI, *Pts*I or *Hind*III) and a frequent cutter (4-bp recognition site, *Mse*I or *Taq*I). The adaptors are then ligated to both ends of the fragments to provide known sequences for PCR amplification. The double-stranded oligonucleotide adaptors are designed in such a way that the initial restriction site is not restored after ligation. Therefore, only the fragments which have been cut by the frequent cutter and rare cutter will be amplified. This property of AFLP makes it very reliable, robust and immune to small variations in PCR amplification parameters (e.g., thermal cycles, template concentration), and it also can produce a high marker density. The AFLP products can be separated in high-resolution electrophoresis systems. The fragments in gel-based or capillary DNA sequencers can be detected by dye-labeling primers radioactively or fluorescently. The number of bands produced can be manipulated by the number of selective nucleotides and the nucleotide motifs used.

A typical AFLP fingerprint (restriction fragment patterns generated by the technique) contains 50-100 amplified fragments, of which up to 80% may serve as genetic markers. In general, AFLP assays can be conducted using relatively small DNA samples (1-100 ng per individual). AFLP has a very high multiplex ratio and genotyping throughput, and is relatively reproducible across laboratories. Another advantage is that it does not require sequence information or probe collection prior to generating the fingerprints, and a set of primers can be used for different species. This is especially useful when DNA markers are rare. However, AFLP assays have some limitations also. For instance, polymorphic information content for bi-allelic markers is low (the maximum is 0.5). High quality DNA is required for complete restriction enzyme digestion. AFLP markers usually cluster densely in centromeric regions in some species with large genomes (e.g., barley and sunflower). In addition, marker development is complicated and not cost-efficient, especially for locus-specific markers. The applications of AFLP markers include biodiversity studies, analysis of germplasm collections, genotyping of individuals, identification of closely linked DNA markers, construction of genetic DNA marker maps, construction of physical maps, gene mapping, and transcript profiling.

SSR markers: SSRs, also called microsatellites, short tandem repeats (STRs) or sequence-tagged microsatellite sites (STMS), are PCR-based markers. They are randomly tandem repeats of short nucleotide motifs (2-6 bp/nucleotides long). Di-, tri- and tetra-nucleotide repeats, e.g. (GT)n, (AAT)n and (GATA)n, are widely distributed throughout the genomes of plants and animals. The copy number of these repeats varies among individuals and is a source of polymorphism in plants. Because the DNA sequences flanking microsatellite regions are usually conserved, primers specific for these regions are designed for use in the PCR reaction. One of the most important attributes of microsatellite loci is their high level of allelic variation, thus making them valuable genetic markers. The unique sequences bordering the SSR motifs provide templates for specific primers to amplify the SSR alleles via PCR. SSR loci are individually amplified by PCR using pairs of oligonucleotide primers specific to unique DNA sequences flanking the SSR sequence. The PCR-amplified products can be separated in high-resolution electrophoresis systems (e.g. AGE and PAGE) and the bands can be visually recorded by fluorescent labeling or silver-staining.

SSR markers are characterized by their hyper-variability, reproducibility, co-dominant nature, locus-specificity, and random genome-wide distribution in most cases. The advantages of SSR markers include that they can be readily analyzed by PCR and easily detected by PAGE or AGE. SSR markers can be multiplexed, have high throughput genotyping and can be automated. SSR assays require only very small DNA samples (~100 ng per individual) and low start-up costs for manual assay methods. However, SSR technique requires nucleotide information for primer design, labor-intensive marker development process and high start-up costs for automated detections. Since the 1990s SSR markers have been extensively used in constructing genetic linkage maps, QTL mapping, marker-assisted selection and germplasm analysis in plants. In many species, plenty of breeder-friendly SSR markers have been developed and are available for breeders. For instance, there are over 35,000 SSR markers developed and mapped onto all 20 linkage groups in soybean, and this information is available for the public (Song et al., 2010).

SNP markers: An SNP is a single nucleotide base difference between two DNA sequences or individuals. SNPs can be categorized according to nucleotide substitutions either as transitions (C/T or G/A) or transversions (C/G, A/T, C/A or T/G). In practice, single base variants in cDNA (mRNA) are considered to be SNPs as are single base insertions and deletions (indels) in the genome. SNPs provide the ultimate/simplest form of molecular markers as a single nucleotide base is the smallest unit of inheritance, and thus they can provide maximum markers. SNPs occur very commonly in animals and plants. Typically, SNP frequencies are in a range of one SNP every 100-300 bp in plants (Edwards et al., 2007; Xu, 2010). SNPs may present within coding sequences of genes, non-coding regions of genes or in the intergenic regions between genes at different frequencies in different chromosome regions.

Based on various methods of allelic discrimination and detection platforms, many SNP genotyping methods have been developed. A convenient method for detecting SNPs is RFLP (SNP-RFLP) or by using the CAPS marker technique. If one allele contains a recognition site for a restriction enzyme while the other does not, digestion of the two alleles will produce different fragments in length. A simple procedure is to analyze the sequence data stored in the major databases and identify SNPs. Four alleles can be identified when the complete base sequence of a segment of DNA is considered and these are represented by A, T, G and C at each SNP locus in that segment. There are several SNP genotyping assays, such as allele-specific hybridization, primer extension, oligonucleotide ligation and invasive cleavage based on the molecular mechanisms (Sobrino et al., 2005), and different detection methods to analyze the products of each type of allelic discrimination reaction, such as gel electrophoresis, mass spectrophotometry, chromatography, fluorescence polarization, arrays or chips, etc. At the present, SNPs are also widely detected by sequencing. Detailed procedures are described in the review by Gupta at el. (2001) and the book Molecular Plant Breeding by Xu (2010).

SNPs are co-dominant markers, often linked to genes and present in the simplest/ultimate form for polymorphism, and thus they have become very attractive and potential genetic markers in genetic study and breeding.

Moreover, SNPs can be very easily automated and quickly detected, with a high efficiency for detection of polymorphism. Therefore, it can be expected that SNPs will be increasingly used for various purposes, particularly as whole DNA sequences become available for more and more species (e.g., rice, soybean, maize, etc.). However, high costs for start-up or marker development, high-quality DNA required and high technical/equipment demands limit, to some extent, the application of SNPs in some laboratories and practical breeding programs.

The features of the widely used DNA markers discussed above are compared in Table 1. The advantages or disadvantages of a marker system are relevant largely to the purposes of research, available genetic resources or databases, equipment and facilities, funding and personnel resources, etc. The choice and use of DNA markers in research and breeding is still a challenge for plant breeders. A number of factors need to be considered when a breeder chooses one or more molecular marker types (Semagn et al., 2006a). A breeder should make an appropriate choice that best meets the requirements according to the conditions and resources available for the breeding program.

3. PRE-REQUISITES AND GENERAL ACTIVITIES OF MARKER-ASSISTED BREEDING

3.1. Prerequisites For An Efficient Marker-Assisted Breeding Program

Compared with conventional breeding approaches, molecular breeding, mainly referred to as DNA marker-assisted breeding, needs more complicated equipment and facilities. In general, the pre-requisites listed below are essential for marker-assisted breeding (MAB) in plants.

a. Appropriate marker system and reliable markers: For a plant species or crop, a suitable marker system and reliable markers available are critically important to initiate a marker-assisted breeding program. As discussed above, suitable markers should have following attributes:

- Ease and low-cost of use and analysis;
- Small amount of DNA required;
- Co-dominance;
- Repeatability/reproducibility of results;

TABLE 1. Comparison of most widely used DNA marker systems in plants; Adapted from Collard et al. (2005), Semagn et al. (2006a), Xu (2010), and others.

Feature and description	RFLP	RAPD	AFLP	SSR	SNP
Genomic abundance	High	High	High	Moderate to high	Very high
Genomic coverage	Low copy coding region	Whole genome	Whole genome	Whole genome	Whole genome
Expression/inheritance	Co-dominant	Dominant	Dominant / co-dominant	Co-dominant	Co-dominant
Number of loci	Small (<1,000)	Small (<1,000)	Moderate (1,000s)	High (1,000s – 10,000s)	Very high (>100,000)
Level of polymorphism	Moderate	High	High	High	High
Type of polymorphism	Single base changes, indels	Single base changes, indels	Single base changes, indels	Changes in length of repeats	Single base changes, indels
Type of probes/primers	Low copy DNA or cDNA clones	10 bp random nucleotides	Specific sequence	Specific sequence	Allele-specific PCR primers
Cloning and/or sequencing	Yes	No	No	Yes	Yes
PCR-based	Usually no	Yes	Yes	Yes	Yes
Radioactive detection	Usually yes	No	Yes or no	Usually no	No
Reproducibility/reliability	High	Low	High	High	High
Effective multiplex ratio	Low	Moderate	High	High	Moderate to high
Marker index	Low	Moderate	Moderate to high	High	Moderate
Genotyping throughput	Low	Low	High	High	High
Amount of DNA required	Large (5 – 50 μg)	Small (0.01 – 0.1 μg)	Moderate (0.5 – 1.0 μg)	Small (0.05 – 0.12 μg)	Small (≥ 0.05 μg)
Quality of DNA required	High	Moderate	High	Moderate to high	High
Technically demanding	Moderate	Low	Moderate	Low	High
Time demanding	High	Low	Moderate	Low	Low
Ease of use	Not easy	Easy	Moderate	Easy	Easy
Ease of automation	Low	Moderate	Moderate to high	High	High
Development/start-up cost	Moderate to high	Low	Moderate	Moderate to high	High
Cost per analysis	High	Low	Moderate	Low	Low
Number of polymorphic loci per analysis	1.0 – 3.0	1.5 – 5.0	20 – 100	1.0 – 3.0	1.0
Primary application	Genetics	Diversity	Diversity and genetics	All purposes	All purposes

- High levels of polymorphism; and
- Occurrence and even distribution genome wide

In addition, another important desirable attribute for the markers to be used is close association with the target gene(s). If the markers are located in close proximity to the target gene or present within the gene, selection of the markers will ensure the success in selection of the gene. Although they can also be used in plant breeding programs, the number of classical markers possessing these features is very small. DNA markers for polymorphism are available throughout the genome, and their presence or absence is not affected by environments and usually do not directly affect the phenotype. DNA markers can be detected at any stage of plant growth, but the detection of classical markers is usually limited to certain growth stages. Therefore, DNA markers are the predominant types of genetic markers for MAB. Each type of markers has advantages and disadvantages for specific purposes. Relatively speaking, SSRs have most of the desirable features and thus are the current marker of choice for many crops. SNPs require more detailed knowledge of the specific, single nucleotide DNA changes responsible for genetic variation among individuals. However, more and more SNPs have become available in many species, and thus they are also considered an important type for marker-assisted breeding.

a. Quick DNA extraction and high throughput marker detection: For most plant breeding programs, hundreds to thousands of plants/individuals are usually screened for desired marker patterns. In addition, the breeders need the results instantly to make selections in a timely manner. Therefore, a quick DNA extraction technique and a high throughput marker detection system are essentially required to handle a large number of tissue samples and a large-scale screening of multiple markers in breeding programs. Extracting DNA from small tissue samples in 96- or 384-well plates and streamlined operations are adopted in many labs and programs. High throughput PAGE and AGE systems are commonly used for marker detection. Some labs also provide marker detection services using automated detection systems, e.g. SNP chips based on thousands to ten thousands of markers.

b. Genetic maps: Linkage maps provide a framework for detecting marker-trait associations and for choosing markers to use in marker-assisted breeding. Therefore, a genetic linkage map, particularly high-density linkage map is very important for MAB. To use markers and select a desired trait present in a specific germplasm line, a proper population of segregation for the trait is required to construct a linkage map. Once a marker or a few markers are found to be associated with the trait in a given population, a dense molecular marker map in a standard reference population will help identify makers that are close to (or flank) the target gene. If a region is found associated with the desired traits of interest, fine mapping also can be done with additional markers to identify the marker(s) tightly linked to the gene controlling the trait. A favorable genetic map should have an adequate number of evenly-spaced polymorphic markers to accurately locate desired QTLs/genes (Babu et al., 2004).

c. Knowledge of marker-trait association: The most crucial factor for marker-assisted breeding is the knowledge of the associations between markers and the traits of interest. Only those markers that are closely associated with the target traits or tightly linked to the genes can provide sufficient guarantee for the success in practical breeding. The more closely the markers are associated with the traits, the higher the possibility of success and efficiency of use will be. This information can be obtained in various ways, such as gene mapping, QTL analysis, association mapping, classical mutant analysis, linkage or recombination analysis, bulked segregant analysis, etc. In addition, it is also critical to know the linkage situation, i.e. the markers are linked in cis/trans (coupling or repulsion) with the desired allele of the trait. Even if some markers have been reported to be tightly linked with a QTL, a plant breeder still needs to determine the association of alleles in his own breeding material. This makes QTL information difficult to directly transfer between different materials.

d. Quick and efficient data processing and management: In addition to above-mentioned pre-requisites, quick and efficient data process and management may provide timely and useful reports for breeders. In a marker-assisted breeding program, not only are large numbers of samples handled, but multiple markers for each sample also need to be screened at the same time. This situation requires an efficient and quick system for labeling, storing, retrieving, processing and analyzing large data sets, and even integrating data sets available from other programs. The development of bioinformatics and statistical software packages provides a useful tool for this purpose.

3.2. Activities Of Marker-Assisted Breeding

Marker-assisted breeding involves the following activities provided the prerequisites are well equipped or available:

a. Planting the breeding populations with potential segregation for traits of interest or polymorphism for the markers used.

b. Sampling plant tissues, usually at early stages of growth, e.g. emergence to young seedling stage.

c. Extracting DNA from tissue sample of each individual or family in the populations, and preparing DNA samples for PCR and marker screening.

d. Running PCR or other amplifying operation for the molecular markers associated with or linked to the trait of interest.

e. Separating and scoring PCR/amplified products, by means of appropriate separation and detection techniques, e.g. PAGE, AGE, etc.

f. Identifying individuals/families carrying the desired marker alleles.

g. Selecting the best individuals/families with both desired marker alleles for target traits and desirable performance/phenotypes of other traits, by jointly using marker results and other selection criteria.

h. Repeating the above activities for several generations, depending upon the association between the markers and the traits as well as the status of marker alleles (homozygous or heterozygous), and advancing the individuals selected in breeding program until stable superior or elite lines that have improved traits are developed.

4. MARKER-ASSISTED SELECTION

4.1. Mas Procedure And Theoretical And Practical Considerations

Marker-assisted selection (MAS) refers to such a breeding procedure in which DNA marker detection and selection are integrated into a traditional breeding program. Taking a single cross as an example, the general procedure can be described as follow:

a. Select parents and make the cross, at least one (or both) possesses the DNA marker allele(s) for the desired trait of interest.

b. Plant F_1 population and detect the presence of the marker alleles to eliminate false hybrids.

c. Plant segregating F_2 population, screen individuals for the marker(s), and harvest the individuals carrying the desired marker allele(s).

d. Plant $F_{2:3}$ plant rows, and screen individual plants with the marker(s). A bulk of F_3 individuals within a plant row may be used for the marker screening for further confirmation in case needed if the preceding F_2 plant is homozygous for the markers. Select and harvest the individuals with required marker alleles and other desirable traits.

e. In the subsequent generations (F_4 and F_5), conduct marker screening and make selection similarly as for $F_{2:3}$s, but more attention is given to superior individuals within homozygous lines/rows of markers.

f. In $F_{5:6}$ or $F_{4:5}$ generations, bulk the best lines according to the phenotypic evaluation of target trait and the performance of other traits, in addition to marker data.

g. Plant yield trials and comprehensively evaluate the selected lines for yield, quality, resistance and other characters of interest.

A frequently asked question about marker-assisted selection is that "how many QTLs should be selected for MAS?" Theoretically, all the QTLs contributing to the trait of interest could be taken into account. For a quantitatively-inherited character like yield, numerous QTLs or genes are usually involved. It is almost impossible to select all QTLs or genes simultaneously so that the selected individuals incorporate all the desired QTLs due to the limitation of resources and facilities. The number of individuals in the population increases exponentially with the increase of target loci involved. The relative efficiency of MAS decreases as the number of QTLs increases and their heritability decreases (Moreau et al., 1998). In other words, MAS will be less effective for a highly complex character governed by many genes than for a simply inherited character controlled by a few genes. The number of genes/QTLs not only impacts the efficiency of MAS, but also the breeding design and implement scheme (detail will be discussed below). Typically no more than three QTLs are regarded as an appropriate and feasible choice (Ribaut and Betran, 1999), although five QTLs were used in improvement of fruit quality traits in tomato via marker-assisted introgression (Lecomte et al., 2004). With development of SNP markers (especially rapid automated detection and genotyping technologies), selection of more QTLs at the same time might be preferred and practicable (Kumpatla et al., 2012).

For MAS for multiple genes/QTLs, it was suggested to limit the number of genes undergoing selection to three to four if they are QTLs selected on the basis of linked markers, and to five to six if they are known loci selected directly (Hospital, 2003). Only the multi-environmentally verified QTLs that possess medium to large effects are selected. The first priority should be given to the major QTLs that can explain greatest proportion of phenotypic variation and/or can be consistently detected across a range of environments and different populations. In addition, an index for selection that weights markers differently could be constructed, depending on their relative importance to the breeding objectives. Flint-Garcia et al. (2003) presented an example of such an index used to select for QTLs with different effect magnitudes.

Another question that is commonly asked also is that "how many markers should be used in MAS?" The more markers associated with a QTL are used, the greater opportunity of success in selecting the QTL of interest will be ensured. However, efficiency is also important for a breeding program, especially when the resources and facilities are limited. From the point of both effectiveness and efficiency, for a single QTL it is usually suggested to use two markers (i.e. flanking markers) that are tightly linked to the QTL of interest. The markers to be used should be close enough to the gene/QTL of interest (<5cM) in order to ensure that only a minor proportion of the selected individuals will be recombinants. If a marker (e.g. the peak marker) is found to be located within the region of gene sequence of interest or in such a close proximity to the QTL/gene that no recombination occurs between the marker and the QTL/gene, such a marker only should be preferable. However, if a marker is not tightly linked to a gene of interest, recombination between the marker and gene may reduce the efficiency of MAS because a single crossover may alternate the linkage association and leads to selection errors. The efficiency of MAS decreases as the recombination frequency (genetic distance) between the marker and gene increases. Use of two flanking markers rather than one may decrease the chance of such errors due to homologous recombination and increase the efficiency of MAS. In this case, only a double crossover (i.e. two single crossovers occurring simultaneously on both sides of the gene/QTL in the region) may result in selection errors, but the frequency of a double crossover is considerably rare. For instance, if two flanking markers with an interval of 20cM or so between them are used, there will be higher probability (99%) for recovery of the target gene than only one marker used.

In practical MAS, a breeder is also concerned about how the markers should be detected, how many generations of MAS have to be conducted, and how large size of the population is needed. In general, detection of marker polymorphism is performed at early stages of plant growth. This is true especially for marker-assisted backcrossing and marker-assisted recurrent selection, because only the individuals that carry preferred marker alleles are expected to be used in backcrossing to the recurrent parent and/or inter-mating between selected individuals/progenies. The generations of MAS required vary with the number of markers used, the degree of association between the markers and the QTLs/genes of interest, and the status of marker alleles. In many cases, marker screening is performed for two to four consecutive generations in a segregating population. If fewer markers are used and the markers are in close proximity to the QTL or gene of interest, fewer generations are needed. If homozygous status of marker alleles of interest is detected in two consecutive generations, marker screening may not be performed in their progenies. Bonnett et al. (2005) discussed the strategies for efficient implementation of MAS involving several issues, e.g. breeding systems or schemes, population sizes, number of target loci, etc. Their strategies include F_2 enrichment, backcrossing, and inbreeding.

In MAS, phenotypic evaluation and selection is still very helpful if conditions permit to do so, and even necessary in cases when the QTLs selected for MAS are not so stable across environments and the association between the selected markers and QTLs is not so close. Moreover, one should also take the impact of genetic background into consideration. The presence of a QTL or marker does not necessarily guarantee the expression of the desired trait. QTL data derived from multiple environments and different populations help a better understanding of the interactions of QTL x environment and QTL x QTL or QTL x genetic background, and thus help a better use of MAS. In addition to genotypic (markers) and phenotypic data for the trait of interest, a breeder often pays considerable attention to other important traits, unless the trait of interest is the only objective of breeding.

There are several indications for adoption of molecular markers in the selection for the traits of interest in practical breeding. The situations favorable for MAS include:

- The selected character is expressed late in plant development, like fruit and flower features or adult characters with a juvenile period (so that it is not necessary to wait for the plant to become fully developed before propagation occurs or can be arranged)
- The target gene is recessive (so that individuals which are heterozygous positive for the recessive allele can be selected and/or crossed to produce some homozygous offspring with the desired trait)
- Special conditions are required in order to invoke expression of the target gene(s), as in the case of breeding for disease and pest resistance (where inoculation with the disease or subjection to pests would otherwise be required), or the expression of target genes is highly variable with the environments.
- The phenotype of a trait is conditioned by two or more unlinked genes. For example, selection for multiple genes or gene pyramiding may be required to develop enhanced or durable resistance against diseases or insect pests.

4.2. Mas For Major Genes Or Improvement Of Qualitative Traits

In crop plants, many economically important characteristics are controlled by major genes/QTLs. Such characteristics include resistance to diseases/pests, male sterility, self-incompatibility and others related to shape, color and architecture of whole plants and/or plant parts. These traits are often of mono- or oligogenic inheritance in nature. Even for some quality traits, one or a few major QTLs or genes can account for a very high proportion of the phenotypic variation of the trait (Bilyeu et al., 2006; Pham et al., 2012). Transfer of such a gene to a specific line can lead to tremendous improvement of the trait in the cultivar under development. The marker loci which are tightly linked to major genes can be used for selection and are sometimes more efficient than direct selection for the target genes. In some cases, such advantages in efficiency may be due to higher expression of the marker mRNA in such cases that the marker is actually within a gene. Alternatively, in such cases that the target gene of interest differs between two alleles by a difficult-to-detect SNP, an external marker of which polymorphism is easier to detect, may present as the most realistic option.

Soybean cyst nematode (SCN) (*Heterodera glycines* Inchinoe) may be taken as an example of MAS for major genes. This pathogen is the most economically significant soybean pest. The principal strategy to reduce or eliminate damage from this pest is the use of resistant cultivars (Cregan et al., 1999). However, identifying resistant segregants in breeding populations is a difficult and expensive process. A widely used phenotypic assay takes five weeks, requires a large greenhouse space, and about 5 to 10 h of labor for every 100 plant samples processed (Young, 1999). Fortunately, the SSR marker Satt309 has been identified to be located only 1–2 cM away from the resistance gene *rhg1* (Cregan et al., 1999), which forms the basis of many public and commercial breeding efforts. In a direct comparison, genotypic selection with Satt309 was 99% accurate in predicting lines that were susceptible in subsequent greenhouse assays for two test populations, and 80% accurate in a third population, each with a different source of SCN resistance (Young, 1999). In soybean, Shi et al. (2009) reported that using molecular markers in a cross J05 x V94-5152, they developed five $F_{4:5}$ lines that were homozygous for all eight marker alleles linked to the genes/loci of resistance to soybean mosaic virus (SMV). These lines exhibited resistance to SMV strains G1 and G7 and presumably carried all three resistance genes (*Rsv1*, *Rsv3* and *Rsv4*) that would potentially provide broad and durable resistance to SMV.

4.3. Mas For Improvement Of Quantitative Traits

Most of the important agronomic traits are polygenic or controlled by multiple QTLs. MAS for the improvement of such traits is a complex and difficult task because it is related to many genes or QTLs involved, QTL x E interaction and epistasis. Usually, each of these genes has a small effect on the phenotypic expression of the trait and expression is affected by environmental conditions. Phenotyping of quantitative traits becomes a complex endeavor consequently, and determining marker-phenotype association becomes difficult as well. Therefore, repeated field tests are required to accurately characterize the effects of the QTLs and to evaluate the stability across environments. The QTL x E interaction reduces the efficiency of MAS and epistasis can result in a skewed QTL effect on the trait.

Despite a tremendous amount of QTL mapping experiments over the past decade, application and utilization of the QTL mapping information in plant breeding has been constrained by a number of factors (Collard and Mackill, 2008):

1. Strong QTL-environmental interaction which make phenotyping difficult since expression may vary from one location/year to another;
2. Lack of universally valid QTL-marker associations applicable across populations. The notion that QTL mapping to identify new QTL markers whenever a new germplasm is used, puts some people off and they lose interest in MAS;
3. Deficiencies in QTL statistical analysis which lead to either overestimation or underestimation of the number of QTLs involved and their effect on the trait;
4. Often times, there are no QTLs with major effects on the trait and this means a large number of QTLs have to be identified and in many cases this becomes a tough goal to achieve and further complicates identification of marker-QTL association.

In order to improve the efficiency of MAS for quantitative traits, appropriate field experimental designs and approaches have to be employed. Attention should be given to replications both over time and space, consistency in experimental techniques, samplings and evaluations, robust data processing and statistical analysis. For example, composite interval mapping (CIM) allows the integration of data from different locations for joint analysis to estimate QTL-environment interaction so that stable QTLs across environments can be identified. A saturated linkage map enables accurate identification of both targeted QTLs as well as linked QTLs in coupling and repulsion linkage phases. In practical breeding for improvement of a quantitative trait, usually not many minor QTLs are considered but only a few major QTLs are used in MAS. In case many QTLs especially minor-effect QTLs are involved, a breeder would prefer to consider the strategy of gene pyramiding (see the later section).

Fusarium head blight (FHB) caused by *Fusarum* species is one of the most destructive diseases in wheat and barley worldwide. To combat this disease, a great effort from multiple fields, including plant breeding and genetics, molecular genetics and genomics, plant pathology, and integrated management, has been dedicated since 1990s. Resistance to HFB in both wheat and barley is quantitatively inherited, and many QTLs have been identified from different resources of germplasm (Buerstmayr et al., 2009). Use of MAS to improve the resistance has become a choice for many breeding programs. In wheat, a major QTL designated as *Fhb1* was consistently detected across multiple environments and populations, and explained 20-40% of phenotypic variation in most cases (Buerstmayr et al., 2009; Jiang et al., 2007a, 2007b). Thus wheat breeders would especially prefer to use this major QTL to develop new cultivars with FHB resistance. Pumphrey et al. (2007) compared 19 pairs of NIL for *Fhb1* derived from an ongoing breeding program and found that the average reduction in disease severity between NIL pairs was 23% for disease severity and 27% for kernel infection. Later investigation from the group also demonstrated successful implementation of MAS for this QTL (Anderson et al. 2007).

In addition, researchers also tried to incorporate multiple QTLs by MAS. Miedaner et al. (2006) demonstrated that MAS for three FHB resistance QTLs simultaneously was highly effective in enhancing FHB resistance in German spring wheat. FHB resistance was the highest in recombinant lines with multiple QTLs combined, especially 3B plus 5A. Jiang et al. (2007a) made a comparison of multiple-locus combinations in a RIL population derived from the cross "Veery x CJ 9306". For three loci, the average levels of resistance from low to high in genotypes were: no favorable allele – one favorable allele – two favorable alleles – three favorable alleles, except for the non-reciprocal comparisons. When four or five loci carrying favorable alleles from the resistant parent CJ 9306 were considered simultaneously, the coefficients of determination between the accumulated effects of alleles for different combinations and the averages of number or percentage of diseased spikelets for the corresponding RILs were 0.33-0.41 ($P<0.01$) (Jiang et al., 2007a). Therefore, the authors concluded that the effects of FHB resistance QTLs could be accumulated and the resistance could be feasibly enhanced by selection of favorable marker alleles for multiple loci in breeding programs.

In the U.S., the Coordinated Agricultural Projects (CAPs) with aims to encourage collaborative efforts in applied plant genomics and molecular research have been implemented in several crops, such as rice, wheat, barley, beans, potato, tomato, etc. An important strategy CAPs take is applying marker-assisted selection to plant breeding and efficiently using genetic resources and facilities available, including thousands and ten thousands of DNA markers and plant introductions, to facilitate development of crop cultivars with improved yield, resistance and quality.

5. MARKER-ASSISTED BACKCROSSING

5.1. Mabc Procedure And Theoretical And Practical Considerations

Marker-assisted or marker-based backcrossing (MABC) is regarded as the simplest form of marker-assisted selection, and at the present it is the most widely and successfully used method in practical molecular breeding. MABC aims to transfer one or a few genes/QTLs of interest from one genetic source (serving as the donor parent and maybe inferior agronomically or not good enough in comprehensive performance in many cases) into a superior cultivar or elite breeding line (serving as the recurrent parent) to improve the targeted trait. Unlike traditional backcrossing, MABC is based on the alleles of markers associated with or linked to gene(s)/QTL(s) of interest instead of phenotypic performance of target trait. The general procedure of MABC is as follow, regardless of dominant or recessive nature of the target trait in inheritance:

a. Select parents and make the cross, one parent is superior in comprehensive performance and serves as recurrent parent (RP), and the other one used as donor parent (DP) should possess the desired trait and the DNA markers allele(s) associated with or linked to the gene for the trait.

b. Plant F_1 population and detect the presence of the marker allele(s) at early stages of growth to eliminate false hybrids, and cross the true F_1 plants back to the RP.

c. Plant BCF_1 population, screen individuals for the marker(s) at early growth stages, and cross the individuals carrying the desired marker allele(s) (in heterozygous status) back to the RP. Repeat this step in subsequent seasons for two to four generations, depending upon the practical requirements and operation situations as discussed below.

d. Plant the final backcrossing population (e.g. BC_4F_1), and screen individual plants with the marker(s) for the target trait and discard the individuals carrying homozygous markers alleles from the RP. Have the individuals with required marker allele(s) selfed and harvest them.

e. Plant the progenies of backcrossing-selfing (e.g. BC_4F_2), detect the markers and harvest individuals carrying homozygous DP marker allele(s) of target trait for further evaluation and release.

Theoretically, the proportion of the RP genome after n generations of backcrossing is given by $1 - (1/2)^{n+1}$ for a single locus and $[1 - (1/2)^{n+1}]^k$ for k loci, respectively, for a population large enough in size (or with adequate individuals) and no selection being made during backcrossing (i.e. "blind" backcrossing only). The percentage of the RP genome is the average of the population, with some individuals possessing more of the RP genome than others. To fully recover the genome of the RP, 6-8 generations of backcrossing is needed typically in case no selection is made for the RP. However, this process is usually slower than expected for the target gene-carrier chromosome, i.e. linkage drag, especially in case a linkage exists between the target gene and other undesirable traits. On the other hand, the process of introgression of QTLs/genes and recovery of the RP genome may be accelerated by selection using markers flanking QTLs and evenly spaced markers from other chromosomes (i.e. unlinked to QTLs) of the RP (Collard et al., 2005) or selection for the performance of the RP conducted simultaneously. For MABC program, therefore, there are two types of selection recognized: Foreground selection and background selection (Hospital, 2003).

In foreground selection, the selection is made only for the marker allele(s) of donor parent at the target locus to maintain the target locus in heterozygous state until the final backcrossing is completed. Then the selected plants are selfed and the progeny plants with homozygous DP allele(s) of selected markers are harvested for further evaluation and release. As described above, this is the general procedure of MABC. The effectiveness of foreground selection depends on the number of genes/loci involved in the selection, the marker-gene/QTL association or linkage distance and the undesirable linkage to the target gene/QTL.

In background selection, the selection is made for the marker alleles of recurrent parent in all genomic regions of desirable traits except the target locus, or selection against the undesirable genome of donor parent. The objective is to hasten the restoration of the RP genome and eliminate undesirable genes introduced from the DP. The progress in recovery of the RP genome depends on the number of markers used in background selection. The more markers evenly located on all the chromosomes are selected for the RP alleles, the faster recovery of the RP genome will be achieved but larger population size and more genotyping will be required as well. In addition, the linkage drag also can be efficiently addressed by background selection using DNA markers, although it is difficult to overcome in a traditional backcrossing program.

Foreground selection and background selection are two respective aspects of MABC with different foci of selection. In practice, however, both foreground and background selection are usually conducted in the same program, either simultaneously or successively. In many cases, they can be performed alternatively even in the same generation. The individuals that have the desired marker alleles for target trait are selected first (foreground selection). Then the selected individuals are screened for other marker alleles again for the RP genome (background selection). It is understandable to do so because selection of the target gene/QTL is the essential and only critical point for backcrossing program, and the individuals that do not have the allele of target gene will be discarded and thus it is not necessary to genotype them for other traits.

The efficiency of MABC depends upon several factors, such as the population size for each generation of backcrossing, marker-gene association or the distance of markers from the target locus, number of markers used for target trait and RP background, and undesirable linkage drag. Based on simulations of 1000 replicates, Hospital (2003) presented the expected results of a typical MABC program, in which heterozygotes were selected at the target locus in each generation, and RP alleles were selected for two flanking markers on target chromosome each located 2 cM apart from the target locus and for three markers on non-target chromosomes. As shown in Table 2, a faster recovery of the RP genome could be achieved by MABC with combined foreground and background selection, compared to traditional backcrossing. Therefore, using markers can lead to considerable time savings compared to conventional backcrossing (Frisch et al., 1999; Collard et al., 2005).

TABLE 2. Expected results of a MABC program with combined foreground and background selection used; Adapted from Hospital (2003).

Backcross generation	Number of individuals	% homozygosity of recurrent parent alleles at selected markers		% recurrent parent genome	
		Chromosome with target locus	All other chromosomes	Marker-assisted backcross	Conventional backcross
BC_1	70	38.4	60.6	79.0	75.0
BC_2	100	73.6	87.4	92.2	87.5
BC_3	150	93.0	98.8	98.0	93.7
BC_4	300	100.0	100.0	99.0	96.9

In a MABC program, the population to be analyzed should contain at least one genotype that has all favorable alleles for a particular QTL. Later, the number of QTLs may be increased progressively, but not beyond six QTLs in most cases because of prohibitive difficulty in handling all QTLs (Hospital, 2003). In addition, the more QTLs/genes are transferred, the larger the proportion of unwanted genes would be due to linkage drag. In general, most of the unwanted genes are located on non-target chromosomes in early BC generations, and are rapidly removed in subsequent BC generations. On the contrary, the quantity of DP genes on the target chromosome decreases much more slowly, and even after generation BC_6 many of the unwanted donor genes are still located on the target chromosome in segregating state (Newbury, 2003). Given a total genome length is 3000 cM, 1% donor DNA fragments after six backcrosses represents a 30 cM chromosomal segment or region, which may host many unwanted genes, especially if the DP is a wild genetic resource. Young and Tanksley (1989) genotyped a collection of tomato varieties in which the resistance gene was previously transferred at the *Tm*-2 locus with RLFP markers. Their data indicated that the size of chromosomal segment retained around the *Tm*-2 locus during backcross breeding was very variable, with one line exhibiting a donor segment of 50 cM after 11 backcrosses and other one possessing 36 cM donor segment after 21 backcrosses. This clearly demonstrates the need for background selection.

As discussed above, linkage drag can be reduced by performing background selection. Typically, two markers flanking the target gene are used, and the individuals (or double recombinants) that are heterozygous at the target locus and homozygous for the recipient (RP) alleles at both flanking markers are selected. Use of closer flanking markers leads to more effective and faster reduction of linkage drag compared to distant markers. However, less distance between two flanking markers implies less probability of double recombination, and thus larger populations and more genotyping are needed. In order to optimize genotyping effort (i.e. the cost of the program), therefore, it is important to determine the minimal population sizes necessary to ensure the desired genotypes can be obtained. Hospital and Decoux (2002) developed a statistical software for determining the minimum population size required in BC program to identify at least one individual that is double-recombinant with heterozygosity at target locus and homozygosity for recurrent parent alleles at flanking marker loci. In addition, for closely-linked flanking markers, it is unlikely to obtain double recombinant genotypes through only one generation of backcrossing. Therefore, additional backcrossing should be conducted. For instance, in one BC generation (e.g. BC_1) single recombination on one side of the target gene is selection, and single recombination on the other side may be selected in another BC generation (e.g. BC_2) (Young and Tanksley 1989). In this way, individuals with desired RP alleles at two flanking markers and donor allele at target locus can be finally obtained.

To accelerate the recovery of RP genome on non-target chromosomes, scientists suggested using markers in backcrossing and discussed how many makers should be used (Tanksley et al., 1989; Hospital et al. 1992; Visscher et al. 1996). In background selection, the approaches involve selecting individuals that are of homozygous recipient type at a collection of markers located on non-carrier chromosomes. From a point of both effectiveness and efficiency, it is important to determine an appropriate number of markers to be used. More markers do not necessarily mean better benefits in practice. Generally, several markers are involved and MABC should be performed over two or more generations. It is unlikely that the selection objective can be realized in a single BC generation.

Dense marker coverage of non-target chromosomes is not mandatory to increase the overall proportion of recurrent parent genome, unless fine-mapping of specific chromosome regions is highly important. An appropriate number of markers and optimal position on chromosomes are important. Computer simulation suggested that for a chromosome of 100 cM, two to four markers are sufficient, and selection based on markers would be most efficient if the markers are optimally positioned along the chromosomes (Servin and Hospital, 2002). In practice, at least two or three markers per chromosome are needed, and every chromosome should be involved. In such a MABC scheme, three to four generations of backcrossing is generally enough to achieve more than 99% of the recurrent parent genome. With respect to the time necessary to release new varieties, the gain due to background selection can be economically valuable. In addition, background selection is more efficient in late BC generations than in early BC generations. For example, if a BC breeding scheme is conducted over three successive BC generations and yet the preference is to genotype individuals only once, then it is more efficient to genotype and select the individuals in BC_3 generation rather than in the BC_1generation (Hospital et al. 1992, Ribaut et al. 2002).

5.2. Application Of Mabc

Success in integrating MABC as a breeding approach lies in identifying situations in which markers offer noticeable advantages over conventional backcrossing or valuable complements to conventional breeding effort. MABC is essential and advantageous when:

1. Phenotyping is difficult and/or expensive or impossible;
2. Heritability of the target trait is low;

3. The trait is expressed in late stages of plant development and growth, such as flowers, fruits, seeds, etc.;
4. The traits are controlled by genes that require special conditions to express;
5. The traits are controlled by recessive genes; and
6. Gene pyramiding is needed for one or more traits.

Among the molecular breeding methods, MABC has been most widely and successfully used in plant breeding up to date. It has been applied to different types of traits (e.g. disease/pest resistance, drought tolerance and quality) in many species, e.g. rice, wheat, maize, barley, pear millet, soybean, tomato, etc. (Collard et al., 2005; Dwivedi et al., 2007; Xu, 2010). In maize, for example, *Bacillus thuringiens* is a bacterium that produces insecticidal toxins, which can kill corn borer larvae when they ingest the toxins in corn cells (Ragot et al. 1995). The integration of the *Bt* transgene into various corn genetic backgrounds has been achieved by using MABC. Aroma in rice is controlled by a recessive gene which is due to an eight base-pair deletion and three single nucleotide polymorphism in a gene that codes for betaine aldehyde dehydrgenase 2 (Bradbury et al., 2005a). This discovery allows identification of the aromatic and non-aromatic rice varieties and discriminates homozygous recessive and dominant as well as heterozygous individuals in segregating population for the trait. MABC has been used to select for aroma in rice (Bradbury et al. 2005b). High lysine *opaque2* gene in corn was incorporated using MABC (Babu et al. 2005). However, the rate of success decreases when large numbers of QTL_s are targeted for introgression. Sebolt et al. (2000) used MABC for two QTL for seed protein content in soybeans. However, only one QTL was confirmed in $BC_3F_{4:5}$. When that QTL was introduced in three different genetic backgrounds, it had no effect in one background. In tomato, Tanksley and Nelson (1996) proposed a MABC strategy, called advanced backcross-QTL (AB-QTL), to transfer resistance genes from wild relative/unadapted genotype into elite germplasm. The strategy has proven effective for various agronomically important traits in tomato, including fruit quality and black mold resistance (Tanksley and Nelson, 1996; Bernacchi et al., 1998; Fulton et al., 2002). In addition, AB-QTL has been used in other crop species, such as rice, barley, wheat, maize, cotton and soybean, collectively demonstrating that this strategy is effective in transferring favorable alleles from the wild/unadapted germplasm to elite germplasm (Wang and Chee, 2010; Concibido et al., 2003).

In barley, a marker linked (0.7 cM) to the *Yd2* gene for resistance to barley yellow dwarf virus was successfully used to select for resistance in a backcrossing scheme (Jefferies et al., 2003). Compared to lines without the marker, the BC_2F_2-derived lines carrying the linked marker had lighter leaf symptoms and higher yield when infected by the virus. In maize, marker-facilitated backcrossing was also successfully employed to improve complex traits such as grain yield. Using MABC, six chromosomal segments each in two elite lines, Tx303 and Oh43, were transferred into two widely used inbred lines, B73 and Mo17, through three generations of backcrossing followed by two selfing generations. Then the enhanced lines with better performance were selected based on initial evaluations of testcross hybrids. The single-cross hybrids of enhanced B73 x enhanced Mo17 out-yielded the check hybrids by 12-15% (Stuber et al., 1999). Zhao et al. (2012) reported that a major quantitative trait locus (named *qHSR1*) for resistance to head smut in maize was successfully integrated into ten high-yielding inbred lines (susceptible to head smut). Each of the ten high-yielding lines was crossed with a donor parent Ji 1037 that contains *qHSR1* and is completely resistant to head smut, followed by five generations of backcrossing to the respective recurrent parents. In BC_1 through BC_3 only phenotypic selection was conducted to identify highly resistant individuals after artificial inoculation. In BC_4phenotypic selection, foreground selection and recombinant selection were conducted to screen for resistant individuals with the shortest *qHSR1* donor regions. In BC_5, phenotypic selection, foreground selection and background selection were performed to identify resistant individuals with the highest proportion of the recurrent parent genome, followed by one generation of self-pollination to obtain homozygous genotypes at the *qHSR1* locus. The ten improved inbred lines all showed substantial resistance to head smut, and the hybrids derived from these lines also showed a significant increase in the resistance. Semagn et al. (2006b) provided a detail review on the progress and prospects of MABC in crop breeding.

Currently, a cooperative marker-based backcrossing project for high-oleic acid in soybean has been initiated among multiple U.S. land-grant universities and USDA-ARS. Backcrossing and selection will be performed using the markers tightly linked to the high-oleic genes/loci. Hopefully, the high-oleic (80% or higher) traits will be successfully transferred from mutant lines or derived lines into other locally superior cultivars/lines, or combined with other unique traits like low linolenic acid (Pham et al., 2012).

6. MARKER-ASSISTED GENE PYRAMIDING AND MARKER-ASSISTED RECURRENT SELECTION

Marker-assisted gene pyramiding (MAGP) is one of the most important applications of DNA markers to plant breeding. Gene pyramiding has been proposed and applied to enhance resistance to disease and insects by selecting for two or more than two genes at a time. For example in rice such pyramids have been developed against bacterial blight and blast (Huang et al., 1997; Singh et al., 2001; Luo et al., 2012). Castro et al. (2003) reported a success in pyramiding qualitative gene and QTLs for resistance to stripe rust in barley. The advantage of using markers in this case allows selecting for QTL-allele-linked markers that have the same phenotypic effect. To enhance or improve a quantitatively inherited trait in plant breeding, pyramiding of multiple genes or QTLs is recommended as a potential strategy (Richardson et al., 2006). The cumulative effects of multiple-QTL pyramiding have been proven in crop species like wheat, barley and soybean (Richardson et al., 2006; Jiang et al., 2007a, 2007b; Li et al., 2010; Wang et al., 2012). Pyramiding of multiple genes/QTLs may be achieved through different approaches: multiple-parent crossing or complex crossing, backcrossing, and recurrent selection. A suitable breeding scheme for MAGP depends on the number of genes/QTLs required for improvement of traits, the number of parents that contain the required genes/QTLs, the heritability of traits of interest, and other factors (e.g. marker-gene association, expected duration to complete the plan and relative cost). Assuming three or four desired genes/QTLs exist separately in three or four lines, pyramiding of them can be realized by three-way, four-way or double crossing. They may also be integrated by convergent backcrossing or stepwise backcrossing. However, if there are more than four genes/QTLs to be pyramided, complex or multiple crossing and/or recurrent selection may be often preferred.

For MABC-based gene pyramiding, in general, there may be three strategies or breeding schemes: stepwise, simultaneous/synchronized and convergent backcrossing or transfer. Supposing one cultivar W is superior in comprehensive performance but lack of a trait of interest, and four different genes/QTLs contributing to the trait have been identified in four germplasm lines (e.g. P1, P2, P3 and P4). Three MABC schemes for pyramiding the genes/QTLs can be described as follow.

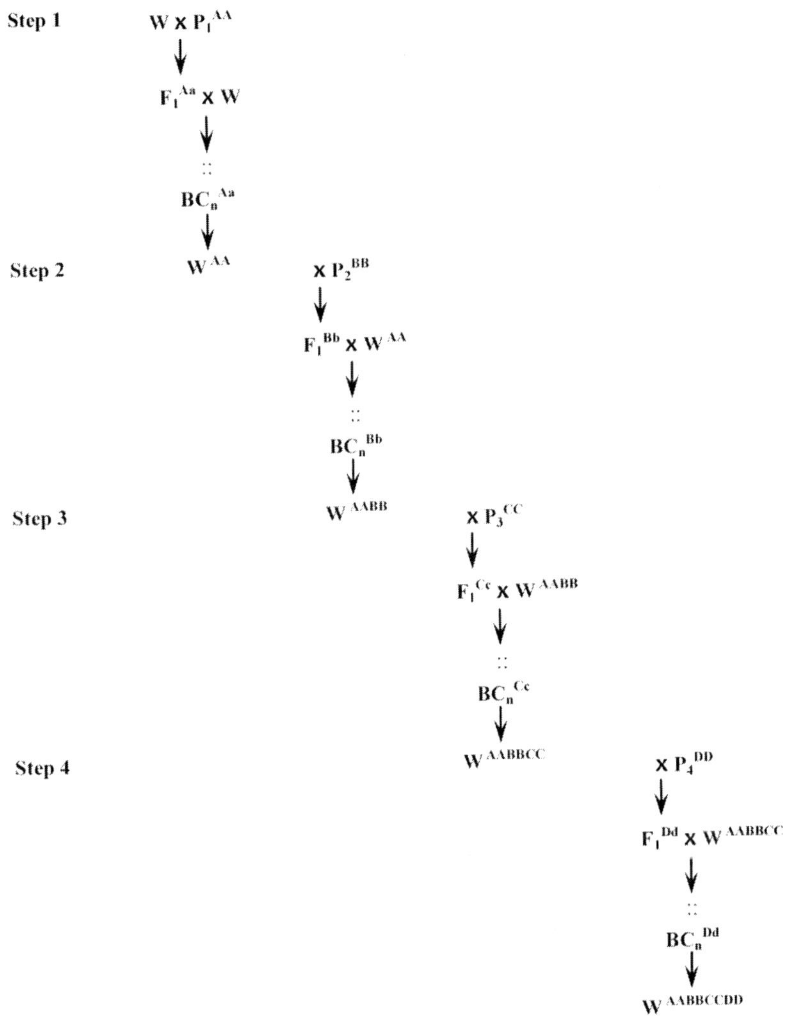

SCHEME 1. Stepwise Backcrossing

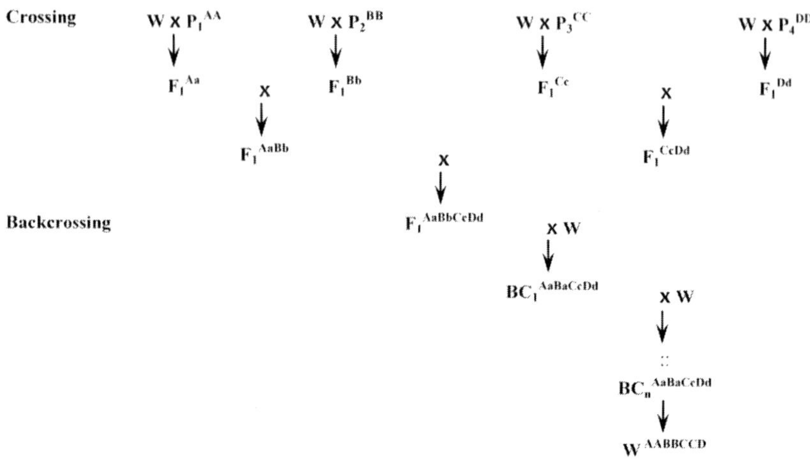

SCHEME 2.Simultaneous/Synchronized Backcrossing

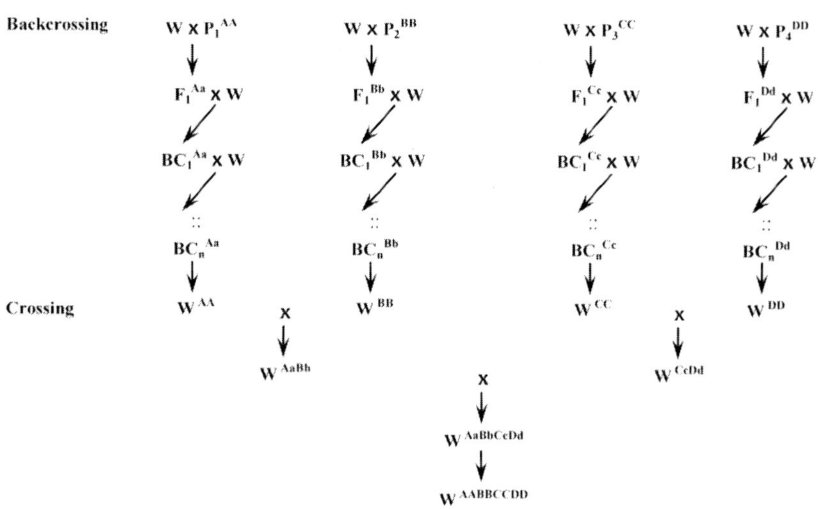

SCHEME 3.Convergent Backcrossing

In the stepwise backcrossing, four target genes/QTLs are transferred into the recurrent parent W in order. In one step of backcrossing, one gene/QTL is targeted and selected, followed by next step of backcrossing for another gene/QTL, until all target genes/QTLs have been introgressed into the RP.

The advantage is that gene pyramiding is more precise and easier to implement as it involves only one gene/QTL at one time and thus the population size and genotyping amount will be small. The improved recurrent parent may be released before the final step as long as the integrated genes/QTLs (e.g. two or three) meet the requirement at that time. The disadvantage is that it takes a longer time to complete. In the simultaneous or synchronized backcrossing, the recurrent parent W is first crossed to each of four donor parents to produce four single-cross F_1s. Two of the four single-cross F_1s are crossed with each other to produce two double-cross F_1s, and these two double-cross F_1s are crossed again to produce a hybrid integrating all four target genes/QTLs in heterozygous state. The hybrid and/or progeny with heterozygous markers for all four target genes/QTLs is subsequently crossed back to the RP W until a satisfactory recovery of the RP genome, and finalized by one generation of selfing. The advantage of this method is that it takes the shortest time to complete. However, in the backcrossing all target genes/QTLs are involved at the same time and thus it requires a large population and more genotyping. Convergent backcrossing is a strategy combining the advantages of stepwise and synchronized backcrossing. First the four target gene/QTLs are transferred separately from the donors into the recurrent parent W by single crossing followed by backcrossing based on markers linked to the target genes/QTLs, to produce four improved lines (W^{AA}, W^{BB}, W^{CC}, and W^{DD}). Two of the improved lines are crossed with each other and the two hybrids are then intercrossed to integrate all four genes/QTLs together and develop the final improved line with all four genes/QTLs pyramided (i.g. $W^{AABBCCDD}$). Relatively speaking, convergent backcrossing is more acceptable because in this scheme not only is time reduced (compared to stepwise transfer) but gene fixation and/or pyramiding is also more easily assured (compared to simultaneous transfer).

Theoretical issues and efficiency of MABC for gene pyramiding have been investigated through computer simulations (Ribaut et al., 2002; Servin et al., 2004; Ye and Smith, 2008). Practical application of MABC to gene pyramiding has been reported in many crops, including rice, wheat, barley, cotton, soybean, common bean and pea, especially for developing durable resistance to stresses in crops. However, there is very limited information available about the release of commercial cultivars resulted from this strategy. Somers et al. (2005) implemented a molecular breeding strategy to introduce multiple pest resistance genes into Canadian wheat. They used high throughput SSR genotyping and half-seed analysis to process backcrossing and selection for six FHB resistance QTLs, plus orange blossom wheat midge resistance gene *Sm1* and leaf rust resistance gene *Lr21*. They also used 45-76 SSR markers to perform background selection in backcrossing populations to accelerate the restoration of the RP genetic background. This strategy resulted in 87% fixation of the elite genetic background at the BC_2F_1 on average and successfully introduced all (up to 4) of the chromosome segments containing FHB, *Sm1* and *Lr21* resistance genes in four separate crosses(Somers et al., 2005). Joshi and Nayak (2010) and Xu (2010) recently reviewed the techniques and practical cases in marker-based gene pyramiding.

Similar to the simultaneous/synchronized backcrossing scheme, marker-assisted complex or convergent crossing (MACC) can be undertaken to pyramid multiple genes/QTLs. In particular, MACC is a proper option of breeding schemes for gene pyramiding if all the parents are improved cultivars or lines with good comprehensive performance and have different or complementary genes or favorable alleles for the traits of interest. The difference from simultaneous backcrossing is that selfing hybrid and progenies replaces backcrossing hybrid to the recurrent parent. In MACC, the hybrid of convergent crossing is subsequently self-pollinated and marker-based selection for target traits is performed for several consecutive generations until genetically stable lines with desired marker alleles and traits have been developed. In order to reduce population size and to avoid loss of most important genes/QTLs, different markers may be used and selected in different generations, depending on their relative importance. The markers for the most important genes/QTLs can be detected and selected first in early generations and less important markers later. Once homozygous alleles of the markers for a gene/locus are detected, they may not be necessarily detected again in the subsequent generations. Instead, phenotypic evaluation should be conducted if conditions permit.

Using markers to select or pyramid for multiple genes/QTLs is more complex and less proven. Recurrent selection is widely regarded as an effective strategy for the improvement of polygenic traits. However, the effectiveness and efficiency of selection are not so satisfactory in some cases because phenotypic selection is highly dependent upon environments and genotypic selection takes a longer time (2-3 crop seasons at least for one cycle of selection). Marker-assisted recurrent selection (MARS) is a scheme which allows performing genotypic selection and intercrossing in the same crop season for one cycle of selection (Fig. 1). Therefore, MARS could enhance the efficiency of recurrent selection and accelerate the progress of the procedure (Jiang et al., 2007a), particularly helps in integrating multiple favorable genes/QTLs from different sources through recurrent selection based on a multiple-parental population.

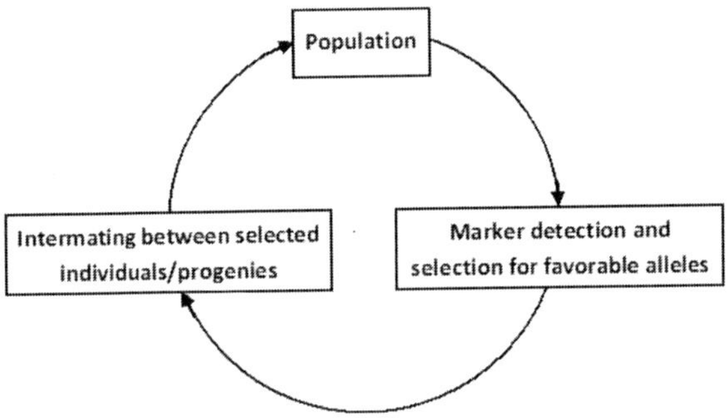

FIGURE 1. General procedure of marker-assisted recurrent selection (MARS)

For complex traits such as grain yield, biotic and abiotic resistance, MARS has been proposed for "forward breeding" of native genes and pyramiding multiple QTLs (Ragot et al., 2000; Ribaut et al., 2000, 2010; Eathington, 2005; Crosbie et al., 2006). As defined by Ribaut et al. (2010), MARS is a recurrent selection scheme using molecular markers for the identification and selection of multiple genomic regions involved in the expression of complex traits to assemble the best-performing genotype within a single or across related populations. Johnson (2004) presented an example to demonstrate the efficiency of MARS for quantitative traits. In their maize MARS programs, a large-scale use of markers in bi-parental populations, first for QTL detection and then for MARS on yield (i.e. rapid cycles of recombination and selection based on associated markers for yield), could allow increased efficiency of long-term selection by increasing the frequency of favorable alleles (Johnson, 2004). Eathington (2005) and Crosbie et al. (2006) also indicated that the genetic gain achieved through MARS in maize was about twice that of phenotypic selection (PS) in some reference populations. In upland cotton, Yi et al. (2004) reported significant effectiveness of MARS for resistance to *Helicoverpa armigera*. The mean levels of resistance in improved populations after recurrent selection were significantly higher than those of preceding populations.

7. GENOMIC SELECTION

Genomic selection (GS) or genome-wide selection (GWS) is a form of marker-based selection, which was defined by Meuwissen (2007) as the simultaneous selection for many (tens or hundreds of thousands of) markers, which cover the entire genome in a dense manner so that all genes are expected to be in linkage disequilibrium with at least some of the markers. In GS genotypic data (genetic markers) across the whole genome are used to predict complex traits with accuracy sufficient to allow selection on that prediction alone. Selection of desirable individuals is based on genomic estimated breeding value (GEBV) (Nakaya and Isobe, 2012), which is a predicted breeding value calculated using an innovative method based on genome-wide dense DNA markers (Meuwissen et al., 2001). GS does not need significant testing and identifying a subset of markers associated with the trait (Meuwissen et al., 2001). In other words, QTL mapping with populations derived from specific crosses can be avoided in GS. However, it does first need to develop GS models, i.e. the formulae for GEBV prediction (Nakaya and Isobe, 2012). In this process (training phase), phenotypes and genome-wide genotypes are investigated in the training population (a subset of a population) to predict significant relationships between phenotypes and genotypes using statistical approaches. Subsequently, GEBVs are used for the selection of desirable individuals in the breeding phase, instead of the genotypes of markers used in traditional MAS. For accuracy of GEBV and GS, genome-wide genotype data is necessary and require high marker density in which all quantitative trait loci (QTLs) are in linkage disequilibrium with at least one marker.

GS can be possible only when high-throughput marker technologies, high-performance computing and appropriate new statistical methods become available. This approach has become feasible due to the discovery and development of large number of single nucleotide polymorphisms (SNPs) by genome sequencing and new methods to efficiently genotype large number of SNP markers. As suggested by Goddard and Hayes (2007), the ideal method to estimate the breeding value from genomic data is to calculate the conditional mean of the breeding value given the genotype at each QTL. This conditional mean can only be calculated by using a prior distribution of QTL effects, and thus this should be part of the research to implement GS. In practice, this method of estimating breeding values is approximated by using the marker genotypes instead of the QTL genotypes, but the ideal method is likely to be approached more closely as more sequence and SNP data are obtained (Goddard and Hayes, 2007).

Since the application of GS was proposed by Meuwissen et al. (2001) to breeding populations, theoretical, simulation and empirical studies have been conducted, mostly in animals (Goddard and Hayes, 2007; Jannink et al., 2010). Relatively speaking, GS in plants was less studied and large-scale empirical studies are not available in public sectors for plant breeding (Jannink et al., 2010), but it has attracted more and more attention in recent years (Bernardo, 2010; Bernardo and Yu, 2007; Guo et al., 2011; Heffner et al., 2010, 2011; Lorenzana and Bernardo, 2009; Wong and Bernardo,

2008; Zhong et al., 2009). Studies indicated that in all cases, accuracies provided by GS were greater than might be achieved on the basis of pedigree information alone (Jannink et al., 2010). In oil palm, for a realistic yet relatively small population, GS was superior to MARS and PS in terms of gain per unit cost and time (Wong and Bernardo, 2008). The studies have demonstrated the advantages of GS, suggesting that GS would be a potential method for plant breeding and it could be performed with realistic sizes of populations and markers when the populations used are carefully chosen (Nakaya and Isobe, 2012).

GS has been highlighted as a new approach for MAS in recent years and is regarded as a powerful, attractive and valuable tool for plant breeding. However, GS has not become a popular methodology in plant breeding, and there might be a far way to go before the extensive use of GS in plant breeding programs. The major reason might be the unavailability of sufficient knowledge of GS for practical use (Nakaya and Isobe, 2012). Statistics and simulation discussed in terms of formulae in GS studies are most likely too specific and hard for plant breeders to understand and to use in practical breeding programs. From a plant breeder's point of view, GS can be practicable for a few breeding populations with a specific purpose, but may be impractical for a whole breeding program dealing with hundreds and thousands of crosses/populations at the same time. Therefore, GS must shift from theory to practice, and its accuracy and cost effectiveness must be evaluated in practical breeding programs to provide convincing empirical evidence and warrant a practicable addition of GS to a plant breeder's toolbox (Heffner et al., 2009). Development of easily understandable formulae for GEBVs and user-friendly software packages for GS analysis is helpful in facilitating and enhancing the application of GS in plant breeding. Kumpatla et al. (2012) recently presented an overall review on the GS for plant breeding.

8. MARKER-BASED BREEDING AND CONVENTIONAL BREEDING: CHALLENGES AND PERSPECTIVES

Marker-assisted breeding became a new member in the family of plant breeding as various types of molecular markers in crop plants were developed during the 1980s and 1990s. The extensive use of molecular markers in various fields of plant science, e.g. germplasm evaluation, genetic mapping, map-based gene discovery, characterization of traits and crop improvement, has proven that molecular technology is a powerful and reliable tool in genetic manipulation of agronomically important traits in crop plants. Compared with conventional breeding methods, MAB has significant advantages:

a. MAB can allow selection for all kinds of traits to be carried out at seedling stage and thus reduce the time required before the phenotype of an individual plant is known. For the traits that are expressed at later developmental stages, undesirable genotypes can be quickly eliminated by MAS. This feature is particularly important and useful for some breeding schemes such as backcrossing and recurrent selection, in which crossing with or between selected individuals is required.

b. MAB can be not affected by environment, thus allowing the selection to be performed under any environmental conditions (e.g. greenhouse and off-season nurseries). This is very helpful for improvement of some traits (e.g. disease/pest resistance and stress tolerance) that are expressed only when favorable environmental conditions present. For low-heritability traits that are easily affected by environments, MAS based on reliable markers tightly linked to the QTLs for traits of interest can be more effective and produce greater progress than phenotypic selection.

c. MAB using co-dominance markers (e.g. SSR and SNP) can allow effective selection of recessive alleles of desired traits in the heterozygous status. No selfing or test crossing is needed to detect the traits controlled by recessive alleles, thus saving time and accelerating breeding progress.

d. For the traits controlled by multiple genes/QTLs, individual genes/QTLs can be identified and selected in MAB at the same time and in the same individuals, and thus MAB is particularly suitable for gene pyramiding. In traditional phenotypic selection, however, to distinguish individual genes/loci is problematic as one gene may mask the effect of additional genes.

e. Genotypic assays based on molecular markers may be faster, cheaper and more accurate than conventional phenotypic assays, depending on the traits and conditions, and thus MAB may result in higher effectiveness and higher efficiency in terms of time, resources and efforts saved.

The research and use of MAB in plants has continued to increase in the public and private sectors, particularly since 2000s. However, MAS and MABC were and are primarily constrained to simply-inherited traits, such as monogenic or oligogenic resistance to diseases/pests, although quantitative traits were also involved (Collard and Mackill, 2008; Segmagn et al., 2006; Wang and Chee, 2010). The application of molecular markers in plant breeding has not achieved the results as expected previously in terms of extent and success (e.g. release of commercial cultivars). Collard and Mackill (2008) listed ten reasons for the low impact of MAS and MAB in general. Improvement of most agronomic traits that are of complicated inheritance and economic importance like yield and quality is still a great challenge for MAB including the newly developed GS. From the viewpoint of a plant breeder, MAB is not universally or necessarily advantageous. The application of molecular technologies to plant breeding is still facing the following drawbacks and/or challenges:

a. Not all markers are breeder-friendly. This problem may be solved by converting of non-breeder-friendly markers to other types of breeder-friendly markers (e.g. RFLP to STS, sequence tagged site, and RAPD to SCAR, sequence characterized amplified region).

b. Not all markers can be applicable across populations due to lack of marker polymorphism or reliable marker-trait association. Multiple mapping populations are helpful in understanding marker allelic diversity and genetic background effects. In addition, QTL positions and effects also need to be validated and re-estimated by breeders in their specific germplasm (Heffner et al., 2009).

c. False selection may occur due to recombination between the markers and the genes/QTLs of interest. Use of flanking markers or more markers for the target gene/QTL can help.

d. Imprecise estimates of QTL locations and effects result in slower progress than expected. The efficiency of QTL detection is attributed to multiple factors, such as algorithms, mapping methods, number of polymorphic markers, and population type and size (Wang et al., 2012). High marker density fine mapping with large populations and well-designed phenotyping across multiple environments may provide more accurate estimates of QTL location and effects

e. A large number of breeding programs have not been equipped with adequate facilities and conditions for a large-scale adoption of MAB in practice.

f. The methods and schemes of MAB must be easily understandable, acceptable and implementable for plant breeders, unless they are not designed for a large scale use in practical breeding programs.

g. Higher startup expenses and labor costs.

With a long history of development, especially since the fundamental principles of inheritance were established in the late 19th and early 20th centuries, plant breeding has become an important component of agricultural science, which has features of both science and arts. Conventional breeding methodologies have extensively proven successful in development of cultivars and germplasm. However, subjective evaluation and empirical selection still play a considerable role in conventional breeding. Scientific breeding needs less experience and more science. MAB has brought great challenges, opportunities and prospects for conventional breeding. As a new member of the whole family of plant breeding, however, MAB, as transgenic breeding or genetic manipulation does, cannot replace conventional breeding but is and only is a supplementary addition to conventional breeding. High costs and technical or equipment demands of MAB will continue to be a major obstacle for its large-scale use in the near future, especially in the developing countries (Collard and Mackill, 2008; Ribaut et al., 2010). Therefore, integration of MAB into conventional breeding programs will be an optimistic strategy for crop improvement in the future. It can be expected that the drawbacks of MAB will be gradually overcome, as its theory, technology and application are further developed and improved. This should lead to a wide adoption and use of MAB in practical breeding programs for more crop species and in more countries as well.

REFERENCES

1. Anderson, J.A., S. Chao, and S. Liu, 2007: Molecular breeding using a major QTL for Fusarium head blight resistance in wheat. Crop Sci. 47, S-112–119.
2. Babu, R., S.K. Nair, B.M. Prasanna, and H.S. Gupta. 2004. Integrating marker-assisted selection in crop breeding – Prospects and challenges. Current Science 87: 607-619.
3. Babu, R., S.K. Nair, A. Kumar, S. Venkatesh, J.C. Sekhar, N.N Singh, G. Srinivasan, and H.S. Gupta. 2005. Two-generation marker-aided backcrossing for rapid conversion of normal maize lines to quality protein maize (QPM). Theor. Appl. Genet. 111: 888-897.
4. Bernacchi, D., T. Beck-Bunn, D. Emmatty, Y. Eshed, S. Inai, J. Lopez, V. Petiard, H. Sayama, J. Uhlig, D. Zamir, and S.D. Tanksley. 1998. Advanced backcross QTL analysis of tomato: II. Evaluation of near-isogenic lines carrying single-donor introgressions for desirable wild QTL-alleles derived from *Lycopersicon hirsutum* and *L. pimpinellifolium*. Theor. Appl. Genet. 97: 170-180.
5. Bernardo, R. 2010. Genomewide selection with minimal crossing in self-pollinated crops. Crop Sci. 50: 624-627.
6. Bernardo, R., and J. Yu. 2007. Prospects for genomewide selection for quantitative traits in maize. Crop Sci. 47: 1082-1090.
7. Bilyeu, K., L. Palavalli, D.A. Sleper, and P. Beuselinck. 2006. Molecular genetic resources for development of 1% linolenic acid soybeans. Crop Sci. 46: 1913-1918.
8. Bonnett, D.G., G.J. Rebetzke, and W. Spielmeyer. 2005. Strategies for efficient implementation of molecular markers in wheat breeding. Mol. Breeding 15: 75-85.
9. Botstein, D., R.L. White, M. Skolnick, and R.W. Davis. 1980. Construction of a genetic linkage map in man using restriction fragment length polymorphisms. Am. J. Hum. Genet. 32: 314-331.
10. Bradbury, L.M.T., T.L. Fitzgerald, R.J. Henry, Q. Jin, and Waters, D.L.E. (2005a). The gene for fragrance in rice. Plant Biotech. J. 3: 363–370.

11. Bradbury, L.M.T., R.J. Henry, Q. Jin, R.F. Reinke, and Waters, D.L.E. (2005b). A perfect marker for fragrance genotyping in rice. Mol. Breeding 16:279-283.
12. Buerstmayr, H., T. Ban, and J.A. Anderson. 2009. QTL mapping and marker-assisted selection for Fusarium head blight resistance in wheat: a review. Plant Breeding 128: 1-26.
13. Castro, A.J., F. Capettini, A.E. Corey, T. Filichkina, P.M. Hayes, A. Kleinhofs, D. Kudrna, K. Richardson, S. Sandoval-Islas, C. Rossi, and H. Vivar. 2003. Mapping and pyramiding of qualitative and quantitative resistance to stripe rust in barley. Theor. Appl. Genet. 107: 922-930.
14. Collard, B.C.Y., and D.J. Mackill. 2008. Marker-assisted selection: an approach for precision plant breeding in the twenty-first century. Phil. Trans. R. Soc. B 363: 557-572.
15. Collard, B.C.Y., M.Z.Z. Jahufer, JB Brouwer, and E.C.K. Pang. 2005. An introduction to markers, quantitative trait loci (QTL) mapping and marker-assisted selection for crop improvement: the basic concepts. Euphytica 142: 169-196.
16. Concibido, V.C., B. La Vallee, P. Mclaird, N. Pineda, J. Meyer, L. Hummel, J. Yang, K. Wu, and X. Delannay. 2003. Introgression of a quantitative trait locus for yield from *Glycine soja* into commercial soybean cultivars. Theor. Appl. Genet. 106: 575-582.
17. Cregan, P.B., J. Mudge, E.W Fickus, D. Danesh, R. Denny, and N.D. Young. 1999. Two simple sequence repeat markers to select for soybean cyst nematode resistance conditioned by the *rhg1* locus. Theor. Appl. Genet. 99: 811-818.
18. Crosbie, T.M., S.R. Eathington, G.R. Johnson, M. Edwards, R. Reiter, S. Stark, R.G. Mohanty, M. Oyervides, R.E. Buehler, A.K. Walker, R. Dobert, X. Delannay, J.C. Pershing, M.A. Hall, and K.R. Lamkey. 2006. Plant breeding: past, present and future. In: K.R. Lamkey and M. Lee (eds.), Plant Breeding: the Arnel R. Hallauer International Symposium, Blackwell Publishing, Oxford, UK, pp. 3-50.
19. Dwivedi, S.L., J.H. Crouch, D.J. Mackill, Y. Xu, M.W. Blair, M. Ragot, H.D. Upadhyaya, and R. Oritiz. 2007. The molecularization of public sector crop breeding: progress, problems and prospects. Advances in Agronomy 95: 163-318.
20. Eathington, S.R. 2005.Practical applications of molecular technology in the development of commercial maize hybrids. In: Proceedings of the 60th Annual Corn and Sorghum Seed Research Conferences, American Seed Trade Association, Washington, D.C.
21. Edwards, D., J.W. Forster, D. Chagne, and J. Batley. 2007. What is SNPs? In: N.C. Oraguzie, E.H.A. Rikkerink, S.E. Gardiner and H.N. de Silva (eds.), Association Mapping in Plants. Springer, Berlin, pp 41-52.
22. Farooq, S., and F. Azam. 2002a. Molecular markers in plant breeding-I: Concepts and characterization. Pakistan J Biol. Sci. 5: 1135-1140.
23. Farooq, S., and F. Azam. 2002b. Molecular markers in plant breeding-II: Some pre-requisites for use. Pakistan J Biol. Sci. 5: 1141-1147.
24. Flint-Garcia, S.A., L.L. Darrah, M.D. McMullen, and B.E. Hibbard. 2003. Phenotypic versus marker-assisted selection for stalk strength and second-generation European corn borer resistance in maize. Theor. Appl. Genet. 107: 1331-1336.
25. Frisch, M., M. Bohn, and A.E. Melchinger. 1999. Comparison of selection strategies for marker-assisted backcrossing of a gene. Crop Sci. 39: 1295-1301 (See also Errata, Frisch et al., 1999. Crop Sci. 39: 1903).
26. Fulton, T.M., P. Bucheli, E. Voirol, J. Lopez, V. Peetiard, and S.D. Tanksley. 2002. Quantitative trait loci (QTL) affecting sugars, organic acids and other biochemical properties possibly contributing to flavor, identified in four advanced backcross populations of tomato. Euphytica 127: 163-177.
27. Goddard, M.E., and B.J. Hayes. 2007. Genomic selection. J. Anim. Breed. Genet..124: 323-330.
28. Guo, Z., D.M. Tucker, J. Liu, V. Kishore, and G. Gay. 2011. Evaluation of genomewide selection efficiency in maize nested association mapping populations. Theor. Appl. Genet. 124: 261-275.

29. Gupta, P.K., J.K. Roy, and M. Prasad. 2001. Single nucleotide polymorphisms: a new paradigm for molecular marker technology and DNA polymorphism detection with emphasis on their use in plants. Current Science 80: 524-535.
30. Heffner, E.L., M.E. Sorrells, and Jannink, J.-L. 2009. Genomic selection for crop improvement. Crop Sc. 49: 1-12.
31. Heffner, E.L., A.J. Lorenz, Jannink, J.-L. and M.E. Sorrells. 2010. Plant breeding with genomic selection: gain per unit time and cost. Crop Sci. 50: 1681-1690.
32. Heffner, E.L., Jannink, J.-L., H. Iwata, E. Souza, and M.E. Sorrells. 2011. Genomic selection accuracy for grain quality traits in biparental wheat populations. Crop Sci. 51: 2597-2606.
33. Hospital, F. 2003.Marker-assisted breeding. In: H.J. Newbury (ed.), Plant molecular breeding. Blackwell Publishing and CRC Press, Oxford and Boca Raton, pp. 30-59.
34. Hospital, F., C. Chevalet, and P. Mulsant. 1992. Using markers in gene introgression breeding programs. Genetics 231: 1199-1210.
35. Hospital, F., and G. Decoux. 2002. Popmin: a program for the numerical optimization of population sizes in marker-assisted backcross breeding programs. Journal of Heredity 93: 383-384.
36. Huang, N., E.R. Angeles, J. Domingo, G. Magpantay, S. Singh, G. Zhang, N. Kumaravadivel, J. Bennet, and G.S. Khush. 1997. Pyramiding of bacterial blight resistance genes in rice: marker-assisted selection using RFLP and PCR. Theor. Appl. Genet. 95: 313-320.
37. Jannink, J.-L., A.J. Lorenz, and H. Iwata. 2010. Genomic selection in plant breeding: from theory to practice. Briefings in Functional Genomics 9: 166-177.
38. Jefferies, S.P., B.J. King, A.R. Barr, P. Warner, S.J. Logue, and P. Langridge. 2003. Marker-assisted backcross introgression of the *Yd2* gene conferring resistance to barley yellow dwarf virus in barley. Plant Breeding 122: 52-56.
39. Jiang, G.-L., Shi, J., & Ward, R. W. (2007a). QTL analysis of resistance to Fusarium head blight in the novel wheat germplasm CJ 9306. I. Resistance to fungal spread. Theor. Appl. Genet. 116: 3-13.
40. Jiang, G.-L., Dong, Y., Shi, J., & Ward, R. W. (2007b). QTL analysis of resistance to *Fusarium* head blight in the novel wheat germplasm CJ 9306. II. Resistance to deoxynivalenol accumulation and grain yield loss. Theor. Appl. Genet. 115: 1043-1052.
41. Johnson, R. 2004. Marker assisted selection. In: J. Jannick (Ed.), Plant Breed. Rev. Vol. 24, Part 1, 293-309.
42. Joshi, R.K., and S. Nayak. 2010. Gene pyramiding – A broad spectrum technique for developing durable stress resistance in crops. Biotechol. Mol. Biol. Rev. 5: 51-60.
43. Konieczny, A., and F. Ausubel. 1993. A procedure for mapping Arabidopsis mutations using co-dominant ecotype-specific PCR based markers. The Plant Journal 4: 403-410.
44. Kumpatla, S.P., R. Buyyarapu, I.Y. Abdurakhmonov, and J.A. Mammadov. 2012. Genomics-assisted plant breeding in the 21st century: technological advances and progress. In: I.Y. Abdurakhmonov (ed.), Plant Breeding, InTech, pp 131-184.
45. Lecomte, L., P. Duffé, M. Buret, B. Servin, F. Hospital, and M. Causse. 2004. Marker-assisted introgression of five QTLs controlling fruit quality traits into three tomato lines revealed interactions between QTLs and genetic backgrounds. Theor. Appl. Genet. 109: 568-668.
46. Li, X., Y. Han, W. Teng, S. Zhang, K. Yu, V. Poysa, T. Anderson, J. Ding, and W. Li. 2010. Pyramided QTL underlying tolerance to Phytophthora root rot in mega-environment from soybean cultivar 'Conrad' and 'Hefeng 25'.Theor. Appl. Genet. 121: 651-658.
47. Liu, B.-H. 1991. Development and prospects of dwarf male-sterile wheat. Chinese Science Bulletin, 36(4), 306.
48. Lorenzana, R.F., and R. Bernardo. 2009. Accuracy of genotypic predictions for marker-based selection in biparental plant populations. Theor. Appl. Genet. 120: 151-161.

49. Luo, Y., J.S. Sangha, S. Wang, Z. Li, J. Yang, and Z. Yin. 2012. Marker-assisted breeding of *Xa4*, *Xa21* and *Xa27* in the restorer lines of hybrid rice for broad-spectrum and enhanced disease resistance to bacterial blight. Mol. Breeding DOI 10.1007/s11032-012-9742-7.
50. Meuwissen, T. 2007. Genomic selection: marker assisted selection on a genome wide scale. J. Anim. Breed. Genet. 124: 321-322.
51. Meuwissen, T.H.E., B.J. Hayes, and M.E. Goddard. 2001. Prediction of total genetic value using genome wide dense marker maps. Genetics 157: 1819-1829.
52. Miedaner, T., F. Wilde, B. Steiner, H. Buerstmayr, V. Korzun, and E. Ebmeyer, 2006. Stacking quantitative trait loci (QTL) for Fusarium head blight resistance from non-adapted sources in an European elite spring wheat background and assessing their effects on deoxynivalenol (DON) content and disease severity. Theor. Appl. Genet. 112: 562-569.
53. Moreau, L., A. Charcosset, F. Hospital, and A. Gallais. 1998. Marker-assisted selection efficiency in populations of finite size. Genetics 148: 1353-1365.
54. Mullis, K. 1990. The unusual origin of the polymerase chain reaction. Scientific American 262 (4): 56–61, 64–65.
55. Nakaya, A., and S.N. Isobe. 2012. Will genomic selection be a practical method for plant breeding? Annals of Botany,
56. Newbury, H.J. 2003. Plant Molecular Breeding. Blackwell Publishing, CRC Press, Birmingham, UK.
57. Payne, P.I., M.A. Nigtingale, A.F. Krattiger, and L.M. Holt. 1987. The relationship between HMW glutenin subunit composition and the bread-making quality of British-grown wheat varieties. Journal of the Science of Food and Agriculture 40: 51-65.
58. Pham, A.-T., J.G. Shannon, and K.D. Bilyeu. 2012. Combinations of mutant FAD2 and FAD3 genes to produce high oleic acid and low linolenic acid soybean oil. Theor. Appl. Genet. 125: 503–515.
59. Pumphrey, M. O., R. Bernardo, and J. A. Anderson, 2007: Validating the *Fhb1* QTL for Fusarium head blight resistance in near-isogenic wheat lines developed from breeding populations. Crop Sci. 47: 200-206.
60. Ragot, M., M. Biasiolli, M.F. Dekbut, A. Dell'orco, L. Margarini, P. Thevenin, J. Vernoy, J. Vivant, R. Zimmermann, and G. Gay. 1995. Marker-assisted backcrossing: a practical example. In: A. Berville and M. Tersac (eds.), Les Colloques, No. 72, INRA, Paris, 45-46.
61. Ragot, M., G. Gay, J.P. Muller, and J. Durovray. 2000. Efficient selection for the adaptation to the environment through QTL mapping and manipulation in maize. In: J.M. Ribaut and D. Poland (eds.), Molecular Approaches for the Genetic Improvement of Cereals for Stable Production in Water-limited Environments. CIMMYT, Mexico, pp 128-130.
62. Ribaut, J.M., and J. Betran. 1999. Single large-scale marker-assisted selection (SLS-MAS). Mol. Breeding 5: 531-541.
63. Ribaut, J.M., G. Edmeades, E. Perotti, and D. Hoisington. 2000. QTL analysis, MAS results and perspectives for drought-tolerance improvement in tropical maize. In: J.M. Ribaut and D. Poland (eds.), Molecular Approaches for the Genetic Improvement of Cereals for Stable Production in Water-limited Environments. CIMMYT, Mexico, pp 131-136.
64. Ribaut, J.M., C. Jiang, and D. Hoisington. 2002. Simulation experiments on efficiencies of gene introgression by backcrossing. Crop Sci. 42: 557-565.
65. Ribaut, J.M., M.C. de Vicente, and X. Delannay. 2010. Molecular breeding in developing countries: challenges and perspectives. Current Opinion in Plant Biology 13: 1-6.
66. Richardson, K.L., M.I. Vales, J.G. Kling, C.C. Mundt, and P.M. Hayes. 2006. Pyramiding and dissecting disease resistance QTL to barley stripe rust. Theor. Appl. Genet. 113: 485-495.

67. Rogers, W.J., P.I. Payne, and Harinder K. (1989). The HMW glutenin subunit and gliadin compositions of German-Grown wheat varieties and their relationship with bread-making quality. Plant Breeding 103: 89-100.
68. Sebolt, A.M., R.C. Shoemaker, and B.W. Diers. 2000. Analysis of a quantitative trait locus allele from wild soybean that increases seed protein concentration in soybean. Crop Sci. 40: 1438-1444.
69. Semagn, K., A. Bjornstad, and M.N. Ndjiondjop. (2006a). An overview of molecular marker methods for plants. Afr. J. Biotechnol. 5: 2540-2568.
70. Semagn, K., A. Bjornstad, and M.N. Ndjiondjop. (2006b). Progress and prospects of marker assisted backcrossing as a tool in crop breeding programs. Afr. J. Biotechnol. 5: 2588-2603.
71. Servin, B., and F. Hospital. 2002. Optimal positioning of markers to control genetic background in marker-assisted backcrossing. The Journal of Heredity 93: 214-217.
72. Servin, B., O.C. Martin, M. Mezard, and F. Hospital. 2004. Toward a theory of marker-assisted gene pyramiding. Genetics 168: 513-523.
73. Shi, A., P. Chen, D. Li, C. Zheng, B. Zhang, and A. Hou. 2009. Pyramiding multiple genes for resistance to soybean mosaic virus in soybean using molecular markers. Mol. Breeding 23: 113-124.
74. Singh, S., J.S. Sidhu, N. Huang, Y. Vikal, Z. Li, D.S. Brar, H.S. Dhaliwal, and G.S Khush. 2001. Pyramiding three bacterial blight resistance genes (*xa5, xa13* and *Xa21*) using marker-assisted selection into indica rice cultivar PR106. Theor. Appl. Genet. 102: 1011-1015.
75. Sobrino, B., M. Briona, and A. Carracedoa. 2005. SNPs in forensic genetics: a review on SNP typing methodologies. Forensic Science International 154: 181-194.
76. Somers, D.J., J. Thomas, R. DePauw, S. Fox, G. Humphreys, and G. Fedak. 2005. Assembling complex genotypes to resist *Fusarium* in wheat (*Triticum aestivum* L.). Theor. Appl. Genet. 111: 1623-1631.
77. Song, Q., G. Jia, Y. Zhu, D. Grant, R.T. Nelson, , Hwang, E.-Y., D.L. Hyten, and P.B. Cregan. 2010. Abundance of SSR motifs and development of candidate polymorphic SSR markers (BARCSOYSSR_1.0) in soybean. Crop Sci. 50: 1950-1960.
78. Southern, E. M. (1975). Detection of specific sequences among DNA fragments separated by gel electrophoresis. Journal of Molecular Biology 98: 503-517.
79. Stuber, C.W., M. Polacco, and M.L. Senior. 1999. Synergy of empirical breeding, marker-assisted selection and genomics to increase crop yield potential. Crop Sci. 39: 1571-1583.
80. Tanksley, S.D., and J.C. Nelson. 1996. Advanced backcross QTL analysis: a method for the simultaneous discovery and transfer of valuable QTLs from unadapted germplasm into elite breeding lines. Theor. Appl. Genet. 92: 191-203.
81. Tanksley, S.D., N.D. Young, A.H. Paterson, and M.W. Bonierbale. 1989. RFLP mapping in plant breeding: new tools for an old science. Bio. Technology 7: 257-263.
82. Visscher, P.M., R. Thompson, and C.S. Haley. 1996. Confidence intervals in QTL mapping by booststrapping. Genetics 143: 1013-1020.
83. Wang, B., and P.W. Chee. 2010. Application of advanced backcross quantitative trait locus (QTL) analysis in crop improvement. Journal of Plant Breeding and Crop Science 2: 221-232.
84. Wang, X., Jiang, G.-L., M. Green, R.A. Scott, D.L. Hyten, and P.B. Cregan. 2012. Quantitative trait locus analysis of saturated fatty acids in a population of recombinant inbred lines of soybean. Mol. Breeding, DOI 10.1007/s11032-012-9704-0.
85. Wong, C.K., and R. Bernardo. 2008. Genome wide selection in oil palm: increasing selection gain per unit time and cost with small populations. Theor. Appl. Genet. 116: 815-824.
86. Xu, Y. 2010. Molecular plant breeding. CAB International.
87. Ye, G., and K.F. Smith. 2008. Marker-assisted gene pyramiding for inbred line development: Basic principles and practical guidelines. Intern. J. Plant Breed. 2: 1-10.

88. Yi, C., W. Guo, X. Zhu, L. Min, and T. Zhang. 2004. Pyramiding breeding by marker-assisted recurrent selection in upland cotton II. Selection effects on resistance to *Helicoverpa armigera*. Scientia Agricultura Sinica 37: 801-807.
89. Young, N.D. 1999. A cautiously optimistic vision for marker-assisted breeding. Mol. Breeding 5: 505-510.
90. Young, N.D., and S.D. Tanksley. 1989. RFLP analysis of the size of chromosomal segments retained around the *Tm-2* locus of tomato during backcross breeding. Theor. Appl. Genet. 77: 353-359.
91. Zhao, X., G. Tan, Y. Xing, L. Wei, Q. Chao, W. Zuo, T. Lübberstedt, and M. Xu. 2012. Marker-assisted introgression of *qHSR1* to improve maize resistance to head smut. Mol. Breeding, DOI 10.1007/s11032-011-9694-3
92. Zhong, S., J.C.M. Dekkers, R.L. Fernando, and Jannink, J.-L. (2009). Factors affecting accuracy from genomic selection in populations derived from multiple inbred lines: a barley case study. Genetics 182: 355-364.

CHAPTER 5

Breeding Strategies for Improving Rice Yield—A Review

Mudasir Hafiz Khan[1*], Zahoor Ahmad Dar[2], Sher Ahmad Dar[2]

[1]*Department of Agriculture, Lal Mandi, Srinagar (J&K)*

[2]*Division of Plant Breeding and Genetics, S. K. University of Agricultural Sciences & Technology of Kashmir, Shalimar Campus, Srinagar (J&K)*

ABSTRACT

An increase in productivity is always one of the main goals of any crop breeding program including rice. However, many goals can be identified for this crop varying in importance from region to region, country to country, and even within a given country. Increase in grain yield potential is the major goal of almost all rice breeders programs. The major impacts are related to the development of new strategies to increase the genetic grain yield potential of the varieties. Rice breeders have been very successful in improving the crop. Some milestones are the contribution to the green revolution with the semi-dwarf varieties, the new rice plant type and hybrid rice. The main breed- ing method used to improve rice is the pedigree, but development of hybrids and population improvement are added to the breeder's portfolio. Breeders have been taking advantage of biotechnology tools to enhance their breeding capacity; however, many programs are still struggling on how to integrate them into the breeding programs and how to balance the allocation of resources between conventional and modern tools.

Keywords: Rice, Hybridization, Ideotype, Heterosis, Male Sterility, Wide Hybridization, Genetic Engineering, MAS Breeding

1. INTRODUCTION

Rice (Oryza sativa L., 2n = 2x = 24) is second most widely grown cereal crop and the staple food for more than half of world's population, providing two thirds of calorie intake for more than three billion people in Asia and one-third of calorie intake of nearly 1.5 billion people in Africa and Latin America. It is probably a descendent of wild grass that was most likely cultivated in the foothills of the far Eastern Himalayas. Some believes that the rice plant may have originated in southern India, then spread to the north of the country and then onwards to China. It then arrived in Korea, the Philippines (about 2000 B. C.) and then Japan and Indonesia (about 1000 B. C.). When Alexander the Great invaded India in 327 B. C., it is believed that he took rice back to Greece. Arab travellers took it to Egypt, Morocco and Spain and that is how it travelled all across Europe. Portugal and Netherlands took rice to their colonies in West Africa and then it travelled to America through the "Columbian Exchange" of natural resources. But as is traditionally known, rice is a slow starter and this is also true to the fact that it took close to two centuries after the voyages of Columbus for rice to take root in the Americas. Thereafter the journey of rice continues with the Moors taking it to Spain in 700 A. D. and then the Spanish brought rice to South America at the beginning of 17th century.

The average growth rate of rice yield was 3.68% per year in the early 1980's but has decreased to 0.74% per year in the late 1990's. Top ten rice producing countries are given in Table 1. China and India together produced 364.73 m·t from an area of 74.60 m·ha. However, the average productivity of China (6.72 t/ha) is much higher than that of India (3.61 t/ha).

In India rice provides 43% of calorie requirement for more than 70% of population. Ten states of India contributed 77.49% (816.11 Lt) of total rice production during 2012 (Table 2). The largest production of 146.06 Lt came from West Bengal followed by 140.22 Lt from UP and 128.95 Lt from Andhra Pradesh.

Table 1. Top ten rice producing countries of world.

S No.	Country	Rice Production (000 t)	Area (000 ha)	Yield (t/ha)
1	China	205,714	30,600	6.72
2	India	159,016	44,000	3.61
3	Indonesia	59,370	12,160	4.88
4	Vietnam	44,480	7780	5.72
5	Thailand	31,061	10,900	2.85
6	Myanmar	18,984	7065	2.69
7	Philippines	19,365	4890	3.96
8	Brazil	12,500	2450	5.10
9	Japan	10,577	1590	6.65
10	USA	9661	1154	8.37

Source: USDA 2014.

Table 2. Top ten rice producing states of India.

S.No	Country	Rice Production (Lt)	Area (Lha)	Yield (t/ha)
1	West Bengal	146.06	54.34	2.69
2	Andhra Pradesh	128.95	40.96	3.15
3	Uttar Pradesh	140.22	59.47	2.36
4	Punjab	105.42	28.18	3.74
5	Orissa	58.07	44.45	1.31
6	Tamil Nadu	74.59	19.04	3.92
7	MP	22.27	16.62	1.34
8	Bihar	71.63	33.24	2.15
9	Haryana	37.59	12.35	3.04
10	Jharkhand	31.31	14.69	2.13
	Total (All India)	1053.122	440.068	2393

Rice Production in Jammu and Kashmir

In the State of Jammu & Kashmir, rice being the principal food crop, is cultivated across diverse agro-ecological situations extending from sub-tropical region (>1000 m·asl) of Jammu to temperate cold high altitude areas (1800 - 2500 m·asl) of Kashmir. The trends in area, production and productivity over years are presented in Figure 1. During 2012, 5.45 Lt of rice was grown from 2.63 L·ha with an average yield of about 2.07 t/ha. In Kashmir the production environment is typically temperate, with cold winters and mild summers, characterized by short growing season, long days, cool nights and occurrence of wide differences between maximum and minimum temperatures. Compared with average yield of temperate regions of world (5.8 t/ha), the rice productivity is disappointingly very low (2.5 t/ha) in Kashmir. The rice area in Kashmir can be broadly categorized into two distinct zones with entirely different agro-climatic conditions. The higher belts, comprising areas between 1800 - 2500 m·asl, constitute 10% - 12% of total rice area. The growing season is restricted to 125 - 135 days. Besides cool soil and atmospheric temperatures cold water from melting snow is directly used for irrigation purposes leading to an environmentally handicapped ecology. The lower belt or valley basin enjoys relatively favourable growing conditions of 140 - 145 days and lies between 1500 - 1800 m·asl. The varietal requirement of two zones is different with Japonica rice being more predominantly grown in upper belts and Indicas in the valley basin.

The ultimate goal of crop breeding is to develop varieties with high yield potential and desirable agronomic characteristics. In rice breeding, the most important qualities sought by breeders have been high yield potential; resistance to major diseases and insects; and improved grain and eating quality. However, there seems to be some conflict between these aims. Emphasis on high grain quality tends to result in unstable yields. Conversely, too much emphasis on disease and insect resistance and stable yields leads to poor grain quality. Hence, breeding efforts should concentrate on varieties with the potential to minimize yield losses under unfavorable conditions, and to maximize yields when conditions are favorable. Various breeding strategies for increasing the yield potential include:

1) Conventional Hybridization

2) Ideotype Breeding

3) Heterosis Breeding

4) Male Sterility

5) Wide Hybridization

6) Genetic Engineering

7) Molecular Marker Assisted Breeding

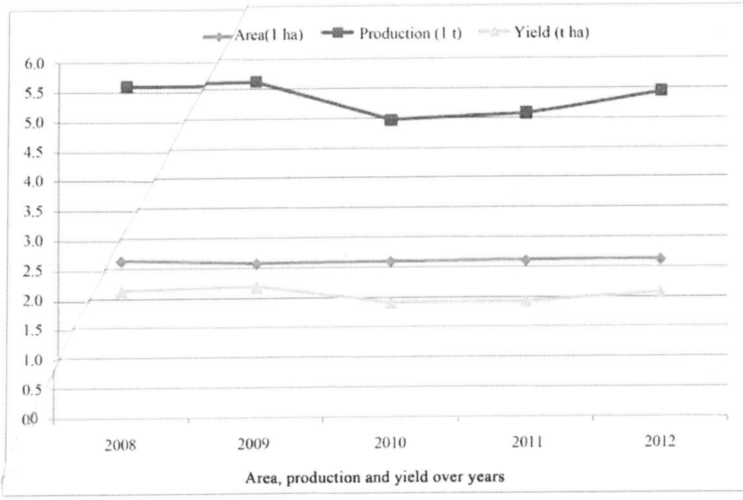

Figure 1. Trends in area, production and yield of rice over years.

2. CONVENTIONAL HYBRIDIZATION AND SELECTION PROCEDURES

This is the time tested strategy for selecting crop cultivars with higher yield potential. It has two phases. The first phase involves the creation of variability through hybridization between diverse parents. In the second phase desirable individuals are selected on the basis of field observations and yield trials. It has been estimated that on the average about 1.0% increase has occurred per year in the yield potential of rice over a 35 year period since the development of first improved variety of rice, IR8 [1] . The yields of crops where there is enough investment in research have been continuously increased and there is no reason why further increases cannot be attained.

Submergence Tolerance through Conventional Breeding Approach

The traditional rice varieties have evolved suitable mechanism(s) to survive under this stress. Systematic screening of rice germplasm has shown that there are excellent flood-tolerant rice types locally available. Among these are especially the "FR13A" and "FR43B" of India, "Kurkaruppan" of Sri

Lanka and "Goda Heenati" of Indonesia. Detailed study on such contrasting rice types has indicated that flooding tolerance is controlled by one or a few genes with major effects and additional genes with smaller, modifying effects. Breeding efforts at IRRI led to evolution of tolerant lines with good agronomic traits. The line "IR 49830-7-2-2" combines high tolerance levels with higher yield potential and resistance to diseases and insect pests and it has been extensively used as a donor parent in the breeding programme. "Sudhir" is another variety which has been developed from the "FR13A" × "Biraj" crosses. This variety has been released by the "Central Variety Release Committee, Indian Council of Agricultural Research, Government of India" in 1999.

In spite of the above developments, till date no variety has been developed which combines desirable levels of flooding tolerance with grain yield. It is attributed to the fact that "FR 13A" is a poor combiner for yield and other agronomic traits. Apart from "FR 13A", "BKNFR 76106-16-0-1-0" and "Kurkaruppan" have major dominant gene for submergence tolerance. Flooding tolerant rice "Goda Heenati" apparently does not have the same submergence tolerance gene. It has been suggested that there are at least three submergence tolerance genes in "FR13A", "Kurkaruppan" and "Goda Heenati". Further, though "FR13A" is considered to be an excellent source of submergence tolerance especially at young seedling stage (10 days), "CN 540 (Suresh)" is found to be more resistant than "FR13A" and "FR43B" might result in genotypes with better submergence tolerance.

Major attempts have also been made in breeding for improved submergence tolerance through the use of double haploid lines (DHLs), developed using crosses between submergence tolerant and sensitive rice cultivars [2] . Two DHL populations for submergence tolerance have been developed at IRRI using the cross combination of 1) "IR 49830 × CT 6241" and 2) "FR13A × IR 42". These populations were evaluated simultaneously in Thailand and Philippines. In Thailand, a distinct bimodal distribution of lines for survival after submergence was observed supporting the concept of a single gene for submergence tolerance [3] . However, the result from Philippines did not support this theory.

Use of DHLs has recently been reported for identification of two RFLP (restriction fragment length polymorphism) markers for submergence tolerance of rice which are mapped to a segment of chromosome 9. This chromosome segment [Sub 1(t)] accounted for 70% of the phenotypic variance in submergence tolerance of this population [4] . It would be rewarding if the Sub 1(t) locus is cloned through map-based cloning technique. Recently, a high resolution map has been constructed around the Sub 1(t) locus through the use of RFLP and AFLP (amplified fragment length polymorphism) markers. Present attempts are focused on cloning of tightly-linked AFLP markers for eventual screening of gene libraries [5] .

3. IDEOTYPE BREEDING

Ideotype breeding aimed at modifying the plant architecture is a time tested strategy to achieve increases in yield potential. Thus selection for short statured cereals such as wheat, rice, and sorghum resulted in doubling of yield potential. Yield potential is determined by the total dry matter or biomass and the harvest index (HI). Tall and traditional rice had HI of around 0.3 and total biomass of about 12 tons per hectare. Thus their maximum yield was 4 tons per hectare. Their biomass could not be increased by application of nitrogenous fertilizers as the plants grew excessively tall, lodged badly and the yield decreased instead of increasing. To increase the yield potential of topical rice it was necessary to improve the harvest index and nitrogen responsiveness by increasing the lodging resistance. This was accomplished by reducing the plant height through incorporation of a recessive gene sd1 for short stature.

The first short statured variety IR8, developed at the International Rice Research Institute (IRRI) also had a combination of other desirable traits such as profuse tillering, dark green and erect leaves for good canopy architecture and sturdy stems. It responded to nitrogenous fertilizer much better and had a higher biomass (about 18 tons) and HI of 0.45. Its yield potential was 8 - 9 tons per hectare [6] .

To increase the yield potential of rice further, a new plant type was conceptualized in 1988. Modern semi- dwarf rice produce a large number of unproductive tillers and excessive leaf area which cause mutual shading and reduce canopy photosynthesis and sink size, especially when they are grown under direct sowing conditions. To increase the yield potential of these semi-dwarf rice, IRRI scientists proposed further modifications of plant architecture with following characteristics:

Ø Low tillering, (9 - 10 tillers for transplanted conditions)

Ø No unproductive tillers

Ø 200 - 250 grains per panicle

Ø Dark green, thick and erect leaves

Ø Vigorous and deep root system

This proposed ideotype became the "New Plant Type" (NPT) highlighted in IRRI's strategic plan [7] and breeding efforts to develop NPT were initiated in 1990. The objective was to develop improved germplasm with 15% - 20% higher yield than the existing high yielding varieties. Numerous breeding lines with desired ideotype were developed [8] and shared with the national rice improvement programs. Three NPT lines have been released in China and one in Indonesia. Other NARS are evaluating and further improving NPT lines.

Reference [9] reported that First-generation new plant type (NPT) lines developed from tropical japonica at IRRI did not yield well because of limited biomass production and poor grain filling. Progress has been made in second-generation NPT lines developed by crossing elite indica with improved tropical japonica. Several second-generation NPT lines out yielded the first-generation NPT lines and indica check varieties. China's "super" rice breeding project has developed many F1 hybrid varieties using a combination of the ideotype approach and inter-subspecific heterosis. These hybrid varieties produced grain yield of 12 $t{\cdot}ha^{-1}$ in on-farm demonstration fields, 8% - 15% higher than the hybrid check varieties. The success of China's "super" hybrid rice was partially the result of assembling the good components of IRRI's NPT design in addition to the use of inter-subspecific heterosis.

The following plant type has been proposed by Chineses scientists to be combined with heterosis breeding to produce super rice hybrids capable of yielding 100 kg/ha yield per day.

Ø Plant height about 100 cm, with culum length of 70 cm.

Ø Top three leaves;

Ø Long: Flag leaf is 50 cm and 2nd and 3rd comes 55 cm in length. The top two leaves are higher than the top of the panicle.

Ø Erect: Leaf angle of the flag, –2nd and –3rd leaves are around 50, 100 and 200, respectively.

Ø Narrow and V shaped: The leaves look narrow but they are wide when flattened.

Ø Thick: 55 $g{\cdot}m^{-2}$ of specific leaf weight for the top three leaves.

Ø Plant type: Moderately compact type with moderate tillering capacity; droopy panicles, the panicle top about 60 cm from the ground after filled; erect leaf callopy without appearance of panicle.

Ø Panicle weight and number; grain weight 5 g per panicle, 270 - 300 panicles m^{-2}.

Ø Leaf area index (LAI): The LAI of top three leaves is about 6

Ø Harvest index: about 0.55

4. HETEROSIS BREEDING

Rice hybrids with a yield advantage of about 10% - 15% over best inbred varieties were introduced in China in mid 1970s and are now planted to about 45% of the rice land in that country. Rice hybrids adapted to tropics have now been bred at IRRI and by NARS and show similar yield advantage. The increased yield advantage of tropical rice hybrids is due to increased biomass, higher spikelet number and to some extent higher grain weight. Increased adoption of hybrids in the tropics should contribute to increased productivity.

4.1. Improving Parental Lines to Increase Efficiency of Hybrid Rice Breeding

Because of the potential of hybrid rice to increase rice production and productivity, many countries are currently working to exploit the benefits of this technology. The major problems in hybrid rice breeding are the limited number of parental lines with specific desirable traits, a lower frequency of maintainers and restorers among elite breeding lines, the narrow genetic base, a lack of resistance to biotic stresses, and poor grain quality of some parental lines. Therefore, improving parental lines must be an integral part of hybrid rice breeding to develop heterotic hybrids and improve breeding efficiency. The frequency of maintainers and restorers has been significantly increased after initiating specific maintainer and restorer breeding programs. Diverse parental lines with better grain quality and multiple resistances have been developed in China and at IRRI. Random mating of composite populations of maintainers and restorers developed at IRRI has helped to widen the genetic base of parental lines. Transferring restorer genes into tropical japonicas and incorporating wide compatibility genes into promising elite lines are considered essential to developing indica/japonica hybrids. Thermosensitive genic male sterile lines are being generated in indica, tropical japonica, and basmati genetic backgrounds. Anther culture and marker-aided selection can be deployed to expedite the parental line improvement program.

4.2. Using Tropical Japonica Germplasm to Enhance Heterosis in Rice

The current level of heterosis (15% - 20% or 0.75 - 1.0 $t \cdot ha^{-1}$) in indica rice hybrids developed for the tropics is economically viable, but a higher level would be more attractive. Studies at IRRI have showed a higher heterosis for yield in tropical japonica/indica crosses than in indica/indica crosses.

Increased heterosis for yield in tropical japonica/indica crosses is possible only if they exhibit normal spikelet fertility. Several tropical japonica rice cultivars that possess the wide compatibility (WC) gene have been identified. Suitable parental lines in improved tropical japonica germplasm possessing the WC gene are being bred at IRRI. These parental lines would help to develop heterotic tropical japonica/indica hybrids.

4.3. The Wide Compatibility System

The genetic basis for hybrid sterility is presented with methods to analyze it. Several independent loci have been identified that cause partial abortion of female gametes carrying an allele in heterozygotes. Similar genetic mechanisms were also identified for male gametes that affect pollen fertility. The hybrid sterility in indica/japonica crosses was mostly attributed to an S-5 locus. Therefore, a single wide compatibility allele can be used to obtain fertile hybrids between the two groups. But hybrid sterility in crosses between rice cultivars from the Indian subcontinent and other areas was controlled by several loci, as detected in Basmati 370 crosses. In hybrid rice breeding, hybrid sterility genes in cytoplasmic male sterile lines affect the screening of potential restorers but not maintainers. It is important to analyze hybrid sterility genes in maintainers to achieve enhanced heterosis in hybrid rice breeding.

5. MALE STERILITY

In rice, there are three major types of CMS/restoration systems, including CMS-BT (borotype), CMS-WA (wild abortive), and CMS-HL (Honglian). The first commercially used CMS-WA germplasm was discovered by Chinese scientist Long Ping Yuan in 1970's and was used to develop the three-line system hybrid rice. As a gametophytic system, CMS-BT is the most widely investigated rice CMS system at the genetic level, and is originally derived from the cytoplasm of an indica rice variety Chinsurash Boro II. Recently, Dr. Yaoguang Liu's laboratory published the results of a study, in which they cloned the CMS and two Rf genes in the CMS-BT/restora- tion system and elucidated the molecular mechanism for male sterility and fertility restoration [10] . This is a significant contribution to hybrid rice dedicated by Chinese scientists. A number of fertility restorer genes that are involved in other CMS systems have been studied genetically. Two Rf loci, Rf5 and Rf6 (t), responding to CMS-HL, have been located in different regions of chromosome 10 [11] . Both the CMSDT restorer gene Rf-d1 (t) and the CMS-DA restorer gene Rf-d (t) were also detected on chromosome 10 [12] [13] .

Two CMS-WA restorer genes Rf3 (t) and Rf4 (t) have been located on chromosomes 1 and 10, respectively [14] .

5.1. Diversifying the CMS System to Improve the Sustainability of Hybrid Rice Technology

Cytoplasmic male sterility (CMS) resulting from nuclear-cytoplasm interaction has been commercially exploited for the production of F1 hybrid seed in a number of crops such as maize, sorghum, sunflower and sugarbeet. More recently, major progress has been made in developing high-yielding hybrid rice varieties based on the CMS system, occupying over 18 million hectares in China. The first CMS line used to develop commercial F1 hybrids were developed in China in 1973 from a single male sterile plant (Oryza sativa f. spontanea) designated as wild abortive (WA). Since then, more than 20 CMS sources have been developed from various accessions of cultivated rice and wild species. A large number of CMS lines have been produced in China, at IRRI, and in several other countries from these cytosterility systems. But 95% of the total area planted to hybrid rice has a single CMS source, WA cytoplasm. This cytoplasmic uniformity of hybrid varieties could result in disease epidemics, such as the outbreak of Southern corn blight due to Helminthosporium maydis on US maize hybrids carrying T-type cytoplasm. To overcome the danger of genetic vulnerability of hybrid rice to diseases and insects, we urgently need to diversify the cytoplasm of the male sterility sources. Conventional backcrossing and somatic cell hybridization are possible avenues to achieve diversification using cultivated and wild species as CMS donors. Two major problems are 1) the lack of efficient techniques for characterizing CMS sources and 2) non-availability of effective restorers, particularly when the cytoplasmic donors (wild species) and recipient parents are distantly related. Under such situations, the donor species itself should be explored as a possible restorer. CMS sources are distinguished from each other on the basis of the fertility of crosses of CMS lines with restorers.

These procedures are laborious and fertility is affected by environment. Advances in molecular biology offer potential to develop precise, reliable, and quick molecular techniques to characterize different CMS sources. Fertile and male sterile lines including different sources of CMS can be distinguished from each other using molecular diagnostic probes based on the restriction endonuclease fragment pattern of mtDNA, including the nature of mitochondrial protein products and ultrastructure of the mitochondrial genome. But the role of these elements in producing cytosterility is not well understood. Protoplast fusion offers promise to produce cybrids and new alloplasmic male sterile lines. Moreover, this technology can further enhance efficiency through the quick transfer of

CMS from one source to other elite breeding lines. Otherwise, this requires 5 - 6 generations of conventional backcrossing. Both two-line (photoperiod-sensitive and thermosensitive genic male sterility) and one-line (apomixis) systems offer a good alternative to the current three-line (CMS) system for producing hybrid seed. But until these systems become available for commercial production of hybrid seed, CMS continues to be the best strategy and thus needs diversification for the sustainability of hybrid rice technology.

5.2. Advances in Two-Line Hybrid Rice Breeding

The discovery of environment-sensitive genic male sterility in rice led to the development of a simpler and more efficient two-line hybrid breeding system compared with the cytoplasmic male sterility or three-line system. Several elite photoperiod-sensitive (PGMS) and thermosensitive genic male sterile (TGMS) lines have been developed in China. The commercial two-line hybrids developed using these lines occupied about 330,000 ha in 1996. Multiplying PGMS and TGMS lines in a pure form requires some special handling. These methods have been developed in China. Similar methods of seed production have been adopted for both two-line and three-line hybrids. The key point is to determine the time when a PGMS or TGMS line will show complete sterility for about 1 mo at a given location. Seed yield of two-line hybrids in China is 2.25 - 3.0 $t \cdot ha^{-1}$. Under tropical conditions, in which day length differences are marginal, the TGMS system is considered to be more useful than the PGMS system. Genetic analysis at IRRI confirmed the monogenic recessive control of the TGMS trait. The TGMS gene of the IRRI mutant, IR32364 TGMS, was found to be non-allelic to the TGMS genes identified in China (tms 1) and Japan (tms 2). IRRI has developed some TGMS lines in indica rice that possess the tms 2 gene; these lines are being evaluated in national agricultural systems. As expected, the TGMS system gave a higher frequency of heterotic hybrids than the CMS system.

5.3. Breeding and Characterizing Indica PGMS and TGMS Lines in China

Hubei photoperiod-sensitive genic male sterile (PGMS) rice Nongken 58S, which was used as the donor of genic male sterile genes, was crossed, backcrossed, and multi-crossed with various target indica varieties. Sterile plants were selected under different environments. In the past 15 years, eight thermosensitive genic male sterile (TGMS) lines and four PGMS lines were developed. The characteristic response of these lines to photoperiod and temperature was studied under various environments in the field and in phytotrons. The results showed that most of them had a completely stable sterile period of about 45 days at Wuhan. The temperature-sensitive stage ranged from the differentiating stage of the pollen mother cell to the early ripe stage of pollen. The critical temperature points (CTP) of fertility alteration varied in different TGMS lines: 26.5°C (daily mean) for W6154S, W6184S, W6111S, W6417S, and W8103S; 25.5°C for W9046S and W9056S; and 24°C for W91607S. The fertility expression of PGMS lines was controlled simultaneously by photoperiod and temperature. PGMS line W7415S had a CTP of 26°C and a critical photoperiod point (CPP) of 13.5 h. The CTP and CPP for W9451S and W9461S were 24°C and 14.0 h, and for W9593S they were 24°C and 13.0 h, respectively. A practical and effective procedure for breeding PGMS and TGMS lines has been established.

6. WIDE HYBRIDIZATION

Crop gene pools are widened through hybridization of crop cultivars with wild species, weedy races as well as intra-subspecific crosses. Such gene pools are exploited for improving many traits including yield. Reference [15] reported that a quarter of lines from BC2-BC4 segregants from the Avena sativa × Avena sterilis crosses were significantly higher in grain yield than the cultivated recurrent parent. Nine lines from this study when tested over years and sites had agronomic traits similar to the recurrent parent and 10% - 29% higher grain yield. The higher yield potential of these inter-specific derivatives was attributed to higher vegetative growth rates or early seedling vigor.

Reference [16] reported that some backcross derivates from a cross between an Oryza rufipogon accession from Malaysia and cultivated rice, out yielded the recurrent parent by as much as 18%. They identified two QTL from wild species with major contribution to yield increase. These QTL are now being transferred to several modern semi-dwarf varieties.

Wide Hybridization for Disease and Insect Resistance

Wild species of rice are a rich source of genes for resistance breeding. For example, none of the cultivated rice was found to be resistant to grassy stunt. Oryza nivara, a wild species closely related to cultivated rice was found to be resistant and the dominant gene for resistance was transferred to improved germplasm through backcrossing. This gene for resistance has been incorporated into many widely grown varieties. When genes are to be transferred from more distantly related species, special techniques such as embryo rescue are employed to reproduce inter-specific hybrids. Reference [17] transferred genes for resistance to three biotypes of brown plant hopper from O. officinalis to an elite breeding line. Reference [18] transferred genes for resistance to brown plant hopper from O. australiensis to cultivated rice. Similarly genes for resistance to blast and bacterial blight have been transferred from O. minuta to improved rice germplasm [19] .

7. GENETIC ENGINEERING

Protocols for rice transformation have been developed which allow transfer of foreign genes from diverse biological systems into rice. Direct DNA transfer methods such as protoplast based [20] and biolistic as well as Agrobacterium-mediated [21] are being used for rice transformation. Major targets for rice improvement through transformation are disease and insect resistance.

As early as 1987, genes encoding for toxins from Bacillus thuringiensis (BT) were transferred to tomato, tobacco and potato, where they provided protection against Lepidoptern insects. A major target for BT deployment in transgenic rice is the yellow stem borer. This pest is widespread in Asia and causes substantial crop losses. Improved rice cultivar are either susceptible to the insect or have only partial resistance. Thus BT transgenic rice has much appeal for controlling the stem borer. Codon optimized BT genes have been introduced into rice and show excellent levels of resistance in the laboratory and greenhouse [22] . Bt rices have also been tested under field conditions in China [23] and have excellent resistance to diverse populations of yellow stem borer. Besides BT genes, other genes for insect resistance such as those for proteinase inhibitors, α-amylase inhibitors and lectins are also beginning to receive attention. Insects use diverse proteolytic or hydrolytic enzymes in their digestive gut for the digestion of food proteins and other food components. Plant derived proteinase inhibitors or α-amylase inhibitors are of particular interest because these inhibitors are a part of the natural plant defense system against insect predation. Reference [24] reported tansgenic rice carrying cowpea trypsin inhibitor (Cpti) gene with enhanced resistance against striped stem borer and pink stem borer.

Several viral diseases cause serious yield losses in rice. A highly successful strategy termed coat protein (CP) mediated protection has been employed against certain viral diseases such as tobacco mosaic virus in tobacco and tomato. A coat protein gene from rice strip virus was introduced into two japonica varieties by electroporation of protoplasts [25] . The resultant transgenic plants expressed CP at high level and exhibited a significant level of resistance to virus infection and the resistance was inherited to the progenies.

Transgenic Rice for Improved Nutritional Quality—Golden Rice

Malnutrition disorders are the cause of 24,000 deaths a day. "Golden Rice" represents a genetic engineering concept for the development of nutrient-dense staple crops which can make an important contribution to the reduction of malnutrition in developing countries. Major micronutrient deficiency disorders concern protein, energy, iron, zinc, vitamin A and iodine. These deficiencies are especially severe in countries where rice is the major staple.

Rice contains neither b-carotene (provitamin A) nor C40 carotenoid precursors in its endosperm. Rice in its milled form (as it is usually consumed) is therefore entirely without vitamin A and its carotenoid precursors. Millions of rice consumers who depend on rice for a large proportion of their calories suffer from vitamin A deficiency. Reference [26] produced transgenic rice (Golden Rice) with the provitamin A (b-carotene) biosynthetic pathway engineered into its endosperm. Agrobacterium-mediated transformation was applied to introduce three genes: phytoene synthase (psy), phytoene desaturase (crt1) and lycopene cyclase (lcy). HPLC (high performance liquid chromatography) analysis revealed the presence of b-carotene in transgenic seeds. The transformed rice Taipei 309 is no longer cultivated. Efforts are now underway at IRRI to transfer the genes for b-carotene into widely grown varieties, such as IR 64, through conventional backcrossing and transformation.

Reference [27] introduced the entire coding sequence of the soybean ferritin gene into kita-ake, a rice cultivar via Agrobacterium-mediated transformation. The introduced ferritin gene was regulated by the rice seed storage protein glutelin promoter, GluB-1, and terminated by the Nos polyadenylation signal. Synthesis of soybean ferritin protein was confirmed in each of the transformed rice seeds by western blot analysis, and specific accumulation in endosperm was determined by immunological tissue printing. The iron content of T1 seeds was up to three times higher than in untransformed seeds.

8. MOLECULAR MARKER ASSISTED BREEDING

Numerous genes for disease and insect resistance are repeatedly transferred from one varietal background to the other. Most genes behave in dominant or recessive manner and require time consuming efforts to transfer. Sometimes the screening procedures are cumbersome and expensive and require large field space. If such genes can be tagged by tight linkage with molecular markers, time and money can be saved in transferring these genes from one varietal background to another. The presence or absence of the associated molecular marker indicates at an early stage, the presence or absence of the desired target gene. A molecular marker very closely linked to the target gene can act as a "tag" which can be used for indirect selection of target gene.

Two of the most serious and widespread diseases in rice production are rice blast caused by the fungus Pyrcularia oryzae, and bacterial blight caused by Xanthomonas oryzae pv. oryzae. Development of durable resistance to these diseases is the focus of a coordinated effort at IRRI using molecular marker technology. Efforts to detect markers closely linked to bacterial blight resistance genes have taken advantage of the availability of near isogenic lines having single genes for resistance. Segregating populations were used to confirm co-segregation between RFLP markers and genes for resistance. Protocols for converting RFLP markers into PCR based markers and using the PCR markers in marker-aided selection have been established [28] . The PCR markers were also used for pyramiding genes for resistance to bacterial blight. Thus xa4, x5, xa13, and Xa21 were combined into same breeding line [29] . The pyramided lines showed a wider spectrum and higher level of resistance than lines with only a single gene for resistance. MAS has also been employed for moving genes from pyramided lines into new plant type [30] as well as into improved varieties grown in India [31] .

Novel alleles and genetic diversity widely exist in wild relatives of cultivated plants. For example, wild relatives of rice within the genus Oryza are not only a rich source of information on the origins of variation within the genus but also a viable source of a wide variety of agronomically important germplasm for future breeding. To fill the gulf between national research programs and breeding applications in developing countries, an international program, Challenge Programs was established to unlock the genetic diversity existing in a wide spectrum of germplasm collections. Molecular markers have been proven particularly useful for accelerating the backcrossing of a gene or QTL from exotic cultivars or wild relatives into an elite cultivar or breeding line. Favorable genes or alleles from wild species of rice have been detected after backcrossing to elite cultivars [32] . Similarly, this approach can identify alleles from exotic cultivars that result in improved phenotype, even though the parent may not possess inferior phenotype for this trait. This approach is thought to be promising in rice because a number of rice cultivars are widely grown for their adaptation, stable performance, and desirable grain quality. Reference [33] used such an approach to transfer the bacterial blight resistance gene Xa21 into Minghui 63, a widely used parent for hybrid rice production in China. Reference [34] used a similar approach to introgress two QTLs controlling resistance to rice yellow mottle virus into the cultivar IR64. Such approaches, however, can only sample a small number of accessions.

8.1. Breeding for Durable Resistance

Full yield potential of modern rice varieties is not realized because of the toll taken by the attack of disease and insect organisms. It is estimated that diseases and insects cause yield losses of up to 25% annually. Genetic improvement to incorporate durable resistance to pests is the preferred strategy to minimize these losses. There is no cost to farmers and resistant cultivars are easily adopted and disseminated unlike "knowledge based" technologies. Also concern for the environment has become an important public policy issue and pest management methods that minimize the use of crop protection chemicals are increasingly finding favor.

8.2. Biotic Stress Resistance

To date, two blight-resistant genes, Xa26 and Xa13, have been cloned by Chinese scientists [35] . Xa26 on chromosome 11 encodes an leucine rich repeat (LRR) receptor kinase-like protein [36] . xa13, a recessive allele conferring disease resistance against bacterial blight, plays a key role in both disease resistance and pollen development, and represents a new type of plant disease resistance. The dominant allele, Xa13, is required for both bacterial growth and pollen development. Promoter mutations in Xa13 cause down-regulation of expression during host pathogen interaction, resulting in the fully recessive xa13 that confers race-specific resistance [35] . Also five other genes (Xa4, Xa2, Xa-7, Xa-23 and Xa-25) with resistance to bacterial blight have been mapped or delimited to a narrow physical interval [37] [38] .

Insect-resistance studies are mainly carried out on brown plant hopper (BPH) and white brown plant hopper (wbph). Reference [39] fine-mapped a resistance gene, which was designated tentatively as bph19 (t), against BPH on chromosome 3. Reference [40] delimited bph15 to a region of 47 kb on chromosome 4. Another three BPH-resistant genes, Bph12 (t), bph2, bph9 and a WBPH-resistant gene wbph6 (t) were located on chromosomes 4, 12, 12 and 11, respectively [41] .

9. CONCLUSION

Mudasir Hafiz Khan, Zahoor Ahmad Dar, Sher Ahmad Dar (MAS) has become integral component of germplasm improvement. A large number of genes for various traits have been tagged with molecular markers to apply MAS for trait improvement. Map-based cloning has resulted in isolation of several genes for resistance to biotic and abiotic stresses as well as yield-related traits. This has opened the possibility of applying MAS for yield enhancement. Recombinant DNA technology has resulted in production of transgenic rice with new genetic traits and for resistance to biotic and abiotic stresses.

REFERENCES

1. Peng, S.R., Laza, C., Visperas, R.M., Sanico, A.L., Cassman, K.G. and Khush, G.S. (2000) Grain Yield of Rice Cultivars and Lines Developed in the Philippines since 1966. Crop Science, 40, 307-314.
2. Sarkarung, S., Singh, O.N., Roy, J.K., Vanavichit, A. and Bhekasut, P. (1995) Breeding Strategies for Rainfed Lowland Ecosystem. Proc. Int. Rice Res. Conf., Manila, IRRI, 709-720.
3. Program Report for 1994 IRRI (1995) International Rice Research Institute, Manila, 309.
4. Mackill, D.J. and Xu, K. (1996) Genetics of Seedling-Stage Submergence Tolerance in Rice; Rice Genetics III. In: Khush, G.S. Ed., Proceedings of Third International Rice Genetics Symposium, IRRI, Manila, 607-612.
5. Kenong, X., Ronald, P.C. and Mackill, D.J. (1997) Development of High Resolution Map around the Rice Submergence Tolerance Locus Sub I. General Meeting of the International Program on Rice Biotechnology, Malacca, 158.
6. Chandler Jr., R.F. (1969) Plant Morphology and Stand Geometry in Relation to Nitrogen. In: Eastin, J.D., Haskin, F.A., Sullivan, C.Y. and Van Baul, C.H.M., Eds., Physiological Aspects of Crop Yield, ASA Publication, Madison, 265-285.
7. IRRI (1989) IRRI towards 2000 and beyond. International Rice Research Institute, P.O. Box 933, Manila.
8. Khush, G.S. (1995) Breaking the Yield Barrier of Rice. GeoJournal, 35, 329-332.
9. Peng, S., Khush, G.S., Virk, P., Tang, Q. and Yingbin, Z. (2008) Progress in Ideotype Breeding to Increase Rice Yield Potential. Field Crops Research, 108, 32-38.
10. Wang, Z.H., Zou, Y.J., Li, X.Y., Zhang, Q.Y., Chen, L.T. and Wu, H. (2006) Cytoplasmic Male Sterility of Rice with Boro II Cytoplasm Is Caused by a Cytotoxic Peptide and Is Restored by Two Related PPR Motif Genes via Distinct Modes of mRNA Silencing. Plant Cell, 18, 676-687.
11. Liu, X.Q., Xu, X., Tan, Y.P., Li, S.Q., Hu, J. and Huang, J.Y. (2004) Inheritance and Molecular Mapping of Two Fertility-Restoring Loci for Honglian Gametophytic Cytoplasmic Male Sterility in Rice (Oryza sativa L.). Molecular Genetics and Genomics, 271, 586-594.
12. Xie, J.K., Zhuang, J.Y., Fan, Y.Y., Tu, G.Q., Xia, Y.W. and Zheng, K.L. (2001) Mapping of a Fertility-Restoring Gene for Dwarf-Wild-Abortive CMS in Rice. Chinese Journal of Rice Science, 15, 161-164.
13. Tan, X.L., Tan, Y.L., Zhao, Y.H., Zhang, X.M., Hong, R.K. and Jin, S.L. (2004) Identification of the Rf Gene Conferring Fertility Restoration of the CMS Dian-Type 1 in Rice by Using Simple Sequence Repeat Markers and Advanced Inbred Lines of Restorer and Maintainer. Plant Breeding, 123, 338-341.
14. Yang, H.Y., Ren, X., Weng, Q.M., Zhu, L.L. and He, G.C. (2002) Molecular Mapping and Genetic Analysis of a Rice Brown Planthopper (Nilaparvata lugens) Resistance Gene. Hereditas, 136, 39-43.

15. Lawerence, P.L. and Frey, K.J. (1976) Backcross Variability for Grain Yield in Species Crosses (Avena sativa L. × A. sterilis L.). Euphytica, 24, 77-85.
16. Xiao, J., Grandillo, S., Ahn, S.N., McCouch, S.R. and Tanksley, S.D. (1996) Genes from Wild Rice Improve Yield. Nature, 38, 1223-224.
17. Jena, K.K. and Khush, G.S. (1990) Introgression of Genes from Oryza officinalis Well ex Watt to Cultivated Rice, O. sativa L. Theoretical and Applied Genetics, 80, 737-745.
18. Multani, D.S., Jena, K.K., Brar, D.S., Delos Reyes, B.G., Angeles, E.R. and Khush, G.S. (1994) Development of Monosomic Alien Addition Lines and Introgressions of Genes from Oryza australiensis Domin. to Cultivated Rice O. sativa L. Theoretical and Applied Genetics, 88, 102-109.
19. Brar, D.S. and Khush, G.S. (1997) Wide Hybridization for Rice Improvements: Alien Gene Transfer and Molecular Characterization of Introgression. In: Jones, M.P., Dingkhun, M., Johnson, D.E. and Fagade, S.O., Eds., Inter-Specific Hybridization: Progress and Prospect WARDA, 01BP2551, Cote d'voire, Bouake, 21-29.
20. Datta, S.K., Peterhaus, A., Datta, K. and Potrykus, I. (1990) Genetically Engineered Fertile Indica Rice Recovered from Protoplasts. Nature Biotechnology, 8, 736-740.
21. Hei, Y., Ohta, S., Komari, T. and Kumashiro, T. (1994) Efficient Transformation of Rice (Oryza sativa L.) Mediated by Agrobacterium and Sequence Analysis of the Boundaries of T-DNA. The Plant Journal, 6, 271-282.
22. Datta, S.K., Torrizo, L., Tu, J., Oliva, N. and Datta, K. (1997) Production and Molecular Evaluation of Transgenic Rice Plants. IRRI Discussion Paper Series No. 21, International Rice Research Institute, Manila.
23. Tu, J., Zhang, G., Datta, K., Xu, C., He, Y., Zhang, Q., Khush, G.S. and Datta, S.K. (2000) Field Performance of Transgenic Elite Commercial Hybrid Rice Expressing Bacillus thuringensis Endoprotein. Nature Biotechnology, 18, 1101-1104.
24. Xu, D., Xue, Q., McElroy, D., Mawal, Y., Hilder, V.A. and Wu, R. (1996) Constitutve Expression of a Cowpea Trypsin Inhibitor Gene CpTi in Transgenic Rice Plants Confers Resistance to Two Major Rice Insect Pests. Molecular Breeding, 2, 167-173.
25. Hayakawa, T., Zhu, Y., Ito, K. and Kimura, Y. (1992) Genetically Engineered Rice Resistant to Rice Stripe Virus, an Insect Transmitted Virus. Proceedings of the National Academy of Sciences of the United States of America, 89, 9865- 9869.
26. Ye, X., Al-Babili, S., Kloti, A., Zhang, A., Lucca, P., Beyer, P. and Potrykus, I. (2000) Engineering the Provitamin A (β-Carotene) Biosynthetic Pathway into (Carotenoid-Free) Rice Endosperm. Science, 287, 303-305.
27. Goto, F., Yoshihara, T., Shigemoto, N., Toki, S. and Takaiwa, F. (1999) Iron Fortification of Rice Seed by the Soybean Ferritin Gene. Nature Biotechnology, 17, 282-286.
28. Zheng, K., Huang, N., Bennett, J. and Khush, G.S. (1995) PCR-Based Marker-Assisted Selection in Rice Breeding. IRRI Discussion Paper Series No. 12, International Rice Research Institute, Manila.
29. Huang, N., Angeles, E.R., Domingo, J., Magpantay, G., Singh, G., Zhang, G., Kumaravadivel, N., Bennett, J. and Khush, G.S. (1997) Pyramiding of Bacterial Blight Resistance Genes in Rice: Marker Assisted Selection Using RFLP and PCR. Theoretical and Applied Genetics. 95, 313-320.
30. Sanchez, A.C., Brar, D.S., Huang, N., Li, Z. and Khush, G.S. (2000) STS Marker-Assisted Selection for Three Bacterial Blight Resistance Genes in Rice. Crop Science, 40, 792-797.
31. Singh S., Sidhu, J.S., Huang, N., Vikal, Y., Li, Z., Brar, D.S., Dhaliwal, H.S. and Khush, G.S. (2001) Pyramiding Three Bacterial Blight Resistance Genes (xa5, xa13 and Xa21) Using Marker-Assisted Selection into Indica Rice Cultivar PR106. Theoretical and Applied Genetics, 102, 1011-1015.
32. Moncada, P., Martinez, C.P., Borrero, J., Chatel, M., Gauch, H., Guimaraes, E., Tohme, J. and McCouch, S.R. (2001) Quantitative Trait Loci for Yield and Yield Components in an Oryza sativa × Oryza rufipogon BC2F2 Population Evaluated in an Upland Environment. Theoretical and Applied Genetics, 102, 41-52.

33. Chen, S., Lin, X.H., Xu, C.G. and Zhang, Q.F. (2000) Improvement of Bacterial Blight Resistance of "Minghui 63", an Elite Restorer Line of Hybrid Rice, by Molecular Marker-Assisted Selection. Crop Science, 40, 239-244.
34. Ahmadi, N., Albar, L., Pressoir, G., Pinel, A., Fargette, D. and Ghesquiere, A. (2001) Genetic Basis and Mapping of the Resistance to Rice Yellow Mottle Virus. III. Analysis of QTL Efficiency in Introgressed Progenies Confirmed the Hypothesis of Complementary Epistasis between Two Resistance QTLs. Theoretical and Applied Genetics, 103, 1084- 1092.
35. Chu, Z.H., Yuan, M., Yao, J.L., Ge, X.J., Yuan, B. and Xu, C.G. (2006) Promoter Mutations of an Essential Gene for Pollen Development Result in Disease Resistance in Rice. Genes & Development, 20, 1250-1255.
36. Sun, L.H., Wang, C.M., Su, C.C., Liu, Y.Q., Zhai, H.Q. and Wan, J.M. (2006). Mapping and Marker-Assisted Selection of a Brown Planthopper Resistance Gene Bph2 in Rice (Oryza sativa L.). Journal of Genetics and Genomics, 33, 717-723.
37. He, Q., Li, D.B., Zhu, Y.S., Tan, M.P., Zhang, D.P. and Lin, X.H. (2006) Fine Mapping of Xa2, a Bacterial Blight Resistance Gene in Rice. Molecular Breeding, 17, 1-6.
38. Fan, Y.L., Chen, X.W., Wang, C.L., Zhu, L.H., Zhang, Q. and Zhao, K.J. (2006) Mapping the Rice Bacterial Blight Resistance Gene Xa23 with RFLP Markers and Converting RFLP to STS Marker. Acta Genetica Sinica, 32, 931-935.
39. Chen, J.W., Wang, L., Pang, X.F. and Pan, Q.H. (2006) Genetic Analysis and Fine Mapping of a Rice Brown Planthopper (Nilaparvata lugens Stål) Resistance Gene Bph19 (t). Molecular Genetics and Genomics, 275, 321-329.
40. Yang, H.Y., You, A.Q., Yang, Z.F., Zhang, F.T., He, R.F. and Zhu, L.L. (2004) High-Resolution Genetic Mapping at the Bph15 Locus for brown Planthopper Resistance in Rice (Oryza sativa L.). Theoretical and Applied Genetics, 110, 182-191.
41. Su, C.C., Zhai, H.Q., Wang, C.M., Sun, L.H. and Wan, J.M. (2006) SSR Mapping of Brown Planthopper Resistance Gene Bph9 in Kaharamana, an Indica Rice (Oryza sativa L.). Journal of Genetics and Genomics, 33, 262-268.

CHAPTER 6

Breeding for Drought Tolerance in Maize (Zea mays L.)

Abdoul-Raouf Sayadi Maazou*, Jialu Tu*, Ju Qiu, Zhizhai Liu

College of Agronomy and Biotechnology, Southwest University, Chongqing, China

ABSTRACT

Drought, like many other environmental stresses, has adverse effects on crop yield including maize (Zea mays L.). Low water availability is one of the major causes for maize yield reductions affecting the majority of the farmed regions around the world. Therefore, the development of drought-tolerant lines becomes increasingly more important. In maize, a major effect of water stress is a delay in silking, resulting in an increase in the anthesis-silking interval, which is an important cause of yield failures. Diverse strategies are used by breeding programs to improve drought tolerance. Conventional breeding has improved the drought tolerance of temperate maize hybrids and the use of managed drought environments, accurate phenotyping, and the identification and deployment of secondary traits has been effective in improving the drought tolerance of tropical maize populations and hybrids as well. The contribution of molecular biology will be potential to identify key genes involved in metabolic pathways related to the stress response. Functional genomics, reverse and forward genetics, and comparative genomics are all being deployed with a view to achieving these goals. However, a multidisciplinary approach, which ties together breeding, physiology and molecular genetics, can bring a synergistic understanding to the response of maize to water deficit and improve the breeding efficiency.

Keywords: Maize (Zea mays L.), Drought Stress, Anthesis-Silking Interval, Breeding

1. INTRODUCTION

*These authors contribute equally to this work.

Maize (Zea mays L.) along with wheat and rice provides at least 30% of the food calories to more than 4.5 billion people in 94 developing countries where one-third of children are malnourished [1] [2] . By 2050, the demand for maize in the developing world will be almost double to the current demand [3] . However, an estimated 15% to 20% of maize grain yield is lost each year due to drought and such losses may further increase as droughts become more frequent and severe because of climate change [4] . Irrigation is not an option for large numbers of farmers and there is limited potential for any expansion of irrigation in developing countries [5] .

The use of genetics to improve drought tolerance and provide yield stability is an important part of the solution to stabilizing global production. That is why the development of maize varieties with enhanced tolerance to drought stress and higher water use efficiency (WUE) has become a high priority goal for major breeding programs, both in the private and public sectors. The breeding programs improve drought tolerance via diverse strategies such as recurrent selection and evaluation of segregating population under managed and multi-location drought-stress environment, use of secondary traits for selection under drought condition, genomic-based approach and transgenic technology. Understanding the nature of drought response in maize and some major strategies used for improving drought stress-tolerant maize lines will provide opportunities to improve the breeding process.

Although other reviews on plant breeding for drought tolerance are available [6] - [9] , this review will focus on the major strategies or approaches used by major breeding programs for improving drought tolerance in maize.

2. MAIZE RESPONSES TO DROUGHT

Drought is a multidimensional stress, affecting plants at various levels of their organization over space and time, so that the physiological responses to it are complex and often unpredictable. However, in maize, a major effect of water stress is a delay in silking, resulting in an increase in the anthesis-silking interval (ASI), which is an important cause of yield failures [10] . In fact, typical visual symptoms of drought stress in maize are a change in color from green to green-gray, and rolling of the lower leaves followed by those in the upper canopy. At the same time stomas are closing, photosynthesis is being sharply reduced and growth is slowing. When stress coincides with the 7 - 10 days period prior to flowering, ear growth will slow more than tassel growth and there is a delay in silk emergence relative to pollen shed, giving rise to an interval between anther extrusion and silk exposure [11] . This ASI was shown to be highly correlated with grain yield, in particular kernel number and ear number per plant [10] . At the same time leaf senescence begins at the base of the plant and spreads upwards to the ear. Severe stress at flowering may lead to the complete abortion of ears and the plant becomes barren. Drought-affected ears typically have fewer kernels that will be poorly filled if drought extends throughout grain filling [12] .

In many cropping environments, drought stress increases in severity at the end of the growing season. In this situation, assimilate which has accumulated to a high level in the maize stem is re-mobilized, and contributes to yield [13] . From the root to the aerial tissues the action of abscisic acid involves interactions with other growth regulators, particularly ethylene and reactive oxygen species [14] . Moreover, the development of a root system capable of accessing water far down the soil profile is a valuable trait in drought- affected environments [15] . Many species, including maize, respond to water deficit by redirecting growth and dry matter accumulation away from the shoot to the root [16] .

Another potentially valuable stress tolerance mechanism is the osmotic adjustment (OA) because it involves the active accumulation of solutes in the cell [14] . This accumulation enables the retention of water during episodes of low external water potential, limiting turgor loss and damage from cell shrinkage. Under more prolonged or severe moisture deficit, these solutes also are implicated in the stabilization of various macromolecular structures.

Finally, the stomata respond dynamically to changes in the environment and play a vital role in limiting water loss during drought. Maize is an "isohydric" species, so called because the signaling and response system is sensitive to a decrease in water potential, limiting water loss in the early phases of water deficit, so that leaf water potential can be maintained at the well-watered level until more advanced stages of drought [17] . The sensitivity of maize stomata also protects xylem from cavitation [18] , and under high vapor pressure deficit, it curtails water loss in the middle of the day, so that photosynthesis and transpiration is restricted to the cooler morning periods when WUE is higher [14] .

Many of these mechanisms favor survival but may have limited value in enhancing grain yield under a level of water stress that varies spatially and across seasons.

3. SELECTION AND EVALUATION OF SEGREGATING POPULATION UNDER MANAGED AND MULTI-LOCATION DROUGHT-STRESS ENVIRONMENTS

Breeding for drought tolerance in maize is a complex task, not least because drought can affect the crop at any stage of development. The choice of a selection strategy is critical to breeding for stress tolerance. Probably the most widely used strategy is to select for yield under non-stressed conditions, and then evaluate those selections at many sites with variable moisture availability or "random stress" [19] . Underlying assumptions of this approach are that genes for drought tolerance are present in elite high yielding material, even after the number of genotypes has been narrowed to the few evaluated under random stress, and that selection under optimum growing conditions can also increase performance in sub-optimum conditions [20] . Moreover, hybrids usually yield better than varieties under drought with heterosis acting as an important source of stress tolerance [6] .

The choice of the testing environment(s) is critical to the rate of achievable genetic gain. Ideally, the selection environment should mirror the target environment in rainfall distribution, physical and chemical soil properties, water distribution profiles and potential evapotranspiration rates, otherwise significant genotype × environment (G × E) interactions will result in much of the gain achieved in the selection environment not being reproduced in the target environment [14] . The use of uniform soils, along with rigorous experimental design and appropriate statistical analysis will together maximize the precision of genotypic means [14] . Multi-location evaluation is necessary to estimate the importance of G × E. It is especially critical in the context of breeding for drought tolerance, where a consequence of lowered plant vigor is a higher responsiveness to environmental variation.

This methodology was used to develop drought tolerant versions of several elite lowland tropical populations [21] - [26] . For example, Monneveux et al. evaluated direct and correlated responses to recurrent selection for drought tolerance in two CIMMYT maize populations, DTP1 and DTP2, adapted to the lowland and mid-altitude tropics [25] . Cycles C0, C3, and C6 of DTP1 and C0, C3, C5 and C9 of DTP2 were evaluated under drought, low N, and optimal conditions. Evaluation for drought tolerance was based on replicated trials at one or two water stress levels during a rain-free period using controlled irrigation. Severe water stress was induced during flowering and grain filling such that average grain yield in these trials was reduced 15% to 30% relative to unstressed yields. The same germplasm was also grown under well-watered conditions. This selection was based on an index involving grain yield under drought and well-watered conditions and ASI, barrenness, leaf senescence, and leaf rolling under drought. They observed a significant yield gains in both populations under drought conditions. Yield gain from C0 to C6 cycles was 0.20 and 0.12 $Mg\cdot ha^{-1}Cycle^{-1}$ (16.5% and 10.1% $Cycle^{-1}$) for the white and yellow versions, respectively. This gain was explained by a significant increase of both the number of ears per plant and the number of grains per ear. Yield gain, however, was less than that under drought, and was associated mainly with an increase in the number of grains per ear. They also reported that selection for drought tolerance did not affect yield and the number of grains per ear under optimal conditions but did increase 1000-kernel weight in DTP1. Ears per plant were significantly correlated with grain yield across cultivars among three environments (Figure 1). Moreover, they found that effects of recurrent selection on dry matter partitioning at maturity showed a significant increase in ear weight and a significant decrease in tassel, stem, and leaf weight in DTP1 when evaluated under drought [25] .

In the same study, Monneveux et al. observed a positive correlation between tassel and stem weight per plant across cultivars, both at anthesis and maturity (Figure 2) [25] . There was also a significant negative correlation between ear weight and tassel weight at anthesis (Figure 3).

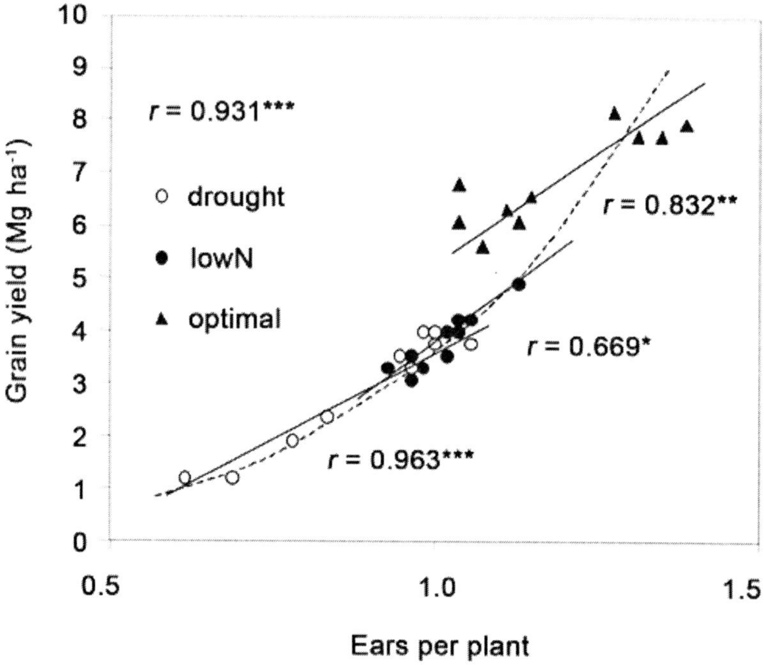

Figure 1. Relationship between ears per plant and grain yield across cultivars within and across environments [25] . *, **, and *** Significant at P = 0.05, 0.01, and 0.001, respectively.

In another study using similar methods, Edmeades et al. observed an increase in grain yield under drought of 0.26 Mg ha^{-1} or 12.6% Cycle^{-1} following S1 recurrent selection using a selection intensity of 5 to 10%, and a lower rate of improvement (0.08 Mg ha^{-1} or 3.8% Cycle^{-1}) following fullsib recurrent selection with a selection intensity of 26 to 32% [24] . They also reported a small but significant increase in grain yield, ears per plant, kernel number per fertile ear, and individual kernel weight under well-wa- tered conditions. Bänziger et al. evaluated contrasting selection cycles of several of these populations under low N and observed similar yield gains under low N as well [27] . They suggested that common mechanisms were responsible for increased partitioning of assimilates to the developing ear and for increased yields under both types of stress.

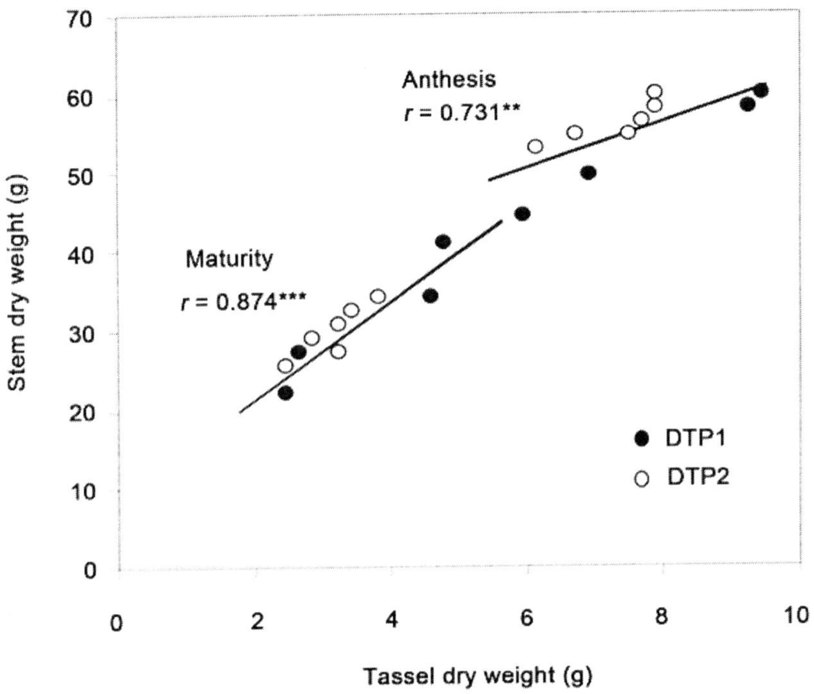

Figure 2. Relationship between tassel and stem dry weight at anthesis and maturity across cultivars, under drought conditions [25] . **, and *** Significant at P = 0.01 and 0.001, respectively.

Despite success achieved using selection and evaluation under managed and multi- location drought-stress environment, the effectiveness of this method is however, largely unknown. For example, using the same approach, Magorokosho et al. compared two maize populations, ZM601 and ZM607 for drought tolerance during flowering, the most drought-vulnerable period for the maize plant [19] . The results did not show much difference in drought tolerance between ZM601 and ZM607. Differences between population means for grain yield, ASI, and number of ears per plant (EPP) were small.

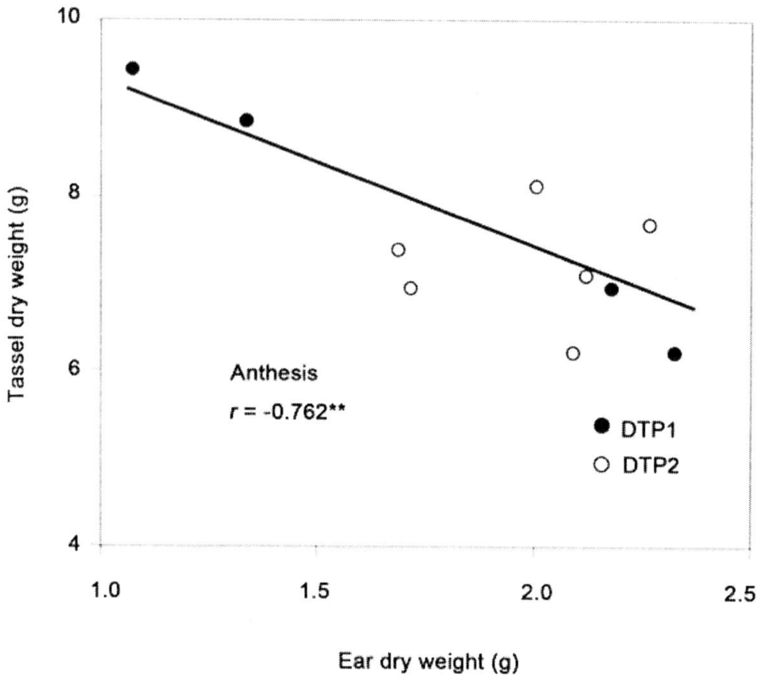

Figure 3. Relationship between ear and tassel dry weight at anthesis across cultivars, under drought conditions [25] . **Significant at P = 0.01.

4. SECONDARY TRAITS FOR SELECTION UNDER DROUGHT CONDITIONS

Secondary traits are those other than economic yield itself which can provide a measure of plant performance [28] . An ideal secondary trait would be genetically correlated with grain yield in the target environment, genetically variable, have a high level of heritability, be simple, cheap, non-destructive and fast to assay, be stable throughout the measurement period and would not be associated with any yield loss under non-stressed conditions [14] . Under drought stress conditions, breeding progress is impeded by a significant level of G × E (both with respect to cropping season and location). Given the poor heritability of grain yield under drought stress conditions, genetic progress is hard to achieve via direct selection. However, because under drought, both the heritability (h^2) of at least some secondary traits remains high and the genetic correlation between grain yield and these traits increases significantly, recourse to indirect selection becomes an attractive strategy [28] [30] . Selection based on secondary traits which reflect the direct effects of drought can improve the response, since it avoids the confounding effects of other stresses, such as poor soil fertility, micronutrient deficiency and pathogen presence [14] . Application of this strategy has generated genetic gains under a range of environmental conditions.

In fact, after evaluation of a total of 3509 inbred progenies (Sl to S3 level) in 50 separate yield trials under two or three water regimes, Bolaños and Edmeades reported a general tendency to decrease with increasing moisture stress in the heritability for grain yield from around 0.60 in well-watered environments to values of 0.40 or less at very low yield levels [29] . The h^2 of kernels ear^{-1} and weight $kernel^{-1}$ was around 0.60 under well-watered conditions, but also decreased with increasing stress. In contrast, the heritability for ASI and ears plant either increased or remained fairly constant with increasing moisture stress and declining yield levels. The h^2 for days to anthesis remained fairly constant across all moisture regimes. They also suggesting that earlier-flowering progenies were associated with high grain yields and similarly, a short ASI was linked to high grain yield under stress. This perception is reinforced by findings of Magorokosho et al. who found that a phenotypic correlation between grain yield and ASI were small ($r = -0.08$ to -0.21*) under adequate moisture conditions and became much larger ($r = -0.40$** to -0.43**) at moisture-stressed sites [19] . Similarly, the relationship between ears per plant (EPP) and grain yield, also became stronger with increasing moisture stress (from 0.08 to 0.24** without stress, to 0.40** to 0.45** with moisture stress).

In an earlier study using similar methods, Ziyomo and Bernardo observed a significant decrease in h^2 for grain yield under drought (0.60 in the control experiment and 0.37 under drought) but not for ASI and leaf senescence [31] . They observed a higher h^2 and genetic variance for ASI and leaf senescence under drought stress. Furthermore, in this drought experiments, the strongest genetic correlation was observed between grain yield and ASI (−0.77). In summary, compared with direct selection for grain yield under drought, indirect selection based on grain yield in the control experiments had a relative efficiency of 0.78, indirect selection based on ASI had a relative efficiency of 1.04, and indirect selection based on leaf senescence had a relative efficiency of 0.98.

These results indicate that ASI, leaf senescence and EPP are useful secondary traits for the selection of grain yield at moisture stressed sites. Therefore, the use of this strategy can increase selection efficiency.

5. GENOMIC-BASED APPROACH

Drought tolerance is a complex quantitative trait controlled by many genes, and is one of the most difficult traits to study and characterize. Compared to conventional approaches, genomics offers unprecedented opportunities for dissecting quantitative traits into their single genetic determinants, the so-called quantitative trait loci (QTL), thus paving the way to marker-assisted selection (MAS) and, eventually, cloning of QTLs and their direct manipulation via genetic engineering [32] . Therefore, it is possible to identify major QTLs regulating specific drought responses and it will provide an efficient way to improve drought tolerance in maize germplasm [33] . The increasing number of studies reporting QTLs for drought-related traits and yield in drought- stressed crops indicates a growing interest in this approach [10] [33] - [37] .

In fact, using a modeling approach combined with field measurements, Ribaut et al. identified a common QTLs for both leaf growth and ASI in a recombinant inbred line population evaluated under water stress conditions [14] . For all common QTL the allele conferring high leaf elongation rate conferred a short ASI, indicating a high silk elongation rate. They also observed unsurprisingly that drought-related QTL are dispersed throughout the maize genome. In another study, Sari-Gorla et al. performed a linkage analysis between the expression of male and female flowering time, ASI, plant height and molecular markers [10] . The experiment was carried out under two environmental conditions, well-watered and water-stressed, on a maize population of 142 recombinant inbred lines obtained by selfing the F1 between lines B73 and H99 and genotyped for a total of 153 loci. Linkage analysis revealed that, for male flowering time and plant height, most of the QTLs detected were the same under control and stress conditions. In contrast, with respect to female flowering time and ASI diverse QTLs appeared to be expressed either under control conditions or under stress. All of the QTLs conferring tolerance to drought were located in different chromosome regions. They suggested that plant tolerance, in its different components, is not attributable to the presence of favorable allelic combinations controlling the trait but is based on physiological characteristics not directly associated with the control of the character. Zhu et al. also found five, five, six, four, and five QTLs for ASI, plant height, grain yield, ear height, and ear setting under full irrigation condition, respectively, and four, seven, six, four, and four QTLs for ASI, plant height, grain yield, ear height, and ear setting under severe late stress conditions, respectively with similar method [33] .

The universal QTLs information generated in these studies will aid in undertaking an integrated breeding strategy for further genetic studies in drought tolerance improvement in maize.

6. TRANSGENIC TECHNOLOGY

Traditional breeding techniques contribute considerably to the popularization and application of drought-resistant lines and cultivars, but the limitations are the long breeding cycle and the extensive time consumption. Thanks to rapid progress in biotechnology and genome sequencing, there is now a diverse choice of tools for the identification of candidates for genes involved in specific processes, including the response to drought. Many drought tolerance candidate genes have been documented in the literature [34] - [41] , but few of them have been validated either via reverse genetics or transgenic approaches, or by the demonstration that directed selection under field conditions in diverse genetic backgrounds produces a clear genetic gain [14] .

However, some reports have showed potential results. For example, Quan et al. reported that glycine betaine plays an important role in some plants, including maize, in conditions of abiotic stress, but different maize varieties vary in their capacity to accumulate glycine betaine [42] . An elite maize inbred line DH4866 was transformed with the betA gene from Escherichia coli encoding choline dehydrogenase, a key enzyme in the biosynthesis of glycine betaine from choline. The transgenic maize plants accumulated higher levels of glycine betaine and exhibited higher tolerance to drought stress than wild-type plants (non-transgenic) at germination and the young seedling stage. Most importantly, the grain yield of transgenic plants was significantly higher than that of wild-type plants after drought treatment. They suggested that enhanced glycine betaine accumulation in transgenic maize provides greater protection of the integrity of the cell membrane and greater activity of enzymes compared with wild-type plants in conditions of drought stress.

In a most recent study, Liu et al. cloned maize ZmSDD1 and dissected its functions and performance to drought stress [43] . Genetically engineered overexpression of ZmSDD1 was employed to transform maize and cultivate new drought-resistant lines. Under an optical microscope, in the visual field × 100 with the adaxial ends, 71 and 51 stomata were found in wild types (WTs) (Figure 4(A-a)) and transgenic plants (TPs) (Figure 4(A-b)), respectively, with the same number of stomata with the abaxial ends, 71 in WTs (Figure 4(A-e)) and 51 in TPs (Figure 4(A-f)). In the fields × 200 of adaxial and abaxial ends, 21 stomata in WTs (Figure 4(A-c) and Figure 4(A-g)) and 15 stomata in TPs (Figure 4(A-d) and Figure 4(A-h)) were observed. A stomatal density of 90/mm^2 was determined in WT, whereas it was 63/mm^2 in TP, with a reduction of 30% in TP (Figure 4(B)). The ZmSDD1 expression level in TPs was 6.68 times higher than that in WTs (Figure 4(C-a)); the expression rate of TMM related to the stomatal development signaling pathway was augmented by 4.29 times (Figure 4(C-b)), while the

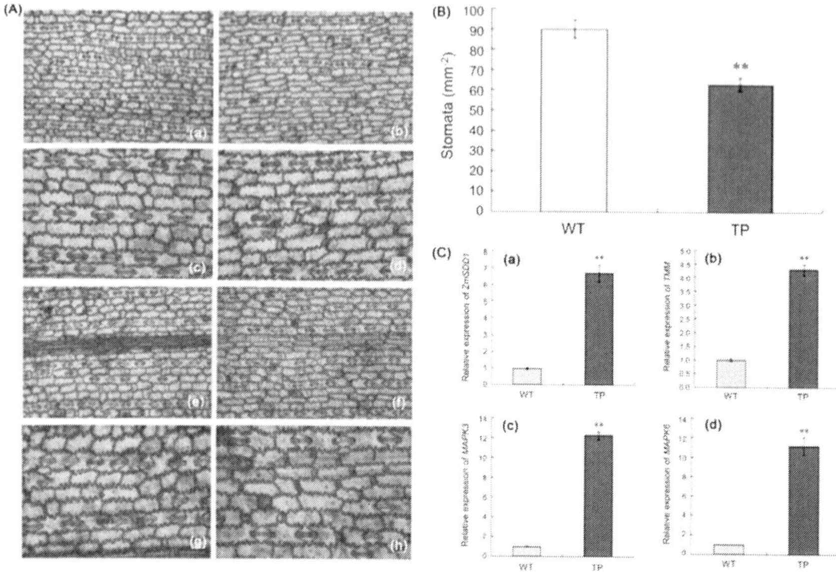

Figure 4. Reduction of stomatal density in TP plants with overexpression of ZmSDD1 [43] . (A) WT and TP maize under optical microscopy; (B) Stomata numbers in TP and WT maize; (C) Genes correlated with stomatal development were regulated by overexpression of ZmSDD1 (a ZmSDD1, b TMM, c MAPK3, d MAPK6). Data represent mean values ± SD from three independent experiments. Asterisks indicate statistical significance (**P < 0.01, Student's t test) of differences between transgenic and WT plants (WT wild type, TP transgenic plant).

degrees of expression of MAPK3 and MAPK6 increased by 12.21 (Figure 4(C-c)) and 11.16 (Figure 4(C-d)) times, respectively. These results indicate that the reduction of stomatal density in the TPs might be attributed to the regulation of ZmSDD1. Therefore, the transpiration is reduced in TPs compared with WTs.

In the same study, The TP and WT maize groups were drought-treated 3 times during 45 days, and the survival rates were observed and summarized. The results showed that 19 TP plants survived and 1 plant died, constituting a survival rate of 95 %. In contrast, only 3 WT plants survived, while 17 plants died, forming a survival rate of 15 %. Therefore, the transgenic maize exhibited a survival rate 80 %, higher than that of WT [43] . These results indicate that, in contrast to the effect on WTs, the drought tolerance of TPs was significantly enhanced.

7. CONCLUSION AND PERSPECTIVES

Breeding for drought tolerance in maize is a complex task, not least because drought can affect the crop at any stage of development. Many breeders have focused on alleviating the effects of drought at flowering and during grain filling because maize is most vulnerable to drought at these stages. There is no consensus about the best strategy to use in breeding maize for drought tolerance, and some researchers recommend a combination of two or more of the above mentioned approaches. However, a selection index combining secondary traits such as ASI and EPP with grain yield should result in faster improvement of grain yield under drought stress than selection for grain yield alone. It is desirable that more breeding programs use high-priority abiotic stresses in their mainstream breeding program, so that more experience on breeding approaches that effectively target stress environments can be gained. Such insights are particularly relevant for breeders in low-income countries that target production conditions that are stressed due to both biophysical and socio-economic reasons.

ACKNOWLEDGEMENTS

Financial supports from the General Project of Natural Science Foundation of Chong- qing (cstc2016jcyjA0344), the National University Student Innovation Program (2014- 10635042), the Fundamental Research Funds for the Central Universities (XDJK2014- C149), and Maize Germplasm Resources Protection Project (2016NWB036-04-10) are thankfully acknowledged.

CITE THIS PAPER

Maazou, A.-R.S., Tu, J.L., Qiu, J. and Liu, Z.Z. (2016) Breeding for Drought Tolerance in Maize (Zea mays L.). American Journal of Plant Sciences, 7, 1858- 1870.

REFERENCES

1. Hoisington, D., Khairallah, M., Reeves, T., Ribaut, J.M., Skovmand, B., Taba, S. and Warburton, M. (1996) Plant Genetic Resources: What Can They Contribute toward Increased Crop Productivity. Proceedings of the National Academy of Sciences of the United States of America, 96, 5937-5943.

2. Von Braun, J., Byerlee, D., Chartres, C., Lumpkin, T., Olembo, N. and Waage, J.J. (2010) A Draft Strategy and Results Framework for the CGIAR. World Bank, CGIAR, Washington DC. http://www.cgiar.org/www-archive/www. cgiar.org/ changemanagement /pdf/cgiar_srf_june7_2010.pdf
3. Chaudhary, H.K., Kaila, V. and Rather, S.A. (2014) Maize. In: Pratap, A. and Kumar, J., Eds., Alien Gene Transfer in Crop Plants: Achievements and Impacts, Springer, New York, 27-50.
4. FAOSTAT (2010) Statistical Database of the Food and Agriculture Organization of the United Nations. FAO, Rome.
5. World Bank (2005) Agriculture Investment Sourcebook. Agriculture and Rural Development. World Bank, Washington DC. http://documents.worldbank.org /curated/en/2005/06/6425137/agriculture-investment-sourcebook
6. Bruce, W.B., Edmeades, G.O. and Barker, T.C. (2002) Molecular and Physiological Approaches to Maize Improvement for Drought Tolerance. Journal of Experimental Botany, 53, 13-25.
7. Lopes, M.S., Araus, J.L., Van Heerden, P.D.R. and Foyer, C.H. (2011) Enhancing Drought Tolerance in C4 Crops. Journal of Experimental Botany, 62, 3135-3153.
8. Kikuchi, A., Huynh, H.D., Endo, T. and Watanabe, K. (2015) Review of Recent Transgenic Studies on Abiotic Stress Tolerance and Future Molecular Breeding in Potato. Breeding Science, 65, 85-102.
9. Mickelbart, M.V., Hasegawa, P.M. and Bailey-Serres, J. (2015) Genetic Mechanisms of Abiotic Stress Tolerance That Translate to Crop Yield Stability. Nature Reviews Genetics, 16, 237-251.
10. Sari-Gorla, M., Krajewski, P., Di Fonzo, N., Villa, M. and Frova, C. (1999) Genetic Analysis of Drought Tolerance in Maize by Molecular Markers II. Plant Height and Flowering. Theoretical and Applied Genetics, 99, 289-295.
11. Edmeades, G.O. (2013) Progress in Achieving and Delivering Drought Tolerance in Maize: An Update. ISAAA, Ithaca.
12. Edmeades, G.O., Bolanos, J., Elings, A., Ribaut, J.M., Banziger, M. and Westgate, M.E. (2000) The Role and Regulation of the Anthesis-Silking Interval in Maize. In: Westgate, M.E. and Boote, K.J., Eds., Physiology and Modeling Kernel Set in Maize, Vol. 29, CSSA Special Publication, Wisconsin, 43-73.https://dl.sciencesocieties. org/publications/books/pdfs/cssaspecialpubl/physiologyandmo/43
13. Jurgens, S.K., Johnson, R.R. and Boyer, J.S. (1978) Dry Matter Production and Translocation in Maize Subjected to Drought during Grain Fill. Agronomy Journal, 70, 678-682.
14. Ribaut, J.M., Betran, J., Monneveux, P. and Setter, T. (2009) Drought Tolerance in Maize. In: Bennetzen, J. and Hake, S., Eds., Handbook of Maize: Its Biology, Springer, New York, 311-344.
15. Robertson, M.J., Fukai, S., Ludlow, M.M. and Hammer, G.L. (1993) Water Extraction by Grain Sorghum in a Sub-Humid Environment. II. Extraction in Relation to Root Growth. Field Crops Research, 33, 99-112.
16. Hsiao, T.C. and Xu, L.K. (2000) Sensitivity of Growth of Roots versus Leaves to Water Stress: Biophysical Analysis and Relation to Water Transport. Journal of Experimental Botany, 51, 1595-1616.
17. Tardieu, F.T. and Simonneau, T. (1998) Variability among Species of Stomatal Control under Fluctuating Soil Water Status and Evaporative Demand: Modeling Isohydric and Anisohydric Behaviors. Journal of Experimental Botany, 49, 419-432.
18. Cochard, H. (2002) Xylem Embolism and Drought-Induced Stomatal Closure in Maize. Planta, 215, 466-471.
19. Magorokosho, C., Pixley, K.V. and Tongoona, P. (2003) Selection for Drought Tolerance in Two Tropical Maize Populations. African Crop Science Journal, 11, 151-161. http://www.bioline.org.br/pdf?cs03018
20. Russell, W.A. (1984) Agronomic Performance of Maize Cultivars Representing Different Eras of Breeding. Maydica, 29, 375-390.https://www.resear

chgate.net/publication/284582291_Agronomic_performance_of_maize_cultivars_representing_different_eras_of_breeding
21. Bolanos, J. and Edmeades, G.O. (1993) Eight Cycles of Selection for Drought Tolerance in Tropical Maize. I. Responses in Grain Yield, Biomass, and Radiation Utilization. Field Crops Research, 31, 233-252.
22. Bolanos, J. and Edmeades, G.O. (1993) Eight Cycles of Selection for Drought Tolerance in Tropical Maize. II. Responses in Reproductive Behavior. Field Crops Research, 31, 253-268.
23. Chapman, S.C. and Edmeades, G.O. (1999) Selection Improves Drought Tolerance in Tropical Maize Populations: II. Direct and Correlated Responses among Secondary Traits. Crop Science, 39, 1315-1324.
24. Edmeades, G.O., Bolanos, J., Chapman, S.C., Lafitte, H.R. and Banziger, M. (1999) Selection Improves Drought Tolerance in Tropical Maize Populations: I. Gains in Biomass, Grain Yield, and Harvest Index. Crop Science, 39, 1306-1315.
25. Monneveux, P., Sanchez, C., Beck, D. and Edmeades, G.O. (2005) Drought Tolerance Improvement in Tropical Maize Source Populations. Crop Science, 46, 180-191.
26. Banziger, M., Setimela, P.S., Hodson, D. and Vivek, B. (2005) Breeding for Improved Abiotic Stress Tolerance in Maize Adapted to Southern Africa. Agricultural Water Management, 80, 212-224.
27. Banziger, M., Edmeades, G.O. and Lafitte, H.R. (1999) Selection for Drought Tolerance Increases Maize Yields over a Range of Nitrogen Levels. Crop Science, 39, 1035-1040.
28. Lafitte, H.R., Blum, A. and Atlin, G. (2003) Using Secondary Traits to Help Identify Drought-Tolerant Genotypes. In: Fischer, K.S., Lafitte, R.H., Fukai, S., Atlin, G. and Hardy, B., Eds., Breeding Rice for Drought-Prone Environments, International Rice Research Institute, Manila, 37-48.
29. Bolanos, J. and Edmeades, G.O. (1996) The Importance of the Anthesis-Silking Interval in Breeding for Drought Tolerance in Tropical Maize. Field Crops Research, 48, 65-80.
30. Banziger, M. and Lafitte, H.R. (1996) Efficiency of Secondary Traits for Improving Maize for Low-Nitrogen Target Environments. Crop Science, 37, 1110-1117.
31. Ziyomo, C. and Bernardo, R. (2013) Drought Tolerance in Maize: Indirect Selection through Secondary Traits versus Genome Wide Selection. Crop Science, 53, 1269-1275.
32. Tuberosa, R. and Salvi, S. (2006) Genomics-Based Approaches to Improve Drought Tolerance of Crops. Trends in Plant Science, 11, 405-412.
33. Zhu, J.J., Wang, X.P., Sun, C.X., Zhu, X.M., Li, M., Zhang, G.D., Tian, Y.C. and Wang, Z.L. (2011) Mapping of QTL Associated with Drought Tolerance in a Semi-Automobile Rain Shelter in Maize (Zea mays L.). Agricultural Sciences in China, 10, 978-996.
34. Ribaut, J.-M., Edmeades, G. and Hoisington, D. (2000) The Genetic Basis of Drought Tolerance in Maize and Options for Improvement via Marker-Assisted Selection. Developments in Plant Genetics and Breeding, 5, 35-41.
35. Fu, F.-L., Feng, Z.-L., Gao, S.-B., Zhou, S.-F. and Li, W.-C. (2008) Evaluation and Quantitative Inheritance of Several Drought-Relative Traits in Maize. Agricultural Sciences in China, 7, 280-290.
36. Li, W.-J., Liu, Z.-Z., Shi, Y.-S., Song, Y.-C., Wang, T.-Y., Xu, C.-W. and Li, Y. (2010) Detection of Consensus Genomic Region of QTLs Relevant to Drought-Tolerance in Maize by QTL Meta-Analysis and Bioinformatics Approach. Acta Agronomica Sinica, 36, 1457-1467.
37. Messmer, R., Fracheboud, Y., Banziger, M., Stamp, P. and Ribaut, J.M. (2011) Drought Stress and Tropical Maize: QTLs for Leaf Greenness, Plant Senescence, and Root Capacitance. Field Crops Research, 124, 93-103.
38. Bohnert, H.J., Bressan, R.A. and Hasegawa, P.M. (2006) Genetic Basis of Ion Homeostasis and Water Deficit. In: Ribaut, J.M., Ed., Drought Adaptation in Cereals, The Haworth Press, Inc., New York, 551-582.

39. Bohnert, H.J., Gong, Q., Li, P. and Ma, S. (2006) Unraveling Abiotic Stress Tolerance Mechanisms Getting Genomics Going. Current Opinion in Plant Biology, 9, 180-188.
40. Li, M., Xu, M., Yang, W., Kong, Z. and Xue, Y. (2007) Genome-Wide Gene Expression Profiling Reveals Conserved and Novel Molecular Functions of the Stigma in Rice (Oryza sativa L.). Plant Physiology, 144, 1797-1812.
41. Talame, V., Ozturk, N.Z., Bohnert, H.J. and Tuberosa, R. (2007) Barley Transcript Profiles under Dehydration Shock and Drought Stress Treatments: A Comparative Analysis. Journal of Experimental Botany, 58, 229-240.
42. Quan, R., Shang, M., Zhang, H., Zhao, Y. and Zhang, J. (2004) Engineering of Enhanced Glycine Betaine Synthesis Improves Drought Tolerance in Maize. Plant Biotechnology Journal, 2, 477-486.
43. Liu, Y., Qin, L., Han, L., Xiang, Y. and Zhao, D. (2015) Overexpression of Maize SDD1 (ZmSDD1) Improves Drought Resistance in Zea mays L. by Reducing Stomatal Density. Plant Cell, Tissue and Organ Culture, 122, 147-159.

CHAPTER 7

Castor Breeding

Máira Milani[1] and Márcia Barreto de Medeiros Nóbrega[1]

[1] Embrapa Cotton, Campina Grande, Paraíba, Brazil

1. INTRODUCTION

Produced from the seeds of *Ricinus communis*, castor oil is an important feedstock for the chemical industry because it is the only commercial source of ricinoleic acid, a hydroxy fatty acid, which comprises about 90% of the oil. In addition to the traditional uses of ricinoleic acid, there is also a demand for vegetable oil to be used as biofuel and for nem products derived from the castor oil. Due to the increasing demand in the global market, there is a short supply of castor oil and this trend seems to get worst every year. Castor is an ideal candidate for production of high value, industrial oil feedstocks because of the very high oil content (48-60%) of the seed, and the extremely high levels of potential oil production [1] Due to the ricinoleic acid, castor oil and its derivatives arc of great versatility being used in synthesis routes for a large number of products and are increasing rapidly [2,3]. In some places of the world it is used like an ornamental due to their vibrant leaf and floral coloration [4].

All over the world, cultivation is done by small farmers in countries such as India, China, and Brasil, and FAO statistics report seed yield averages of 1,104.8; 911.8; and 701.1kg/ha respectively for these countries (Figure 1). In Brazil and in India, the production is made in arid or semiarid regions. In these environments, the rainfall, is generally erratic and low, and the availability of water is the major factor affecting yield.

In Brazil, the production of castor oil is concentrated in the semi-arid northeast, mainly in the state of Bahia, which accounts for more than 80% of the production and acreage [6, 7]. The culture system used by small producers in Brazil usually involves intercropping with food crops mainly maize and beans, and low adoption of technologies. The whole system of production, from planting to processing is manual [8]. Mostly the use of local varieties with long cycle and uneven seed maturation, little or no soil tillage and fertilization. Using such technology farmers have low-income, and the national seed yield of castor is low as 600 kg/ha [7]. These values are too low to make the production profitable. The cultivars developed by Embrapa Cotton for the traditional areas of cultivation of castor, produce an average of 1500 kg/ha of castor in farmers fields [9].

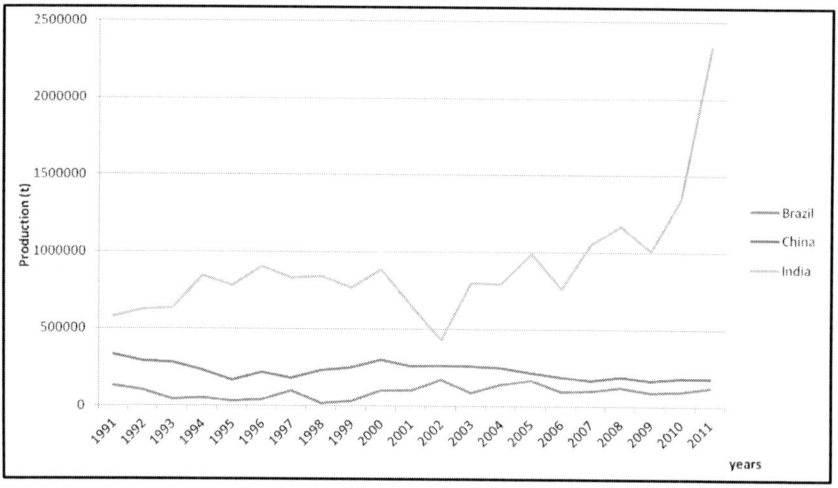

FIGURE 1. Yield of castor seed in the three main producers countries in twenty years. Data Source: [5]

The populations synthesized by the breeding program of Embrapa Cotton have been evaluated along with public and private partners since 1987. The Research Group mainly evaluates selected genotypes based on the behavior per se of individuals or populations. The main objective is to obtain cultivar that are more productive and adapted to the environment of each growing region, to the production system, and the technological level applied.

The program aims to develop cultivars that are indehiscent, short, and easy to harvest. Earliness of seed maturation is also desired.

2. GENETIC RESOURCES

The taxonomy and geographic distribution were thoroughly studied and documented earlier in USSR [10], USA [1], Brazil [11] and India [12,13,14].

Castor is reported to have a polyphyletic origin, both India and Africa were considered as the origin of castor based on its widespread cultivation, documents of its medicinal uses and physical evidences. Due to its widespread survival and perennial nature, all possible transitions from an uncultivated plant to a weedy plant and from semi cultivated to a field crop exist and there is no gap between uncultivated and cultivated castor.

The genetic diversity in castor is restricted due to its monotypic existence. Six subspecies viz., persicus, chinensis, zanzibarinus, sanguineus, Africans, Mexicans were identified based on eco-geographical grouping [12, 10, 15]. However, there is no difference in the chromosome number (2n=20) among the sub-species and they all can cross easily with each other [12, 16].

According [17] and [18], the castor can be classified as:

- Superdivision: *Spermatophyta* - Seed plants
- Division: *Magnoliophyta* - Flowering plants
- Class: *Magnoliopsida* - Dicotyledons
- Subclass: *Rosidae*
- Order: *Euphorbiales*
- Family: *Euphorbiaceae* - Spurge family
- Genus: *Ricinus* L. - Ricinus
- Species: *Ricinus communis* L. - Castor

According [19], castor must be classified as Angiospermae, Eudicotyledone, Rosanae and Malpighiales.

Although generally known as "castor bean", this plant is not a legume, and the term "bean" should be discontinued in favor of castor plant and castor seed [15] 2000). Avoiding to use the term bean is really important because these seed and the whole plant are very poisonous and should not be eaten.

FIGURE 2. Examples of castor bean plants with different height. Photos: Máira Milani, Embrapa Cotton.

A great variation in phenotypic expression is observed due to its cross-pollinated nature. Example of high variability in morphological characters are stem color, epicuticular wax (bloom wax), plant height (Figure 2), presence of spines on capsules (Figure 3), branching pattern, leaf shape, sex expression (Figure 4), seed color, and response to environmental conditions. Wide variation was observed in several morphological traits in the germplasm collections in India, USSR and elsewhere [14, 20, 15]. Also for quantitative traits its genetic polymorphism is exploitable in breeding programs [21, 22, 23, 24].

Germplasm banks are the basic providers of useful genes and genotypes needed to achieve the desirable genetic improvement in breeding programs; however, the resources available in castor germplasm worldwide have been barely tapped for castor genetic improvement and the majority of them have been poorly characterized [25]. The use of genetic resources by the global castor community could be increased if there were characterization of accessions, consolidated reports on available resources, free accession to information on banks, and uniform data collection standards among repositories [25].

FIGURE 3. Examples of different colors in the fruits of castor bean. Photos: Máira Milani, Embrapa Cotton.

These enhancements would allow an estimate of the genetic variability with single collections without the flux of accessions between countries. Germplasm characterization would also be easier if fast, non-destructive, and reliable screening methods were developed. An example is the quick and non-destructive method for estimating ricinoleic fatty acid content by Nuclear Magnetic Resonance in seeds [26].

Normally, castor is monoecious, with pistillate flowers on the upper part of raceme and staminate flowers on the lower part (Fig. 4a). This type is referred to as normal monoecious. Another type, referred as interspersed monoecious, has pistillate and staminate flowers interspersed along the entire raceme axis. The proportion of pistillate and staminate flowers among racemes can vary widely both within and among genotypes. It can also be influenced considerably by environment [27]

FIGURE 4. Arrangement of male and female flowers in racemes of castor: a)monoic normal; b and c) gynodioc; d) androdioic; e)interspersed; f) monoic bearing some perfect flowers.

In normal monoecious varieties, the percentage of pistillate flowers along the raceme axis is usually the highest on the first raceme, with a decreasing percentage on subsequently developed racemes. With the decrease in pistillate flowers, there is a proportional increase in the number of staminate flowers [27]. This within plant variation is generally associated with the seasons. Female tendency is highest in spring and early summer; male tendency is highest in mid and late summer. Temperature is probably the main environmental component affecting sex. Moderate temperatures promote female flowers while high temperature promote male flowers. However, age of plant and nutrition can also influence sex expression. Femaleness is strongest in young plants with a high level of nutrition. Maleness is strongest in old plants with a low level of nutrition [28].

In addition to monoecism, a subtype of dioecism occurs in plants with only pistillate flowers along the entire raceme axis of all racemes [27]. The counterpart, plants with only staminate flowers, can occur in extreme climatic conditions, with high temperature or water deficit (Fig. 4d).

There are three types of pistillate lines (employed for hybrid production): N, S, and NES. In the N type, the femaleness is controlled by a recessive gene (ff). In the S type, femaleness is controlled by a polygenic complex with dominant and epistatic effects in which the plant starts as female, but a reversion can occur at any time. In the NES type, the plant has the recessive gene (ff) that allows it to start as female, but the presence of environmentally sensitive genes triggers a sexual reversion when temperature is higher than 31 °C [29,30, 31].

The development of pistillate lines has allowed breeders to successfully utilize heterosis (hybrid vigor) in castor. Prior to the development of pistillate lines, inbred lines having many female flowers were used as female lines. Commercial exploitation of heterosis in India was instantly adopted after the development of VP-1, a S type stable pistillate line derived from TSP 10 R (Texas Stable Pistillate 10R) introduced from the USA [31]. Several pistillate lines were developed using VP-1 source of pistillate expression [32,33,34]. Other pistillate lines were developed using NES type of sexual expression but GCH-6 is the only commercial hybrid based on that system. Several other sources of pistillate lines were identified by screening 1 250 accessions from the germplasm bank at DOR, India [25, 34].

The adoption of male-sterile lines could be an alternative to pistillate lines for the production of hybrid seeds. Some studies were performed looking for genotypes with male sterility or inducing it through mutation [35], but we did not find any reference to a genotype expressing male-sterility for adoption into hybrid development programs.

A male flower, after opening, releases viable pollen grains for 1 to 2 days. The best environmental conditions for pollen dispersal are at a temperature between 26 °C to 29 °C and relative humidity of 60%, which may vary according to the cultivar. The pollen contain allergenic substances similar to those found in the seeds, which are of protein nature, such as ricin, robina, crotin, and circina Arbina [15] 2000). High temperatures, plant age, and short day length favor the appearence of male flowers.

Regarding the female flowers, the literature shows that stigmas become receptive before the anthesis of male flowers. However the existence of this short protogynous phase [36] is not accepted by most researchers [37] who claim the male flowers reach maturity first and anthesis usually occurs in a short period of time before the opening of the female flowers [1]. In this way, there is a large source of pollen for the first pistillate flowers that open and become receptive. The stigma is fully receptive a few hours after the flower opens, but it is difficult for pollination to occur shortly after the opening of the flower. The stigma remains receptive after anthesis, , for a period of 5 to 10 days depending on environmental conditions [38].

Castor has a mixed mating system generating both selfed and cross fertilized offspring. Under natural conditions, cross pollination in castor can exceed 80% [11], but the actual level of cross pollination is dependent on both genotype and environmental conditions. Since pollination occurs mostly by wind, genetic purity of individual accessions can be maintained by planting in isolation by time or space (usually 1,000 m from other accessions) or covering the inflorescence with a paper bag [39]. This later option is labor intensive and expensive, but usually more practical if breeders need just a few seeds. Storing pollen is another option for germplasm conservation. [40] observed that castor pollen grains were viable after being stored at temperatures of -196° C, -80° C, and -18° C for up 30 days and there is evidence that pollen viability would be retained for long periods with cryopreservation at -80° C.

3. GOALS OF CASTOR BEAN BREEDING

Presently the main objectives of the breeding programs around the world are: earliness of seed maturation, plants architecture for mechanized harvest and disease resistance (root not and gray mold). These should be combined with superior productivity of cultivars and at least of 48% oil content of seed. Most breeding programs have searched genotypes with short height (less than 1.5m), height of primary raceme between 20 and 40cm, less than 150 days for harvesting, erect plant and non-shattering fruits.

In some regions, castor has have been selected for increased tolerance to abiotic stresses such as drought, high temperature, salinity, and exchangeable aluminium. Considering that the castor is not a food crop, it is often considered to be cultivated in marginal areas.

The reduction of the toxicity of castor seeds has also been the target of improvement programs. Ricin is a protein toxin found in the endosperm of castor seed capable of inhibiting protein synthesis by enzymatically blocking the ribosomes of eukaryotes [41]. Castor oil does not contain ricin because this protein is insoluble in oil, and any residual ricin is eliminated in the refining process. Ricin content varies among genotypes. The ricin content varied from 1.9 a 16 g/kg among 263 accessions from the USDA Germplasm Bank [42] and from 3.5 to 32.2 in varieties and accessions from the Embrapa Germplasm Bank [43].

The development of new cultivars with traits of interest and adapted to specific microclimates is only possible when there is available knowledge about the extent of genetic diversity of the species [44]. Despite the recent publication of the castor bean genome [45], little is known about the actual genetic diversity of this species. Genetic diversity analyses of castor bean germplasm collections worldwide have showed low levels of variability and lack of geographically structured genetic populations, regardless of marker system used (e.g. [4, 46, 47]). Thus, the remarkable phenotypic variation observed in castor does not seem to reflect a high genetic diversity, similar to the one reported for physic nut, in which variations in epigenetic mechanisms may have a more important role in the diversity of the species than genetic variability per se [48]. Castor diversity is still poorly characterized by means of molecular marker systems [49, 50, 51, 52, 53]. In fact, the species has been overlooked until the late 2000s, when analyses regarding genetic diversity of germplasm collections were first published [46]. Thus, obtaining the desired genotypes implies the characterization of the germplasm banks and the proper publication of these results.

4. BREEDING METHODS

In early phases of breeding programs, more attention is given to qualitative characters, but in later stages of improvement greater emphasis is shifted to quantitative traits such as yield, plant height, days to flowering, and traits associated with agronomic and economic factors.

Because castor has both self and crossed pollination and most of pollination is made by wind, contamination of varieties during seed production is a constant risk. To prevent contamination, it is necessary to isolate the area, physically (1000 meters) or temporally, or the use of self-fertilization using paper bags (Fig. 4). Both strategies are expansive. The self-fertilization is a hand labor and normally demands many people and time. is practically impossible to keep the distance recommended in areas multiplication of lines, where they are multiplied dozens of strains simultaneously because it would require a large extension of the area.

Heterosis is a option for the development of hybrid cultivars of castor oil, representing an effective way to increase yield. In castor, this technique is possible due to the occurrence of gynodioecious plants whose genetic control is assigned to a recessive allele. However, the maintenance of female lines in castor increases the costs of production of hybrids. Thus, it is believed that the maintenance and propagation of female lines by micro propagation could be performed in vitro and therefore the purity of the female lines could be easily ensured, and manufacturing costs would be lower. Embrapa Cotton has been testing methods of clonal propagation in vitro and ex vitro in the castor, in order to regenerate and increase germplasm bank accessions, including a few female lines.

FIGURE 5. Self fertilization in castor. A) raceme with growing flowers; B) paper bag being placed over the raceme; C) Fixing paper bag; D) Identification. Photos of Marcia B. M. Nobrega, Embrapa Cotton.

The cultivar development is divided into two main phases: pre-breeding and breeding. Both are essential to reaching its ultimate goal,which is to release new productive cultivars with wide adaptation, stability and good acceptance among producers. The pre-breeding, by definition, is the "bridge" between genetic resources and improvement [54]. In addition to the activities of these two steps it is essential to support activities of processes such as evaluation of the rate of outcrossing, asexual multiplication (in vitro and ex vitro), seed multiplication, and others. Each process is very important for the outcome. On average, the development of cultivars takes 10 to 12 years from the selection of germplasm to the legal process of plant variety protection.

Embrapa Cotton with partners developed four castor cultivars using methods applied for inbreed populations: BRS Nordestina, BRS Paraguaçu, BRS Energia e BRS Gabriela. These varieties are recommended for the states of North and Northeast of Brazil.

FIGURE 6. Embrapa's castor cultivars: (A) BRS Nordestina, (B) BRS Paraguaçu, (C) BRS Energia, and (D) BRS Gabriela. Photos of Máira Milani, Embrapa Algodão.

BRS Nordestina stands out from the average height of 1.90 m, greenish stems with the presence of wax, conical racemes, semi-dehiscent fruits, and black seeds. The period between emergence and first flowering raceme is 50 days, on average, while the average weight of 100 seeds is 68 g, and the oil seed content is 48%. The average yield is 1,500 kg/ha under conditions of normal rainfall in the Northeast semiarid region. The period between the emergencee until the last harvest is 250 days. The BRS Paraguaçu has an average height of 1.60 m, purple wax stem, oval raceme, semi-dehiscent fruits and black seeds. The period from emergence to flowering is 54 days, while the average weight weight of 100 seeds is 71g, and the oil seed content is 47%. The average productivity is 1,500 kg/ha under rainfed conditions of the semiarid region of the Northeast. Earliness is a key feature of BRS Energia, whose average cycle is 120 days between emergence and maturation of the last racemes. The appearance of the first raceme occurs about 30 days after germination. The yield of this cultivar is 1.800kg/ha under the same climatic conditions of the others. The average plant height is 1.40 m, 100 seed weight is around 40 g and the seed oil content is 48%. The BRS Gabriela has the highest seed oil content, 50% on average. It has a mean cycle of 150 days, productivity of 1900 kg/ha. The racemes have a round shape, immature green fruit with wax, an average density of spines, average density of fruit and green pink spines. Under extremes of precipitation (high or low), the density of fruits in racemes can be sparse. The fruits are indehiscent.

4.1. Mass Selection

Mass selection consists in the selection of superior types and the discharge of undesirable types within a plant population. It is used for imprioved cultivars or established local types to improve, or standardize traits of economic importance.

Mass selection is the most effective method for characteristics with high heritability in populations with high levels of natural genetic variability. Two procedures are useful in increasing the efficiency of the mass selection in populations of castor: the self-fertilization of the selected plants to prevent cross pollination, and the use of controlled selection techniques to reduce environmental variation [55]. Mass selection was used to develop IAC-38, an important dwarf castor cultivar in Brazil [11]

4.2. Individual Plant Selection With Progeny Tests

This method consists in selecting individual plants and the subsequent study of their offspring in progeny trials. It is based on the principle that the breeding value of a plant may be measured by the performance of its progeny. It is a straightforward procedure to achieve greater uniformity and increased production in castor. The method of progeny test is highly effective for the improvement of populations of castor with high levels of natural genetic variability. This method was successfully used in the development of the high yielding cultivar 'Guarany' [56]

4.3. Methods Involving Sexual Hybridization

When populations of castor with sufficient natural genetic variation for agronomic characteristics are not available, it is necessary to generate variability by producing hybrids between different lines or cultivars [55]. The choice of the parents of these populations must be based on their agronomic performance within the targeted production region, and diallel cross can be used if there are several promising parents or cultivars [55].

The pedigree method is adequate for simultaneous selection of several traits. This methods has been used to develop the cultivar IAC-2028, a dwarf and not-shattering genotype in Brazil [57].

The bulk method is the most effective option when the main objective is to improve the adaptation of castor to stress conditions such as drought, acid soils, high levels of salt and resistance to diseases [55]. The backcross method of selection is the most effective when there is a need to improve some simply inherited, qualitative characteristic in a commercial cultivar or promising elite line. The non-recurrent parent must have the characteristic absent from the recurrent parent. The method of backcrossing is especially effective in castor for the improvement of characteristics such as seed shattering, flower height, and disease resistance [55].

Recurrent selection is defined as successive cycles of selection and recombination of selected lines or individual plants [55]. It is not often used for castor selection, but it has been successful on the reduction of height of the cultivar Guarani [58, 59]. In each of five cycles of selection, plant height was reduced by 28 cm, 13 cm, 19.9 cm, 11.7 cm and 3.4 cm [59].

In the last three decades, India has made significant progress in the development of hybrids [60]. The availability of pistilate lines, like the VP-1, was the base for launching hybrids such as GAUCH-1, GCH-2, and GCH-4 during 1990s and ten more high yielding hybrids later on [60].

The first commercial castor hybrid, 'GCH 3', was developed in India and had high seed yield potential (88% superior to the most planted cultivars at that time), drought tolerance, medium maturity time (140-210 days) and high oil content (46%). Since then, a total of 15 hybrids were released in India, some of them with resistance to fusarium wilt and high seed yield potential [34]. The advantages of hybrids over cultivars resulted in a predominance of hybrids (50 to 60%) in the castor production in India. In the State of Gujarat, the use of hybrid seed is up to 95% of the cultivated area. In Gujarat, where castor is cultivated mostly under irrigated conditions, the adoption of hybrid seed has caused an increase in seed yield from 350 to 1 970 kg/ha within a few years [34].

The intensity of heterosis on castor seed yield depends on both the genetic diversity and individual combining ability of the parents [61, 62, 63, 64, 32, 33, 34]. More studies on genetic diversity and combining ability are necessary for supporting the development of hybrids.

An alternative method for selecting castor hybrids was successfully evaluated by Toppa [65]. The method of cryptic hybrids was proposed by Lonqquist [66] for maize, consisting of simultaneous self-pollination and crossing in the same plant, allowing the selection of the best progeny at each cycle. Because castor has a low endogamic depression and produces more than one raceme per plant, the method can be employed. After four cycles of selection, the 12 cryptic hybrids had higher seed yield (1,675 kg/ha) than the 12 conventional hybrids (1,550 kg/ha) evaluated over two years in two locations [65].

5. CHALLENGES

The scarcity of labor that has been observed in rural areas over the past decade, has raised the costs of operation management and cultivation. A research group in Brazil has been focusing on the research for indehiscent cultivars with shorter plants to facilitate mechanical harvesting operations. Also earlier genotypes have been obtained to reduce the residence time of the crop in the field.

In a review of the challenges to reach greater productivity with the castor [67], the authors mention that the main challenge in developing cultivars is the castor plant adaptation to Combine harvesting. Both cultivars as machines, require further adjusted in order to obtain more efficiency in the process of Combine harvesting.

Nowadays, breeders look for plants that could be adapted for a variety environmental conditions including the increased ambient temperature caused by Global Climate Change.

REFERENCES

1. Brigham RD. Castor: Return of an old crop. In Janick J; Simon JE (ed.) Progress in New Crops. New York: Wiley, 1993. 380-383.
2. Chierice GO; Claro Neto S. Aplicação industrial do óleo. In: Azevedo DMP; Beltrão NEM (Ed.). O Agronegócio da mamona no Brasil. Brasília, DF:Embrapa Informação Tecnológica, Campina Grande: Embrapa Algodão, 2007, 417-447.
3. CASTOROIL.IN Comprehensive Castor Oil Report: A report on castor oil & castor oil derivatives, in: http://www.castoroil.in/reference/report/report.html. (Accessed 31 january 2013)

4. Foster JT, Allan GJ, Chan AP, Rabinowicz PD, Ravel J, Jackson PL. Single nucleotide polymorphisms forasses sing genetic diversity in castor bean (Ricinus communis). BMC Plant Biol. , 201010:13-23.
5. FAOSTAT http://faostatfao.org/site/567/DesktopDefault.aspx?PageID=567#ancor 2012 (accessed 12 november 2012)
6. Carvalho BCL. Manual do cultivo da mamona. Salvador: EBDA. 2005
7. CONAB. Série histórica: mamona. Brasília, DF: Central de Informações Agropecuárias. http://www.conab.gov.br/conteudos.php?a=1382&t=2. (Accessed 31jan2013)
8. Savy Filho A. Mamona: tecnologia agrícola. Campinas: Emopi, 2005
9. Freire EC; Lima EF; Andrade FP; Milani M; Nóbrega MBM. Melhoramento genético. In: Azevedo DMP; Beltrão NEM (Ed.). O Agronegócio da mamona no Brasil. Brasília, DF:mbrapa Informação Tecnológica, Campina Grande: Embrapa Algodão, 2007. 169-194
10. Moshkin VA. Castor. New Delhi: Amerind; 1986
11. Savy Filho A. Castor bean breeding. In: Borém A (Ed.) Improvement of Cultivated Species. Viçosa: Federal University of Viçosa, 2005
12. Kulkarni LG, Ramanamurthy GV. Castor. New Delhi: Indian Council of Agric Res., 1977
13. DOR. Castor in India. Hyderabad: Directorate of Oilseeds Research. 2003
14. Anjani K. Extra-early maturing germplasm for utilization in castor improvement. Ind. Crops Products 2010, 31, 139-144
15. Weiss EA. Oilseed crops. 2nd ed. Oxford: Blackwell Science, 2000
16. Atsmon D. Castor. In: Röbbelen G, Downey RK, Ashri A (Eds.) Oil Crops of the World. New York: McGraw Hill, 1989; p 348–447,
17. Vidal WN, Vidal MRR. Fitossistemática: famílias de angiospermas.Viçosa: UFV, 1980. 59.
18. Popova GM, Moshkin VA. Botanical classification. In: Moshkin VA (Ed.). Castor. New Delhi: Amerind, 1986
19. Angiosperm Phylogeny Group III. An update of the Angiosperm Phylogeny Group classification for the orders and families of flowering plants: APG III. Botanical Journal of the Linnean Society, London, 2009, 161: 105 – 121
20. Moshkin VA, Dvoryadkina AG. Cytology and genetics of qualitative characters. In: Moshkin VA Castor. New Delhi: Amerind, 1986. 93-102
21. Uguru MI, Abuka IN. Hybrid vigour and genetic actions for two qualitative traits in castor plant (Ricinus communis L.). Ghana Journal. Agric. Sci. 1998; 31: 81 – 82
22. Bahia HF, Silva SA, Fernandez LG, Ledo CAS, Moreira RFC. Divergência genética entre cinco cultivares de mamoneira. Pesquisa Agropecuária Brasileira, 2008. 43, 357-362.
23. Bezerra Neto FV, Leal NR, Gonçalves LSA, Rêgo Filho LM, Amaral Júnior AT. Descritores quantitativos na estimativa da divergência genética entre genótipos de mamoneira utilizando análises multivariadas Rev. Ciênc. Agron., 2010, 41, 294-299
24. Nóbrega MBM, Geraldi IO, Carvalho ADF. Avaliação de cultivares de mamona em cruzamentos dialélicos parciais. Bragantia 2010. 69. 281-288
25. Anjani K. Castor genetic resources: a primary gene pool for exploitation. Ind. Crops. Products 2012.35:1-14
26. Berman P, Nizri S, Parmet Y, Wiesman Z. Large scale-scale screening of intact castor seeds by viscosity using time domain NMR and chemometrics. J. Am. Oil Chem. Soc. 2010. 87:1247-1254.
27. Zimmerman LH, Smith JD. Production of F1 seed in castorbean by use of sex genes sensitive to environment. Crop Science, 1966, 6: 406-409
28. Shifriss O. Sex Instability in Ricinus. Genetics, 1956, 41: 265–280.
29. L. H Zimmerman, Castorbeans: a new crop for mechanized production. Adv. Agron. 1958X:257288
30. O Shifriss, Conventional and unconventional systems controlling sex variations in *Ricinus*. J. Genet. 1960

31. Ankineedu G, Rao GP. Development of pistillate castor. Indian J. Genet. Plant Breeding 1973, 33:416-422
32. Lavanya C, Ramanarao PV, Gopinath VV. Studies on combining ability and heterosis in castor hybrids. Journal of Oilseeds Research 2006, 23:174-177.
33. Pathak HC. Crop Improvement in Castor. In D.M. Hegde (ed.) Vegetable Oils Scenario: Approaches to Meet the Growing Demands. Hyderabad: Indian Society of Oilseeds Research, 2009. 82-94
34. Lavanya C, Solanki SS. Crop improvement of castor. The challenges ahead. In Hegde DM (ed.). Research and Development in Castor. Present status and future strategies. Hyderabad: Indian Society of Oilseeds Research, 2010, p 36-55
35. Chauhan SVS, Singh KP, Kinoshit T. Gamma-ray induced pollen sterility in castor. J. Fac. Agric. Hokkaido Univ. 1990.64, 229-234
36. Gurgel JT. Estudos sobre a mamoneira (Ricinus communis L.). Thesis for teaching profession. ESALQ/USP; 1945
37. Tavora FJAF. A cultura da mamona. Fortaleza: Epace, 1982
38. Moreira JAN, Lima EF, Farias FJC, Azevedo, DMP. Melhoramento da mamoneira (Ricinus communis L.). Campina Grande. Embrapa/CNPA, 1996
39. Rizzardo RAG. O papel de Apis mellifera L. como polinizador da mamoneira (Ricinus communis L.):avaliação da eficiência de polinização das abelhas e incremento de produtividade da cultura, MS thesis.Universidade Federal do Ceará; 2007
40. Vargas DP, Souza SAM, Silva SDA, Bobrowski VL. Pollen grain analysis of some cultivars of castor-oil plant (Ricinus communis L., Euphorbiaceae): conservation and viability. Arquivos do Instituto Biologico 2009. 76, 115-120
41. Khvostova IV. Ricin: the toxic protein of seed. In: Moshkin VA (Ed.) Castor. New Delhi: Amerind Publ., 1986. p 85–92
42. Pinkerton SD, Rolfe RD, Auld DL, Ghetie V, Lauterbach BF. Selection of castor with divergent concentrations of ricin and Ricinus communis agglutinin. Crop Sci. 1999; 39:353-357
43. Baldoni AB, Carvalho MH, Sousa NL, Nobrega MBM, Milani M, Aragão FJL. Variability of ricin content in mature seeds of castor bean. Pesquisa Agropecuária Brasileira, 46,, 2011. 776-779
44. P Gepts, Crop domestication as a long-term selection experiment. In: Janick J. (Ed.) Plant Breeding Reviews, Habaken: John Wiley & Sons, Inc., 200424144
45. Chan AP, Crabtree J, Zhao Q, Lorenzi H, Orvis J, Puiu D, Melake-Berhan A, Jones KM, Redman J, Chen G, Cahoon EB, Gedil M, Stanke M, Haas BJ, Wortman JR, Fraser-Liggett CM, Ravel J, Rabinowicz PD. Draft genome sequence of the oilseed species Ricinus communis. Nat. Biotechnol. 2010; 28:951-956.
46. Allan G, Williams A, Rabinowicz PD, Chan AP, Ravel J, Keim P. Worldwide genotyping of castor bean germplasm (Ricinus communis L.) using AFLPs and SSRs. Genet. Resour. Crop Evol. 2008; 55:365-378.
47. Qiu LJ, Yang C, Tian B, Yang JB, Liu AZ. Exploiting EST databases for the development and characterization of EST-SSR markers in castor bean (Ricinus communis L.). BMC Plant Biol. 2010; 10:278-287.
48. Yi C, Zhang S, Liu X, Bui H, Hong Y. Does epigenetic polymorphism contribute to phenotypic variances in Jatropha curcas L.? BMC Plant Biology 2010; 10: 1-9
49. Billotte N, Jourjon M, Marseillac N, Berger A, Flori A, Asmady H, Adon B, Singh R, Nouy B, Potier F, Cheah S, Rohde W, Ritter E, Courtois B, Charrier A, Mangin B. QTL detection by multi-parent linkage mapping in oil palm (Elaeis guineensis Jacq.). Theoretical and Applied Genetics, 2010; 120, 1673-1687
50. Feng S, Li W, Huang H, Wang J, Wu Y. Development, characterization and cross-species/genera transferability of EST-SSR markers for rubber tree (Hevea brasiliensis). Molecular Breeding, 2009; 23: 85-97
51. Sayama T, Hwang TY, Komatsu K, Takada Y, Takahashi M, Kato S, Sasama H, Higashi A, Nakamoto Y, Funatsuki H, Ishimoto M. Development and application of a whole-genome simple sequence repeat panel for high-throughput genotyping in soybean. DNA Research, 2011; 18:107-115

52. Sraphet S, Boonchanawiwat A, Thanyasiriwat T, Boonseng O, Tabata S, Sasamoto S, Shirasawa K, Isobe S, Lightfoot DA, Tangphatsornruang S, Triwitayakorn K. SSR and EST-SSR-based genetic linkage map of cassava (Manihot esculenta Crantz). Theoretical and Applied Genetics, 2011; 122: 1161-1170
53. Talia P, Nishinakamasu V, Hopp HE, Heinz RA, Paniego N. Genetic mapping of EST-SSRs, SSR and In Dels to improve saturation of genomic regions in a previously developed sunflower map. Electronic Journal of Biotechnology, 2010; 13: 6
54. Nass LL. Utilização de recursos genéticos no melhoramento. In: Nass LL, Valois ACC, Melo IS, Valadares-Inglis MC. Recursos genéticos e melhoramento de plantas. Rondonópolis: Fundação MT, 2001; 29-56
55. Auld DL, Zanotto MD, Mckeon T, Morris JB. Castor. In: Vollmann J, Rajcan I. (ed.) Oil Crops - Handbook of Plant Breeding. New York: Springer, 2009, 316-332
56. Amaral JGC Genetic variability for agronomic characteristics between self pollinated lines of castor (*Ricinus communis* Lcv. AL Guarany. Ph. Dissertation, College of Agronomic Sciences, São Paulo State University; 2003
57. Savy Filho A, Amorim EP, Ramos NP, Martins ALM, Cavichioli JC. IAC 2028: nova cultivar de mamona. Pesq. agropec. bras., 2007, 42:449-452
58. Zanotto MD, Amaral JGC, Poletine JP. Recurrent selection with use of self pollinated lines for reduction of the height of plants of castor (Ricinus communis L.) in a common Guarani population. Proc. Congresso Brasileiro de Mamona, 1st, 2004, 1:1-5. Campina Grande. Embrapa Algodão, Campina Grande, 2004
59. Oliveira IJ, Zanotto MD. Efficiency of recurrent selection for reduction of the stature of plants in castor (Ricinus communis L.). Cienc. Agrotecnol. 2008, 32:1107–1112.
60. Sujatha M, Reddy TP, Mahasi MJ. Role of biotechnological interventions in the improvement of castor (Ricinus communis L.) and Jatropha curcas L. Biotechnol. Adv. 2008; 26:424–435.
61. Costa MN, Pereira WE, Bruno RLA, Freire EC, Nóbrega MBM, Milani M, Oliveira AP. Genetic divergence on castor bean accesses and cultivars through multivariate analysis. Pesquisa Agropecuária Brasileira, 2006; 41:1617-1622.
62. Golakia PR, Madaria RB, Kavani RH, Mehta DR. Gene effects, heterosis and inbreeding depression in castor, Ricinus communis L. Journal of Oilseeds Research, 2004; 21:270-273
63. Lavanya C, Chandramohan Y. Combining ability and heterosis for seed yield and yield components in castor. Journal of Oilseeds Research, 2003;20:220-224
64. Ramana PV, Lavanya C, Ratnasree P. Combining ability and heterosis studies under rainfed conditions in castor (Ricinus communis L.,). Indian Journal of Genetics and Plant Breeding, 2005, 65:325-326
65. Toppa EVB. Análise comparativa da produtividade de híbridos de mamoneira (Ricinus communis L.) obtidos por meio da hibridação convencional e do método dos híbridos crípticos. Ph dissertation, FCA/Unesp. 2011
66. Lonnquist JH, Williams NE. Development of maize hybrids through among full-sib families. Crop Sci. 1967; 7:369-370
67. Severino LS, Auld DL, Baldanzi M, Cândido MJD, Chen G, Crosby W, Tan D, et al. A Review on the Challenges for Increased Production of Castor. Agronomy Journal, 2012; 104(4):853–879.

CHAPTER 8

Breeding to Improve Symbiotic Effectiveness of Legumes

Vladimir A. Zhukov[1], Oksana Y. Shtark[1], Alexey Y. Borisov[1] and Igor A. Tikhonovich[1]

[1] *All-Russia Research Institute for Agricultural Microbiology, St.-Petersburg, Russia*

1. INTRODUCTION

Symbioses with beneficial microorganisms constitute the universal and ecologically highly effective strategy of adaptation of plants towards nearly all types of environmental challenges. Representatives of many groups of fungi and bacteria participate in plant-microbial symbioses (PMS) wherein they can colonize the plant surfaces, tissues or intra-cellular compartments using two basic adaptive strategies: nutritional and defensive. Construction of niches for hosting the symbiotic microbes involves the complicated developmental programs implemented under the joint control by plant and microbial partners and based on the cross-regulation of their genes.

Legume plants (family Fabaceae) are known to form symbioses with extremely broad range of beneficial soil microorganisms (BSM), representing examples of almost all plant-microbe mutualistic systems. Different groups of beneficial microbes improve host mineral nutrition, acquisition of water, promote the plant development and offer protection from pathogens and pests. For ecology and agriculture, the most important beneficial legume symbioses are arbuscular mycorrhiza (AM) and root nodule (RN) symbiosis. These symbioses demonstrate high level of genetic and metabolic integrity, compared with other interactions of legumes with plant growth-promoting rhizosphere bacteria (PGPR) and/or beneficial endophytic bacteria.

High integrity of AM and RN symbioses implies highly specific mutual recognition of partners, formation of special complex symbiotic compartments and integration of partners' metabolic pathways. In the symbioses, legume plant plays a role of the organizing center of the system as it performs functions of coordination and regulation of all developmental processes. During last decade, a significant progress has been achieved in revealing the genetic bases of symbioses formation and functioning, so the knowledge of the plant genetic control over symbioses can effectively facilitate breeding new varieties of legumes that are needed for modern sustainable agriculture. In this chapter, we describe the present state of the developmental genetics of legume symbioses and depict the potential to organize the multi-component symbioses to be used for optimizing the broad spectrum of plant adaptive functions and to improve the sustainability of legume crop production.

2. MECHANISMS OF POSITIVE EFFECT OF BSM ON THE ENVIRONMENT AND HEALTH AND YIELD OF THE LEGUME PLANT

2.1. Legume-Rhizobia Root-Nodule (Rn) Symbiosis

Leguminous plants are able to grow in the soil/substrate without any combined nitrogen due to the fixation of atmospheric nitrogen by symbiotic nodule bacteria (collectively called rhizobia). In collaboration with rhizobia, legumes make a large contribution to the global nitrogen balance in natural and agricultural ecosystems [1]. Nitrogen fixation occurs within special plant organs, root nodules (in some associations stem nodules are also formed). Development of these organs represents a well-organized process based on the tightly coordinated expression of specialized symbiotic plant and bacterial genes. The legume nodules provide an ecological niche for bacteria, as well as structure for metabolic/signal exchange between the partners and for the control of symbionts by the hosts [2].

Family Fabaceae contains 17000-19000 species divided between three sub-families (Caesalpinioideae, Mimosoideae and Papilionoideae) and more than 700 genera of world-wide distribution [3]. With a single exception (*Parasponia*: Ulmaceae), the ability for symbioses with rhizobia is restricted to Fabaceae, although in eight related dicotyledonous families (Rosid I clade) an ability to form nodules with the nitrogen-fixing actinomycete *Frankia* is known [4].

By contrast to legumes, their nitrogen-fixing microsymbionts do not constitute a taxonomically coherent group of organisms. The majority of rhizobia belong to the α-proteobacteria assigned into the Rhizobiaceae family solely on the basis of their ability to nodulate the legumes (e.g. *Azorhizobium, Bradyrhizobium, Mesorhizobium, Rhizobium, Sinorhizobi um*). Recently some β-proteobacteria (close to *Burkholderia, Cupriavidus, Pseudomonas* and *Ralstonia*) and even some γ-proteobacteria have been discovered that can form nitrogen-fixing nodules with the legumes [5]. All these bacteria vary enormously in their overall genome organization, location of "symbiotic" (*sym*) genes and their molecular organization and regulation [6, 7].

Root-nodule symbiosis is well known as highly specific plant-microbe interaction. According to the early surveys of symbiotic specificity [8], legumes were suggested to comprise a range of taxonomically restricted cross-inoculation groups within which the free cross inoculation occurs, while the species from different groups do not cross-inoculate. The best studied examples of this classification are represented by four cross-inoculation groups: "*Trifolium – Rhizobium leguminosarum* bv. *trifolii*", "*Pisum, Vicia, Lathyrus, Lens – R. leguminosarum* bv. *viciae*", "*Galega – R. galegae*", "*Medicago, Melilotus, Trigonella – Sinorhizobium meliloti, S. medicae*". However, it was demonstrated later [9, 10] that such strictly defined specificity is limited to the herbage papilionoid legumes growing in temperate zones and representing the "Galegoid complex".

The specificity of legume-rhizobia interactions is expressed just during the pre-infection stage when rhizobia recognize the roots of appropriate host plants and colonize their surfaces. The interaction starts when the root-excreted signals, in particular, flavonoids, activate the rhizobial nodulation genes (*nod/nol/noe*) [11]. These genes control the synthesis of lipochitooligosaccharidic (LCO) nodulation factors (Nod factors, NFs) which induce the early stages of RN symbiosis development [12-14]. NFs represent the unique group of bacterial signal molecules not known outside legume-rhizobia symbiosis. They are among the most potent developmental regulators: their effect is expressed at concentrations of 10^{-8} – 10^{-12} M only. The core structure of these molecules, common for all rhizobia species, consists of 3-6 residues of N-acetylglucosamine and of a fatty acid (acyl) chain. The type of symbiotic specificity is dependent mainly on the chemical modifications in NF structures [11]. However, a sufficient role in the host specificity of RN symbiosis may also be implemented by the interactions between bacterial surface molecules (some polysaccharides and proteins) [15, 16] and the lectins located on the root hair surfaces [17].

The main enzyme of nitrogen fixation in nodules is a nitrogenase that has a complex structure [18, 19]. Synthesis of nitrogenase (the enzyme catalysing reduction of N_2 into NH_4^+) and other proteins involved in nitrogen fixation is induced in bacterial cells after they differentiate into a specific form called bacteroids. Bacteroids are embedded into a membrane structure named symbiosome, which formation as well as bacteroid differentiation is induced by plant [20]. These symbiosomes are organelle-like units of plant cell responsible for nitrogen fixation [21, 22]. Peri-bacteroid membrane (PBM) that surrounds bacteroids is an active interface of RN symbiosis where exchange of metabolites between symbionts occurs [23].

A pronounced differentiation is typical for rhizobia-infected plant cells, such as an increase in internal membrane structures participating in the PBM formation and biosynthetic processes. Polyploidization and chromatin decondensation are typical for these cells correlating with an elevated transcription activity [24]. Biochemically plant cell differentiation is expressed as a *de novo* synthesis of many proteins including leghaemoglobin and nodule-specific isozymes of carbon and nitrogen metabolism [25]. Leghaemoglobin binds oxygen actively ensuring its transport towards symbiosomes (which are characterised by the intensive respiration necessary to support energy consuming nitrogen fixation) and microaerobic conditions inside the nodules (required for the nitrogenase activity). The carbon and nitrogen metabolic enzymes responsible for the energy supply to nitrogenase and for the assimilation of fixed nitrogen are nodule specific [26]. Organic nitrogenous compounds formed from N_2 fixation are transported to the upper parts of the plant either as amides (mainly asparagine (Asn), but also glutamine (Gln)) or as ureides (allantoin and allantoate), so that legumes can be classified as amide or ureide exporters according to the compounds they use for the mobilization of fixed nitrogen [27].

Rhizobial cells also undergo differentiation, but its level varies in different legume species. The terminal bacteroid differentiation (when bacteroids increase their size and DNA content and lack the capacity to divide) is specific for legumes belonging to the inverted repeat–lacking clade (IRLC) such as *Medicago*, *Pisum*, or *Trifolium*, whereas bacteroids in the non-IRLC legumes, such as *Lotus*, show no sign of terminal differentiation as they maintain their normal bacterial size, genome content, and reproductive capacity [28]. The same rhizobia strains that form symbiosis with both IRLC and non-IRLC legumes have different bacteroid differentiation fates in the two legume types. It was demonstrated that in *Medicago* and probably in other IRLC legumes, the nodule-specific NCR peptides act as symbiotic plant effectors to direct the bacteroids into a terminally differentiated state [29]. Possibly, IRLC legumes use nodule-specific NCR peptides to dominate the endosymbionts: NCR peptides interfere with many aspects of the bacteroid metabolism to allow the efficiency of the nitrogen fixation process to be optimized, for example, by stimulation enlargement and polyploidization of bacteroids [30]. Also, the peptides could be part of a mechanism to avoid the "cheating" of rhizobia that could use host resources to accumulate carbon storage compounds instead of fixing nitrogen [29], which is often observed in the non-IRLC legumes but not in the IRLC [30].

It was also found that nodules where terminal bacteroid differentiation takes place are more efficient in terms of energy use. Oono and Denison [31] reported that legume species with terminal bacteroid differentiation (such as peas (*Pisum sativum* L.) and peanuts (*Arachis hypogaea* L.) invest less in nodule construction but have greater fixation efficiency when compared to species with reversible bacteroid differentiation (such as beans (*Phaseolus vulgaris* L.) and cow peas (*Vigna unguiculata* (L.) Walp.). This effect is probably due to genomic endoreduplication of the bacteroids and full contact of single undivided bacteroid with peribacteroid membrane (some reproductive bacteroids can lose contact with PBM after they divide). Still, this is not known if these useful features of terminal bacteroids differentiation in some legumes could be transferred into other legume species. In work of van de Velde et al. [29], expression of NCR genes in nodules of *Lotus japonicus* (Regel.) K. Larsen (with normally reversible bacteroid differentiation) was sufficient to induce bacteroid morphologies reminiscent of terminally differentiated bacteroids of *Medicago truncatula* Gaertn. But, no positive effect on nitrogen fixation efficiency was reported, probably because there are much more regulatory genes needed to make bacteroids work propertly in such a heterologous system.

2.2. Arbuscular Mycorrhiza (AM)

Arbuscular mycorrhiza (AM) is formed by at least 80% of contemporary terrestrial plants with fungi of phylum Glomeromycota. The Glomeromycota are unique as the only monophyletic mycorrhizal fungus lineage that has co-evolved with land plants throughout their history. They are obligate biotrophs that colonize plant roots obtaining photosynthates, such as carbohydrates (hexoses), and niches for both their growth and reproduction. The AM is evolved more than 400 million years ago and was considered to play a decisive role in plants achieving a terrestrial existence [32-34]. The AM is supposed to be "the mother of plant root endosymbioses" [35]. Since legumes originated long after AM, about 60 million years ago [36], it may be assumed that all of them have the potential to produce this type of symbioses. *Lupinus* is the only known genus where this ability had apparently been lost [37-39].

Specificity of AM symbiosis is relatively low [34]. Symbiosis establishment starts with molecular dialogue between the partners. Plant roots release sesquiterpenes (also known as inducers of parasitic plant seed germination) as well as different phenolic compounds, including flavonoids, which induce fungal hypha growth and branching [40, 41]. Similar to rhizobia, AM-fungi produce signal molecules termed Myc factors (mycorrhization factors) [42], which can be recognized by the plant. They are a mixture of several lipochitooligosaccarides, the structure of which is close to that of rhizobial Nod factors, but is presumably more universal for different plant-fungus combinations [43]. Both sesquiterpenes and Myc factors are released constitutively and in the absence of physical contact with symbiotic partner [44].

The AM-fungi penetrate the root to colonize inner cortical cells. Plant plays an active role in fungus hosting inside the root tissues using cellular mechanisms similar to those used during rhizobial invasion, such as nucleus reposition, cytoplasm aggregation, special cytoskeletal tunnel assembly and symbiotic membrane formation (reviewed in: [45]). A special intracellular compartment of AM providing tight metabolic exchange between the partners is arbuscule, which is highly branched fungal hypha surrounded by membrane of plant origin [34] similar to symbiosome of RN symbiosis [35, 46].

Inner-root and outer parts of mycelium remain bound with arbuscules and are a single continuum *via* which the fungus is able to translocate mineral nutrient and water from the soil into the root system [47]. Thus, well developed AM-symbiosis allows plant growing well in nutrient-poor and drought-affected soils, increases its resistance against pathogens and pests and heavy metals, and improves soil structure (see below).

Phosphorous (P) is one of the mineral nutrients essential for the plant growth (constituting up to 0.2% of the dry weight of the plant cell) and development. It plays the diverse regulatory, structural, and energy transfer roles and consequently is required in significant amounts [48, 49]. The plants can acquire soluble forms of phosphorous directly from soil through the plant specific phosphate transporters (PTs). The dominant available forms in soil (orthophosphate ions, P_i) are very poorly mobile because of the abundance of cations such as Ca^{2+}, Fe^{3+} and Al^{3+} [50]. In such environments where inorganic phosphorous is the predominant form in soil, a range of root adaptations, most of them primarily involved in mobilization and assimilation of phosphorous, are described including plant dependence on arbuscular mycorrhizas (see for review: [34, 51]). In most cases there is a preferential uptake *via* fungal hyphae (the mycorrhizal uptake pathway) [52]. Studies employing radioactive tracers to track hyphal P_i uptake from soil have shown considerable AM contributions to phosphorous uptake [53-55]. The process involves several fungal transport systems some of which have an extremely high affinity for P_i [56].

After transporting into hyphae, the major part of P_i is polymerized by polyphosphate kinase into polyphosphates (poly-P), the linear chains of P_i. The granules rich in poly-P together with phosphorous-containing esters are packed into the cylindrical vacuoles which are transported along the hyphae by tubulin fibrils. After reaching the arbuscules, phosphorous compounds are destroyed by phosphatases and the released P_i cross the partners' interface [56-58]. The arbuscule is the site of phosphate transfer from fungus to plant. It is well documented that plants possess many classes of phosphate transport proteins, including those which are expressed only in AM symbiosis [59-61]. It was discovered that five plant and one fungal PT genes are consistently expressed inside the arbusculated cells [60]. A plant phosphate transporter MtPt4 was shown to be expressed specifically on the peri-arbuscular membrane in *Medicago truncatula* [62].

The mycorrhizal P_i uptake pathway is controlled by the plant host. Many results suggest that the plant phosphorous status is a major regulator controlling induction/repression of plant PT genes at both the soil-root interface and the inner-root symbiotic interface [63-65]. It was shown that high phosphorous concentrations counteract the induction of the mycorrhizal P_i transporter genes by phospholipid extracts from mycorrhizal roots containing the mycorrhiza signal lysophosphatidylcholine [65]. The efflux of P_i probably occurs in coordination with its uptake and the fungus, on its side, might exert the control over the amount of P_i delivered to the plant [66].

Although P_i acquisition receives more attention, the important advances in investigations on nitrogen uptake by AM-fungi have been made in recent years. AM-fungi directly uptake ammonium (NH_4^+), nitrate and amino acids [67] with preference to NH_4^+ [68]. The first step in the nitrogen uptake requires the activity of specific transporters located at the interface between the soil and extraradical mycelium. A fungal transporter gene (*GintAMT1*) involved in the process and having high affinity with NH_4^+ was characterized [68].

Inorganic nitrogen that was taken up by the extraradical mycelium should then be incorporated into the amino acids and translocated to the intraradical mycelium, mainly as arginine (Arg) since this is the predominant free amino acid in the external hyphae [69]. The glutamine synthetase/glutamate synthase (GS/GOGAT) cycle is possibly responsible for a subsequent NH_4^+ assimilation in AM extraradical hyphae [70, 71], although the involvement of glutamate dehydrogenase has not been experimentally excluded. Arg similar to Poly-P is stored and is translocated along hyphae in vacuoles and is later released to the plant apoplast [66].

A mycorrhizal-specific NH_4^+ transporter *LjAMT2;2* has been revealed recently in transcriptomic analysis of *Lotus japonicus* roots upon colonization with *Gigaspora margarita*. The gene has been characterized as a high-affinity AMT belonging to the AMT2 subfamily. It is strongly up-regulated and exclusively expressed in the mycorrhizal roots, but not in the nodules, and transcripts have preferentially been located in the arbusculated cells [72].

The plants colonized by AM-fungi have been demonstrated to manifest an increased resistance to attack of some pathogenic microorganisms, such as fungi, nematodes, bacteria, phytoplasma, and plant viruses (reviewed in: [73]) as well as to plant feeding insects [74, 75]. However, it is still unknown whether such increased resistance to pathogens is a consequence of improved plant overall fitness or it is due to the specific defense responses induced by AM-fungi.

Actually, a range of processes occurring as a result of pathogen invasion (plant defense responses) also takes place in mycorrhized root tissues. They include the signal perception, signal transduction and defense-related gene activation [76-80]. The elements of hypersensitive responses have been observed to take place at both compatible and non-compatible combinations of plants with AM-fungi; reactions similar to the "oxidative burst" are typical for AM during fungus penetration into the epidermal cell [81]. In AM, as in other compatible biotrophic interactions, the defense-like response appears to be weak and occurs transitorily during the early phases of colonization, suggesting that the suppression of plant defense responses by the fungal signals may contribute to successful, compatible AM fungal colonization [76, 82]. AM-fungi are known to alter both constitutive and induced defenses in foliar tissues [83-85].

Drought stress is a major agricultural constraint in the semi-arid tropics. In most cases symbiosis with AM-fungi has been shown to increase host plant growth rates during drought stress and plant resistance to drought. Several mechanisms explaining this phenomenon have been proposed: an influence of AM on plant hormone profiles, increasing intensity of gaseous exchange and photosynthesis in leaves, direct water transport via fungal hyphae from soil into the host plant, enhanced water uptake through improved hydraulic conductivity and increasing leaf conductance and photosynthetic activity, nitrate assimilation by fungal hyphae, enhanced activity of plant enzymes involved in defence against oxidative stress, plant osmosis regulation, and changes in cell-wall elasticity (reviewed in: [86-89]).

The AM fungal hyphae grow into the soil matrix and create conditions conducive to the formation of microaggregates and then their packing into macroaggregates due to production copious amounts of the glycoprotein glomalin [90, 91]. Through AM-fungi-mediated effects on soil structure, it seems logical to suggest that AM colonization of a soil might affect its moisture retention properties and, in turn, the behaviour of plants growing in the soil, particularly when it is relatively dry [88].

AM-fungi were found to play an important role in heavy metal detoxification and the establishment of vegetation in strongly polluted areas (see for review: [92]). Fungal strains isolated from old zinc wastes also decrease heavy metal uptake by plants growing on metal rich substrates, limiting the risk of increasing the levels of these elements in the food chain [93]. Phytoremediation of metal contaminated areas attracts the increasing interest as a cheaper alternative to chemical methods, more friendly for environment and nondestructive to soil biota. The effectiveness of the bioremediation techniques depends on the appropriate selection of both the plant and the fungal partners. Plants conventionally introduced in contaminated areas disappear relatively soon, while those appearing during natural succession are better adapted to harsh conditions. Much more stable are plants that appear on the wastes spontaneously, but, it takes a long time till they establish and form stable communities. Symbiotic partners selected on the basis of such research are often the best choice for future phytoremediation technologies [93-96]. Introduction of plants from xerothermic grasslands into the soils contaminated with industrial metal rich wastes is supposed to be a new solution for waste revegetation [97]. Further improvements can be obtained by optimization of diverse microbiota including various groups of rhizospheric bacteria and shoot endophytes [92].

2.3. Associations Of Roots With Plant Growth-Promoting Rhizobacteria (PGPR)

Plant Growth Promoting Rhizobacteria (PGPR) are the taxonomically diverse group including different bacteria (*Arthrobacter, Azospirillum, Bacillus, Enterobacter, Pseudomonas, Paenibacillus, Streptomyces*) and even some archaea [98]. The PGPR are inhabitants of soil in the vicinity of plant roots and are dependent on consuming root exudates. Many PGPR are able to attach to root surfaces and to AM and other fungal hyphae. The PGPR provide several benefits affecting the host plant either directly (due to mineral nutrient improvement and stimulation of root development) or indirectly (due to defence of plants from soil-borne pathogens and improving host tolerance to abiotic stresses).

Similar to rhizobia, *Azospirillum* possess nitrogenase and therefore is able to fix atmospheric nitrogen. In the early papers, plant growth promoting activity was attributed mainly to associative nitrogen fixation. A broad distribution of cereal-*Azospirillum* associations was identified (reviewed in: [99]). It was demonstrated later, however, that a partial role in these plant–PGPR associations was due to phytohormone auxin (indole-3-acetic acid, IAA) synthesis [100] which improves the root growth and assimilatory capabilities and hence aids nitrogen uptake by plants. In spite of absence of the visible anatomic differentiation in root-*Azospirillum* associations, its development involves a range of molecular interactions some of which may be common to endosymbiotic associations with rhizobia. Moreover, there is a visible taxonomic relatedness between *Azospirillum* and *Bradyrhizobium* genera. Thus, azospirilla and those slow-growing rhizobia might originate from a common *Azospirillum*-like ancestor (see for review: [101, 102]).

Many PGPR are able to solubilize sparingly soluble phosphates, usually by releasing chelating organic acids. Phosphate solubilizing bacteria (PSB) have been identified, but their effectiveness in the soil-plant system is unclear. The ability of an inoculated PSB to supply phosphopous to plant may be limited, either because the compounds released by PSB to solubilize phosphate are rapidly degraded or because the solubilized phosphate is re-fixed before it reaches the root surface [103].

The best studied examples of bacteria providing efficient defense from phytopathogens are: *Pseudomonas* (*P. fluorescens, P. chlororaphis, P. putida*), *Bacillus* (*B. cereus, B. subtilis*) and some *Serratia* (e.g., *S. marcescens*) species. Many of these bacteria are capable of preventing attacks by pathogenic fungi, nematodes and bacteria [98, 104, 105]. Diverse mechanisms may be involved in host protection offered by PGPR.

The best studied mechanism is the competitive exclusion of pathogens often related to their direct suppression by the bacterial antibiotic substances: phenazine-1-carboxamide, 2,4-diacetylphloroglucinol, kanosamine, oligomycin A, oomycin A, pyoluterin, pyrrolnitrin, xanthobaccin, zwittermycin A, volatile dyes (HCN) and cyclic lipopeptides [98, 104, 105].

An important mechanism for the suppression of pathogens by biocontrol microbes may result from competition for iron or other metals, that involves bacterial siderophores which may possess much greater affinities for ferric ions than those for fungal siderophores [104, 105]. The value of siderophores in biocontrol effects under natural conditions is predominantly associated with their ability to induce forms of systemic resistance in plant [98, 106].

Competitive exclusion of pathogens by PGPR is best achieved when the bacteria exhibit high root-colonizing activity. Application of the technique of genetic labeling with Green Fluorescent Protein (GFP) suggested that these bacteria do not regularly colonize the root interiors, and only rarely they can be observed inside the outer root tissues [98]. Most PGPR cells are concentrated on the root surface where the micro-colonies [98, 104] or bio-films are formed [107]. Since the interactions of plants with root-associated bacteria are not specific, the bacteria will colonize the roots of a broad spectrum of hosts. Specificity of the defensive association may be expressed however, at the point when antimicrobial compounds are being synthesized and this does not always correlate with bacteria taxonomy; many strains of *Bacillus* and *Pseudomonas* which possess plant-protective properties have close relatives amongst phytopathogenic types [106, 108, 109].

Microscopic observations demonstrated that suppression of fungi may be correlated with bacterial attachment to pathogen hyphae. As a result of this attachment, some PGPR strains commence their biocontrol functions by behaving as hyper-parasites of pathogenic fungi. This suppression may be related to the production of bacterial enzymes which destroy the pathogen cell walls [104, 105, 110, 111].

Sometimes the biocontrol activities of PGPR do not correlate with intensive colonization of host roots and plant protection results from only a small number of bacteria cells. This occurs when PGPR inoculation induces the systemic resistance mechanisms that make the root non-accessible by pathogens. Initially this effect of PGPR was called ISR (Induced Systemic Resistance) and was attributed exclusively to nonpathogenic systems [112]; SAR (Systemic Acquired Resistance) reactions, by contrast, were considered to be typical for the interactions with plant pathogens. Nevertheless, it was later found that the reactions of both types occur in either pathogen or nonpathogen systems and are distinguished by the nature of their endogenous elicitors (reviewed in: [106, 113]). The conventional SAR reaction is characterized by an accumulation of salicylic acid as signal molecules and pathogenesis-related proteins (PR-proteins), whereas ISR reaction is based on signal transduction pathways regulated by jasmonates and ethylene. The systemic defence responses of both types may be elicited exogenously by PGPR cells attached to the roots or penetrating their outer tissues. Some molecules produced by PGPR (cell wall and cyclic lipopolysaccharides, flagella components, exoenzymes, phytohormones, type III secretion system (TTSS) effectors, siderophores, salicylic acid, and toxins) may be perceived by the plant and elicit a defensive response [106].

It has been reported that PGPR which produce an enzyme which is involved in the catabolism of 1-aminocyclopropane-1-carboxylate (ACC) – the ACC deaminase, can lower ethylene concentration in a developing or stressed plant, protecting it against the deleterious effects of ethylene induced stress and facilitating the formation of longer roots [114, 115]. This demonstrates that ethylene is negative regulator of plant interaction with PGPR.

Despite relatively low specificity of plant associations with PGPR, plant genotype has been shown to influence their effectiveness (i.e. genetic integration exists between the partners), and a series of genome loci (QTL) was identified controlling its quantitative variation [106, 116, 117]. The most pronounced plant species-specificity has been observed in the manifestation of ISR reactions caused by PGPR [118, 119].

Both highly effective direct promotion of plant growth and biocontrol may be due to an ability of the host to regulate PGPR functions by modulating the composition of root exudates. Root-excreted organic acids, but not sugars, are optimal for support of different types of PGPR [120-122]. Additionally, some plants (including the legumes, pea and alfalfa) regulate their PGPR functions by exudating specialised signals from the roots which mimic the bacterial "quorum sensing" regulators required for root colonization and antifungal activities [123]. These observations suggest that improvement of biocontrol functions in root-PGPR associations may be achieved *via* manipulations with the bacterial and plant host genotypes.

2.4. Mutually Beneficial Associations Of Plants With Endophytic Bacteria

Healthy naturally propagated plants grown in the field or in pot cultures are colonized by populations of endophytic bacteria. The spectrum of endophytic bacteria isolated from the roots of various plants covers a wide range of species; representatives of the genera *Pseudomonas*, *Bacillus* or *Streptomyces* are most frequently encountered as endophytes (reviewed in: [124]). Newly developed molecular methods enable complete analyses of the diversity of culturable and non-culturable bacteria [125]. Most of the known genera include some phytopathogenic endophytes. Endophytes and pathogens both possess many similar virulence factors (reviewed in: [124]).

Some endophytes are seed-borne, but others have mechanisms for colonizing plants that have yet to be elucidated [126]. Although there are occasional poorly substantiated reports of intracellular colonization of bacteria providing a consistent and effective increase in the productivity of crops, it is still considered that the intercellular apoplastic space is the most suitable niche for endophytes [127]. It is suggested that many bacterial 'endophytes' may not colonize the living tissues, but occupy protective niches in dead surface tissues or closely adhering soil of rhizosheaths. Consistent entry of endophytes into living root tissues in the field is supposed to require a bacterial capability to hydrolyse the hydrophobic incrustations of the walls of epidermal, hypodermal, endodermal, and other cortical cells [128].

Plant associations with endophytic bacteria can increase plant growth and promote general development or improve plant resistance to pathogens and other environmental stresses enhancing the host's ability to acquire nutrients, or by production of plant growth-regulating, allelopathic or antibiotic compounds [127, 129]. Sometimes improved plant resistance can be linked to induced systemic resistance caused by bacterial elicitors coming from the endophyte [130].

It is necessary to study the natural associations between bacterial endophytes and their hosts for the purposes of employing such systems most productively in sustainable agriculture [127]. Delivery of endophytes to the environment or agricultural fields should be carefully evaluated to avoid introducing plant, animal and human pathogens [131].

2.5. Synergistic Effect of Microbes In Rhizosphere

Microorganisms in the rhizosphere are under the influence of root exudates and plant as a whole as well as of interspecies interactions with each other. Many fungi including AM-fungi can interact with different bacterial species which frequently attach to fungal mycelium (reviewed in: [103, 132, 133]). For those bacteria known to stimulate mycelial growth of mycorrhizal fungi and/or enhance root mycorrhization the term 'mycorrhiza-helper-bacteria' has been proposed [133, 134]. Particularly, the bacteria may encourage growth of AM-fungi at the perisymbiotic stage of development, which precedes the establishment of a direct contact of the microsymbiont with the plant root [103, 135]. On the other hand, AM-fungi directly modify the environment due to mycelial exudation [136], forming the so-called 'mycorrhizosphere' [137]. In addition, the stimulation of root exudation as a result of interactions with AM-fungi leads to qualitative and quantitative changes of the bacterial community in the rhizosphere (reviewed in: [103, 132]).

Synergistic effect between RN and AM symbioses of legumes was described by many authors [95, 138-142]. AM formation is known to promote nodule development and nitrogen fixation by rhizobia, In particular, by means of improvement of mineral (predominantly phosphorous) host plant nutrition (see for review: [103]). AM-fungi also manifest synergism during interactions with PGPR (both indigenous and introduced), which perform biocontrol, nitrogen fixation, and phosphate mobilization during double and complex inoculation [103, 132]. The synergetic effect of plant inoculation by rhizobia and PGPR (*Azospirillum*, *Bacillus*, and *Pseudomonas*) is well known. In particular, it is associated with PGPR production of indole-3-acetic acid, which encourages nodule formation [103, 143]. Triple inoculation of a model legume *Anthyllis cytisoides* with PGPR, AM-fungi and rhizobia was shown to be the most effective approach for revegetation in mediterranean semi-arid ecosystems [94].

Thus, the potential of microbial synergism allows us to speak about high prospects of biotechnologies focused on creation multicomponent symbioses (MCS) that increase the fertility and quality of agricultural legume and nonlegume crops. At the same time, the results of experiments with plant symbioses with AM-fungi, rhizobia and PGPR, including multimicrobial systems, show the important role of physiological and genetic adaptation of microorganisms to local environmental conditions [92, 94, 95, 144]. Hence, during the development of such biotechnologies, it is recommended to use a complex of local microbial isolates adapted to particular environmental conditions.

3. PLANT GENETIC CONTROL OVER DEVELOPMENT AND FUNCTIONING OF MUTUALISTIC SYMBIOSES OF LEGUMES

3.1. Legume Genes Involved In Development Of Rn And Am Symbioses

The complex developmental processes which lead to the formation of intercellular and sub-cellular symbiotic compartments in RN and AM symbioses are controlled by both macro- and microsymbiont. Genetic systems of the symbionts are highly integrated, because some genes and gene products of one partner can switch certain genetic programs in another partner, still the development and function of the symbioses is reliant to the greatest extent on the plant. Developmental genetics of RNS is now well described because both plants and nodule bacteria can be subjected to genetic analysis during nitrogen-fixing nodule formation and functioning. There has been less investigation of AM systems. Mainly this is due to the difficulties encountered in culturing AM-fungi, caused by their obligate symbiotic lifestyle and impossibility of using selective media. Additionally, genetic analysis of AM fungi is more complex because of their heterokaryotic nature and lack of sexual process [34, 145].

The plant genes involved in development of RN and AM symbioses may be divided into two groups, according to approach which was used for the gene identification. The first group, *Sym* genes [146], had been identified with the use of formal genetic analysis (started from selection of plant mutants defective in nodule development). The other group of genes comprises nodulins (from *nodule*) [147], mycorrhizins (from *mycorrhiza*) [148, 149], and symbiosins (from *symbiosis*) [149, 150]. These genes were identified by molecular genetic methods, through identification of proteins and/or RNAs synthesized *de novo* in root nodules (nodulins) or roots colonized by AM fungi (mycorrhizins). The genes whose expression is induced during the development of both endosymbioses, RNS and AM are called symbiosins [150].

Genes of these groups are suspected to play different roles in the processes which may be referred collectively as "management of microsymbionts" inside plant roots. Specifically, the products of some nodulin genes represent the structural elements of newly constructed temporary compartments developed during symbiosis [151]. The other nodulin genes may play essential roles in modulating the hormonal status of the developing nodules [152, 153]. Resently, *in silico* and microarray-based transcriptome profiling approaches have allowed identification of nodulins, mycorrhizins and symbiosins, which are being activated in response to an AM fungal signal, or by either rhizobial or AM fungal stimulus, respectively [149]. Several hundred genes were found to be activated at different stages of either symbiosis, with almost 100 genes being co-induced during nodulation and in AM formation. These co-induced genes representing the common evolutionary bases of AM and nodular symbioses can be associated with those cellular functions which are required for symbiotic efficiency, such as the facilitation of nutrient transport across the perisymbiotic membranes that surround the endosymbiotic bacteroids in root nodules and the arbuscules in AM roots [150]. However, it should be remembered that although most of the nodulins/mycorrhizins/symbiosins were already cloned and sequenced, functions for many of them have been identified only preliminary using the sequence data of the encoding genes and location of the gene products in the symbiotic compartments.

Still, most of nodulins, mycorrhizins and symbiosins seem to play a subordinate role in the regulatory scheme of symbiosis, nevertheless being indispensable for its functionality and stability. In turn, the major, regulatory role in realization of symbiotic programmes is to be assigned to *Sym* genes. These genes, in contrast to nodulin genes, are usually not expressed outside symbiotic structures and there are many examples of the high functional and sequence homologies between them in different legumes. First genes of this group had been identified using the spontaneous mutants from natural legume populations [154] and afterwards using the experimentally induced mutants defective in nodulation or nitrogen fixation (Nod$^-$ and Fix$^-$ phenotypes) [155]. Afterwards, it was demonstrated that mutations in some of these genes also affect the ability of plant to form AM [148]. The presence of such common genes necessary for both AM and RN symbioses development suggested that both endosymbioses were more closely related than it was suspected before. The cloning and sequencing of the common symbiotic genes helped to understand that AM and RN symbioses share the overlapping signaling pathways, which probably were established during evolution the AM symbiosis and was recruited afterwards into the RN symbiosis development [35].

The large sizes of genomes of crop legumes (e.g. soybean *Glycine max* (L.) Merr. and pea) in which the formal genetics of symbioses was initially developed, as well as low capability for genetic transformation, complicate greatly the cloning of symbiotic genes, analysis of their primary structures and the gene manipulations. Therefore, in early 1990s the new legume species, *Lotus japonicus* [156] and *Medicago truncatula* [157, 158] have been introduced in studies as model plants. These species are characterized by small genomes (470 – 500 Mb; [159]) and can easily be genetically transformed [158, 160-162]. In addition, the short lifecycle and high seed productivity made them attractive and convenient model objects for studying molecular bases of RN and AM symbioses.

Genetic analysis in model legumes as well as in crop legumes was started with experimental mutagenesis. Large-scale programs of insertion, chemical and X-rays mutagenesis, performed by different research groups, resulted in generation of numerous symbiotic mutants in *L. japonicus* and *M. truncatula* [163, 164] which allowed researchers to identify and characterize a series of *Sym* genes. The genes involved at the initial stages of nitrogen-fixing symbiosis (early *Sym* genes) were of primary interest, allowing dissection of the mechanisms by which the NF signal is perceived and transduced by host plants [165]. It turned to be that after the perception of NF, the nodulation process follows the same signalling pathway as AM does, with slight differences, though.

3.2. Common Symbiosis Pathway (CSP)

The data obtained during the last fifteen years allowed reconstruction the symbiotic signaling pathway which starts in RN and AM symbioses with recognition the Nod and Myc factors, respectively, and goes on as the signal transduction inside the root. In legumes Nod and Myc factors are most likely perceived by specific receptor complexes [35]. The receptor for NF is considered to be a heterodimer composed of at least two LysM containing receptor kinases [166-168]. Alike, receptor for Myc factor (which is not known yet) also supposed to be consisted of similar receptor kinases, or even include one or more kinases participating in Nod factor signaling [169]. The system of receptor kinases perceiving signal molecules of microsymbionts seems to be complicated, with some receptor complexes being necessary not only for starting the interactions, but also on later stages, during penetration bacteria into the root cortex through root hair. Moreover, some receptor kinases could non-specifically bind Nod and Myc factors, which results in intensified growth of lateral roots [43, 169].

Probably, the diversity of receptor kinases should complement the variability of soil microorganisms and increase the specificity of interactions with mutually beneficial ones.

After the first step of reception of Nod or Myc factor, the symbiotic signal is being transmitted to the common pathway, named Common Symbiosis Pathway [170]. The first player in this pathway is LRR-receptor kinase, or SymRK (symbiotic receptor kinase), which is required for both RN and AM symbioses development [171, 172]. The ligand for this receptor kinase is not known yet. Interestingly, the activity of this kinase is also required for proper progression of late symbiotic stages, at least for rhizobial infection [173]. SymRK kinase domain has been shown to interact with 3-hydroxy-3-methylglutaryl CoA reductase 1 (HMGR1) from *M. truncatula* [174], and an ARID-type DNA-binding protein [175]. These results suggest that SymRK may form protein complex with key regulatory proteins of downstream cellular responses. Symbiotic Remorin 1 (SYMREM1) from *M. truncatula* and SymRK-interacting E3 ligase (SIE3) from *L. japonicus* have also been shown to interact with SymRK [176, 177].

The symbiosis receptor kinase SymRK acts upstream of the Nod and Myc factor-induced Ca^{2+} spiking in the perinuclear region of root hairs within a few minutes after NF application [178]. Perinuclear calcium spiking involves the release of calcium from a storage compartment (probably the nuclear envelope) through as-yet-unidentified calcium channels. To date it is known that the potassium-permeable channels might compensate for the resulting charge imbalance and could regulate the calcium channels in plants [179-183]. Also, nucleoporins NUP85 and NUP133 (to date described only in *Lotus*) are required for calcium spiking, although their mode of involvement is currently unknown. Probably, NUP85 and NUP133 might be a part of specific nuclear pore subcomplex that plays a crucial role in the signal process requiring interaction at the cell plasma membrane and at nuclear and plastid organelle-membranes to induce a Ca^{2+} spiking [184, 185]. Recently, the third constituent of a conserved subcomplex of the nuclear pore scaffold, NENA, was identified as indispensable component of AM and RN endosymbiotic development [186].

The calcium spiking is characteristic for both RN and AM symbioses formation [187]. These Ca^{2+}spikes are supposed to activate a calcium- and calmodulin-dependent protein kinase (CCaMK) that is also required for NF signaling and AM development [188]. This kinase contains an autoinhibition domain, removing of which leads to a spontaneous activation of downstream transcription events and induction of nodule formation in the absence of rhizobia [189]. Thus, CCaMK appears to be a general manager for both symbioses activating different cascades of signaling for N_2-fixing symbiosis and AM in response to different Ca^{2+} spiking, because the next steps of nodulation signaling are independent from those of AM:

the mutations in downstream *Sym* genes do not change the mycorrhizal phenotype of the legume. Interestingly, mutations in any *Sym* genes do not influence the defense reactions, suggesting that signaling pathways of mutualistic symbioses and pathogenesis are sufficiently different.

The calcium-calmodulin-dependent protein kinase (CCaMK) is supposed to be also involved in legume interactions with PGPR and/or endophytic bacteria as it was shown using inoculation of *M. truncatula* by *P. fluorescens* that *MtDMI3* gene (encoding for CCaMK) regulates intercellular root colonization by bacteria as well as expression of some plant housekeeping genes known earlier as mycorrhizins [190].

The CCaMK is known to form a complex with CYCLOPS, a phosphorylation substrate, within the nucleus [35]. *cyclops* mutants of *Lotus* severely impair the infection process induced by the bacterial or fungal symbionts, and are also defective in arbuscule development [149]. During RNS, *cyclops* mutants exhibit the specific defects in infection-thread initiation, but not in the nodule organogenesis [191], indicating that CYCLOPS acts in an infection-specific branch of the symbiotic signaling network [35]. *Cyclops* encodes a protein with no overall sequence similarity to proteins with known function, but containing a functional nuclear localization signal and a carboxy-terminal coiled-coil domain.

It is supposed that CCaMK with help of CYCLOPS probably phosphorilates the specific transcription factors already present in cell: NSP1 and NSP2, which influence the changes of expression in several genes related to the symbiosis development [192, 193]. The activity of these proteins leads to the transcriptional changes in root tissues, for instance, increasing the level of early nodulins ENOD40, ENOD11, ENOD12, ENOD5, which are known to be the potential regulators of infection thread growth and nodule primordium formation [165, 194, 195). Also, the changes in cytokinin status of plant are detected, followed by up-regulation of genes encoding for RN symbiosis-specific cytokinin receptors [196-198]. Moreover, transcription regulators NIN and ERN are to be induced specifically downstream of the early NF signaling pathway in order to coordinate and regulate the correct temporal and spatial formation of root nodules [199-202].

The presented genes are responsible for the signal cascade which is aimed to induce the nodulin, mycorrhizin and symbiosin genes responsible for building the symbiotic structures and implementing their biochemical functions. It is supposed that the signaling pathway did not appear *de novo* in legumes when they become able to form nodules, but was developed from already existed system of AM formation into which the novel, nodule-specific genes were recruited. Still, new genes had been involved in RN symbiosis development, especially those encoding the receptors recognizing hormones (e.g. cytokinins) and hormone-like molecules (Nod factors).

3.3. Autoregulation Of Symbioses Formation

Autoregulation of symbiosis development is an important process that takes place after successful mutual partners' recognition and signal exchange. For RN symbiosis, it is considered that legume host controls the root nodule numbers by sensing the external and internal cues. A major external cue is the soil nitrate, whereas a feedback regulatory system in which earlier formed nodules suppress further nodulation through shoot-root communication is an important internal cue. The latter is known as the autoregulation of nodulation (AON), and is believed to consist of two long-distance signals: a root-derived signal that is generated in infected roots and transmitted to the shoot; and a shoot-derived signal that inhibits nodulation systemically [203, 204]. Therefore, AON represents a strategy through which the host plant can balance the symbiotrophic N nutrition with the energetically more "cheap" combined N nutrition.

Recent findings on autoregulation of nodulation suggest that the root-derived ascending signals to the shoot are short peptides belonging to the CLE peptide family [205, 206]. The leucine-rich repeat receptor-like kinase HAR1 of *Lotus* and its homologues in *M. truncatula* and *P. sativum* (*SUNN* and *Sym29*, respectively) mediate AON and also the nitrate inhibition of nodulation, presumably by recognizing the root-derived signal [207-210]. Other genes, like *ASTRAY*, *KLAVIER* and *TML* in *Lotus*, and *RDN1* in *M. truncatula*, are also supposed to play a sufficient role in AON [211-214].

Very little is still known about the plant regulation of mycorrhization process. In split-root systems on alfalfa (*Medicago sativa*), inoculation of one half of a split-root system with the fungus *Glomus mosseae* significantly reduced later AM colonization on the other half. A similar suppressive effect on mycorrhization was observed after inoculation with *Sinorhizobium meliloti* [215]. Furthermore, prior addition of purifed rhizobial Nod factors on one half signifcantly reduced mycorrhization on the other half of the split-root system, and reciprocally, prior mycorrhization on one side suppressed nodule formation on the other side of the split-root system. Together these data point to a common autoregulation circuit for both symbioses [210]. It was suggested that Nod factor signaling, as well as mycorrhizal Myc factor signaling, induces expression or post-translation processing of CLE peptides, which likely function as ascending long-distance signals to the shoot [210]. Also, it was demonstrated that mutations in *HAR1* of *Lotus* and corresponding orthologues in other legumes increase both nodulation and mycorrhization suggesting the shared role of these orthologous genes in controlling the rate of root colonization by microsymbionts. Thereby, not only the local signal transduction (CSP) but the systemic autoregulation is common for the RN and AM symbioses.

3.4. Next Stage Of Development The Genetics Of Symbioses

The next-coming step of development the genetics of symbioses is studying gene networks on intergenomic level, i.e. the coordinated expression of plant and microbe genes. For AM, with use of the new molecular approaches, in particular transcriptomics, a series of AM fungal genes has been identified, having altered expression levels during the AM formation [216-219]. Still it is not well studied at which stages of fungal-plant interaction the complementary partners' genes are induced or repressed, and so the use of plant mutants impaired at different steps of AM development might be a challenging approach to reveal the pattern of plant and fungal genetic cooperation [220]. The same research aimed at identification of plant-rhizobial gene interactions with the use of plant and microbe mutants is also in progress.

It has been recently observed that *Medicago truncatula* showed significantly lower efficiency of nitrogen fixation than its close relative *Medicago sativa* L. [221]. The number of nodules formed on the roots of *M. truncatula* was less than that of *M. sativa*, and the nitrogen fixation measured on plants at the beginning of flowering (as well as specific N_2 fixation (μgN h^{-1} mg $nodule^{-1}$)) was significantly lower. The reasons for the low efficiency in nitrogen fixation were partially a result of low relative efficiency (electron allocated to N_2 versus H^+), and slow nitrogen export from nodules in *M. truncatula* when compared to *M. sativa*. This might be connected with a low malate concentration in the nodule tissue of *M. truncatula*, and thus insufficient carbon provision for asparagine formation (fixed nitrogen is to be added to malate to form asparagine) [221]. Therefore, Sulieman and Schulze [221] suggest that improvement the malate formation in *M. truncatula* nodules could help improving the effectiveness of nitrogen fixation.

According to these data, genes encoding for enzymes of malate synthesis should be good candidates for markers to be used as selection and breeding aimed at improvement of symbiotic properties in *M. truncatula*. But, in different species potential markers of symbiotic effectiveness could be found among genes of different functional groups. Our original data on sequencing alleles of symbiotic genes in pea (*Pisum sativum*) varieties with different symbiotic effectiveness suggest that polymorphism of genes belonging to CSP does not correspond to symbiotic properties of pea varieties analysed (Zhukov V.A., unpublished results). Perhaps, good candidates for markers of symbiotic effectiveness could be found during large-scale screening by transcriptome sequencing in different pea genotypes, which is now underway in our laboratory.

4. NEW APPROACHES OF APPLICATION OF MUTUALLY BENEFICIAL PLANT-MICROBE SYSTEMS IN SUSTAINABLE AGRICULTURE

4.1. Development Of Multicomponent Inocula Containing Bsm

An existence of plant genes [148, 150, 222] and their molecular products [223] common for both AM and RN symbioses led to a conclusion that system of legume symbiotic genes should be considered as a single whole, controlling the development of a tripartite symbiosis (legume plant + AM fungi + rhizobia). This fact along with the demonstration of synergistic activity in beneficial soil microbes (reviewed in: [103]) and a suggestion that plant genetic systems controling the development of RN and, probably, of some other beneficial plant-microbe associations evolved on the basis of that of AM [35] have great importance for the application of tripartite or even multi-partite symbiotic systems in low-input sustainable environmentally-friendly agrotechnologies.

The use in sustainable agriculture of inocula based on beneficial soil microbes as described above allows the improvement crop productivity with decreased doses of mineral fertilizers and pesticides (reviewed in: [224, 225]). These days the majority of commercial inocula contain pure cultures of single microorganisms and only occasionally multiple combinations. There are several objections to the use of mono-inoculation. Firstly, endemic microbial communities are stable and the introduced microbe may be allowed to occupy a very small niche in the whole community or even get lost in a first week after introduction. Secondly, genetic material in microbes is very plastic, and consequently strains introduced into natural ecosystems can rapidly lose their beneficial traits. Thirdly, the existence of microbial cooperation in the rhizosphere [103] as well as in natural synergistic associations of different microbes including those between AM fungi and their endocellular or superficial symbionts [103, 132] question the possibility and expediency of applying mono-inoculants and even use of the term 'mono-inoculation' itself. Finally, plants possess relatively stable genomes and this fact contributes significantly to the effectiveness of symbiosis [226]. Therefore, for industrial plant production in sustainable systems we should use plants having highly effective interactions with all kinds of beneficial soil microbes, which can encourage the development of multiple niches hosting microbes and regulating their activity. For this it is necessary to develop new multi-component microbial inocula which increase the content and biodiversity of beneficial soil microbes in agricultural land.

There is experimental evidence of the effectiveness of simultaneous inoculation of legumes with AM fungi and nodule bacteria leading to increased productivity and quality of the yield, e.g. groundnut [138], pea [139-141, 227], albaida (*Anthyllis cytisoides*) [95], and soybean [142]. The effect achieved equalled or exceeded that achieved with mineral fertilizers [140, 141, 227]. The effect also exceeded that of mono-inoculation with AM fungi or with rhizobia either in model experiments or under field conditions [139, 140, 142]. In long-term experiments in a desertified Mediterranean ecosystem, it was found that simultaneous inoculation with AM fungi and rhizobia enhanced the establishment of key plant species and increased soil fertility and quality; increased soil nitrogen content, organic matter content, and soil aggregate hydrostability and enhanced nitrogen transfer from nitrogen-fixing to non-fixing species associated with the natural succession of the plants [95].

There is an example of application of triple inoculum (AM fungi, rhizobia and PGPR) to the legume *A. cytisoides* which was successful only when the microorganisms used were isolated from local environment [94]. In collaboration with an innovation company "Bisolbi-Inter" (Russian Federation) the All-Russia Research Institute for Agricultural Microbiology (ARRIAM), Saint-Petersburg, Russian Federation, has developed technology for the production and application of a new multifunctional biopreparation BisolbiMix [228] containing a complex of the most effective isolates of endosymbiotic microbes (AM fungi and rhizobia) and associative bacteria (PGPR) from the collection held at ARRIAM. A non-sterile substrate-carrier which is derived from washing-filtration by-products of a sugar-beet factory contains its own microbial community including all the above groups of beneficial microbes. The preparation can be formulated into a seed dressing (not effective for all the crop plants tested) or granules. The efficacy of BisolbiMix was demonstrated in field trials with legumes, e.g. pea [227] or non-legumes such as wheat, pumpkin and potato (Chebotar V.K. et al., unpublished results). The use of microbial formulations containing rhizobia for non-legumes seems to be sensible because it is known that nodule bacteria which do not form nodules on a non-host legume as well as non-legume roots can operate as PGPR [229, 230]. Thus, the selection of rhizobia with both PGPR activity and efficient symbiotic nitrogen fixation should be advantageous in crop rotations or intercropping systems using legumes and non-legumes.

It is possible, therefore, to develop effective multi-microbial inoculants, but it is necessary to use local communities of beneficial microbes because this exploits the natural biological and genetical adaptations of the partners to their environment [94, 231].

4.2. Breeding For Improving Legume Symbiotic Effectiveness

During development of plant-microbe systems for low-input sustainable ecologically friendly plant cultivation it is necessary to be guided by conclusions of EC experts about global productivity of legumes (http://www.grainlegumes.com/aep/; http://ec.europa.eu/research/biosociety/food_quality/projects/002_en.html) for sustainable agriculture. The use of legumes in agriculture is leading to: improved soil fertility and increased diversity of crops and soil microbial communities; reductions in the use of non-renewable natural resources; decreased negative effects from intensive agrotechnologies on the natural environment due to decreased requirement for mineral fertilizers and pesticides and decreased production of animal protein and associated wastes; local production of pollution-free food and forage; and a more stable income for the agricultural producers. This is why it is necessary to breed legumes which have highly effective interactions with beneficial soil microbes.

For more than twenty five years the authors' laboratory has specialized in the genetics of plant-microbe interactions using pea (*P. sativum* L.) as a model plant. Our experience for improving the effectiveness of beneficial plant-microbe systems with pea is consequently given as an example. At the same time, the authors' team knows only single record of other activity of this nature: genetic variability of onion (*Allium* spp.) has been shown with respect to its responsiveness to AM fungi inoculation which indicate that onion breeding for improving efficacy of associations with AM fungi is possible [232]. The necessity for this sort of plant breeding is also considered, mainly with respect to the effectiveness of RN symbiosis [233-236].

4.3. Analysis Of Genetic Variability Of Pea With Respect To Its Effectiveness Of Interactions With Beneficial Soil Microbes

A high level of genetic variability was demonstrated in analyses of the symbiotic effectiveness under double inoculation with AM fungi and nodule bacteria of 99 land-races and outclassed heritage cultivars of *P. sativum* from the collection N.I. Vavilov's All-Russia Research Institute of Plant Industry, Saint-Petersburg, Russian Federation, of different geographical origin [139, 141]. In a few genotypes considerable increases in plant dry weight (about 300 %), seed productivity (more than 650%), phosphorus and nitrogen content (more than 900 and more than 300 %, respectively) were observed. The most promising highly symbiotically effective genotypes and those with low symbiotic potential were included in the Pea Genetic Collection (ARRIAM) to be used for experiments studying the functioning of tripartite/multipartite symbiosis. Types identified as highly symbiotically effective genotypes were involved in breeding programmes to create commercial pea cultivars with great potential for interactions with beneficial soil microbes (in collaboration with All-Russia Institute of Leguminous and Groat Crops (ARILGC), Orel, Russian Federation).

The most promising highly symbiotically effective pea genotypes previously selected and different commercial pea cultivars created without consideration of symbiotic effectiveness were involved in three-year field trials (Orel district) [227]. Seed productivity and plant dry weight were chosen as the main criteria for the evaluation of symbiosis effectiveness in legume crops. The double (actually multiple, see above comments on the nature of AM fungi) inoculation was shown to increase seed productivity and plant dry weight in most of the pea genotypes studied and sometimes this could exceed the effect of mineral fertilizers. The effectiveness of legume breeding to improve the symbiotic potential of legume cultivars was proven therefore under field conditions and the genotypes to be used in such breeding programmes were identified. The genotype K-8274 (non-commercial) was selected as a standard of symbiotic effectiveness. Additionally, it was demonstrated that highly effective genotypes can be also found among commercial pea cultivars created without consideration for effectiveness of interactions with beneficial soil microbes. Taking into account that most commercial legume cultivars have accidentally lost their abilities for symbiotrophic nutrition without selective pressure during breeding of intensive crops, the latter constitutes a very important finding for plant breeders and gives them the possibility for concurrent generation of cultivars with required pea plant architecture, other agriculturally important traits and high effectiveness of interactions with all types of beneficial soil microbes in a single breeding programme.

4.4. Breeding To Improve Pea Symbiotic Effectiveness

In order to cultivate plants with improved symbiotic potential a special breeding nursery was created in the experimental trials ground of ARILGC on land where for the 5 years before nursery establishment mineral fertilizers had not been applied. To reduce the incidence and severity of root pathogens a 6-field crop rotation was used where cultivation of winter wheat was followed by peas. The multi-component preparation BisolbiMix was used for the inoculation of test plants.

Using the breeding nursery as well as a breeding protocol developed from long-term collaboration of ARRIAM with ARILGC the first (in the whole history of legume breeding) pea cultivar "Triumph" having increased potential of interactions with beneficial soil microbes was intentionally created [237]. It arose as a result of crossing a commercial cultivar 'Classic' (donor of agriculturally important traits) and the genotype K-8274 (donor of symbiotic effectiveness trait) and subsequent individual selection of genotypes with high productivity and capacity for supporting various beneficial microbes.

The cultivar "Triumph" is of middle stem height, semi-leafless and has stable productivity under different climate conditions, it is comparatively resistant to root rots and pests. Its productivity is not lower than those of the productivity standards for Orel district using the conventional production technologies and 10% greater in comparison with the standard cultivars when inoculated with BisolbiMix. As a result of two-year state trials (2007-2008) the productivity of "Triumph" was shown to be comparable with those of standard regional cultivars enabling recommendation for commercial cultivation in the Central region of Russian Federation (unpublished results). Thus, the innovative concept of the authors' research team for plant breeding (applicable not only for legumes, but also for non-legumes) is bearing its first fruits.

5. CONCLUSIONS

Intimate associations of beneficial soil microbes with the host plants described above in detail are applicable in sustainable crop production if taken either separately or in combination. Many authors are now recognizing the need for using the multi-microbial plant inoculants and the advantages of using the indigenous plants (or varieties of local breeding) and microbes.

The authors' team proposes its own concept which offers fundamentally new approaches to plant production. Firstly, it is necessary to consider plant genetic systems controlling interactions with different beneficial soil microbes in unison. Secondly, plants used as a component of this complex plant-microbe system controlling its effectiveness should be bred to improve the effectiveness of interactions with all types of beneficial soil microbes. Increases of plant biomass production due to plant-microbe symbiosis should be used as the main parameter for an evaluation of plant effectiveness in interactions with beneficial soil microbes. The plant production should be done with inoculation composed of multi-component microbial inocula consisting of AM fungi, rhizobia, PGPR and/or beneficial endophytic bacteria. Finally, taking into consideration the importance of legumes for global agriculture, greater emphasis should be placed on plant-microbial systems in the development of low-input agro-biotechnologies enabling wider cultivation of leguminous crops.

Molecular markers are considered to be a convenient tool to facilitate breeding via MAS (marker-assisted selection) approach. But, search for suitable markers that are associated with symbiotic effectiveness trait is rather complicated problem. To our knowledge, there was no direct link between sequences of symbiotic genes and symbiotic effectiveness, and there are only a few examples of successful use of QTL analysis in legumes to trace loci associated with some symbiotic traits in pea [238] and *Lotus* [239]. So there's a gap between molecular genetic bases of symbioses development, from one side, and effective functioning the symbiotic systems in field conditions, from the other side. In our opinion, substantial improvement of methods of molecular genetics and bioinformatics, such as next-generation sequencing and proteome analysis, could help to build a bridge between fundamental and applied science in this area, and to improve the sustainability of the legume crop production.

NAN. ACKNOWLEDGEMENTS

We thank Dr. Margarita A. Vishnyakova (N.I. Vavilov All-Russia Research Institute of Plant Industry, St-Petersburg, Russian Federation), Dr. Tatiana S. Naumkina (All-Russia Institute of Legumes and Groat Crops, Orel, Russian Federation) and Dr. Vladimir K. Chebotar ("Bisolbi-Inter" Ltd., St-Petersburg, Russian Federation) for the long-term collaboration in the fields of legume breeding and microbial inocula development. We are also grateful to Victoria Semenova (Komarov Botanical Institute, St-Petersburg, Russian Federation) for critical reading the manuscript. This work was supported by the grants of RFBR (10-04-00961, 10-04-01146, 12-04-01867), Grant to support leading Russian science school (HIII–337.2012.4) and Governmental contracts for research with Ministry of Science and Education of Russian Federation (II1304, 16.512.11.2155, #8109).

REFERENCES

1. Vance CP. Symbiotic Nitrogen Fixation and Phosphorus Acquisition. Plant Nutrition in the World of Declining Renewable Resources. Plant Physiol 2001;127(2): 390-397.
2. Brewin NJ. Plant Cell Wall Remodeling in the Rhizobium-Legume Symbiosis. Crit Rev Plant Sci 2004;23: 1-24.
3. Allen ON, Allen EK. The Leguminosae. A Source Book of Characteristics, Uses and Nodulation. Madison: The University of Wisconsin Press; 1981.
4. Wall LG. The Actinorhizal Symbiosis. J Plant Growth Regul 2000;19(2): 167-182.
5. Balachandar D, Raja P, Kumar K, Sundaram SP. Non-Rhizobial Nodulation in Legumes. Biotechnol Molec Biol Rev 2007;2(2): 49-57.
6. Spaink HP, Kondorosi A, Hooykaas PJJ., editors. The Rhizobiaceae. Molecular Biology of Model Plant-Associated Bacteria. Dordrecht/Boston/London: Kluwer; 1998.
7. MacLean AM, Finan T, Sadowsky MJ. Genomes of the Symbiotic Nitrogen-Fixing Bacteria of Legumes. Plant Physiol 2007;144(2): 615-622.
8. Fred EB, Baldwin IL, McCoy E. Root Nodule Bacteria and Leguminous Plants. Madison: Univ Wisconsin Stud Sci; 1932.
9. Provorov NA. The Interdependence between Taxonomy of Legumes and Specificity of Their Interaction with Rhizobia in Relation to Evolution of the Symbiosis. Symbiosis 1994;17: 183-200.
10. Broughton WJ, Perret X. Genealogy of Legume-Rhizobium Symbiosis. Curr Opin Plant Biol 1999;2(4): 305-311.
11. Ovtsyna AO, Staehelin C. Bacterial signals required for the Rhizobium-legume symbiosis. In: Pandalai SG. (ed) Recent Research Developments in Microbiology, Vol 7 (Part II). Trivandrum, India: Research Signpost; 2003. p631-648.
12. Spaink HP. The Molecular Basis of Infection and Nodulation by Rhizobia: the Ins and Outs of Sympathogenesis. Ann Rev Phytopathol 1995;33: 345-368.
13. Schultze M, Kondorosi A. Regulation of Symbiotic Root Nodule Development. Annu Rev Genet 1998;32: 33-57.
14. D'Haeze W, Holsters M. Nod Factor Structures, Responses, and Perception During Initiation of Nodule Development. Glycobiology 2002;12(6): 79-105.
15. Becker A, Pühler A. Production of Exopolysaccharides. In: Spaink HP, Kondorosi A, Hooykaas PJJ. (eds) The Rhizobiaceae. Molecular Biology of Model Plant-Associated Bacteria. Dordrecht/Boston/London: Kluwer; 1998. p87-118.
16. Lugtenberg BJJ. Outer Membrane Proteins. In: Spaink HP, Kondorosi A, Hooykaas PJJ. (eds) The Rhizobiaceae. Molecular Biology of Model Plant-Associated Bacteria. Dordrecht/Boston/London: Kluwer; 1998. p45-53.
17. Jones KM, Sharopova N, Lohar DP, Zhang JQ, VandenBosch KA, Walker GC. Differential Response of the Plant *Medicago truncatula* to its Symbiont *Sinorhizobium meliloti* or an Exopolysaccharide-Deficient Mutant. Proc Natl Acad Sci USA 2008;105(2): 704-709.
18. Hirsch AM. Developmental Biology of Legume Nodulation. New Phytol 1992;122: 211-237.

19. Sprent JI. Nodulation in Legumes. Royal Botanical Gardens, Kew: Cromwell Press Ltd; 2001.
20. Oke V, Long SR. Bacteroid Formation in the Rhizobium-Legume Symbiosis. Curr Opin Microbiol 1999;2(6): 641-646.
21. Brewin NJ (1998) Tissue and Cell Invasion by Rhizobium: the Structure and Development of Infection Threads and Symbiosomes. In: Spaink HP, Kondorosi A, Hooykaas PJJ. (eds) The Rhizobiaceae. Molecular Biology of Model Plant-Associated Bacteria. Dordrecht/Boston/London: Kluwer; 1998. p417-429.
22. Tsyganov VE, Voroshilova VA, Herrera-Cervera JA, Sanjuan-Pinilla JM, Borisov AY, Tikhonovich IA, Priefer UB, Olivares J, Sanjuan J. Developmental Down-Regulation of Rhizobial Genes as a Function of Symbiosome Differentiation in Symbiotic Root Nodules of *Pisum sativum* L. New Phytol 2003;159(2): 521-530.
23. Mylona P, Pawlowski K, Bisseling T. Symbiotic Nitrogen Fixation. Plant Cell 1995;7(7): 869-885.
24. Pawlowski K, Bisseling T. Rhizobial and Actinorhizal Symbioses: What Are the Shared Features? Plant Cell 1996;8(10): 1899-1913.
25. Vance CP, Heichel GH. Carbon in N2 Fixation: Limitation or Exquisite Adaptation? Annu Rev Plant Physiol 1991;42: 373-392.
26. Fedorova M, Tikhonovich IA, Vance CP. Expression of C-assimilating Enzymes in Pea (*Pisum sativum* L.) Root Nodules. In situ Localization in Effective Nodules. Plant Cell Environ 1999;22(10): 1249-1262.
27. Schubert KR. Products of Biological Nitrogen Fixation in Higher Plants: Synthesis, Transport, and Metabolism. Annu Rev Plant Physiol 1986;37: 539-574.
28. Mergaert P, Uchiumi T, Alunni B, Evanno G, Cheron A, Catrice O, Mausset AE, Barloy-Hubler F, Galibert F, Kondorosi A, Kondorosi E. Eukaryotic Control on Bacterial Cell Cycle and Differentiation in the Rhizobium-Legume Symbiosis. Proc Natl Acad Sci USA 2006;103(13): 5230-5235.
29. Van de Velde W, Zehirov G, Szatmari A, Debreczeny M, Ishihara H, Kevei Z, Farkas A, Mikulass K, Nagy A, Tiricz H, Satiat-Jeunemaître B, Alunni B, Bourge M, Kucho K, Abe M, Kereszt A, Maroti G, Uchiumi T, Kondorosi E, Mergaert P. Plant Peptides Govern Terminal Differentiation of Bacteria in Symbiosis. Science 2010;327(5969): 1122-1126.
30. Oono R, Denison RF, Kiers ET. Controlling the Reproductive Fate of Rhizobia: How Universal Are Legume Sanctions? New Phytol 2009;183(4): 967-979.
31. Oono R, Denison RF. Comparing symbiotic Efficiency Between Swollen Versus Nonswollen Rhizobial Bacteroids. Plant Physiol 2010;154(3): 1541-1548.
32. Schüßler A, Schwarzott D, Walker C. A New Fungal Phylum, the Glomeromycota: Phylogeny and Evolution. Mycol Res 2001;105: 1413-1297.
33. Brundrett MC Coevolution of Roots and Mycorrhizas of Land Plants. New Phytol 2002;154: 275-304.
34. Smith SE, Read DJ. Mycorrhizal Symbiosis (3rd ed.). London: Academic Press; 2008.
35. Parniske M. Arbuscular Mycorrhiza: the Mother of Plant Root Endosymbioses. Nat Rev Microbiol 2008;6(10): 763-775.
36. Lavin M, Herendeen PS, Wojciechowski MF Evolutionary Rates Analysis of Leguminosae Implicates a Rapid Diversification of Lineages during the Tertiary. Syst Biol 2005;54: 574-594.
37. Sprent JI. Evolving Ideas of Legume Evolution and Diversity: A Taxonomic Perspective on the Ocurrence of Nodulation. New Phytologist 2007;174: 11-25.
38. Sprent J. Evolution and Diversity of Legume Symbiosis. In: Dilworth MJ, James EK, Sprent JI, Newton WE (eds) Nitrogen-Fixing Leguminous Symbioses. Dordrecht: Springer; 2008. P 363-394.
39. Sprent JI, James EK. Legume Evolution: Where Do Nodules and Mycorrhizas Fit in? Plant Physiol 2007;144: 575-581.
40. Akiyama K, Matsuzaki K, Hayashi H. Plant Sesquiterpenes Induce Hyphal Branching in Arbuscular Mycorrhizal Fungi. Nature 2005;435: 824-827.

41. Nagahashi G, Douds DD, Ferhatoglu Y. Functional Categories of Root Exudate Compounds and their Relevance to AM Fungal Growth. In: Koltai H, Kapulnik Y (eds) Arbuscular Mycorrhizas: Physiology and Function. Dordrecht: Springer; 2010. P 33-56.
42. Kosuta S, Chabaud M, Lougnon G, Gough C, Denarie J, Barker DG, Becard G. A Diffusible Factor From Arbuscular Mycorrhizal Fungi Induces Symbiosis-Specific *Mtenod11* Expression in Roots of *Medicago truncatula*. Plant Physiol 2003;131: 952-962.
43. Maillet F, Poinsot V, André O, Puech-Pagès V, Haouy A, Gueunier M, Cromer L, Giraudet D, Formey D, Niebel A, Martinez EA, Driguez H, Bécard G, Dénarié J. Fungal Lipochitooligosaccharide Symbiotic Signals in Arbuscular Mycorrhiza. Nature 2011;469(7328): 58-63.
44. Navazio L, Moscatiello R, Genre A, Novero M, Baldan B, Bonfante P, Mariani P. A Diffusible Signal From Arbuscular Mycorrhizal Fungi Elicits a Transient Cytosolic Calcium Elevation in Host Plant Cells. Plant Physiol 2007;144: 673-681.
45. Genre A, Bonfante P. The Making of Symbiotic Cells in Arbuscular Mycorrhizal Roots. In: Koltai H, Kapulnik Y (eds) Arbuscular Mycorrhizas: Physiology and Function. Dordrecht: Springer; 2010. P 57-81.
46. Parniske M. The *Lotus japonicus* LjSym4 Gene is Required for the Successful Symbiotic Infection of Root Epidermal Cells. Molec Plant-Microbe Iinteract 2000;13: 1109-1120.
47. Genre A, Bonfante P. Building a Mycorrhiza Cell: How to Reach Compatibility Between Plants and Arbuscular Mycorrhizal Fungi. J Plant Interact 2005;1: 3-13
48. Bieleski RL Phosphate Pools, Phosphate Transport and Phosphate Availability. Annu Rev Plant Physiol 1973;24: 225-252.
49. Schachtman DP, Reid RJ, Ayling SM. Phosphorus Uptake by Plants: From Soil to Cell. Plant Physiol 1998;116: 447-453.
50. Tinker PB, Nye PH. Solute Movement in the Rhizosphere. Oxford, UK: Oxford University Press; 2000.
51. Neumann E, George E. Nutrient Uptake: The Arbuscular Mycorrhiza Fungal Symbiosis as a Plant Nutrient Acquisition Strategy In: Koltai H, Kapulnik Y (eds) Arbuscular Mycorrhizas: Physiology and Function. Dordrecht: Springer; 2010. P 137-167.
52. Smith SE, Smith FA, Jakobsen I. Mycorrhizal Fungi Can Dominate Phosphate Supply to Plants Irrespective of Growth Responses. Plant Physiol 2003;133: 16–20
53. Schweiger P, Jakobsen I. Laboratory and Field Methods for Measurement Of Hyphal Uptake of Nutrients in Soil. Plant Soil 2000,226: 237–244.
54. Nielsen JS, Joner EJ, Declerck S, Olsson S, Jakobsen I. Phospho-Imaging as a Tool for Visualization and Noninvasive Measurement of P Transport Dynamics in Arbuscular Mycorrhizas. New Phytol 2002;154: 809–820.
55. Rufyikiri G, Declerck S, Thiry Y. Comparison of 233U and 33P Uptake and Translocation by the Arbuscular Mycorrhizal Fungus *Glomus intraradices* in Root Organ Culture Conditions. Mycorrhiza 2004;14: 203–207.
56. Ezawa T, Smith SE, Smith FA. P Metabolism and Transport in AM Fungi. Plant Soil 2002;244: 221–230.
57. Ashford A. Tubular Vacuoles in Arbuscular Mycorrhizas. New Phytol 2002;54: 545–547.
58. Ezawa T, Cavagnaro TR, Smith SE, Smith FA, Ohtomo R. Rapid Accumulation of Polyphosphate in Extraradical Hyphae of an Arbuscular Mycorrhizal Fungus as Revealed by Histochemistry and a Polyphosphate Kinase/Luciferase System. New Phytol 2003;161: 387–392.
59. Karandashov V, Bucher M. Symbiotic Phosphate Transport in Arbuscular Mycorrhizas. Trends Plant Sci 2005;10: 22-29.
60. Balestrini R, Gomez-Ariza J, Lanfranco L, Bonfante P. Laser Microdissection Reveals that Transcripts for Five Plant and One Fungal Phosphate Transporter Genes are

Contemporaneously Present in Arbusculated Cells. Molec Plant-Microbe Interact 2007;20: 1055-1062.

61. Javot H, Pumplin N, Harrison MJ. Phosphate in the Arbuscular Mycorrhizal Symbiosis: Transport Properties and Regulatory Roles. Plant Cell Environ 2007;30: 310-322

62. Harrison MJ, Dewbre GR, Liu J. A Phosphate Transporter from *Medicago truncatula* Involved in the Acquisition of Phosphate Released by Arbuscular Mycorrhizal Fungi. Plant Cell 2002;14: 2413–2429.

63. Maeda D, Ashida K, Iguchi K, Chechetka SA, Hijikata A, Okusako Y, Deguchi Y, Izui K, Hata S. Knockdown of an Arbuscular Mycorrhiza-Inducible Phosphate Transporter Gene of *Lotus japonicus* Suppresses Mutualistic Symbiosis. Plant Cell Physiol 2006;47: 807-817.

64. Javot H, Penmetsa RV, Terzaghi N, Cook DR, Harrison MJ. A *Medicago truncatula* Phosphate Transporter Indispensable For The Arbuscular Mycorrhizal Symbiosis. Proc. Natl. Acad. Sci. USA 2007;104: 1720-1725.

65. Nagy R, Drissner D, Amrhein N, Jakobsen I, Bucher M. Mycorrhizal Phosphate Uptake Pathway in Tomato Iis Phosphorusrepressible and Transcriptionally Regulated. New Phytol 2009;181: 950–959.

66. Bonfante P, Balestrini R, Genre A, Lanfranco L. Establishment and Functioning Of Arbuscular Mycorrhizas. In: Deising H (ed) The Mycota V. Plant Relationship (2nd ed.). Berlin, Heidelberg: Springer-Verlag; 2009. P 259-274.

67. Hawkins HJ, Johansen A, George E. Uptake and Transport of Organic and Inorganic Nitrogen by Arbuscular Mycorrhizal Fungi. Plant Soil 2000;226: 275-285.

68. Lopez-Pedrosa A, Gonzalez-Guerrero M, Valderas A, Azcon-Aguilar C, Ferrol N. *GintAMTi* Encodes a Functional High-Affinity Ammonium Transporter that is Expressed in the Extraradical Mycelium of *Glomus intraradices*. Fungal Genet Biol 2006;43: 102-110.

69. Jin H, Pfeffer PE, Douds DD, Piotrowski E, Lammers PJ, Shachar-Hill Y. The Uptake, Metabolism, Transport and Transfer of Nitrogen in an Arbuscular Mycorrhizal Symbiosis. New Phytol 2005;168: 687-696.

70. Johansen A, Finlay RD, Olsson PA. Nitrogen Metabolism of External Hyphae of the Arbuscular Mycorrhizal Fungus *Glomus intraradices*. New Phytol 1996;133: 705-712.

71. Breuninger M, Trujillo CG, Serrano E, Fischer R, Requena N. Different Nitrogen Sources Modulate Activity but not Expression of Glutamine Synthetase in Arbuscular Mycorrhizal Fungi. Fungal Genet Biol 2004;41: 542–552.

72. Guether M, Neuhauser B, Balestrini R, Dynowski M, Ludewig U, Bonfante P. A Mycorrhizal-Specific Ammonium Transporter from Lotus japonicus Acquires Nitrogen Released by Arbuscular Mycorrhizal Fungi. Plant Physiol 2009;150: 73–83.

73. Akhtar MS, Siddiqui ZA. Arbuscular Mycorrhizal Fungi as Potential Bioprotectants Against Plant Pathogens. In: Siddiqu ZA et al (eds) Mycorrhizae: Sustainable Agriculture and Forestry. Dordrecht: Springer; 2008. P 61–97.

74. Koricheva J, Gange AC, Jones T. Effects of Mycorrhizal Fungi on Insect Herbivores: a Metaanalysis. Ecology 2009;90: 2088–2097.

75. Currie AF, Murray PJ, Gange AC. Is a Specialist Root-Feeding Insect Affected by Arbuscular Mycorrhizal Fungi? Appl Soil Ecol 2011;47: 77–83.

76. Harrison MJ. Molecular and Cellular Aspects of the Arbuscular Mycorrhizal Symbiosis. Ann Rev Plant Physiol Plant Mol Biol 1999;50: 361-389.

77. Blilou I, Bueno P, Ocampo JA, Garcia-Garrido JM. Induction of Catalase And Ascorbate Peroxidase Activities in Tobacco Roots Inoculated with the Arbuscular Mycorrhizal Fungus *Glomus mossea*. Mycol Res 2000;104: 722-725.

78. Blilou I, Ocampo JA, Garcia-Garrido JM. Resistance of Pea Roots to Endomycorrhizal Fungus or *Rhizobium* Correlates with Enhanced Levels of Endogenous Salicylic Acid. J Exp Bot 1999;50: 1663-1668.

79. Lambais MR. Regulation of Plant Defence-Related Genes in Arbuscular Mycorrhizae. In: Podila GK, Douds DD (eds) Current Advances in Mycorrhizae research The American Phytopathological Society, Minnesota; 2000. P 45-59.
80. Bonanomi A, Wiemken A, Boller T, Salzer P. Local Induction of a Mycorrhiza-Specific Class III Chitinase Gene in Cortical Root Cells of *Medicago truncatula* Containing Developing or Mature Arbuscules. Plant Biol 2001;3: 94-199.
81. Salzer P, Corbiere H, Boller T. Hydrogen Peroxide Accumulation in *Medicago Truncatula* Roots Colonized by the Arbuscular Mycorrhiza-Forming Fungus *Glomus mosseae*. Planta 1999;208: 319-325.
82. Garcia-Garrido JM, Ocampo JA. Regulation of the Plant Defence Response in Arbuscular Mycorrhizal Symbiosis. J Exp Bot 2002;53: 377-1386
83. Liu J, Maldonado-Mendoza I, Lopez-Meyer M, Cheung F, Town CD, Harrison MJ. Arbuscular Mycorrhizal Symbiosis is Accompanied by Local and Systemic Alterations in Gene Expression and an Increase in Disease Resistance in the Shoots. Plant J 2007;50: 529-544.
84. Bennett AE, Bever JD, Bowers MD. Arbuscular Mycorrhizal Fungal Species Suppress Inducible Plant Responses and Alter Defensive Strategies Following Herbivory. Oecologia 2009;160: 711–719.
85. Kempel A, Schmidt AK, Brandl R, Schadler M. Support from the Underground: Induced Plant Resistance Depends on Arbuscular Mycorrhizal Fungi. Funct Ecol 2010;24: 293–300.
86. Quilambo OA. The Vesicular-Arbuscular Mycorrhizal Symbiosis. African J Biotechnol 2003;2: 539-546.
87. Augé RM. Water Relations, Drought and Vesicular-Arbuscular Mycorrhizal Symbiosis Mycorrhiza 2001;11: 3-42.
88. Augé RM, Moore JL, Sylvia DM, Cho K. Mycorrhizal Promotion of Host Stomatal Conductance in Relation to Irradiance and Temperature. Mycorrhiza 2004;14: 85-92.
89. Ruiz-Lozano JM, Aroca R. Host Response to Osmotic Stresses: Stomatal Behaviour and Water Use Efficiency of Arbuscular Mycorrhizal Plants. In: Koltai H, Kapulnik Y (eds) Arbuscular Mycorrhizas: Physiology and Function. Dordrecht: Springer; 2010. P 239-256.
90. Celik I, Ortas I, Kilic S. Effects of Compost, Mycorrhiza, Manure and Fertilizer on Some Physical Properties of a Chromoxerert Soil. Soil and Tillage Res 2004;78: 59-67.
91. Rillig MC. Arbuscular Mycorrhizae, Glomalin and Soil Aggregation. Can J Soil Sci 2004;84: 355-363.
92. Turnau K, Ryszka P, Wojtczak G. Metal Tolerant Mycorrhizal Plants: A Review from the Perspective on Industrial Waste in Temperate Region In: Koltai H, Kapulnik Y (eds) Arbuscular Mycorrhizas: Physiology and Function. Dordrecht: Springer; 2010. P 257-279.
93. Ryszka P, Turnau K. Arbuscular Mycorrhiza of Introduced and Native Grasses Colonizing Zinc Wastes: Implications for Restoration Practices. Plant Soil 2007;298: 219–229.
94. Requena N, Jimenez J, Toro M, Barea JM. Interactions Between Plant-Growth-Promoting Rhizobacteria (PGPR), Arbuscular Mycorrhizal Fungi and *Rhizobium* spp. in the Rhizosphere of *Anthyllis cytisoides*, a Model Legume for Revegetation in Mediterranean Semi-Arid Ecosystems. New Phytologist 1997;136(4): 667-677.
95. Requena N, Perez-Solis E, Azcón-Aguilar C, Jeffries P, Barea JM. Management of Indigenous Plant-Microbe Symbioses Aids Restoration of Desertified Ecosystems. Appl Environ Microbiol 2001;67(2): 495-498.
96. Turnau K, Orlowska E, Ryszka P, Zubek S, Anielska T, Gawronski S, Jurkiewicz A. Role of Mycorrhizal Fungi In Phytoremediation and Toxity Monitoring of Heavy Metal Rhich Industrial Wastes in Southern Poland. In: Twardowska I, Allen HE, Häggblom MM, Stefaniak S (eds) NATO Science Series. Soil and Water Pollution Monitoring, Protection and Remediation. Dordrecht: Springer; 2007;69: 533-551.

97. Turnau K, Anielska T, Ryszka P, Gawroński S, Ostachowicz B, Jurkiewicz A. Establishment of Arbuscular Mycorrhizal Plants Originating from Xerothermic Grasslands on Heavy Metal Rich Industrial Wastes – New Solution for Waste Revegetation. Plant Soil 2008;305: 267–280.
98. Bloemberg G, Lugtenberg BJJ. Molecular Basis of Plant Growth Promotion and Biocontrol by Rhizobacteria. Curr Opin Plant Biol 2001; 4: 343–350
99. Dobereiner J Isolation and Identification of Root Associated Diazotrophs. Plant Soil 1988;110: 207-212.
100. Costacurta A, Vanderleyden J. Synthesis of Phytohormones by Plant-Associated Bacteria. Crit Rev Microbiol 1995;21: 1-18.
101. Provorov NA, Vorobyov NI. Evolutionary Genetics of Plant-Microbe Symbioses. New York: NOVA Science Publishers; 2010.
102. Shtark O., Provorov N., Mikić A., Borisov A., Ćupina B., Tikhonovich I. Legume Root Symbioses: Natural History and Prospects For Improvement. Ratarstvo i povrtarstvo (Field and Vegetable Crops Research) 2011;48: 291-304.
103. Barea JM, Pozo MJ, Azcon R, Azcon-Aguilar C. Microbial Cooperation in the Rhizosphere. J Exp Botany 2005;56(417): 1761-1778.
104. Haas D, Defago G. Biological Control of Soil-Borne Pathogens by Fluorescent Pseudomonads. Nature Rev Microbiol. AOP,
105. Siddiqui ZA. PGPR: Prospective Biocontrol Agents of Plant Pathogens. In: Siddiqui ZA (ed) PGPR: Biocontrol and Biofertilization. Dordrecht: Springer; 2005. P 111–142.
106. Preston GM Plant Perceptions of Plant Growth-Promoting Pseudomonas. Phil Trans R Soc Lond B 2004;359: 907-918.
107. O'Toole GA, Kolter R. Initiation of Biofilm Formation in *Pseudomonas fluorescens* WCS365 Proceeds via Multiple, Convergent Signalling Pathways: a Genetic Analysis. Mol Microbiol 1998;28(3): 449-61.
108. Stephens C, Murray W. Pathogen Evolution: How Good Bacteria go Bad. Curr Biol 2001;11: 53-56.
109. Catara V. *Pseudomonas corrugata*: Plant Pathogen and/or Biological Resource? Mol Plant Pathol 2007;8: 233-244.
110. Bolwerk A, Lagopodi AL, Wijfjes AHM, Lamers GEM, Lugtenberg BJJ, Bloemberg GV. Interactions between *Pseudomonas* Biocontrol Strains and *Fusarium oxysporum* f.sp. *radicis-lycopersici* in the Tomato Rhizosphere. In: Tikhonovich IA, Lugtenberg BJJ Provorov NA (eds) Biology of Plant-Microbe Interactions. IS-MPMI, St.-Petersburg 2004;4: 323-326.
111. Popova EV, Khatskevich LK. In: Tikhonovich IA, Lugtenberg BJJ Provorov NA (eds) Biology of Plant-Microbe Interactions. IS-MPMI, St.-Petersburg 2004;4: 315-318.
112. Van Loon LC, Bakker PA, Pieterse CMJ. Systemic Resistance Induced by Rhizosphere Bacteria. Annu Rev Phytopathol 1998;36: 453-483.
113. Vallad E, Goodman RM. Systemic Acquired Resistance and Induced Systemic Resistance in Conventional Agriculture. Crop Sci 2004;44: 1920-1934.
114. Penrose DM, Glick BR. Methods for Isolating and Characterizing ACC Deaminase-Containing Plant Growth-Promoting Rhizobacteria. Physiol Plantarum 2003;118: 10-15.
115. Glick BR. The Role of Bacterial ACC Deaminase in Promoting the Plant Growth. In: Tikhonovich IA, Lugtenberg BJJ Provorov NA (eds) Biology of Plant-Microbe Interactions. IS-MPMI, St.-Petersburg 2004;4: 557-560.
116. Wu P, Zang G, Ladha JK, McCouch SR, Huang N. Molecular-Marker-Facilitated Investigation on the Ability to Stimulate N2 Fixation in the Rhizosphere by the Irrigated Rice Plants. Theor Appl Genet 1995;91: 1177-1183.
117. Smith KP, Goodman RM. Host Variation for Interactions with Beneficial Plant-Associated Microbes. Annu Rev Phytopathol 1999;37: 473-491.
118. Liu L, Kloepper JW, Tuzun S. Induction of Systemic Resistance in Cucumber by Plant Growth-Promoting Rhizobacteria: Duration of Protection and Root Colonization. Phytopathol 1995;85: 1064-1068.

119. Van Wees SCM, Pieterse CMJ, Trijssenaar A, van Westende TAM, Hartog F, van Loon LC. Differential Induction of Systemic Resistance in *Arabidopsis* by Biocontrol Bacteria. Mol Plant-Microbe Interact 1997;10: 716-724.

120. Kravchenko LV, Leonova EI. Use of Tryptophane from Root Exometabolites for Biosynthesis of Indolil-3-Acetic Acid by the Root-Associated Bacteria. Russian J Microbiol 1993;62: 453–459.

121. Shtark OY, Shaposhnikov AI, Kravchenko LV. The Production of Antifungal Metabolites by *Pseudomonas chlororaphis* Grown on Different Nutrient Sources. Russian J Microbiol 2003;72(5): 574–578.

122. Kamilova F, Kravchenko LV, Shaposhnikov AI, Makarova N, Lugtenberg B. Effect of Tomato Pathogen *Fusarium oxysporum* f. sp. *radicis-lycopersici* and Biocontrol Bacterium *Pseudomonas fluorescens* WCS635 on the Composition of Organic Acids and Sugars in Tomato Root Exudates. Mol. Plant-Microbe Interact 2006;19: 1121-1126.

123. Teplitski M, Robinson JB, Bauer WD. Plants Secrete Substances that mimic Bacterial N-Acyl Homoserine Lactone Signal Activities and Affect Population Density-Dependent Behaviors in Associated Bacteria. Mol Plant-Microbe Interact 2000;13: 637-648.

124. Hallmann J, Berg G. Spectrum and Population Dynamics of Bacterial Root Endophytes. In: Schulz B, Boyle S, Sieber T (eds) Microbial Root Endophytes. Dordrecht: Springer; 2006. P 15-32.

125. Van Overbeek LS, van Vuurde J, van Elsas JD. Application of Molecular Fingerprinting Techniques to Explore the Diversity of Bacterial Endophytic Communities. In: Schulz B, Boyle S, Sieber T (eds) Microbial Root Endophytes. Dordrecht: Springer; 2006. P 337-354.

126. Schulz B, Boyle S, Sieber T. What are Endophytes? In: Schulz B, Boyle S, Sieber T (eds) Microbial Root Endophytes. Dordrecht: Springer; 2006. P 1-14.

127. Sturz AV, Christie BR, Nowak J. Bacterial Endophytes: Potential Role in Developing Sustainable Systems of Crop Production. Cr Rev Plant Sci 2000;19(1): 1-30.

128. McCully ME Niches for Bacterial Endophytes in Crop Plants: a Plant Biologists's View. Aust J Plant Physiol 2001;28: 983-990.

129. Berg G, Hallmann J. Control of Plant Pathogenic Fungi with Bacterial Endophytes. In: Schulz B, Boyle S, Sieber T (eds) Microbial Root Endophytes. Dordrecht: Springer; 2006. P 53–70.

130. Kloepper JW, Ryu CM. Bacterial Endophytes as Elicitors of Induced Systemic Resistance. In: Schulz B, Boyle S, Sieber T (eds) Microbial Root Endophytes. Dordrecht: Springer; 2006. P 33-52.

131. Rosenblueth M, Martinez-Romero E. Bacterial Endophytes and their Interactions with Hosts. Mol Plant-Microbe Interact 2006;19: 827-837.

132. Artursson V, Finlay RD, Jansson JK. Interactions between Arbuscular Mycorrhizal Fungi and Bacteria and Their Potential for Stimulating Plant Growth. Environ Microbiol 2006;8(1): 1-10.

133. Frey-Klett P, Garbaye J, Tarkka M. The Mycorrhiza Helper Bacteria Revisited. New Phytol 2007;176: 22–36.

134. Garbaye J. Helper Bacteria: a New Dimension to the Mycorrhizal Symbiosis. New Phytol 1994;128: 197-210.

135. Hildebrandt U, Ouziad F, Marner FJ, Bothe H. The Bacterium *Paenibacillus validus* Stimulates Growth of the Arbuscular Mycorrhizal fungus *Glomus intraradices* up to the formation of Fertile Spores. FEMS Microbiol Lett 2006;254: 258-267.

136. Toljander JF, Lindahl BD, Paul LR, Elfstrand M, Finlay RD. Influence of Arbuscular Mycorrhizal Mycelial Exudates on soil Bacterial Growth and Community Structure. FEMS Microbiol Ecol 2007;61: 295–304.

137. Rambelli A. The Rhizosphere of Mycorrhizae. In: Marks, G.L., and Koslowski, T.T. (eds) Ectomycorrhizae. New York, USA: Academic Press; 1973. P 299–343.

138. Ibrahim KK, Arunachalam V, Rao PSK, Tilak KVBR. Seasonal Response of Groundnut Genotypes to Arbuscular Mycorrhiza – *Bradyrhizobium* Inoculation. Microbiol Res 1995;150: 218-224.
139. Jacobi LM, Kukalev AS, Ushakov KV, Tsyganov VE, Provorov NA, Borisov AY, Tikhonovich IA. Genetic Variability of Garden Pea (*Pisum sativum* L.) for Symbiotic Capacities. Pisum Genetics 1999;31: 44-45.
140. Borisov AY, Tsyganov VE, Shtark OY, Jacobi LM, Naumkina TS, Serdyuk VP, Vishnyakova MA. Pea (Symbiotic effectiveness). In: Tikhonovich IA, Vishnyakova MA. (eds) The Catalogue of World-Wide Collection. Issue 728. Saint Petersburg: VIR; 2002.
141. Borisov AY, Shtark OY, Danilova TN, Tsyganov VE, Naumkina TS. Effectiviness of Combined Inoculation of Field Peas with Arbuscular Mycorrhizal Fungi and Nodule Bacteria. Russian Agricultural Sciences (Doklady Rossiiskoi Akademii Sel'skohozyaistvennykh Nauk) 2004;4: 5-7.
142. Labutova NM, Polyakov AI, Lyakh VA, Gordon VL. Influence of Inoculation with Nodule Bacteria and Endomycorrhizal Fungus *Glomus intraradices* on Yield and Seed Protein and Oil Content of Different Soybean Cultivars. Russian Agricultural Sciences (Doklady Rossiiskoi Akademii Sel'skohozyaistvennykh Nauk) 2004;4(2): 2-4.
143. Bakker PAHM, Raaijmakers JM, Bloemberg G et al (eds). New Perspectives and Approaches in Plant Growth-Promoting Rhizobacteria Research (Reprinted from Eur J Plant Pathol 2007;119(2)). Dordrecht: Springer; 2007.
144. Andronov EE, Petrova SN, Chizhevskaya EP, Korostik EV, Akhtemova GA, Pinaev AG. Influence of Introducing the Genetically Modified Strain *Sinorhizobium meliloti* ACH-5 on the Structure of the Soil Microbial Community. Russian J Microbiol 2009;78(4): 474–482.
145. Pawlowska TE, Taylor JW. Organization of Genetic Variation in Individuals of Arbuscular Mycorrhizal Fungi. Nature 2004;427(6976): 733-737.
146. Lie TA. Temperature-Dependent Root-Nodule Formation in Pea cv. Iran. Plant Soil 1971;34(3): 751-752.
147. van Kammen A. Suggested Nomenclature for Plant Genes Involved in Nodulation and Symbiosis. Plant Mol Biol Rep 1984;2: 43-45.
148. Gianinazzi-Pearson V. Plant Cell Responses to Arbuscular Mycorrhizal Fungi: Getting to the Roots of the Symbiosis. Plant Cell 1996;8(10): 1871-1883.
149. Kistner C, Winzer T, Pitzschke A, Mulder L, Sato S, Kaneko T, Tabata S, Sandal N, Stougaard J, Webb KJ, Szczyglowski K, Parniske M. Seven *Lotus japonicus* Genes Required for Transcriptional Reprogramming of the Root During Fungal and Bacterial Symbiosis. Plant Cell 2005;17(8): 2217-2229.
150. Küster H, Vieweg MF, Manthey K, Baier MC, Hohnjec N, Perlick AM. Identification and Expression Regulation of Symbiotically Activated Legume Genes. Phytochemistry 2007;68(1): 8-18.
151. Albrecht C, Geurts R, Lapeyrie F, Bisseling T. Endomycorrhizae and Rhizobial Nod Factors Both Require SYM8 to Induce the Expression of the Early Nodulin Genes PsENOD5 and PsENOD12A. Plant J 1998;15(5): 605-614.
152. Kumagai H, Kinoshita E, Ridge RW, Kouchi H. RNAi Knock-Down of ENOD40s Leads to Significant Suppression of Nodule Formation in *Lotus japonicus*. Plant Cell Physiol 2006;47(8): 1102-1111.
153. Wan X, Hontelez J, Lillo A, Guarnerio C, van de Peut D, Fedorova E, Bisseling T, Franssen H. *Medicago truncatula* ENOD40-1 and ENOD40-2 Are Both Involved in Nodule Initiation and Bacteroid Development. J Exp Bot 2007;58(8): 2033-2041.
154. Nutman PS. Genetical Factors Concerned in the Symbiosis of Clover and Nodule Bacteria. Nature 1946;157: 463-465.
155. Jacobsen E. Modification of Symbiotic Interaction of Pea (*Pisum sativum* L.) and *Rhizobium leguminosarum* by Induced Mutations. Plant Soil 1984;82(3): 427-438.
156. Handberg K, Stougaard J. *Lotus japonicus*, an Autogamous, Diploid Legume Species for Classical and Molecular Genetics. Plant J 1992;2: 487-496.

157. Barker D, Bianchi S, Blondon F, Dattee Y, Duc G, Essad S, Flament P, Gallusci P, Genier G, Guy P, Muel X, Tourneur J, Denarie J, Huguet T. *Medicago truncatula*, a Model Plant for Studying the Molecular Genetics of the Rhizobium-Legume Symbiosis. Plant Mol Biol Rep 1990;8: 40-49.
158. Cook DR. *Medicago truncatula* – a Model in the Making. Curr Opin Plant Biol 1999;2(4): 301-304.
159. Young ND, Mudge J, Ellis THN. Legume Genomes: More Than Peas in a Pod. Curr Opin Plant Biol 2003;6(2): 199-204.
160. Cook DR, Vandenbosch K, de Brujin FJ, Huguet T. Model Legumes Get the Nod. Plant Cell 1997;9: 275-281.
161. Udvardi MK. Legume Models Strut Their Stuff. Mol Plant Microbe Interact 2001;14(1): 6-9.
162. Stougaard J. Genetics and Genomics of Root Symbiosis. Curr Opin Plant Biol 2001;4(4): 328-335.
163. Penmetsa RV, Cook DR. A Legume Ethylene-Insensitive Mutant Hyperinfected By Its Rhizobial Symbiont. Science 1997;275(5299): 527-530.
164. Schauser L, Handberg K, Sandal N, Stiller J, Thykjaer T, Pajuelo E, Nielsen A, Stougaard J. Symbiotic Mutants Deficient in Nodule Establishment Identified After T-DNA Transformation of *Lotus japonicus*. Mol Gen Genet 1998;259(4): 414-423.
165. Albrecht C, Geurts R, Bisseling T. Legume Nodulation and Mycorrhizae Formation; Two Extremes in Host Specificity Meet. EMBO J 1999;18(2): 281-288.
166. Radutoiu S, Madsen LH, Madsen EB, Felle HH, Umehara Y, Gronlund M, Sato S, Nakamura Y, Tabata S, Sandal N, Stougaard J. Plant Recognition of Symbiotic Bacteria Requires Two LysM Receptor-Like Kinases. Nature 2003;425(6958): 585-592.
167. Madsen EB, Madsen LH, Radutoiu S, Olbryt M, Rakwalska M, Szczyglowski K, Sato S, Kaneko T, Tabata S, Sandal N, Stougaard J. A Receptor Kinase Gene of the LysM Type Is Involved in Legume Perception of Rhizobial Signals. Nature 2003;425(6958): 637-640.
168. Limpens E, Franken C, Smit P, Willemse J, Bisseling T, Geurts R. LysM-domain Receptor Kinases Regulating Rhizobial Nod Factor-Induced Infection. Science 2003;302(5645):630-633.
169. Op den Camp R, Streng A, De Mita S, Cao Q, Polone E, Liu W, Ammiraju JS, Kudrna D, Wing R, Untergasser A, Bisseling T, Geurts R. LysM-type Mycorrhizal Receptor Recruited for Rhizobium Symbiosis in Nonlegume Parasponia. Science 2011;331(6019): 909-912.
170. Banba M, Gutjahr C, Miyao A, Hirochika H, Paszkowski U, Kouchi H, Imaizumi-Anraku H. Divergence of Evolutionary Ways Among Common Sym Genes: CASTOR and CCaMK Show Functional Conservation Between Two Symbiosis Systems and Constitute the Root of a Common Signaling Pathway. Plant Cell Physiol 2008;49(11): 1659-1671.
171. Endre G, Kereszt A, Kevei Z, Mihacea S, Kalo P, Kiss GB. A Receptor Kinase Gene Regulating Symbiotic Nodule Development. Nature 2002;417(6892): 962-966.
172. Stracke S, Kistner C, Yoshida S, Mulder L, Sato S, Kaneko T, Tabata S, Sandal N, Stougaard J, Szczyglowski K, Parniske M. A Plant Receptor-Like Kinase Required for Both Bacterial and Fungal Symbiosis. Nature 2002;417(6892): 959-962.
173. Limpens E, Mirabella R, Fedorova E, Franken C, Franssen H, Bisseling T, Geurts R. Formation of Organelle-Like N2-fixing Symbiosomes in Legume Root Nodules Is Controlled by DMI2. Proc Natl Acad Sci USA 2005; 102(29): 10375-10380.
174. Kevei Z, Lougnon G, Mergaert P, Horvath GV, Kereszt A, Jayaraman D, Zaman N, Marcel F, Regulski K, Kiss GB, Kondorosi A, Endre G, Kondorosi, E-Ané JM. 3-Hydroxy-3-methylglutaryl Coenzyme A Reductase 1 Interacts with NORK and Is Crucial for Nodulation in *Medicago truncatula*. Plant Cell 2007;19(12): 3974-3989.
175. Zhu H, Chen T, Zhu M, Fang Q, Kang H, Hong Z, Zhang Z. A Novel ARID DNA-binding Protein Interacts with SymRK and Is Expressed During Early Nodule Development in *Lotus japonicus*. Plant Physiol 2008;148(1): 337-347.

176. Lefebvre B, Timmers T, Mbengue M, Moreau S, Herve C, Tóth K, Bittencourt-Silvestre J, Klaus D, Deslandes L, Godiard L, Murray JD, Udvardi MK, Raffaele S, Mongrand S, Cullimore J, Gamas P, Niebel A, Ott T. A Remorin Protein Interacts with Symbiotic Receptors and Regulates Bacterial Infection. Proc Natl Acad Sci USA 2010;107(5): 2343-2348.
177. Yuan S, Zhu H, Gou H, Fu W, Liu L, Chen T, Ke D, Kang H, Xie Q, Hong Z, Zhang Z. A Ubiquitin Ligase of Symbiosis Receptor Kinase Involved in Nodule Organogenesis. Plant Physiol 2012;
178. Wais RJ, Galera C, Oldroyd G, Catoira R, Penmetsa RV, Cook D, Gough C, Dénarié J, Long SR. Genetic Analysis of Calcium Spiking Responses in Nodulation Mutants of *Medicago truncatula*. Proc Natl Acad Sci USA 2000;97(24): 13407-13412.
179. Ané JM, Kiss GB, Riely BK, Penmetsa RV, Oldroyd GE, Ayax C, Lévy J, Debellé F, Baek JM, Kalo P, Rosenberg C, Roe BA, Long SR, Dénarié J, Cook DR. *Medicago truncatula* DMI1 Required for Bacterial and Fungal Symbioses in Legumes. Science 2004;303(5662): 1364-1367.
180. Imaizumi-Anraku H, Takeda N, Charpentier M, Perry J, Miwa H, Umehara Y, Kouchi H, Murakami Y, Mulder L, Vickers K, Pike J, Downie JA, Wang T, Sato S, Asamizu E, Tabata S, Yoshikawa M, Murooka Y, Wu GJ, Kawaguchi M, Kawasaki S, Parniske M, Hayashi M. Plastid Proteins Crucial for Symbiotic Fungal and Bacterial Entry into Plant Roots. Nature 2005;433(7025):527-531.
181. Edwards A, Heckmann AB, Yousafzai F, Duc G, Downie JA. Structural Implications of Mutations in the Pea SYM8 Symbiosis Gene, the DMI1 Ortholog, Encoding a Predicted Ion Channel. Mol Plant Microbe Interact 2007;20(10): 1183-1191.
182. Riely BK, Lougnon G, Ané JM, Cook DR. The Symbiotic Ion Channel Homolog DMI1 Is Localized in the Nuclear Membrane of *Medicago truncatula* Roots. Plant J 2007;49(2): 208-216.
183. Peiter E, Sun J, Heckmann AB, Venkateshwaran M, Riely BK, Otegui MS, Edwards A, Freshour G, Hahn MG, Cook DR, Sanders D, Oldroyd GE, Downie JA, Ané JM. The *Medicago truncatula* DMI1 Protein Modulates Cytosolic Calcium Signaling. Plant Physiol 2007;145(1): 192-203.
184. Kanamori N, Madsen LH, Radutoiu S, Frantescu M, Quistgaard EMH, Miwa H, Downie JA, James EK, Felle HH, Haaning LL, Jensen TH, Sato S, Nakamura Y, Tabata S, Sandal N, Stougaard J. A Nucleoporin Is Required for Induction of Ca2+ Spiking in Legume Nodule Development and Essential for Rhizobial and Fungal Symbiosis. Proc Natl Acad Sci USA 2006;103(2): 359-364.
185. Saito K, Yoshikawa M, Yano K, Miwa H, Uchida H, Asamizu E, Sato S, Tabata S, Imaizumi-Anraku H, Umehara Y, Kouchi H, Murooka Y, Szczyglowski K, Downie JA, Parniske M, Hayashi M, Kawaguchi M. NUCLEOPORIN85 Is Required for Calcium Spiking, Fungal and Bacterial Symbioses, and Seed Production in *Lotus japonicus*. Plant Cell 2007;19(2): 610-624.
186. Groth M, Takeda N, Perry J, Uchida H, Draxl S, Brachmann A, Sato S, Tabata S, Kawaguchi M, Wang TL, Parniske M. NENA, a *Lotus japonicus* Homolog of Sec13, is Required for Rhizodermal Infection by Arbuscular Mycorrhiza Fungi and Rhizobia But Dispensable for Cortical Endosymbiotic Development. Plant Cell 2010;22(7): 2509-2526.
187. Oldroyd GE, Downie JA. Calcium, Kinases and Nodulation Signalling in Legumes. Nat Rev Mol Cell Biol 2004;5(7): 566-576.
188. Catoira R, Galera C, de Billy F, Penmetsa RV, Journet EP, Maillet F, Rosenberg C, Cook D, Gough C, Denarie J. Four Genes of *Medicago truncatula* Controlling Components of a Nod Factor Transduction Pathway. Plant Cell 2000;12(9): 1647-1665.
189. Gleason C, Chaudhuri S, Yang T, Munoz A, Poovaiah BW, Oldroyd GE. Nodulation Independent of Rhizobia Induced by a Calcium-Activated Kinase Lacking Autoinhibition. Nature 2006;441(7097): 1149-1152.
190. Sanchez L, Weidmann S, Arnould C, Bernard AR, Gianinazzi S, Gianinazzi-Pearson V. *Pseudomonas fluorescens* and *Glomus mosseae* Trigger *DMI3*-Dependent Activation

of Genes Related to a Signal Transduction Pathway in Roots of *Medicago truncatula*. Plant Physiol 2005;139(2): 1065-1077.

191. Yano K, Yoshida S, Muller J, Singh S, Banba M, Vicker K. CYCLOPS, a Mediator of Symbiotic Intracellular Accommodation. Proc Natl Acad Sci USA 2008;105(51): 20540-20545.

192. Kaló P, Gleason C, Edwards A, Marsh J, Mitra RM, Hirsch S, Jakab J, Sims S, Long SR, Rogers J, Kiss GB, Downie JA, Oldroyd GE. Nodulation Signaling in Legumes Requires NSP2, a Member of the GRAS Family of Transcriptional Regulators. Science 2005;308(5729): 1786-1789.

193. Smit P, Raedts J, Portyanko V, Debellé F, Gough C, Bisseling T, Geurts R. NSP1 of the GRAS Protein Family Is Essential for Rhizobial Nod Factor-Induced Transcription. Science 2005;308(5729): 1789-1791.

194. Heckmann AB, Lombardo F, Miwa H, Perry JA, Bunnewell S, Parniske M, Wang TL, Downie JA. *Lotus japonicus* Nodulation Requires Two GRAS Domain Regulators, One of Which Is Functionally Conserved in a Non-Legume. Plant Physiol 2006;142(4): 1739-1750.

195. Murakami Y, Miwa H, Imaizumi-Anraku H, Kouchi H, Downie JA, Kawasaki KMS. Positional cloning identifies *Lotus japonicus* NSP2, a Putative Transcription Factor of the GRAS Family, Required for NIN and ENOD40 Gene Expression in Nodule Initiation. DNA Res 2006;13: 255-265.

196. Gonzalez-Rizzo S, Crespi M, Frugier F. The *Medicago truncatula* CRE1 Cytokinin Receptor Regulates Lateral Root Development and Early Symbiotic Interaction with *Sinorhizobium meliloti*. Plant Cell 2006;18(10): 2680-2693.

197. Murray JD, Karas BJ, Sato S, Tabata S, Amyot L, Szczyglowski K. A Cytokinin Perception Mutant Colonized by *Rhizobium* in the Absence of Nodule Organogenesis. Science 2007;315(5808): 101-104.

198. Tirichine L, Sandal N, Madsen LH, Radutoiu S, Albrektsen AS, Sato S, Asamizu E, Tabata S, Stougaard J. A Gain-of-Function Mutation in a Cytokinin Receptor Triggers Spontaneous Root Nodule Organogenesis. Science 2007;315(5808): 104-107.

199. Schauser L, Roussis A, Stiller J, Stougaard J. A Plant Regulator Controlling Development of Symbiotic Root Nodules. Nature 1999;402(6758): 191-195.

200. Borisov AY, Madsen LH, Tsyganov VE, Umehara Y, Voroshilova VA, Batagov AO, Sandal N, Mortensen A, Schauser L, Ellis N, Tikhonovich IA, Stougaard J. The *Sym35* Gene Required for Root Nodule Development in Pea Is an Ortholog of *Nin* from *Lotus japonicus*. Plant Physiol 2003;131(3): 1009-1017.

201. Marsh JF, Rakocevic A, Mitra RM, Brocard L, Sun J, Eschstruth A, Long SR, Schultze M, Ratet P, Oldroyd GE. *Medicago truncatula* NIN Is Essential for Rhizobial-Independent Nodule Organogenesis Induced by Autoactive Calcium/Calmodulin-Dependent Protein Kinase. Plant Physiol 2007;144(1): 324-335.

202. Middleton PH, Jakab J, Penmetsa RV, Starker CG, Doll J, Kalo P, Prabhu R, Marsh JF, Mitra RM, Kereszt A, Dudas B, VandenBosch K, Long SR, Cook DR, Kiss GB, Oldroyd GE. An ERF Transcription Factor in *Medicago truncatula* That Is Essential for Nod Factor Signal Transduction. Plant Cell 2007;19(4): 1221-1234.

203. Caetano-Anolles G, Gresshoff PM. Plant Genetic Control of Nodulation. Annu Rev Microbiol 1991;45: 345-382.

204. Ferguson BJ, Indrasumunar A, Hayashi S, Lin MH, Lin YH, Reid DE, Gresshoff PM. Molecular Analysis of Legume Nodule Development and Autoregulation. J Integr Plant Biol 2010;52(1): 61-76.

205. Okamoto S, Ohnishi E, Sato S, Takahashi H, Nakazono M, Tabata S, Kawaguchi M. Nod Factor/Nitrate-Induced CLE Genes That Drive HAR1-Mediated Systemic Regulation of Nodulation. Plant Cell Physiol 2009;50(1): 67-77.

206. Mortier V, Den Herder G, Whitford R, Van de Velde W, Rombauts S, D'Haeseleer K, Holsters M, Goormachtig S. CLE Peptides Control *Medicago truncatula* Nodulation Locally and Systemically. Plant Physiol 2010;153(1): 222-237.

207. Krusell L, Madsen LH, Sato S, Aubert G, Genua A, Szczyglowski K, Duc G, Kaneko T, Tabata S,de Bruijn F, Pajuelo E, Sandal N, Stougaard J. Shoot Control of Root

Development and Nodulation Is Mediated by a Receptor-Like Kinase. Nature 2002;420(6914): 422-426.
208. Nishimura R, Hayashi M, Wu GJ, Kouchi H, Imaizumi-Anraku H, Murakami Y, Kawasaki S, Akao S, Ohmori M, Nagasawa M, Harada K, Kawaguchi M. HAR1 Mediates Systemic Regulation of Symbiotic Organ Development. Nature 2002;420(6914): 426-429.
209. Schnabel E, Journet EP, de Carvalho-Niebel F, Duc G, Frugoli J. The *Medicago truncatula* SUNN Gene Encodes a CLV1-like leucine-Rich Repeat Receptor Kinase That Regulates Nodule Number and Root Length. Plant Mol Biol 2005;58(6):809-822.
210. Staehelin C, Xie ZP, Illana A, Vierheilig H. Long-Distance Transport of Signals During Symbiosis: Are Nodule Formation and Mycorrhization Autoregulated in a Similar Way? Plant Signal Behav 2011;6(3): 372-377.
211. Nishimura R, Ohmori M, Fujita H, Kawaguchi M. A *Lotus* Basic Leucine Zipper Protein with a RING-Finger Motif Negatively Regulates the Developmental Program of Nodulation. Proc Natl Acad Sci USA 2002;99(23): 15206-15210.
212. Oka-Kira E, Tateno K, Miura K, Haga T, Hayashi M, Harada K, Sato S, Tabata S, Shikazono N, Tanaka A, Watanabe Y, Fukuhara I, Nagata T, Kawaguchi M. klavier (klv), a Novel Hypernodulation Mutant of *Lotus japonicus* Affected in Vascular Tissue Organization and Floral Induction. Plant J 2005;44(3): 505-515.
213. Magori S, Oka-Kira E, Shibata S, Umehara Y, Kouchi H, Hase Y, Tanaka A, Sato S, Tabata S, Kawaguchi M. Too Much Love, a Root Regulator Associated with the Long-Distance Control of Nodulation in *Lotus japonicus*. Mol Plant Microbe Interact 2009;22(3): 259-268.
214. Schnabel EL, Kassaw TK, Smith LS, Marsh JF, Oldroyd GE, Long SR, Frugoli JA. The ROOT DETERMINED NODULATION1 Gene Regulates Nodule Number in Roots of *Medicago truncatula* and Defines a Highly Conserved, Uncharacterized Plant Gene Family. Plant Physiol 2011;157(1): 328-340.
215. Catford JG, Staehelin C, Lerat S, Piché Y, Vierheilig H. Suppression of Arbuscular Mycorrhizal Colonization and Nodulation in Split-Root Systems of Alfalfa after Pre-Inoculation and Treatment with Nod Factors. J Exp Bot 2003;54(386): 1481-1487.
216. Grunwald U, Nyamsuren O, Tarnasloukht M, Lapopin L, Becker A, Mann P, Gianinazzi-Pearson V, Krajinski F, Franken P. Identification of Mycorrhiza-Regulated Genes with Arbuscule Development-Related Expression Profile. Plant Mol Biol 2004;55(4): 553-566.
217. Gianinazzi-Pearson V, Brechenmacher L. Functional Genomics of Arbuscular Mycorrhiza: Decoding the Symbiotic Cell Programme. Canad J Bot 2004;82(8): 1228-1234.
218. Seddas PMA, Arnould C, Tollot M, Arias CM, Gianinazzi-Pearson V. Spatial Monitoring of Gene Activity in Extraradical and Intraradical Developmental Stages of Arbuscular Mycorrhizal Fungi by Direct Fluorescent in situ RT-PCR. Fungal Gen Biol 2008;45(8): 1155-1165.
219. Seddas PM, Arias CM, Arnould C, van Tuinen D, Godfroy O, Benhassou HA, Gouzy J, Morandi D, Dessaint F, Gianinazzi-Pearson V. Symbiosis-Related Plant Genes Modulate Molecular Responses in an Arbuscular Mycorrhizal Fungus During Early Root Interactions. Mol Plant Microbe Interact 2009;22(3): 341-351.
220. Kuznetsova E, Seddas-Dozolme PM, Arnould C, Tollot M, van Tuinen D, Borisov A, Gianinazzi S, Gianinazzi-Pearson V. Symbiosis-Related Pea Genes Modulate Fungal and Plant Gene Expression During the Arbuscule Stage of Mycorrhiza with *Glomus intraradices*. Mycorrhiza 2010;20(6): 427-443.
221. Sulieman S, Schulze J. The Efficiency of Nitrogen Fixation of the Model Legume *Medicago truncatula* (Jemalong A17) Is Low Compared to *Medicago sativa*. J Plant Physiol 2010;167(9): 683-692.
222. Duc G, Trouvelot A, Gianinazzi-Pearson V, Gianinazzi S. First Report of Non-Mycorrhizal Plant Mutants (Myc–) Obtained in Pea (*Pisum sativum* L.) and Faba Bean (*Vicia faba* L.). Plant Sci 1989;60(2): 215-222.

223. Frühling M, Roussel H, Gianinazzi-Pearson V, Pühler A, Perlick AM. The *Vicia faba* Leghemoglobin Gene VfLb29 Is Induced in Root Nodules and in Roots Colonized by the Arbuscular Mycorrhizal Fungus *Glomus fasciculatum*. Mol Plant Microbe Interact 1997;10(1): 124-131.

224. Xavier IJ, Holloway G, Leggett M Development of Rhizobial Inoculant Formulations. Crop Management (Online) 2004;

225. Rai MK., editor. Handbook of microbial biofertilizers. New York: Haworth Press, Technology & Engineering; 2006.

226. Tikhonovich IA, Provorov NA. Cooperation of Plants and Microorganisms: Getting Closer to the Genetic Construction of Sustainable Agro-Systems. Biotechnol J 2007;2(7): 833-848.

227. Shtark OY, Danilova TN, Naumkina TS, Vasilchikov AG, Chebotar VK, Kazakov AE, Zhernakov AI, Nemankin TA, Prilepskaya NA, Borisov AY, Tikhonovich IA. Analysis of Pea (*Pisum sativum* L.) Source Material for Breeding of Cultivars with High Symbiotic Potential and Choice of Criteria for Its Evaluation. Ecological genetics ("Ekologicheskaja genetika") 2006;4(2): 22-28 (In Russian).

228. Chebotar VK, Kazakov AE, Erofeev SV, Danilova TN, Naumkina TS, Shtark OY, Tikhonovich IA, Borisov AY. Method of Production of Complex Microbial Fertilizer. Patent No 2318784. 2008.

229. Prévost D, Antoun H. Potential Use of Rhizobium as PGPR with Non-Legumes: Abstracts from the Inoculant Forum, March 17-18, 2005. Saskatoon, Saskatchewan, Canada.

230. Hossain MS, Mårtensson A. Potential Use of *Rhizobium* spp. to Improve Fitness of Non-Nitrogen-Fixing Plants. Acta Agriculturae Scandinavica, Section B – Plant Soil Science 2008;58(4): 352-358.

231. Gentili F, Jumpponen A. Potential and Possible Uses of Bacterial and Fungal Biofertilizers. In: Rai MK. (ed.) Handbook of Microbial Biofertilizers. New York: Haworth Press, Technology & Engineering; 2006. p1-28.

232. Galvan GA, Burger-Meijer K, Kuiper TW, Kik C, Scholten OE. Breeding for Improved Responsiveness to Arbuscular Mycorrhizal Fungi in Onion. Proceedings of 3rd International Congress of the European Integrated Project Quality Low Input Food (QLIF) Congress, Hohenheim, Germany, March 20–23, 2007. (Online) http://orgprints.org/view/projects/int_conf_qlif2007.html

233. Herridge D, Rose I. Breeding for Enhanced Nitrogen Fixation in Crop Legumes. Field Crops Res 2000;65: 229-248.

234. Rengel Z. Breeding for Better Symbiosis. Plant Soil 2002;245(1): 147-162.

235. Graham PH, Hungria M, Tlusty B. Breeding for Better Nitrogen Fixation in Grain Legumes: Where Do the Rhizobia Fit In? Crop Management 2004;

236. Howieson JG, Yates RJ, Foster KJ, Real D, Besier RB. Prospects for the Future Uses of Legumes. In: Dilworth MJ, James EK, Sprent JI, Newton WE. (eds) Nitrogen Fixing Leguminous Symbioses. Berlin/Heidelberg: Springer Science+Business Media BV; 2008. p363-394.

237. Borisov AY, Danilova TN, Shtark OY, Solovov II, Kazakov AE, Naumkina TS, Vasilchikov AG, Chebotar VK, Tikhonovich IA. Tripartite Symbiotic System of Pea (*Pisum sativum* L.): Applications in Sustainable Agriculture. In: FD Dakora, BM Chimphango, AJ Valentine, C Elmerich, WE Newton (eds) Biological Nitrogen Fixation: Towards Poverty Alleviation Through Sustainable Agriculture. Proceedings of 15th International Congress on Nitrogen Fixation & 12th International Conference of the African Association for Biological Nitrogen Fixation. Berlin/Heidelberg: Springer Science and Business Media BV; 2008. p15-17.

238. Bourion V, Rizvi SM, Fournier S, de Larambergue H, Galmiche F, Marget P, Duc G, Burstin J. Genetic Dissection of Nitrogen Nutrition in Pea Through a QTL Approach of Root, Nodule, and Shoot Variability. Theor Appl Genet 2010;121(1): 71-86.

239. Tominaga A, Gondo T, Akashi R, Zheng SH, Arima S, Suzuki A. Quantitative Trait Locus Analysis of Symbiotic Nitrogen Fixation Activity in the Model Legume *Lotus japonicus*. J Plant Res 2012;125(3): 395-406.

CHAPTER 9

Advances in transgenic vegetable and fruit breeding

João Silva Dias[1], Rodomiro Ortiz[2]

[1]*University of Lisbon, Instituto Superior de Agronomia, Lisboa, Portugal*

[2]*Swedish University of Agricultural Sciences, Alnarp, Sweden*

ABSTRACT

Vegetables and fruits are grown worldwide and play an important role in human diets because they provide vitamins, minerals, dietary fiber, and phytochemicals. Vegetables and fruits are also associated with improvement of gastrointestinal health, good vision, and reduced risk of heart disease, stroke, chronic diseases such as diabetes, and some forms of cancer. Vegetable and fruit production suffers from many biotic stresses caused by pathogens, pests, and weeds and requires high amounts of plant protection products per hectare. United States vegetables farmers are benefiting from growing transgenic squash cultivars resistant to Zucchini yellow mosaic virus, Watermelon mosaic virus, and Cucumber mosaic virus, which were deregulated and commercialized since 1996. Bt-sweet corn has also proven effective for control of some lepidopteran species and continues to be accepted in the fresh market in the USA, and Bt-fresh-market sweet corn hybrids are released almost every year. Likewise, transgenic Bt-eggplant bred to reduce pesticide use is now grown by farmers in Bangladesh. Transgenic papaya cultivars carrying the coat-protein gene provide effective protection against Papaya ring spot virus elsewhere. The transgenic "Honey Sweet" plum cultivar provides an interesting germplasm source for Plum pox virus control. Enhanc- ed host plant resistance to Xanthomonas campestris pv. musacearum, which causes the devastating banana Xanthomonas wilt in the Great Lakes Region of Africa, was achieved by plant genetic engineering. There are other vegetable and fruit crops in the pipeline that have been genetically modified to enhance their host plant resistance to insects and plant pathogens, to show herbicide tolerance, and to improve features such as slow ripening that extends the shelf-life of the produce. Consumers could benefit further from eating more nutritious transgenic vegetables and fruits. Transgenic plant breeding therefore provides genetically enhanced seed embedded technology that contributes to integrated pest management in horticulture by reducing pesticide sprays as well as improving food safety by minimizing pesticide residues. Furthermore, herbicide-tolerant transgenic crops can help reducing plough in fields, thereby saving fuel because of less tractor use, which also protects the structure of the soil by reducing its erosion. Transgenic vegetable and fruit crops could make important contributions to sustainable vegetable production and for more nutritious and healthy food. Countries vary, however, in their market standards of acceptance of transgenic crops. Biotechnology products will be successful if clear advantages and safety are de- monstrated to both growers and consumers.

Keywords: Biosafety, Fruits, GM-crops, GMO, Health Benefits, IPM, Market, Nutrition, Plant Breeding, Sustainability, Vegetables

1. INTRODUCTION

Vegetables and fruits are grown worldwide and make up a major portion of the diet of humans in many parts of the world. They play a significant role in human nutrition, especially as sources of vitamins (C, A, B1, B6, B9, E), minerals, dietary fiber and phytochemicals [1] . Vegetables and fruits in the daily diet have been strongly associated with improvement of gastrointestinal health, good vision, and reduced risk of heart disease, stroke, chronic diseases such as diabetes, and some forms of cancer [2] .

Vegetable and fruit consumption worldwide is rising, reflecting the consumer's increased income, desire of diversity, and awareness of nutritional benefits. A world vegetable survey showed that 402 vegetable crops are cultivated worldwide, representing 69 families and 230 genera [3] [4] . Leafy vegetables—of which the leaves or young leafy shoots are consumed—were the most often utilized (53% of the total), followed by vegetable fruits (15%), and vegetables with below ground edible organs comprised 17%. Many vegetable crops have more than one part used. Most of the vegetables are marketed fresh with only a small proportion processed because most vegetables are perishable. Consumption shortly after harvest guarantees optimal vegetable quality.

Vegetable and fruit production, due to their cultivation intensity, suffers from many biotic stresses caused by pathogens, pests, and weeds and requires high amounts of pesticides per hectare. Because of the high diversity of vegetable crops, pest loads vary and are complex vis-à-vis field crops. The main method for controlling pathogens, pests, and weeds has been the use of pesticides because vegetables and fruits are high-value commodities with high cosmetic standards. Synthetic pesticides have been applied to vegetable crops since the 1950s, and have been highly successful in reducing crop losses to some insects, other pathogens, and weeds. Vegetables account for a significant share of the global pesticide market. About 20% of the world's annual pesticides expenditures (amounting to US $8.1 billion) are spent for growing vegetables [5] . Only cotton used more insecticides than vegetables on an area basis. Insecticides are regularly applied to control a complex of insect pests that cause damage by feeding directly on the plant or by transmitting pathogens, particularly viruses. Despite pesticide use, insects, pathogens, and weeds continue to cause a heavy toll on world vegetable production. Preharvest losses are globally estimated as 15% for insect pests, 13% for damage by pathogens, and about 12% for weeds [6] . Pesticide residues can affect the health of growers and consumers and contaminate the environment. Vegetables are often consumed in fresh form, so pesticide residue and biological contamination is a serious issue. Consumers worldwide are increasingly concerned about the quality and safety of their food, as well as the social and the environmental conditions under which it is produced. Vegetable prices will therefore increase by enhancing their quality and safety [1] .

Transgenic crops, commonly referred to as genetically modified (GM), crops enable breeders to bring favorable genes, often previously inaccessible, into already elite cultivars, improving their value considerably and offer unique opportunities for controlling insects and other pathogens. Many vegetable and fruit crops have been genetically modified to include resistance to insects, other pathogens (including viruses), and herbicides and for improved features, such as slow ripening, higher nutritional status, seedless fruit, and increased sweetness. Recently, Dias and Ortiz [7] did a review (based on 372 articles) about the status of transgenic vegetables to improve their production. They analyzed the advances and potentials in transgenic research until 2010 on tomato, eggplant, potato, cucurbits, brassicas, lettuce, alliums, sweet corn, cowpea, cassava, sweet potato, and carrots. Some experimental transgenic vegetables show host plant resistance to insects, nematodes, fungi, bacteria, and viruses, extended shelf-life of the produce, herbicide tolerance, enhanced nutritional status, and seedless fruit and better flavor. Host plant resistance or product quality will increase the value of the vegetable throughout the chain, thereby benefiting farmers, traders and consumers. The most promising traits to reach vegetable and fruit growers seem to be host plant resistances to insects and other pathogens, especially for tomato, potato, eggplant, summer squash, sweet corn, papaya, plum and banana.

Breeding transgenic pest resistant and herbicide tolerant vegetable cultivars can decrease management costs and release growers' time for other economic activities, while also contributing to a less toxic production environment. Transgenic vegetables with tolerance to abiotic stresses or enhanced input efficiency could also provide various benefits to farmers and the environment, e.g. Bt cry genes could provide eggplants with host plant resistance to shoot borer while reducing today's insecticide spraying during the crop season: 40 to 80 times in India or 50 times in the Philippines [8] [9] . Consumers could also benefit further from the use of more nutritious transgenic vegetables, e.g. an increase of crop carotenoids by metabolic sink manipulation through a transgenic breeding approach appears feasible in some vegetable crops [7] . Genetically engineering carrots containing increased calcium (Ca) levels may boost Ca uptake, thereby reducing the incidence of Ca deficiencies such as osteoporosis. Fortified transgenic lettuce with zinc will overcome its deficiency that severely impairs organ function. Transgenic lettuce with improve tocopherol, and resveratrol composition may prevent coronary disease and arteriosclerosis and can contribute to cancer chemo preventative activity. Folates deficiency, which is regarded as a global health problem, can also be overcome with transgenic tomatoes with folate levels that provide a complete adult daily requirement [7] . Likewise, food safety can be enhanced through transgenic approaches, e.g. resource-poor people in rural Africa will benefit eating cyanide-free cultivars of cassava—a main staple in their diets [10] - [12] .

Genetic engineering has the potential to address some of the most challenging biotic constraints faced by vegetables and fruit growers, which are not easily addressed through conventional plant breeding alone. However, horticulture remains in its infancy regarding the use of transgenic crop technology because vegetables are considered minor crops (compared to field crops), due to the lower resources invested (especially by the multinational private seed corporations), and derived of the high costs for deregulation [1] . Most biotech research is done in field crops because multinational seed corporations expect the highest rate of return to their investment. While it is becoming less expensive to create transgenic crops for pest management, developing a marketable product and a regulatory package remains costly. Development and regulatory costs can be recouped more readily if the product is grown on an extensive area (as happens with field crops), which is not generally the case for individual vegetable crops. The large multinational seed corporations have, for the most part, abandoned the development of transgenic vegetable and fruit crops because of the high costs associated with product development and deregulation [1] [7] .

There are many cultivars of the same vegetable or fruit crop, and their expected life can be relatively short. Moreover, introducing a transgene into a breeding program can be complicated and cost prohibitive, especially in crops with difficulty for using backcrossing (e.g. banana, cassava, potato or sweet potato). Furthermore, deregulation of a transgenic trait is event specific in many countries. It may not be therefore possible to develop a single transgenic event that can be converted into many different cultivars of a single or closely related group of vegetable and fruit species through conventional breeding. This article highlights advances in breeding transgenic vegetables and fruits, and issues affecting their use, as illustrated by the case studies of tomato, potato, eggplant, squash, sweet corn, papaya, plum and banana and by the contribution of transgenic vegetables and fruits to more nutritious and healthy food.

2. ADVANCES IN BREEDING TRANSGENIC VEGETABLES AND FRUITS: CASE STUDIES

2.1. Tomato: Delaying Fruit Ripening

The first commercially grown transgenic crop was Flavr Savr™ tomato, which was released by Calgene in 1994. This tomato contains an antisense version of the polygalacturonase (PG) gene. The use of this gene ensued after many years of research on several genes involved in fruit development and tomato ripening. They were identified, cloned, and characterized to breed transgenic tomato cultivars [13] . However, Flavr Savr™ tomato failed in the market since this cultivar was considered inferior by growers, and was rapidly withdrawn from the market. Plant genetic engineers learned an important lesson after this failure: the importance of cooperation with plant breeders.

There has been further research conducted to manipulate fruit ripening, texture and nutritional quality using transgenic approaches. Many of the genes targeted include ethylene because of its role in fruit ripening [7] . Enzymes that regulate ethylene biosynthesis in plants are: S-adenosylmethionine (SAM) synthase, 1-aminocyclo- propane-1-carboxylate (ACC) synthase, and ACC oxidase. The genes encoding these enzymes [14] [15] , as well as those that metabolize SAM or ACC [16] [17] , have been targeted in order to manipulate ethylene biosynthesis, thereby regulating fruit ripening. It has been clearly shown that modulation of ethylene biosynthesis using genetic engineering can yield tomato fruits with predictable ripening characteristics. Tomato fruit ripening manipulation has been however achieved by introducing anti-ripening genes (rin and nor) in heterozygous genotypes. These genes have been incorporated in many fresh and processing tomatoes [7] .

2.2. Potato: Host Plant Resistance to Insects and Viruses

After the first Bt-tomato was reported with partial resistance to lepidopteran insects, improvement through truncation of the Bt gene and codon modification were made to optimize protein expression in plants, and Bt potatoes expressing cry 3A gene were developed. Bt-potato cultivars expressing resistance to Colorado potato beetle (Leptinotarsa decemlineata), the most destructive insect pest of potato in North America, and aphids associated with Potato virus Y and Potato leafroll virus, were approved for sale in the United States in 1995. The reported profits in USA were on average US $55 ha^{-1} for Bt-potato [18] . Likewise, an ex-ante assessment of Monsanto's transfer of transgenic virus resistance technology potatoes suggested an average profit of US $117 ha^{-1} for virus-resistant potato in Mexico [19] . Potato growers, either small or large, can benefit from this transgenic potato, as well as consumers because of the lower prices as a result of higher production.

New Leaf®, New Leaf Y®, and New Leaf Plus® were the trade names of the transgenic potato cultivars sold by Nature Mark®, a subsidiary of Monsanto® [20] . About 600 ha were grown commercially in the United States when New Leaf® cultivars were introduced in 1995, and the commercial acreage further reached about 20,000 ha. Market success of the New Leaf®, New Leaf Y®, and New Leaf Plus® potatoes could be attributed to the difficulty in controlling Colorado potato beetle and also high pest populations of aphids and associated virus problems due to mild winters in the Pacific Northwest [21] . This virus resistance benefited seed producers, while commercial growers benefited from higher yields and reduced need for insecticides [21] . The processing industry and consumers benefited from improved quality. Potatoes were one of the first foods from a transgenic crop that was commonly served in restaurants. New Leaf® potato cultivars were the fastest cultivar adoption in the history of the USA potato industry [21] , until potato processors, concerned about anti-biotech organizations, consumer resistance and loss of market share in Europe and Japan, suspended contracts for Bt-potatoes with grow- ers in 2000 and they were taken off the market [22] . The crop area of transgenic insect and virus resistant potatoes did not increase at the same rate as happened in transgenic maize or transgenic cotton, irrespective of being highly effective and growers increasingly using them. The ability of growers to save potato tubers for future plantings, the registration of imidacloprid, a new insecticide that gave excellent control in the same year that Bt-potato became available to growers, the request by consumers to segregate transgenic potatoes, and trade issues driven by international consumers could explain why the market growth for transgenic potatoes was not as rapid as the owners of the proprietary transgenic technology would have liked [21] [22] . International trade barriers were more substantial for transgenic potato than other technology adoptions [23] , which further led to discontinuing its use by the processor and commercial grower, hence the loss of a market for Nature Mark® potatoes. An additional factor that also account for this rapid demise was that Bt-cultivars were only grown in 3% of the USA potato area [23] . The major impact came when the leading fast food McDonald's chain decided to ban transgenic potatoes from its servings. Nature Mark® dissolved after the 2001 season. This initial failure of transgenic potatoes shows that consumer acceptance should be regarded as key for adopting transgenic crop technol- ogy. Guenthner [23] indicated that although growers accepted transgenic potatoes, the processors had little to gain by accepting transgenic raw materials but exposed themselves to significant risk in marketing because they did not have a "consumer-acceptance accelerator".

The larvae of potato tuber moth (PTM, Phthorimaea operculella), the most common and destructive insect pest for this crop in tropical and sub-tropical areas worldwide [24] [25] , mine the foliage, stems, and tubers in the field and also in storage, which leads to significant yield losses. Combining natural resistance mechanisms with Bt cry genes could be a potential solution to improve potato resistance to PTM [7] . Field and storage trials in Egypt, under natural infestation, revealed that Bt-cry5 transgenic potatoes derived from the cv. "Spunta" ("Spunta-G2", "Spunta-G3", and "Spunta-6a3") had the highest host plant resistance to PTM, with almost 100% of their tubers showing no insect damage in the field, and about 90% free of insect damage after 3-month storage [26]. South Africa was selected as the target country for releasing and marketing transgenic PTM-resistance potato for small holder farmers because of its previous in-country experience with genetically modified potato research [27] [28] . Multi-location yield trials and PTM efficacy trials (field and storage) have been conducted at six locations in South Africa since 2001. Complete control of the PTM was found at all locations in all years, and there was no infestation in the field when examining the foliage and tubers of "Spunta-G2" and no infestation was found in tubers stored up to 6 months. The cry1Ia1 transgene was used to transform two popular South African potato cultivars to ensure a greater impact [7] . Further research by Cooper et al. [29] shows that expression of avidin combined with natural host plant resistance (from S. chacoense) in transgenic potato could be also useful for controlling PTM.

Specific issues need to be addressed for the potential use of transgenic potatoes in the center of origin of this crop: The Andean region. Celis et al. [30] assessed therefore the biosafety of transgenic potato in South America. They did not observe any harm in many non-target organisms but found that gene flow occurs to wild relatives growing near potato crops, which suggest that the fitness of these wild species could be altered if stable transgene introgression occurs. In this regard, Scurrah et al. [31] suggest the use of sterile cultivars with scarce flow- er production and lacking seed production to minimize the risk of gene flow from transgenic potato to wild relatives or native cultivars.

2.3. Eggplant: Resistance to Fruit and Shoot Borer

Eggplant is attacked by a number of insects including thrips, cotton leafhopper, jassids and aphids. The most damaging pest is the eggplant fruit and shoot borer (FSB, Leucinodes orbonalis). Losses have been estimated to be between 54% and 70% in India and Bangladesh and up to 50% in the Philippines [9] . There are no known eggplant cultivars resistant to the fruit and shoot borer, so the use of insecticide sprays continues to be the most common control method used by growers. The borer is vulnerable to sprays only for a few hours before they bore into the plant, which explains why growers often spray every other day, particularly during the fruiting stage that contributes to the presence of pesticide residues.

FSB-resistant Bt-eggplant was genetically engineered by Mahyco under a collaborative agreement with Monsanto® and the first Bt-eggplant with resistance to FSB was produced in 2000. This transgenic eggplant incorporates the cry1Ac gene expressing insecticidal protein to confer resistance against FSB. This Bt-eggplant was effective against FSB, with 98% insect mortality in Bt-eggplant shoots and 100% in fruits compared to less than 30% mortality in non-Bt counterparts [32] . Krishna and Qaim [33] indicated that simulations using farm-survey data suggest the aggregate economic surplus gains of Bt-eggplant hybrids could be around US$ 108 million per year in India. Eggplant consumers will capture a large share of these gains, but farmers and the seed company will also benefit. By sharing this technology with the public sector, Bt-open-pollinated eggplant cultivars may be- come available with a certain time lag [34] [35] , thereby making this technology more accessible for resource- poor farmers.

An expert committee was setup in India to examine biosafety issues and data regarding Bt-eggplant in 2006 [9] . They initially indicated that Bt-eggplant was safe and equivalent to its non Bt-counterpart but more research and trials were still needed to confirm these findings and ascertain its benefits regarding existing methods for pest management and for reducing pesticide use. In 2009 a second expert committee examined the data from these new trials and concluded that benefits of the available Bt-eggplant outweighed the perceived and projected risks, thereby advising the Genetic Engineering Appraisal Committee (GEAC) to recommend its commercialization in India [36] . The GEAC cleared Bt-eggplant for commercialization in October 2009 but the Government of India officially announced in February 2010 that more time was needed before its release to farmers due to the successful lobbying made by anti-transgenic crop activists [37] . The Minister of Environment stated that this moratorium for growing Bt-eggplant in India will remain until reaching a political, scientific and societal consensus [38] . Recently in Bangladesh four cultivars of Bt-eggplant for control of eggplant FSB were deregulated and released commercially on October 2013, in order to reduce pesticide use, decrease cost of cultivation and increase marketable yield [39] .

2.4. Summer Squash: Multiple Virus Resistance

Viruses cause 20% to 80% of yield losses in summer squash in the United States [40] . Three of the most important viruses affecting squash production are Zucchini yellow mosaic virus (ZYMV), Watermelon mosaic virus (WMV), and Cucumber mosaic virus (CMV). Summer squash cultivars with satisfactory resistance to CMV, ZYMV, and WMV are yet to become available from cross breeding [41] . Two lines of squash expressing the coat protein (CP) gene of ZYMV, WMV and CMV were deregulated and commercialized in the United States in 1996. Subsequently, many squash types and cultivars have been bred, using crosses and backcrosses with the two initially deregulated lines. This material is highly resistant to infection by one, two or all three of the target viruses [42] - [47] . Virus-resistant transgenic squash limits virus infection rates by restricting challenge viruses, reducing their titers, or inhibiting their replication or cell-to-cell or systemic movement. Therefore, lower virus le- vels reduce the frequency of acquisition by vectors and subsequent transmission within and between fields. Consequently, virus epidemics are substantially limited. The adoption of virus resistant squash cultivars has steadily increased in the United States since 1996. This adoption rate was estimated at 12% (approximately 3100 ha) across the country in 2005 [48] . Virus-resistant transgenic squash has allowed growers to achieve yields comparable to those obtained in the absence of viruses with a net benefit of US $22 million in 2005 [48] . No- wadays United State farmers are benefiting from growing transgenic squash cultivars. Engineered resistance has been so far the only approach to breed squash cultivars with multiple sources of resistance to CMV, ZYMV, and WMV.

2.5. Sweet Corn: from Bt to Triple "Stack" Transgenic Cultivars

Sweet corn, expressing cry1Ab endotoxin, was introduced commercially in the United States in 1998 into an industry that is highly sensitive to damage to corn ears from lepidopteran pests [49] . This endotoxin was very effective against the European corn borer (Ostrinia nubilalis) in the state of New York, providing 100% clean ears when no other lepidopteran species were present and in excess of 97% when the two noctuids, corn earworm (Helicoverpa zea) and fall armyworm (Spodoptera frugiperda), were also present [50] . Studies in other USA states have shown that Bt-sweet corn provided consistently excellent control of the lepidopteran pest complex and the potential for 70% to 90% reductions in insecticide requirement [49] - [54] . About 5% of the 262,196 ha of sweet corn (fresh and processing) grown in the United States in 2006 was with Bt-sweet because corn processors have avoided growing Bt-sweet corn due to concerns about export markets [55] . It has been grown since then only as a fresh market vegetable crop.

An economic assessment in Virginia found a gain of US $1777 ha^{-1} for fresh-market sweet corn vs. non- Bt-sweet corn sprayed up to six times with pyrethroid insecticides [53] . Bt-sweet corn was also much better at preserving major predators of O. nubilalis while controlling the European corn borer than were the commonly used insecticides lambda-cyhalothrin, indoxacarb and spinosad. Bt-sweet corn hybrids can be therefore truly integrated into a biological pest control program.

Monsanto® announced recently the release, through its vegetable seed brand Seminis®, of a "triple-stack" transgenic sweet corn with host plant resistance to insects, which also tolerates glyphosate sprays for weed control [56] . They expect that this transgenic sweet corn provides protection against damage by European corn borers, corn earworms, fall army worms and corn rootworm larvae, and reduces insecticide sprays up to 85% vis- à-vis. a non-transgenic sweet corn [57] . This transgenic sweet corn had below 3% ear damage by fall armyworm or earworm in trials [58] . The use by farmers of this insect-resistant sweet corn that tolerates glyphosate can also result in eco-efficiency because of less tractor trips across the field that help farmers to save fuel, thereby reduc- ing greenhouse gas emissions and decreasing the carbon footprint per ear of corn grown.

2.6. Papaya: Pathogen-derived Resistance Save the Crop in Hawaii

Papaya ringspot virus (PRSV) reduces both fruit quality and edible yield. Transgenic papaya with pathogen- derived resistance carrying the coat-protein gene provides effective protection against PRSV strains over a significant period of the cycle of this perennial fruit crop [59] . Seeds of the transgenic cultivars "SunUp" and "Rainbow" became available to Hawaii's farmers in 1998, and growing cultivar "Raibow" led to a significant increase in papaya production [60] . Prior to the introduction of PRSV-resistant transgenic papaya, growing papaya in Hawaii was no longer viable despite area-wide efforts to erradicate infected trees in order to limit the propagation of the virus. The cultivation of PRSV-resistant transgenic papaya cultivars dramatically reduced the incidence of PRSV in many areas allowing some growers to return to growing non-transgenic papaya, which is important for the high value Japanese market in which transgenic papaya is not allowed. Growers have been able to use the PRSV-resistant transgenic papaya cultivars as a trap crop [61] by growing it as a border around the non-transgenic crop and allowing it to cleanse viruliferous aphids of PRSV [62] . Thus, the Hawaian papaya industry can now produce and market both transgenic and conventional papaya in the same field. The draft genome of the cultivar "Sun Up" makes this papaya cultivar the best-characterized transgenic crop [63] , which may assist to overcome regulatory barriers.

2.7. Plum: Resistance to Plum Pox Virus

There are few sources of natural resistance to Plum pox virus (PPV), the causative agent of Sharka, one of the worldwide most destructive diseases of plum and other stone fruits. Over the past 20 years an intensive international research project between USA, France, Poland, Romania, Spain, and Czech Republic has focused on the development of transgenic resistance to PPV [64] . The PPV coat protein (CP) gene was isolated, sequenced, and cloned by Ravelonandro et al. [65] and was used for Agrobacterium-mediated transformation of plum. This gene was transferred to plum through genetic engineering. As a result, the highly PPV-resistant cultivar "Honey Sweet" plum was bred. This cultivar has shown its host plant resistance after 15 years of field testing in Europe under heavy infection pressure [66] - [68] and in greenhouse tests in the USA (since field inoculations were not allowed because PPV was not in this country until 1999, and is still a quarantine organism under a containment and eradication program). Its transgenic resistance was further transferred to other seedlings through crossbreeding [69] [70] , thus developing additional PPV-resistant cultivars. The efficacy and safety of "Honey Sweet" was assessed through laboratory, greenhouse and field research during 12 years, which led to the regulatory approval of this PPV resistant transgenic plum cultivar in the USA in 2011. “Honey Sweet” is the culmination of years of research and development. In anticipation of the potential spread of PPV to other US regions, breeding with “Honey Sweet” is underway to transfer the PPV resistance transgene into new cultivars adapted to different US growing locations. Breeding with “Honey Sweet” is relatively rapid since an easy molecular marker assay provides means for detecting its resistance transgene in “Honey Sweet” derived-seedlings. This resistance is transferred to the offspring as a single dominant gene. The transgene resistance technology and post-transcriptional gene silencing tested in “Honey Sweet” can be used to develop other resistant stone fruits, such as peach, apricot, Japanese plum, and cherry, which are susceptible to PPV. Studies show that “Honey Sweet” plum fruit quality is excellent, and productivity is very good [64] . The international collaborative program that bred “Honey Sweet” shows the potential of public funding for research and development, which provides safe, efficacious genetic engineering products for the benefit of growers and consumers. “Honey Sweet” plum provides an interesting germplasm source for PPV control elsewhere.

2.8. Banana: Resistance to Xanthomonas Wilt in the Great Lakes Region of Africa

Banana and plantain (Musa sp.) are among the most important food crops in the tropics. They are grown in more than 120 countries over an area of about 10 million hectares and an annual production of 130 million tonnes [71] . Approximately one third of the bananas produced globally are grown in sub-Saharan Africa where the crop provides more than 25% of food energy requirements for more than 100 million people. Xanthomonas wilt caused by the bacterium Xanthomonas campestris pv. musacearum is one of the most important diseases and currently considered as the biggest threat to banana production in the Great Lakes Region of Africa including Uganda, Democratic Republic of Congo, Rwanda, Kenya, Tanzania and Burundi. The pathogen is highly contagious and its spread has endangered the livehood of millions of farmers who rely on banana for food and income. The development of disease resistant bananas remains a high priority since farmers are reluctant to employ labor-inten- sive disease control measures and there is no host plant resistance among banana cultivars. The International Institute of Tropical Agriculture (IITA), in a partnership with Uganda’s National Agricultural Research Organisation (NARO) and other institutions have developed transgenic bananas constitutively expressing the hypersensitivity response-assisting protein (Hrap) and the plant ferredoxin-like protein (Pflp) genes from sweet pepper (Capsicum annuum). These transgenic bananas have exhibited enhanced resistance to X. campestris pv. musacearum [72] . The best 65 transgenic plants expressing Hrap and Pflp and not showing any infection symptoms after artificial inoculation of potted plants with banana Xanthomonas wilt in the screenhouse were evaluated in a confined field trial in Uganda, against this bacterial wilt for two successive crop cycles [73] . Transgenic bananas did not show significant changes in plant morphology versus the non-transgenic banana cultivar. Most transgenic bananas had significantly higher host plant resistance than the non-transgenic control. After testing them as mother plants and first ratoon plants, 11 transgenic bananas were rated as having absolute resistance; i.e., similar to the reference, resistant wild not-edible banana species Musa balbisiana. Their plant phenotype and the bunch weight and size were similar to the non-transgenic counterparts. Transgenic bananas showing Xanthomonas wilt resistance will undergo multi-environment testing in Uganda. They will be further evaluated for environmental and food safety according to the country's biosafety regulations, risk assessment and management, plus procedures for seed registration and release [74] .

3. TRANSGENIC VEGETABLES AND NUTRITIONAL HEALTH BENEFITS

As we saw above transgenic crops, enable plant breeders to bring favorable genes, often previously inaccessible, into already elite cultivars, improving their value considerably and offering unique opportunities for extending shelf-life of the produce and for controlling insects, viruses and other pathogens. Conventional plant breeding that utilizes non-transgenic approaches will remain the backbone of vegetable and fruit genetic improvement strategies. However, transgenic crop cultivars should not be excluded as products capable of contributing to more nutritious and healthy food. In this chapter we will describe several advances of transgenic vegetables to nutritional quality and health benefits, very important for consumers.

Tomato fruit and its processed products are the principal dietary sources of carotenoids such as lycopene and β-carotene. Therefore, there is considerable interest in elevating the levels of carotenoids in tomato fruit by genetic manipulation and thereby improving the nutritional quality of the crop. To enhance the carotenoid content and profile of tomato fruit, Romer et al. [75] have produced transgenic lines containing a bacterial carotenoid gene (crtI) encoding the enzyme phytoene desaturase, which converts phytoene into lycopene. Expression of this gene in transgenic tomato plants of the cultivar "Ailsa Cray" did not elevate total carotenoid levels. However, the β-carotene content increased about threefold, up to 45% of the total carotenoid content. The alteration in carotenoid content of these transgenic tomatoes did not affect growth and development. Levels of noncarotenoid isoprenoids were unchanged in the transformants. The phenotype has been found to be stable and reproducible over at least four generations. Fraser et al. [76] reported also an increase in tomato fruit carotenoids phytoene, lycopene, β-carotene and lutein in cultivar "Ailsa Craig". Phytoene synthase from the bacterium Erwinia uredovora (crtB) has been overexpressed in tomato cultivar. Fruit-specific expression was achieved by using the tomato polygalacturonase promoter, and the CRTB protein was targeted to the chromoplast by the tomato phytoene synthase-1 transit sequence. Total fruit carotenoids of primary transformants (T(0)) were two–to fourfold higher than the controls, whereas phytoene, lycopene, β-carotene, and lutein levels were increased 2.4-, 1.8-, and 2.2-fold, respectively. The biosynthetically related isoprenoids, tocopherols plastoquinone and ubiquinone were unaffected by changes in carotenoid levels. The progeny (T(1) and T(2) generations) inherited both the transgene and phenotype. Ripe tomato fruits accumulate large amounts of lycopene and small amounts of β-carotene (pro-vitamin A). Lycopene is transformed into β-carotene by the action of lycopene beta-cyclase (beta-Lcy). Rosati et al. [77] introduced, via Agrobacterium-mediated transformation, DNA constructs aimed at upregulating (OE construct) or downregulating (AS construct) the expression of the beta-Lcy gene in a fruit-specific fashion. Three tomato transformants containing the OE construct show a significant increase in tomato fruit β-caro- tene content. The tomato fruits from these plants display different colour phenotypes, from orange to orange-red, depending on the lycopene/beta-carotene ratio. Fruits from AS transformants show up to 50% inhibition of beta- Lcy expression, accompanied by a slight increase in lycopene content. Leaf carotenoid composition is unaltered in all transformants. In most transformants, an increase in total carotenoid content is observed with respect to the parental line. This increase occurs in the absence of major variations in the expression of endogenous carotenoid genes.

Current advances in genetic engineering of brassicas have enabled the production of plants with alterations in a range of vitamins or amino acids for improved human nutrition. Lu et al. [78] developed transgenic cauliflower with high levels of β-carotene accumulation, and Wahlroos et al. [79] produced oilseed Brassica rapa with increased histidine content. It is likely in the future that more transgenic vegetable brassicas with altered vitamin or amino acid content also will be developed.

Vitamin E, which includes tocopherols, is a lipid-soluble antioxidant. There are α, β, γ, and δ isoforms of to- copherol with relative vitamin E potencies of 100%, 50%, 10%, and 3%, respectively. Conversion of γ-toco- pherol to α-tocopherol in vegetable crops could increase their value and importance in human health because vi- tamin E reduces the risk of several serious disorders (e.g., cardiovascular diseases and cancer), slows aging and enhances the function of the immune system. Cho et al. [80] developed transgenic lettuce plants of the cultivar "Chungchima" expressing a cDNA encoding γ-tocopherol methyltransferase from Arabidopsis thaliana to improve tocopherol composition. Transgene inheritance and expression in transformed plants increased enzyme activity and conversion of γ-tocopherol to the more potent α form.

Folate deficiency, regarded as a global health problem, causes neural tube defects and other human diseases. Folates are synthesized from pteridine, p-aminobenzoate (PABA), and glutamate precursos. Díaz de la Garza et al. [81] [82] developed trangenic tomatoes by engineering fruit-specific overexpression of GTP cyclohydrolase I that catalyzes the first step of pteridine synthesis and amynodeoxychorismate synthase that catalyzes the first step of PABA synthesis. Vine-ripened fruits contained on average 25-fold more folate than controls by combining PABA- and pteridine-overproduction traits through crossbreeding of transgenic tomato plants. The achieved folate level provides a complete adult daily requirement with less than one standard serving.

Vegetables also offer consumers a diverse mixture of nutrients that promote human health more beneficially than dietary supplements. However, the ingestion of plant-based diets rather than diets that rely primarily on animal products could limit the intake of essential nutrients such as calcium (Ca). Consequently, genetically engineering vegetables containing increase Ca levels may boost Ca uptake, thereby reducing the incidence of Ca deficiencies such as osteoporosis. In this regard, Park et al. [83] modified carrots to express increased levels of the plant Ca transporter sCAX1. These carrot lines were fertile and displayed no adverse phenotypes. Further, mice and human feeding trials demonstrated increased Ca absorption from sCAX1-expressing transgenic carrots vis-à-vis controls [84] . This research supports alternative means of biofortifying vegetables with bioavailable Ca. Zinc is also an essential element in human nutrition, as its deficiency severely impairs organ function. In experiments to fortify lettuce with this element, Zuo et al. [85] used Agrobacterium-mediated gene delivery of a mouse metallothionein mutant β-cDNA in the lettuce cultivar "Salinas 88". The concentration of zinc in the lettuce transgenic plants increased to 400 μg/g dry weight, which is considerably higher than in wild-type plants.

Flavonoids are polyphenols whose dietary intake has the potential to prevent chronic diseases. Schijlen et al. [86] introduced heterologous, flavonoid pathway genes—stilbene synthase, chalcone synthase, chalcone reductase, chalcone isomerase, and flavone synthase—to produce novel flavonoids in tomato fruit. These novel flavonoids-flavones and flavonols increased threefold, mostly in the Q12 peel, which had higher total antioxidant capacity. These findings add further support to the potential of engineering tomato fruit for accumulation of high levels of beneficial nutrients. Similarly, the poliphenol resveratrol, a stilbene, shows cancer chemo-preventative activity and may prevent coronary heart disease and arteriosclerosis. Liu et al. [87] in a quantitative analysis showed that resveratrol in transgenic lettuce plants was 56.0 ± 5.52 μg/g leaf fresh weight, which is comparable to that in the skin of grape fruit (Citrus × paridisi Macfad.).

Flavonoids such as anthocyanins are known as antioxidants in vitro and can reduce the risk of many diseases related to aging. However, some vegetable brassicas, such as cauliflower, are low in anthocyanins. In an attempt to manipulate pigment biosynthesis to increase the health benefits of brassica vegetables, the effect of a regulatory locus of flavonoid content was assessed. Agrobacterium tumefaciens–mediated transformation of a Brassica oleracea line, selected for high transformation ability by Sparrow et al. [88] , was used to produce plants transgenic for the maize lc (leaf color) locus. lc is a regulatory gene in the anthocyanin pathway, and it is expected that its presence will increase the flavonoid content. Seedling explants were co-cultivated with Agrobacterium tumefaciens strain LBA4404 containing a binary vector Q27 with a neomycin phosphotransferase II (NPTII) gene. Under tissue culture conditions, lc-containing plants were green with no visible increase in anthocyanin production. However, after transfer to the greenhouse, the exposure to high light intensity led to visi- ble signs of pigmentation within one week. Increased pigmentation was apparent in stems, petioles, main leaf veins, and sepals. lc-containing lines had 10 to 20 times higher levels of total anthocyanins than controls. In ad- dition, antioxidant activity of lc-containing lines was 1.5 times higher than that of controls [89] .

The unique flavor and odor of alliums is derived from the hydrolysis of organosulfur compounds, which pro- duces pyruvate, ammonia, and volatile sulfur compounds [90] . This reaction is catalyzed by the enzyme alliinase, which is contained in vacuoles within cells and released upon disruption of the tissue [91] . Variations in the ratios of these volatile sulfur compounds are responsible for the difference in flavors and odors between Allium species [90] . Along with health and nutritional benefits associated with these compounds, these thiosulfides are also major contributors to the bitter taste of some onions [90] [92] . Three sets of transgenic onion plants con- taining antisense alliinase gene constructs (a CaMV 35S-driven antisense root alliinase gene, a CaMV 35S-dri- ven antisense bulb alliinase, and a bulb alliinase promoter-driven antisense bulb alliinase) have been recently produced [93] . Results from the antisense bulb alliinase lines have been much more encouraging, and three lines were produced with barely detectable bulb alliinase levels and activity. Progress has been confounded by the poor survival of transgenic plants. Crossing a non-transgenic open-pollinated parental line with a transgenic parental plant carrying a single transgene in the hemizygous state has conducted to a transgenic hybrid onion seed from these transgenic lines. Some resulting seed produced by the non-transgenic parents will be hemizygous for the transgene and can be selected to give F1 heterozygous individuals containing the transgene. Self-fertilization of these individuals produces homozygous, hemizygous, and null F_2progeny for the transgene locus. These homozygous individuals can then be used to generate the bulk seed required for the production of commercial transgenic onion lines with less bitter taste. In attempts to reduce bitterness in lettuce, Sun et al. [94] cloned the gene for the sweet and taste, modifying protein miraculin from the pulp of berries of Richadella dulcifica, a West African shrub. This gene, with the CaMV 35S promoter, was introduced into the lettuce cultivar "Kaiser" using A. tumefaciens GV2260. Expression of this gene in transgenic plants led to the accumulation of significant concentrations of the sweet-enhancing protein. People suffering diabetes may use Miraculin as a food sweetener, which is active at extremely low concentrations. The first successful study conducted to engineer the taste of tomato fruit involved transformation of tomato with the thaumatin gene from the African plant katemfe (Thaumatococcus daniellii) [95] . Thaumatin is a sweet-tasting protein. Fruit fromT2 transgenic plants tasted sweeter than the control plants, leaving a unique and sweet-specific aftertaste.

Food safety can also be enhanced through transgenic approaches. The first transgenic cassava plants became available in the mid-1990s [96] - [98] as plants with reduced cyanogenic content [99] - [101] , which can benefit resource-poor people in rural Asia and Africa where this starchy root crop is the base of their diet.

Some vegetables, mainly tomato, are also genetically modified to become a vaccine delivery. Plant delivery of oral vaccines has attracted much attention because this strategy offers several advantages over vaccine delivery by injection [102] - [104] . Oral vaccines also offer the hope of more convenient immunization strategies and a more practical means of implementing universal vaccination programs worldwide. Tomato has been tested for expression of vaccines that can address human health issues of the developing world. Transgenic tomato plants potentially can bring several positive effects and improve human health. McGarvey et al. [105] engineered tomato plants of cultivar "UC82b" to express a gene encoding a glycoprotein (G-protein), which coats the outer surface of the rabies virus. The recombinant constructs contained the G-protein gene from the environmental risk assessment strain of rabies virus. The G-protein was expressed in leaves and fruit of the transgenic plants, and it was found localized in Golgi bodies, vesicles, plasmalemma, and cell walls of vascular parenchyma cells.

Ma et al. [106] overexpressed hepatitis E virus (HEV) open reading frame 2 partial gene (HEV-E2) in tomato plants to investigate its expression in transformants, the immunoactivity of expressed products, and explore the feasibility of developing a new type of plant-derived HEV oral vaccine. The recombinant protein was produced at 61.22 ng/g fresh weight in tomato fruits and 6.37 - 47.9 ng/g fresh weight in the leaves of the transformants. It was concluded that the HEV-E2 gene was correctly expressed in transgenic tomatoes and that the recombinant antigen derived had normal immunoactivity. These transgenic tomato plants are valuable tools for the development of edible oral vaccines. Chen et al. [107] developed an effective antiviral agent against enterovirus 71 (EV71), which causes seasonal epidemics of hand, foot, and mouth disease associated with fatal neurological complications in young children, by transforming the gene for VP1 protein—a previously defined epitope and also a coat protein of EV71—in tomato plant. VP1 protein was first fused with sorting signals to enable it to be retained in the endoplasmic reticulum of tomato plant, and its expression level increased to 27 mg/g in fresh tomato fruit. Transgenic tomato fruit expressing VP1 protein was then used as an oral vaccine, and the development of VP1- specific fecal IgA and serum IgG were observed in BALB/c mice. Additionally, serum from mice fed transgenic tomato could neutralize the infection of EV71 to rhabdomyosarcoma cells, indicating that tomato fruit expressing VP1was successful in orally immunizing mice. Moreover, the proliferation of spleen cells from orally immunized mice was stimulated by VP1 protein and provided further evidence of both humoral and cellular immunity. Results of this study not only demonstrated the feasibility of using transgenic tomato as an oral vaccine to generate protective immunity in mice against EV71 but also the probability of enterovirus vaccine development. The Gram-negative bacterium Yersinia pestis causes plague, which has affected human health since ancient times. It is still endemic in Africa, Asia, and the American continent. There is the urgent need for a safe and cheap vaccine due to the increasing reports of the incidence of antibiotic-resistant strains and concern with the use of Y. pestis as an agent of biological warfare. Out of all the Y. pestis antigens tested, only F1 and Vinduce present a good protective immune response against a challenge with the bacterium [108] . Alvarez et al. [109] reported the expression in tomato of the Y. pestis F1-Vantigen fusion protein. The immunogenicity of the F1-V transgenic tomatoes was confirmed in mice that were injected subcutaneously with bacterially produced F1-V fusion protein and boosted orally with transgenic tomato fruit. Expression of the plague antigens in the tomato fruit allowed producing an oral vaccine candidate without protein purification and with minimal pro- cessing technology, offering a good system for a large-scale vaccination programs in developing countries. The future of edible plant-based vaccines through transgenic approaches will depend on producing them safely on sufficient amounts.

4. TRANSGENIC VEGETABLES AND INTEGRATED PEST MANAGEMENT

The production of vegetables worldwide tends to be on smaller areas and in more diversified holdings than field crops such as cotton, canola, cereals and soybeans. Vegetables are often in more complex agricultural systems where insects may move from one crop to the next within the same farm. How this will impact the use and effects of transgenic vegetable plants in the agricultural landscape can be complex. Growing multiple insect-resistant transgenic vegetable plants in the same area and exposure of a polyphagous insect to the same Bt protein expressed in the different vegetable species will challenge conventional strategies developed for transgenic cotton or maize cultivars. Thoughtful consideration therefore will be needed before choosing what toxins vegetable plants should express. The selection should be based not only on what will be an effective toxin against the target insect but which toxins are already in use in other vegetable crops that may be hosts for the target insect. Additionally, the difficulty of sampling insect populations for resistant alleles will take on a higher level of complexity in a diversified vegetable system.

Further consideration should also be given to the effects on nontarget organisms within diversified transgenic vegetable plantings. Hoheisel and Fleischer [110] investigated the seasonal dynamics of coccinellids and their food (aphids and pollen) in a vegetable farm system containing plantings of Bt-sweet corn, Bt-potato, and transgenic insect-resistant squash in northeastern USA. Their results indicated that the transgenic vegetable crops provided conservation of coccinellids and resulted in a 25% reduction in insecticide use. In a similar study with these same crops, Leslie et al. [111] compared the soil surface dwelling communities of Coleoptera and Formicidae in the transgenic crops and their isolines and found no differences in species richness and species composition but found that the transgenic vegetables required fewer insecticide applications. Such results make clear that transgenic technology can be introduced within vegetable integrated pest management (IPM) systems and that transgenic vegetables can offer novel and effective ways of controlling insects and the pathogens they transmit.

Virus-resistant transgenic plants are particularly valuable if no germplasm source of resistance has been identified or if host plant resistance is difficult to transfer into elite cultivars by traditional breeding methods due to incompatibility barriers or links to undesired traits. In such cases, engineered host plant resistance may be the only viable option to develop virus resistant cultivars. Growers can also use virus-resistant transgenic vegetables as a trap crop by growing it as a border around the non-transgenic vegetable crop and allowing it to cleanse viruliferous aphids.

In small, diversified vegetable plantings typical of those found throughout the developing world, the challenges for regulatory oversight of transgenic plants are immense. In these countries, farmers will likely save transgenic seed and move transgenic seed between locations, and some transgenic products may move into mar- kets that do not permit these products. These concerns will be lessened if people consume the produce from transgenic vegetables locally after they are released and grown following national biosafety guidelines. However, it is likely that violations will occur, and this will challenge legal systems.

While each vegetable has its own set of one or more key pests, other pests can also be problematic. Traditional broad-spectrum insecticides often control a suite of pest insects. Thus, when transgenic vegetables are introduced into production systems, other methods of control will have to be applied or developed for secondary pests, e.g., biological control of secondary pests or use of selective insecticides, applied either as seed treatments or foliar sprays, may be necessary.

Integrated pest management could benefit from some herbicide-tolerant crops, if alternative nonchemical methods can be applied first to control weeds and the specific herbicide could be used later, only when and where the economic threshold of weeds is surpassed [112] . Herbicide-tolerant transgenic crops can help reducing plough in fields, thereby saving fuel because of less tractor use, which also protects the structure of the soil by reducing its erosion. Repeated use of herbicides in the same area may also create problems of weed herbicide resistance [113] . Although the risk of herbicide-resistant genes in vegetables is globally lower than in field crops because many vegetables are consumed in the vegetative stage.

5. MARKETS, POLITICS AND THE POWER OF ADVOCACY GROUPS

Although transgenic cultivars have proven to be a powerful tool for pest management, and their use has been accompanied by reduced dependence on pesticides, fertilizers and other inputs, and dramatic economic and environmental benefits [114] , many countries are still engaged in discussions about potential negative impacts of these crops on the environment, nontarget organisms, food safety, the unintentional spread of transgenic traits into conventionally bred-crop or landrace gene pools of the same species particularly in centers of crop diversity or origin, and questions of seed ownership. Fear about potential negative effects of transgenic crops has led to the implementation of very stringent regulatory systems in several countries and regulations that are far more restrictive for transgenic crops than for other agricultural technologies [115] . Consumer antagonism has precluded many farmers and consumers from sharing the benefits that these crops can provide. Critics also claim that adop- tion of transgenic crops benefits multinational seed corporations while hurting small farmers because of the additional investments required for growing these crops successfully. These unfounded concerns are present despite the highly successful and rapid adoption of transgenics in maize, soybeans, canola and cotton in many countries of the world.

Farmers and consumers will try to influence political decision making in favor of any technology if they expect to gain from it. The political power of farmers in the developed world was used effectively to gain access to large farm subsidies supported in part by fiscal resources and in part by artificially high consumer prices. Their market power however gradually deteriorated as consumers gained a greater say in the marketplace for food. On the contrary, developing world farmers possess very limited political power and have been taxed rather than subsidized by their governments.

Another factor that has led to differential perspectives between rich and poor countries is the relative power possessed by civil society groups including advocacy activists opposed to genetic engineering in food and agriculture. These activists have been very successful in influencing the debate, and consumer and government attitudes towards genetically modified food in Europe and are also gaining power in selected developing countries. While the European governments have tended to follow the expressed desires by advocacy groups and a majority of consumers opposed to genetically engineered food, the USA and Canadian governments support the farm sector and the private sector engaged in the development and deployment of modern biotechnology-crops. The large majority of the population of China and India (the largest consumers of vegetables) also are in favor of the use of modern biotechnology for both human diseases and pest-resistant crops, while the European countries are much more in favor of the use of modern biotechnology for human diseases than for food and agriculture. Some developing countries are considering the commercialization of genetically modified crops but they have been warned by the European Union that they may lose the European market not only for the commodities that have been genetically modified but possibly also for those that have not been geneticaly modified. This situation happened for vegetables with Africa exporter countries and seems to be also relevant with India government decision respecting Bt-eggplant. Developing countries who might wish to use genetic modification to improve agricultural productivity and nutritional qualities of their foods might therefore be faced with a choice between doing so for the domestic market while losing export opportunities or foregoing the potential benefits to its own people while maintaining export opportunities.

One could argue that the European consumer has gained a great deal of political power over agriculture in her capacity as a consumer, while still agreeing to provide large subsidies to agriculture in her capacity as a taxpayer. The possibility that the application of modern biotechnology in European agriculture could reduce the need for farm subsidies does not seem to enter into the European debate. To the extent this potential contradiction has been considered by consumers and governments, the conclusion seems to be that consumers prefer to pay farmers not to produce genetically modified food either through additional subsidies or through high food prices, at least until now. Such a contradiction is not prevalent in the United States and Canada.

The great opposition made by advocacy groups mainly in Europe and Japan to transgenic crops about food safety and possible health problems is denied by scientific data and various international and national organizations worldwide. Transgenic crops must pass a rigorous assessment based on scientific data of any potential risks. The objective of this appraisal has been to determine whether the transgenic crop is as safe as its conventional counterpart before transgenic modification. For this purpose, scientific data are provided to be reasonably confident that it will not damage consumers' health. Several organizations have stated that transgenic crops approved for commercialization, do not pose more risk to human health than conventional crops. The Food and Agriculture Organization of the United Nations [116] , the Royal Society of London, the US National Academy of Sciences, the Brazilian Academy of Sciences, the Chinese Academy of Sciences, the Indian National Sci- ence Academy, the Mexican Academy of Sciences and the Third World Academy of Sciences [117] , the World Health Organization [118] , the American College of Nutrition [119] , the American Society of Toxicology [120] , the British Medical Association [121] , and the Union of German Academies of Sciences and Humanities [122] , among others, have stated that transgenic crops approved for commercialization, do not pose more risk to human health than conventional crops, and they should be considered as safe as conventional ones.

The World has witnessed a steady increase of transgenic crop area in the last 18 years. The potential impacts on human and animal health have been subject to extensive research, and no evidence has been found against transgenic crops. Some people, however continue to argue the potential long-term risks but without indicating what those risk maybe. Recently, the Federal Office of Consumer Protection and Food Safety of Germany and partners [123] issued the report "Biological and Ecological Evaluation towards Long-term Effects" (also known as the BEETLE report) with the aim of providing scientific data to the European Commission. The BEETLE report reviewed in excess of 100 publications and consulted 52 experts in health issues to assess the possible long-term effect of transgenic crops on the health of consumers and the environment. This report concluded that so far, no adverse effects to human health from eating GM food have been found. The report further stated that although unexpected negative effects are known in conventional crops, none has yet been detected in transgenic crops. The report concludes that there is a negligible probability for adverse effects to consumers' health in the long term.

6. THE CASE FOR CISGENIC VEGETABLES AND FRUITS?

Transgenic vegetables lacking any undesired genes will be useful to overcome any potential resistance for the deployment of such genetically engineered crops [125] . Some systems are available for releasing marker-free transgenic crops, e.g. through co-transformation or highly efficient gene transfer protocols for engineering plants without any marker-aided selection that require the screening to identify transgenic events, and recombination for removal of specific undesired gene sequences from transgenic plants.

Progress in crop genome sequencing, high resolution genetic mapping and precise phenotyping will accelerate the discovery and use of functional alleles and allelic variation that are associated with traits of interest for plant breeding [124] . Genome sequencing and annotation include an increasing range of species such as apple, banana/plantain, (common) bean, cacao, cassava, chickpea, Chinese cabbage, citrus, coffee, (diploid) cotton, cucum- ber, eggplant, grape, model legumes, maize, melon, oil palm, oilseed rape, papaya, peach, pepper, pigeonpea, potato, rice, sorghum, sugarcane, soybean, (woodland) strawberry, tomato and watermelon among other plant species [https://genomevolution.org/wiki/index.php/ Sequenced_plant_genomes#Rate_of_Publication]. Perhaps, one day further research on the genome of a plant species from a drought-prone environment may assist to breed- ing hardier and water efficient related crops due to gene synteny.

The use of plant-derived genes for introducing useful traits and plant-derived promoters will also overcome some concerns about the deployment of genetically-engineered vegetables. In this regard, cisgenesis addresses some negative views regarding the use of genes from non-crossable species for breeding vegetable crops. Cisgenesis involves only genes from the plant itself or from a crossable close relative, and these genes could also be transferred by conventional breeding methods. Crop wild relatives are therefore a valuable source of traits for cisgenesis. For example, wild Solanum species from South America are sources of alleles for improving tomato adaptation to climate change and other abiotic stresses: S. chilense grows in the desert due to its long primary roots and extensive secondary root system, whereas S. pennellii increases its water use efficiency under drought, and S. cheesmanii or S. peruvianum may grow in salty coastal areas. Schouten et al. [125] [126] argued very eloquently that regulations for genetically modified organisms should be altered to exempt cisgenesis because it respects species barriers. They conclude that regulators should treat cisgenic plants as conventionally bred plants regarding biosafety issues. Of course, any cisgenic plants need to be assessed for confirming they only bear the intended modifications and no other foreign gene(s). Very recently, Eriksson et al. [127] warn that cisgenic also poses risks and brings additional costs. They further argue that promoting cisgenics may lead to a lack of trust that public places in science. People may ask why scientists told them for several decades that transgenic crops were not more risky than crossbred crops.

7. OUTLOOK

Vegetables are grown worldwide, on large and small farms, on good and marginal land, and by large commercial growers and small subsistence farmers. Transgenic cultivars can improve nutritional quality and make important contributions to sustainable vegetable production by overcoming some limiting factors in vegetable production, e.g. pathogens (bacteria, fungi, viruses), pests (insects and nematodes), and weeds, which are not easily addressed through conventional breeding alone. They could also help for reducing pesticide residues and human poisoning, and reducing costs in horticulture. However, as noted by Ortiz and Smale [115] , before transgenic cultivars can be made available to vegetables farmers, particularly in the developing world, several hurdles must be overcome. At this time, the most important of these appear to be related to intellectual property management, public awareness, biosafety guidelines, and food safety protocols. Appropriate measures needs to be sought to overcome such hurdles.

The high cost of satisfying biotechnology regulatory requirements is prohibitive for most vegetable crops, restricting the commercial availability of this powerful technology to a few commodities crops such as cotton, maize, soybean and canola. In addition, high regulatory hurdles reduce or eliminate opportunities for small companies and public institutions to commercialize products that might appeal directly to consumers or address human nutritional needs. The lost opportunity costs and reduced innovations due to underutilization of modern biotechnology methods in vegetable cultivar development are significant, particularly in light of the documented benefits of these products. Thus it is necessary to develop cost-efficient benefit/risk analysis systems for products of biotechnology [128] [129] .

Countries vary in their market standards and acceptance of transgenic crops. The advantages of transgenic breeding should provide incentive for integration of this method into the amelioration of challenges for vegetables and horticultural production, if consumer fears are overcome or eased. The future of transgenic vegetables may be in the developing world, especially China and India, which account for 40% of the world population, and grow and consume in excess of 60% of the world's vegetables. They already adopted Bt-cotton, and it is likely that Bt-rice will be reaching soon farmers in China. The acceptance of transgenic vegetables in both countries will likely hasten their adoption in other parts of the world and allow farmers to use these crops.

João Silva Dias, Rodomiro Ortiz Distinct perspectives leading to different policy options and standards may conflict with the current globalization market. Some policy options that may be appropriate for the developed world are sometimes inappropriate for the developing world. Transgenic crops are not silver bullets for achieving food security but coupled with traditional knowledge and conventional agriculture may be a powerful tool for making available affordable food to poor farmers and consumers. Modern biotechnology can provide products with the potential for reducing farming costs, enhancing agricultural productivity, ensuring food supply, alleviating rural poverty, promoting sustainable use of natural resources, protecting the environment and preserving biodiversity. Biotechnology-derived crops will succeed in farming systems if clear advantages and safety are demonstrated to both growers and consumers.

REFERENCES

1. Dias, J.S. and Ryder, E. (2011) World Vegetable Industry: Production, Breeding, Trends. Horticultural Reviews, 38, 299-356.

2. Keatinge, J.D.H., Waliyar, F., Jammadass, R.H., Moustafa, A., Andrade, M., Drechsel, P., Hughes, J.A., Kardivel, P. and Luther. K. (2010) Re-Learning Old Lessons for the Future of Food: By Bread Alone No Longer—Diversifying Diets with Fruit and Vegetables. Crop Science, 50, S51-S62.
3. Kays, S.J. and Dias J.S. (1995) Common Names of Commercially Cultivated Vegetables of the World in 15 Languages. Economic Botany, 49, 115-152.
4. Kays, S.J. (2011) Cultivated Vegetables of the World: A Multilingual Onomasticon. Wageningen Academic Publishers, Wageningen.
5. Krattinger, A. (1998) The Importance of Ag-Biotech to Global Prosperity. ISAAA Briefs No. 6, The International Service for the Acquisition of Agri-biotech Applications, Ithaca, NY, 11.
6. Pimentel, D. (1997) Techniques for Reducing Pesticide Use. Economic and Environmental Benefits. Wiley, New York.
7. Dias, J.S. and Ortiz, R. (2012) Transgenic Vegetable Crops: Progress, Potentials and Prospects. Plant Breeding Reviews, 35, 151-246.
8. Asian Vegetable Research and Development Center (2001) Fact Sheet: Eggplant Fruit and Shoot Borer. AVRDC, Tainan.
9. Choudhary, B. and Gaur, K. (2009) The Development and Regulation of Bt Brinjal in India. ISAAA Brief No. 38, International Service for Acquisition of Agri-Biotech Applications, Ithaca, NY.
10. Siritunga, D. and Sayre, R.T. (2003) Generation of Cyanogen-Free Transgenic Cassava. Planta, 217, 367-373.
11. Siritunga, D. and Sayre, R.T. (2004) Engineering Cyanogen Synthesis and Turnover in Cassava (Manihot esculenta). Plant Molecular Biology, 56, 661-669.
12. Siritunga, D., Arias-Garzon, D., White, W. and Sayre, R.T. (2004) Over-Expression of Hydroxynitrile Lyase in Transgenic Cassava Roots Accelerates Cyanogenesis and Food Detoxification. Plant Biotechnology Journal, 2, 37-43.
13. Kramer, M.G. and Redenbaugh, K. (1994) Commercialization of a Tomato with an Antisense Polygalacturonase Gene: The FLAVR SAVR™ Tomato Story. Euphytica, 79, 293-297.
14. Hamilton, A., Lycett, G. and Grierson, D. (1990) Antisense Gene That Inhibits Synthesis of the Hormone Ethylene in Transgenic Plants. Nature, 346, 284-287.
15. Oeller, P.W., Wong, L.M., Taylor, L.P., Pike, D.A. and Theologis, A. (1991) Reversible Inhibition of Tomato Fruit Senescence by Antisense 1-Aminocyclopropane-1-Carboxylate Synthase. Science, 254, 437-439.
16. Klee, H.J., Hayford, M.B., Kretzmer, K.A., Barry, G.F. and Kishore, G.M. (1991) Control of Ethylene Synthesis by Expression of a Bacterial Enzyme in Transgenic Tomato Plants. Plant Cell, 3, 1187-1193.
17. Good, X., Kellogg, J.A., Wagoner, W., Langhoff, D., Matsumura, W. and Bestwick, R.K. (1994) Reduced Ethylene Syn- thesis by Transgenic Tomatoes Expressing S-Adenosylmethionine Hydrolase. Plant Molecular Biology, 26, 781-790.
18. Gianesi, L. and Carpenter, J. (1999) Agricultural Biotechnology: Insect Control Benefits. National Center for Food and Agricultural Policy, Washington DC.
19. Qaim, M. (1998) Transgenic Virus Resistance Potatoes in Mexico: Potential Social Implications of North-South Biotechnology Transfer. ISAAA Briefing No. 7, International Service for Acquisition of Agri-Biotech Applications, Ithaca, NY.
20. Thomas, P.E., Kaniewski, W.K. and Lawson, E.C. (1997) Reduced Field Spread of Potato Virus in Potatoes Transformed with the Potato Leafroll Virus Coat Protein Gene. Plant Disease, 81, 1447-1453.
21. Thornton, M. (2003) The Rise and Fall of New Leaf Potatoes. NABC Report, 15, 235-243.
22. Grafius, E.J. and Douches, D.S. (2008) The Present and Future Role of Insect-Resistant Genetically Modified Potato Cultivars in IPM. In: Romeis, J., Shelton, A.M. and Kennedy, G.G., Eds., Integration of Insect-Resistant Genetically Modified Crops within IPM Programs, Springer Science + Business Media B.V., Dordrecht, 195-221.

23. Guenthner, J.F. (2002) Consumer Acceptance of Genetically Modified Potatoes. American Journal of Potato Research, 79, 309-316.
24. Visser, D. (2005) Guide to Potato Pests and Their Natural Enemies in South Africa. ARC-Roodeplaat Vegetable and Ornamental Plant Institute, Pretoria.
25. Douches, D.S. and Grafius, E.J. (2005) Transformation for Insect Resistance. In: Razdan, M.K. and Mattoo, A.K., Eds., Genetic Improvement of Solanaceous Crops. Vol. 1: Potato, Science Publishers Inc., Enfield, NH, Plymouth, UK, 235- 266.
26. Douches, D.S., Pett, W., Santos, F., Coombs, J., Grafius, E., Li, W., Metry, E.A., Nasr El-Din, T. and Madkour, M. (2004) Field and Storage Testing Bt Potatoes for Resistance to Potato Tuberworm (Lepidoptera: Gelichiidae). Journal of Economic Entomology, 97, 1425-1431.
27. Douches, D.S., Brink, J.A., Quemada, H., Pett, W., Koch, M., Visser, D., Maredia, K. and Zarka, K. (2007) Commercialization of Potato Tuber Worm Resistant Potatoes in South Africa. Proceedings of 6th World Potato Congress, Boise, 20-26 August 2006. http://www.potatocongress.org/congress/proceedings-2006/
28. Douches, D.S., Brink, J.A., Quemada, H., Pett, W., Koch, M., Visser, D., Maredia, K. and Zarka, K. (2008) Commercialization of Potato Tuber Moth Resistant Potatoes in South Africa. In: Kroschel, J. and Lacey, L.A., Eds., Integrated Pest Management for the Potato Tuber Moth, Phthorimaea operculella (Zeller)—Potato Pest of Global Importance, Tropical Agriculture 20, Advances in Crop Research 10, Margraf Publishers GmbH, Weikersheim, 139-147.
29. Cooper, S.G., Douches, D.S., Zarkas, K. and Grafius, E.J. (2009) Enhanced Resistance to Control Potato Tuberworm by Combining Engineered Resistance, Avidin, and Natural Resistance Derived from Solanum chacoense. American Journal of Potato Research, 86, 24-30.
30. Celis, B.C., Scurrah, M., Cowgill, S., Chumbiauca, S., Green, J., Franco, J., Main, G., Kiezebrink, D., Visser, R. and At- kinson, H.J. (2004) Environmental Biosafety and Transgenic Potato in a Centre of This Crop's Diversity. Nature, 432, 222-225.
31. Scurrah, M., Celis-Gamboa, C., Chumbiauca, S., Salas, A. and Visser, R.G.F. (2008) Hybridization between Wild and Cultivated Potato Species in the Peruvian Andes and Biosafety Implications for Deployment of GM Potatoes. Euphytica, 164, 881-892.
32. ISAAA (2008) Bt Brinjal in India. Pocket K 35. International Service for Acquisition of Agri-Biotech Applications, Ithaca, NY.
33. Krishna, V.V. and Qaim, M. (2008) Potential Impacts of Bt Eggplant on Farmers' Health in India. Agricultural Economics, 38, 167-180.
34 Krishna, V.V. and Qaim, M. (2007) Estimating the Adoption of Bt Eggplant in India: Who Benefits from Public-Pri- vate Partnership? Food Policy, 32, 523-543.
35. Kolady, D. and Lesser, W. (2008) Can Owners Afford Humanitarian Donations in Agbiotech—The Case of Genetically Engineered Eggplant in India. Electronic Journal of Biotechnology, 11, 5. http://www.ejbiotechnology.info /content/ vol11/ issue2/full/5/
36. Kameswara-Rao, C. (2010) Moratorium on Bt Brinjal: A Review of the Order of the Minister of Environment and Forests, Government of India. Foundation for Biotechnology Awareness and Education, Bangalore.
37. Kinetz, E. (2010) India Halts Genetically Modified Eggplant Release. Greenbio. http://seattletimes.com/html/businesstechnology/2011022982_apasindiagmfood.html
38. Jayaraman, K. (2010) Bt Brinjal Splits Indian Cabinet. Nature Biotechnology, 28, 296.
39. Choudhary, B., Nasiruddin, K.M. and Gaur, K. (2014) The Status of Commercialized Bt Brinjal in Bangladesh. ISAAA Brief No. 47, International Service for Acquisition of Agri-Biotech Applications, Ithaca, NY.
40. Gianessi, L.P., Silvers, C.S., Sankula, S. and Carpenter, J.E. (2002) Virus Resistant Squash. In: Plant Biotechnology: Cur- rent and Potential Impact for Improving Pest

Management in US Agriculture. An Analysis of 40 Case Studies, National Center for Food and Agriculture Policy, Washington, DC, 75.
41. Gaba, V., Zelcer, A. and Gal-On, A. (2004) Cucurbit Biotechnology—The Importance of Virus Resistance. In Vitro Cel- lular & Developmental Biology-Plant, 40, 346-358.
42. Ochoa, J.P.A., Dainello, F., Pike, L.M. and Drews, D. (1995) Field Performance Comparison of Two Transgenic Sum- mer Squash Hybrids to Their Parental Hybrid Lineage. HortScience, 30, 492-493.
43. Clough, G.H. and Hamm, P.B. (1995) Coat Protein Transgenic Resistance to Watermelon Mosaic and Zucchini Yellow Mosaic Virus in Squash and Cantaloupe. Plant Disease, 79, 107-1109.
44. Fuchs, M. and Gonsalves, D. (1995) Resistance of Transgenic Squash Pavo ZW-20 Expressing the Coat Protein Genes of Zucchini Yellow Mosaic Virus and Watermelon Mosaic Virus 2 to Mixed Infections by Both Potyviruses. BioTechnology, 13, 1466-1473.
45. Tricoli, D.M., Carney, K.J., Russell, P.F., McMaster, J.R., Groff, D.W., Hadden, K.C., Himmel, P.T., Hubbard, J.P., Boeshore, M.L. and Quemada H.D. (1995) Field Evaluation of Transgenic Squash Containing Single or Multiple Virus Coat Protein Gene Constructs for Resistance to Cucumber Mosaic Virus, Watermelon Mosaic Virus 2, and Zucchini Yellow Mosaic Virus. BioTechnology, 13, 1458-1465.
46. Fuchs, M., Tricoli, D.M., McMaster, J.R., Carney, K.J., Schesser, M., McFerson, J.R. and Gonsalves, D. (1998) Comparative Virus Resistance and Fruit Yield of Transgenic Squash with Single and Multiple Coat Protein Genes. Plant Disease, 82, 1350-1356.
47. Schultheis, J.R. and Walters, S.A. (1998) Yield and Virus Resistance of Summer Squash Cultivars and Breeding Lines in North Carolina. HortScience, 8, 31-39.
48. Shankula, S. (2006) Quantification of the Impacts on US Agriculture of Biotechnology-Derived Crops Planted in 2005. http://www.ncfap.org/
49. Lynch, R., Wiseman, B., Sumner, H., Plaisted, D. and Warnick D. (1999) Management of Corn Earworm and Fall Armyworm (Lepidoptera: Noctuidae) Injury on a Sweet Corn Hybrid Expressing a cryIA (b) Gene. Journal of Economic Entomology, 92, 1217-1222.
50. Musser, F.R. and Shelton, A.M. (2003) Bt Sweet Corn and Selective Insecticides: Their Impacts on Sweet Corn Pests and Predators. Journal of Economic Entomology, 96, 71-80.
51. Burkness, E.C., Hutchison, W.D., Bolin, P.C., Bartels, D.W., Warnock, D.F. and Davis, D.W. (2001) Field Efficacy of Sweet Corn Hybrids Expressing a Bacillus thuringiensis Toxin for Management of Ostrinia nubilalis (Lepidoptera: Crambidae) and Helicoverpa zea (Lepidoptera: Noctuidae). Journal of Economic Entomology, 94, 197-203.
52. Hassel, R. and Shepard, B.M. (2002) Insect Population on Bacillus thuringiensis Transgenic Sweet Corn. Journal of Entomological Science, 37, 285-292.
53. Speese, I.J., Kuhar, T.P., Bratsch, A.D., Nault, B.A., Barlow, V.M., Cordero, R.J. and Shen, Z. (2005) Efficacy and Eco- nomics of Fresh-Market Bt Transgenic Sweet Corn in Virginia. Crop Protection, 24, 57-64.
54. Rose, R. and Dively, G.P. (2007) Effects of Insecticide-Treated and Lepidopteran-Active Bt Transgenic Sweet Corn on the Abundance and Diversity of Arthropods. Environmental Entomology, 36, 1254-1268.
55. NASS (2007) Vegetables: 2006 Annual Summary. National Agricultural Statistics Service, Washington DC.
56. Headrick, J. (2011) The Buzz on Sweet Corn. http://monsantoblog.com/2011/11/08/the-buzz-on-sweet-corn/
57. Stuart, D. (2011) Biotech Sweet Corn Varieties Deliver Sustainable Benefits to Growers. Monsanto Newsroom, 8/8/2011. http://www.monsanto.com/newsviews/Pages/gmo-sweet-corn-variety-coming-soon.aspx
58. Cordero, R., Morjan, W., Fabellar, A., Harty, J. and Subere, C.V. (2011) Potential Impact of Biotech Sweet Corn (MON 89034 × MON 88017) on Pest Management in the Southern US. Proceedings of the 2011 ASHS Annual Conference, Waikoloa, 25-28

September 2011, S288. http://ashs.confex.com/ashs/2011/webprogram/Paper5706.html

59. Lius, S., Manshardt, R.M., Fitch, M.M.M., Slightom, J.L., Sanford, J.C. and Gonsalves, D. (1997) Pathogen-Derived Resistance Provides Papaya with Effective Protection against Papaya Ringspot Virus. Molecular Breeding, 3, 161-168.

60. Susuki, Y.I., Tripathi, S. and Gonsalves, D. (2007) Virus-Resistant Transgenic Papaya: Commercial Development and Regulatory and Environmental Issues. In: Punka, Z.K., De Boer, S.H. and Sanfaçon, H., Eds., Biotechnology and Plant Disease Mmanagement, CAB International, Wallinford, 436-461.

61. Shelton, A.M. and Badenes-Perez, F.R. (2006) Concept and Applications of Trap Cropping in Pest Management. Annual Review of Entomology, 51, 285-308.

62. Fuchs, M. and Gonsalves, D. (2007) Safety of Virus-Resistant Transgenic Plants Two Decades after Their Introduction: Lessons from Realistic Field Risk Assessment Studies. Annual Review of Phytopathology, 45, 173-202.

63. Ming, R., Hou, S., Feng, Y., Yu, Q., Dionne-Laporte, A., Saw, J.H., Senin, P., Wang, W., Ly, B.V., Lewis, K.L.T., Salzberg, S.L., Feng, L., Jones, M.R., Skelton1, R.L., Murray, J.E., Chen, C., Qian, W., Shen, J., Du, P., Eustice1, M., Tong, E., Tang, H., Lyons, E., Paull, R.E., Michael, T.P., Wall, K., Rice, D.W., Albert, H., Wang, M.L., Zhu, Y.J., Schatz, M., Nagarajan, N., Acob, R.A., Guan, P., Blas, A., Wai1, C.M., Ackerman, C.M., Ren, Y., Liu, C., Wang, J., Wang, J., Na, J.K., Shakirov, E.V., Haas, B., Thimmapuram, J., Nelson, D., Wang, X., Bowers, J.E., Gschwend, A.R., Delcher, A.L., Singh, R., Suzuki, J.Y., Tripathi, S., Neupane, K., Wei, H., Irikura, B., Paidi, M., Jiang, N., Zhang, W., Presting, G., Windsor, A., Navajas-Pérez, R.N., Torres, M.J., Feltus, F.A., Porter, B., Li, Y., Burroughs, A.M., Luo, M.C., Liu, L., Christopher, D.A., Mount, S.M., Moore, P.H., Sugimura, T., Jiang, J., Schuler, M.A., Friedman, V., Mitchell-Olds, T., Shippen, D.E., de Pamphilis, C.W., Palmer, J.D., Freeling, M., Paterson, A.H., Gonsalves, D., Wang, L. and Alam, M. (2008) The Draft Genome of the Transgenic Tropical Fruit Tree Papaya (Carica papaya Linnaeus). Nature, 452, 991-996.

64. Scorza, R., Callahan, A., Dardick, C., Cambra, M., Polak, J., Ravelonandro, M., Zagrai, I. and Malinowski, T. (1998) "Honey Sweet"—A Transgenic Plum Pox Virus Resistant Plum—From Laboratory and Experimental Field Plots to Regulatory Approval. Acta Horticulturae, 974, 57-63.

65. Ravelonandro, M., Monsion, M., Teycheney, P.Y., Delbos, R. and Dunez, J. (1992) Construction of a Chimeric Viral Gene Expressing Plum Pox Virus Coat Protein. Gene, 120, 167-173.

66. Hily, J.-M., Scorza, R., Malinowski, T., Zawadzka, B. and Ravelonandro, M. (2004) Stability of Gene Silencing-Based Resistance to Plum Pox Virus in Transgenic Plum (Prunus domestica L.) under Field Conditions. Transgenic Research, 13, 427-436.

67. Malinowski, T., Cambra, M., Capote, N., Zawadzka, B., Gorris, M.T., Scorza, R. and Ravelonandro, M. (2006) Field Trials of Plum Clones Transformed with the Plum Pox Virus Coat Protein (PPV-CP) Gene. Plant Disease, 90, 1012- 1018.

68. Polak, J., Pivalova, J., Kundu, J.K., Jokes, M., Scorza, R. and Ravelonandro, M. (2008) Behavior of Transgenic Plum Pox Virus-Resistant Prunus domestica L. Clone C5 Grown in the Open Field under a High and Permanent Infection Pressure of the PPV-Rec Strain. Journal of Plant Pathology, 90, S1.33-S1.36.

69. Ravelonandro, M., Scorza, R., Renaud, R. and Salesses, G. (1998) Transgenic Plums Resistant to Plum Pox Virus Infection and Preliminary Results of Cross-Hybridization. Acta Horticulturae, 478, 67-71.

70. Scorza, R., Callahan, A., Levy, L., Damsteegt, V. and Ravelonandro, M. (1998) Transferring Potyvirus Coat Protein Genes through Hybridization of Transgenic Plants to Produce Plum Pox Virus Resistant Plums (Prunus domestica L.). Acta Horticulturae, 472, 421-425.

71. Food and Agricultural Organization (2009) Agriculture Data. FAO, Rome. http://faostat.fao.org

72. Tripathi, L., Mwaka, H., Tripathi, J.N. and Tushemereirwe, W. (2010) Expression of Sweet Pepper Hrap Gene in Banana Enhances Resistance to Xanthomonas campestris pv musacearum. Molecular Plant Pathology, 11, 721-731.
73. Tripathi, L., Tripathi, J.N., Kiggundu, A., Korie, S., Shotkoski, F. and Tushemereirwe, W.K. (2014) Field Trial of Xan- thomonas Wilt Disease-Resistant Bananas in East Africa. Nature Biotechnology, 32, 868-870.
74. Tripathi, L. (2012) Transgenics in Crop Improvement Research at IITA. IITA Research for Development (R4D) Review, 8, 58-60.
75. Romer, S., Fraser, P.D., Kiano, J.W., Shipton, C.A., Misawa, N., Schuch, W. and Bramley, P.M. (2000) Elevation of the Pro- vitamin A Content of Transgenic Tomato Plants. Nature Biotechnology, 18, 666-669.
76. Fraser, P.D., Romer, S., Shipton, C.A., Mills, P.B., Kiano, K.W., Misawa, N., Drake, R.G., Schuch, W. and Bramley, P.M. (2002) Evaluation of Transgenic Tomato Plants Expressing an Additional Phytoene Synthase in a Fruit Specific Manner. Proceedings of the National Academy of Sciences of the United States of America, 99, 1092-1097.
77. Rosati, C., Aquilani, R., Dharmapuri, S., Pallara, P., Marusic, C., Tavazza, R., Bouvier, F., Camara, B. and Giuliano, G. (2000) Metabolic Engineering of Beta-Carotene and Lycopene Content in Tomato Fruit. Plant Journal, 24, 413-419.
78. Lu, S., van Eck, J., Zhou, X., Lopez, A.B., O'Halloran, D.M., Cosman, K.M., Conlin, B.J., Paolillo, D.J., Garvin, D.F., Vrebalov, J., Kochian, L.V., Kupper, H., Earle, E.D., Cao, J. and Li, L. (2006) The Cauliflower Or Gene Encodes a DnaJ Cysteine-Rich Domain-Containing Protein That Mediates High Levels of β-Carotene Accumulation. Plant Cell, 18, 3594-3605.
79. Wahlroos, T., Susi, P., Solovyev, A., Dorokhov, Y., Morozov, S, Atabekov, J. and Korpela, T. (2004) Increase of His- tidine Content in Brassica rapa Subsp. oleifera by Over-Expression of Histidine-Rich Fusion Proteins. Molecular Breeding, 14, 455-462.
80. Cho, E.A., Lee, C.A., Kim, Y.S., Baek, S.H., Reyes, B.G. and Yun, S.J. (2005) Expression of γ-Tocopherol Methyltrans- ferase Transgene Improves Tocopherol Composition in Lettuce (Lactuca sativa L.). Molecules and Cells, 19, 16-22.
81. Diaz de la Garza, R.I., Quinlivan, E.P., Klaus, S.M.J., Basset, G.J.C., Gregory, J.F. and Hanson, A.D. (2004) Folate Biofortification in Tomatoes by Engineering the Pteridine Branch of Folate Synthesis. Proceedings of the National Academy of Sciences of the United States of America, 101, 13720-13725.
82. Diaz de la Garza, R.I., Gregory, J.F. and Hanson, A.D. (2007) Folate Biofortification of Tomato Fruit. Proceedings of the National Academy of Sciences of the United States of America, 104, 4218-4222.
83. Park, S., Kim, C.K., Pike, L.M., Smith, R.H. and Hirschi, K.D. (2004) Increased Calcium in Carrots by Expression of an Arabidopsis H^+/Ca^{2+} Transporter. Molecular Breeding, 14, 275-282.
84. Morris, J., Hawthorne, K.M., Hotze, T., Abrams, S.A. and Hirschi, K.D. (2008) Nutritional Impact of Elevated Cal- cium Transport Activity in Carrots. Proceedings of the National Academy of Sciences of the United States of America, 105, 1431-1435.
85. Zuo, X., Zhang, Y., Wu, B., Chang, X. and Ru, B. (2002) Expression of the Mouse Metallothionein Mutant ββ-cDNA in the Lettuces (Lactuca sativa L.). Chinese Science Bulletin, 47, 558-562.
86. Schijlen, E., Ric de Vos, C.H., Jonker, H., van den Broeck, H., Molthoff, J., van Tunen, A., Martens, S. and Bovy, A. (2006) Pathway Engineering for Healthy Phytochemicals Leading to the Production of Novel Flavonoids in Tomato Fruit. Plant Biotechnology Journal, 4, 433-444.
87. Liu, S., Hu, Y., Wang, X., Zhong, J. and Lin, Z. (2006) High Content of Resveratrol in Lettuce Transformed with a Stil- bene Synthase Gene of Partenocissus henryana. Journal of Agricultural and Food Chemistry, 54, 8082-8085.

88. Sparrow, P.A.C., Dale, P.J. and Irwin, J.A. (2004) The Use of Phenotypic Markers to Identify Brassica oleracea Ge- notypes for Routine High-Throughput Agrobacterium Mediated Transformation. Plant Cell Reports, 23, 64-70.
89. Braun, R.H., Morrison, S.C., Schwinn, K.E. and Christey, M.C. (2006) Agrobacterium Mediated Transformation of Brassica oleracea with the Lc Locus. International Association for Plant Tissue Culture and Biotechnology, Beijing, 115.
90. Randle, W.M. and Lancaster, J.E. (2002) Sulphur Compounds in Alliums in Relation to Flavour Quality. In: Brewster, J.L., Ed., Onions and Other Vegetable Alliums, CAB International, Wallingford, Oxfordshire, 329-356.
91. Lancaster, J.E. and Collin, H.A. (1981) Presence of Alliinase in Isolated Vacuoles and of Alkyl Cysteine Sulphoxides in the Cytoplasm of Bulbs of Onion (Allium cepa L.). Plant Science Letters, 22, 169-176.
92. Almeida, D. (2006) Manual de Culturas Hortícolas. Vol. 1, Editorial Presença, Lisboa.
93. Eady, C.C., Davis, S., Farrant, J., Reader, J. and Kenel, F. (2003) Agrobacterium tumefaciens Mediated Transforma- tion and Regeneration of Herbicide Resistant Onion (Allium cepa) Plants. Annals of Applied Biology, 142, 213-217.
94. Sun, H.J., Cui, M.L., Ma, B. and Ezura, H. (2006) Functional Expression of the Taste Modifying Protein, Miraculin, in Transgenic Lettuce. FEBS Letters, 580, 620-626.
95. Bartoszewski, G., Niedziela, A., Szwacka, M. and Niemirowicz-Szczytt, K. (2003) Modification of Tomato Taste in Transgenic Plants Carrying a Thaumatin Gene from Thaumatococcus daniellii BENT. Plant Breeding, 122, 347-351.
96. Li, H.Q., Sautter, C., Potrykus, I. and Pounti-Kaerlas, J. (1996) Genetic Transformation of Cassava (Manihot esculenta Crantz). Nature Biotechnology, 14, 736-740.
97. Raemakers, C.J.J.M., Sofiari, E., Taylor, N.J., Henshaw, G.G., Jacobsen, E. and Visser. R.G.F. (1996) Production of Trans- genic Cassava Plants by Particle Bombardment Using Luciferase Activity as the Selection Marker. Molecular Breeding, 2, 339-349.
98. Schopke, C., Taylor, N.J., Carcamo, R., Konan, N.K., Marmey, P., Henshaw, G.G., Beachy, R.N. and Fauquet, C.M. (1996) Regeneration of Transgenic Cassava Plants (Manihot esculenta Crantz) from Microbombarded Embryogenic Sus- pension Cultures. Nature Biotechnology, 14, 731-735.
99. Siritunga, D. and Sayre, R.T. (2003) Generation of Cyanogen-Free Transgenic Cassava. Planta, 217, 367-373.
100. Siritunga, D. and Sayre, R.T. (2004) Engineering Cyanogen Synthesis and Turnover in Cassava (Manihot esculenta). Plant Molecular Biology, 56, 661-669.
101. Siritunga, D., Arias-Garzon, D., White, W. and Sayre, R.T. (2004) Over-Expression of Hydroxynitrile Lyase in Trans- genic Cassava Roots Accelerates Cyanogenesis and Food Detoxification. Plant Biotechnology Journal, 2, 37-43.
102. Langridge, W.H.R. (2000) Edible Vaccines. Scientific American, 2000, 66-71.
103. Pascual, D.W. (2007) Vaccines Are for Dinner. Proceedings of the National Academy of Sciences of the United States of America, 104, 10757-10758.
104. SunilKumar, G.B., Gananpathi, T.R. and Bapat, V.A. (2007) Production of Hepatitis B Surface Antigen in Recom- binant Plant Systems. Biotechnology Progress, 23, 523-529.
105. McGarvey, P.B., Hammond, J., Dienelt, M.M., Hooper, D.C., Fu, Z.F., Dietzschold, B., Koprowski, H. and Michaels, F.H. (1995) Expression of the Rabies Virus Glycoprotein in Transgenic Tomatoes. Bio/Technology, 13, 1484-1487.
106. Ma, Y., Zhang, J., Lin, S.Q. and Xia, N.S. (2001) Genetic Engineering Vaccines Produced by Transgenic Plants. Jour- nal of Xiamen University, 40, 71-77.
107. Chen, H.F., Chang, M.H., Chiang, B.L. and Jeng, S.T. (2006) Oral Immunization of Mice Using Transgenic Tomato Fruit Expressing VP1 Protein from Enterovirus 71. Vaccine, 24, 2944-2951.
108. Benner, G., Andrews, G., Byrne, W., Strachan, S., Sample, A. and Heath, D. (1999) Immune Response to Yersinia Outer Proteins and Other Yersinia pestis Antigens after Experimental Plague Infection in Mice. Infection and Immunity, 67, 1922-1928.
109. Alvarez, M.L., Pinyerd, H.L., Crisantes, J.D., Rigano, M.M., Pinkhasov, J., Amanda, M., Walmsley, A.M., Masona, H.S. and Cardineau, G.A. (2006) Plant-Made Subunit

Vaccine against Pneumonic and Bubonic Plague Is Orally Immu- nogenic in Mice. Vaccine, 24, 2477-2490.

110. Hoheisel, G.A. and Fleischer, S.J. (2007) Coccinelids, Aphids, and Pollen in Diversified Vegetable Fields with Transgenic and Isoline Cultivars. Journal of Insect Science, 7, 1-12.

111. Leslie, T.W., Hoheisel, G.A., Biddinger, D.J., Rohr, J.R. and Fleisher S.J. (2007) Transgenes Sustain Epigeal Insect Biodiversity in Diversified Vegetable Farm Systems. Environmental Entomology, 36, 234-244.

112. Krimsky, S. and Wrubel, R.P. (1996) Agricultural Biotechnology and the Environment. University of Illinois Press, Urbana.

113. Wrubel, R.P. and Gressel, J. (1994) Are Herbicide Mixtures Useful for Delaying the Rapid Evolution of Resistance? A Case Study. Weed Technology, 8, 635-648.

114. Brookes, G. and Barfoot, P. (2009) GM Crops: Global Socio-Economic and Environmental Impacts 1996-2007. PG Economics, Dorchester.

115. Ortiz, R. and Smale, M. (2007) Transgenic Crops: Pro-Poor or Pro-Rich? Chronica Horticulturae, 47, 9-12.

116. FAO (2000) FAO Statement on Biotechnology. Food and Agriculture Organization of the United Nations, Rome. www.fao.org/biotech/stat.asp?lang=en

117. Royal Society of London, the US National Academy of Sciences, the Brazilian Academy of Sciences, the Chinese Academy of Sciences, the Indian National Science Academy, the Mexican Academy of Sciences and the Third World Academy of Sciences (2000) Transgenic Plants and World Agriculture. National Academies Press, Washington DC.

118. World Health Organization (2002) 20 Preguntas sobre los alimentos modificados genéticamente. WHO, Geneva. http://www.oei.es/salactsi/20oms.htm

119. Chassy, B.M. (2002) Food Safety Evaluation of Crops Produced through Biotechnology. Journal of the American College of Nutrition, 21, S166-S173.

120. Society of Toxicology (2002) The Safety of Genetically Modified Foods Produced through Biotechnology. Society of Toxicology, Virginia. www.toxicology.org/ai/gm/GM_Food.asp

121. British Medical Association (2004) Genetically Modified Foods and Health: A Second Interim Statement. British Me- dical Association, London. http://www.argenbio.org/adc/uploads/pdf/bma.pdf

122. Union of the German Academies of Science and Humanities (2006) Are There Health Hazards for the Consumers from Eating Genetically Modified Foods? Interacademy Panel Initiative on Genetically Modified Organisms, Berlin. http://www.ilsi.org/NorthAmerica/Documents/UGASH.pdf

123. Federal Office of Consumer Protection and Food Safety (Germany) and Partners (2009) Long-Term Effects of Ge- netically Modified (GM) Crops on Health and the Environment (Including Biodiversity): Prioritisation of Potential Risks and Delimitation of Uncertainties. Federal Office of Consumer Protection of Food Safety, Berlin. http://ec.europa.eu/food/food/biotechnology/reports_studies/docs/lt_effects_report_en.pdf

124. Ortiz, R. (2011) Revisiting the Green Revolution: Seeking Innovations for a Changing World. Chronica Horticulturae, 51, 6-11.

125. Schouten, H.J., Krens, F.A. and Jacobsen, E. (2006) Cisgenic Plants Are Similar to Traditionally Bred Plants. EMBO Reports, 8, 750-753.

126. Schouten, H.J., Krens, F.A. and Jacobsen, E. (2006) Do Cisgenic Plants Warrant Less Stringent Oversight? Nature Bio- technology, 24, 753.

127. Eriksson, D., Stymne, S. and Schorring, J.K. (2014) The Slippery Slope of Cisgenesis. Nature Biotechnology, 32, 727.

128. Federoff, N.V. and Brown, N.M. (2004) Mendel in the Kitchen. A Scientist's View of Genetically Modified Foods. Joseph Henry Press, Washington, DC.

129. Bradford, K.J., Van Deynze, A., Gutterson, N., Parrott, W. and Strauss, S.H. (2005) Regulating Transgenic Crops Sen- sibly: Lessons from Plant Breeding, Biotechnology and Genomics. Nature Biotechnology, 23, 439-444.

CHAPTER 10

SRAP Molecular Marker Technology in Plant Science

Genyi Li[1], Peter B. E. McVetty[1] and Carlos F. Quiros[2]

[1] *Department of Plant Science, University of Manitoba, MB, Canada*

[2] *Department of Plant Sciences, University of California, Davis, CA, USA*

1. INTRODUCTION

Molecular markers are commonly used in genetic diversity analysis, genetic map construction, gene mapping and cloning, and marker assisted selection in plant breeding. Based on detection procedure, most molecular marker technologies can be classified into hybridization-based or PCR-based systems. Restriction fragment length polymorphism (RFLP) is the first hybridization-based molecular marker system that was intensively used at the beginning of the molecular biology era in life science while hybridization-based marker methods such as microarrays and diversity array technology (DArT) are used currently to detect single nucleotide polymorphisms (SNP). In contrast, many PCR-based molecular marker detection methods have been developed. For example, amplified fragment length polymorphism (AFLP), random amplified polymorphic DNA (RAPD), simple sequence repeats (SSR) and sequence related amplified polymorphism (SRAP), inter-simple sequence repeat (ISSR), sequence tagged site (STS), and sequence characterized amplification region (SCAR), are commonly used in genomic analysis (Jones et al., 2009).

There are advantages and limitations for all molecular marker detection methods. In particular, RFLP probes can be shared in related species so RFLP is advantageous over other molecular markers in comparative genomics. However, the detection procedure in RFLP is complicated and costly. Additionally, RFLP is not easily automated to analyze thousands of individuals for marker assisted selection. AFLP is a commonly used molecular marker system since it can detect multiple genetic loci in a genome. On the other hand, there are many steps in the AFLP detection procedure, which limits its application in marker assisted selection when thousands of individual DNA samples need to be analyzed in a short time. SSRs often have a high level of polymorphism in plant genomes and are commonly used in most genomic applications. Since SSR technology only detects sequence repeats, the number of SSRs in a genome is relatively limited compared with numerous SNPs. RAPD is easily performed in one round of PCR, however, a low level of reproducibility of RAPD amplification limits its wide use in genomic analysis.

As next generation sequencing (NGS) technologies dramatically increase capacity and throughput of DNA sequencing, whole genome sequencing of many plant species has been accomplished and most economically important crop species such as rice, maize, soybean, sorghum, potato, tomato and Chinese cabbage have been fully sequenced. Although it is still challenging to use NGS for assembling a whole complex genome such as barley and wheat, there are thousands of SNPs identified in NGS that can be used to develop molecular markers in species with complex genomes. Furthermore, NGS is directly used in SNP discovery and a few dozen genotypes can be sequenced simultaneously to assemble ultradense genetic maps. Additionally, different strategies are used to produce partial genomes that can be used to directly sequence SNPs using next generation technologies.

2. SRAP TECHNOLOGY

We developed and published information on the SRAP marker system in 2001 (Li and Quiros). The original thinking was to simplify the AFLP detection procedure and increase throughput and improve reproducibility compared to RAPD. To produce a simple detection procedure, we skipped restriction enzyme digestion and ligation of target DNA fragments and adapters in the AFLP detection protocol. We designed SRAP primers in sizes similar to those in AFLP, but ran one round of PCR instead of two rounds in AFLP. To detect multiple loci with a pair of SRAP primers, we designed a special PCR running program (94^{0}C for 1 min, 35^{0}C for 1 min and 72^{0}C for 1 min for the first 5 cycles and followed by 30 cycles at the raised annealing temperature of 50^{0}C). At the beginning of PCR, the 35^{0}C annealing temperature allowed SRAP primers to anneal to multiple loci in target DNA so that the multiple loci were amplified to produce a profile that is similar to that in AFLP. Similar to AFLP, most SRAP markers are dominant while most SNPs and SSRs are co-dominant. Compared with RAPD, SRAP used a pair of primers with 16 to 22 nucleotides instead of 10-mer short primers in RAPD, which gives SRAP a big advantage over RAPD so one SRAP primer can combine with unlimited number of other primers. Although SRAP PCR starts at 35^{0}C annealing temperature in the first five cycles, the larger sizes of SRAP primers allowed the increase of annealing temperature to 50^{0}C in the following cycles, which significantly improves the reproducibility in SRAP. In contrast, a low level of reproducibility in RAPD is a limitation factor. In addition, SRAP primers can be fluorescently labelled and combined with unlabeled SRAP primers so SRAP PCR products can be separated in capillary instruments such as ABI genetic analyzers.

In general, there is a difference of GC content between gene coding sequences and other sequences in plant genomes. We used this difference to design two sets of SRAP primers. The forward primers contained a GGCC cassette closing the 3' end of SRAP primers that might preferentially anneal to the GC-rich regions while the reverse SRAP primer set was incorporated with an AATT cassette that would preferentially anneal SRAP primers to introns and gene spacers so that SRAP could preferentially amplify gene-rich regions in a genome. After sequencing SRAP fragments and constructing a SRAP genetic map in *B. oleracea*, it was found that SRAP indeed amplified more sequences from genes and more SRAP markers fell into chromosome arm regions and produced fewer markers in the centromeres that were filled with AFLP markers.

There is wide flexibility in the design of SRAP primers. After testing the primers we used for gene cloning, we found that most of these primers worked well in SRAP amplification. In the construction of an ultra-dense genetic map in *B. napus*, we used 12 fluorescently labeled forward primers and 442 unlabeled primers to assemble a genetic map with 13,351 molecular markers (Sun et al., 2007). Based on the SRAP markers on the genetic map, the efficiency of each labeled primer was checked by counting the average numbers of SRAP markers produced by individual labeled primers. Some labeled SRAP primers such as FC1, BG23 and SA7 produced more than ten polymorphic loci while EM2 and DC1 produced less than six polymorphic loci. If we checked individual SRAP primer combinations, we found that there was a big difference in the numbers of polymorphic loci detected by individual SRAP primers. Efficient SRAP primer combinations produced over 20 polymorphic loci while less efficient SRAP primer combinations produced one to three mapped loci. Actually, we tested over a thousand SRAP primers in *B. napus* and *B. rapa* and found that most primers produced good profiles with over 30 strong bands (unpublished data). After testing a large set of SRAP primer combinations, we selected a set of SRAP primer combinations that showed the best performance in SRAP amplification in different accessions and populations in various Brassica species. These SRAP primer combinations serve as a standard set and are routinely used in genetic map construction, mapping of quantitative loci (QTL) and gene cloning in our lab. Therefore, if SRAP is frequently used in a lab, it is worthwhile to select a set of SRAP primer combinations which will enhance the effectiveness and efficiency of SRAP marker detection.

To enhance the capacity and effectiveness of SRAP technology, we combined SRAP with Illumina's Solexa sequencing to directly integrate genetic loci on the *B. rapa* genetic map based on paired-end Solexa sequences (Li et al., 2011). To achieve this objective, we used two rounds of PCR to prepare SRAP products that were pooled and sequenced with Illlumina's Solexa sequencing. The first round of PCR produced SRAP fragments using individual DNA samples from a mapping population with the same set of SRAP primer combinations while the second round was used to tail the SRAP PCR products that allowed identifying the original DNA for producing paired-end Solexa sequences. After Illumina's Solexa sequencing, paired-end sequences were sorted using the sequences of tag primers and the numbers of unique paired-end sequences from each DNA samples were obtained. These numbers represent Solexa sequence frequencies in each DNA sample which was used to integrate Solexa sequences onto the SRAP genetic map. To integrate paired-end sequences on the genetic map, we adopted a QTL mapping strategy by using Windows QTL Cartographer software 2.5. When a paired-end sequence was found to have only one significant LOD score in a bin of the genetic map, this sequence was assigned into this bin. In total, 1737 unique paired-end sequences representing the same number of genetic loci were integrated on the genetic map. Eventually, we constructed a high density *B. rapa* genetic map consisting of 1,737 paired-end Solexa sequences, 9,177 SRAP markers and 46 SSR markers.

3. GENETIC MAP CONSTRUCTION

Genetic maps are extensively used in gene mapping, QTL mapping and assembly of whole genome sequence. High density molecular markers in genetic maps are advantageous and necessary in most applications. The detection of multiple loci in a SRAP PCR reaction can be automated through fluorescently labelled SRAP primers so it is feasible to construct a high density genetic map using SRAP technology.

In *B. napus*, we used a five-color fluorescent dye set including '6-FAM', 'VIC', 'Pet', 'NED' and 'LIZ' to perform SRAP fragment analysis with the ABI genetic analyzer (Sun et al., 2007). 'LIZ' was used as the internal standard while the other four fluorescent dyes were used to label SRAP primers and combined with unlabeled primers. After obtaining SRAP products with four fluorescently labeled and four unlabeled primers, all the products from four SRAP primer combinations were pooled to increase the detection throughput by four fold. We used 1,634 SRAP primer combinations selected from 12 labeled and 442 unlabeled SRAP primers to produce 13,472 SRAP markers. Together with 79 SSR markers, we assembled currently the most saturated genetic map in *B. napus*.

Using cDNA-SRAP technology, we first constructed a transcriptome map based on *B. oleracea* cDNAs obtained from leaf tissue (Li et al., 2003). In cDNA-SRAP, one step PCR allows the amplification of single strand cDNAs after the first strand cDNA is synthesized using reverse transcriptase. Since most cDNA-SRAP markers come from differences in gene sequence, these markers are considered to be functional markers. SRAP products can be easily isolated from polyacrylamide gels for sequencing so we sequenced 190 fragments that corresponded to 190 polymorphic loci from cDNA-SRAP. Through analysis of sequence similarity, 169 out of 190 cDNA marker sequences were homologous to genes reported in Arabidopsis, which allowed the identification of extensive colinearity between the two genomes according to the gene-for-gene alignment. Later, we developed over 1,000 SRAP markers using genomic DNA from the same mapping population and assembled these SRAP markers from both genomic DNA and cDNA samples on the same genetic map (Gao et al., 2007). In addition, we integrated 10 SCAR markers using sequences of genes with known functions in the biosynthesis of glucosinolates and inflorescence architecture, and one SCAR marker flanking a resistance gene to downy mildew.

SRAP technology can be combined with other markers to construct genetic maps. For example, Yu et al., (2007) constructed a high-density genetic map in a cultivated allotetraploid cotton population using SSR, SRAP, AFLP, and target region amplification polymorphism (TRAP). This high density cotton genetic map consists of 697 SSR, 171 TRAP, 129 SRAP, 98 AFLP, and two morphological markers, covering a genetic distance of 4,536.7 cM with the average genetic distance of 4.1 cM per marker. Gulsen et al., (2010) reported a new citrus linkage map using SRAP, RAPD, SSR, ISSR, peroxidase gene polymorphism (POGP), resistant gene analog (RGA), and a morphological marker, Alternaria brown spot resistance gene. In total, they assembled 385 SRAP, 97 RAPD, 95 SSR, 18 ISSR, 12 POGP, and 2 RGA markers on the citrus genetic map.

In the Cucurbitaceae family, Yeboah et al., (2007) constructed genetic maps in cucumber using SRAP and ISSR markers. They developed pseudo-testcross F1 segregating populations from a cross between two diploid parents and constructed male and female parental genetic maps separately with 164 SSR and 108 SRAP markers. More recently, Zhang et al., (2012) constructed a high density consensus genetic map in an inter-subspecific mapping population in cucumber. The consensus map contained over a thousand molecular markers including 1,152 SSR, 192 SRAP, 21 SCAR and one STS. In another cucurbit species, Levi et al., (2006) constructed an extended genetic map for watermelon using five PCR-based molecular markers SRAP, AFLP, SSR, ISSR and RAPD. As suggested by the authors, low polymorphism is often observed in watermelon cultivars, combining several marker systems is necessary to construct a high density genetic map covering the whole genome.

SRAP markers have been used to construct genetic maps in a wide range of plant species. In Dendrobium plants that are used as Chinese herbs, Xue et al., (2010) constructed two genetic maps in two Dendrobium species, *D. officinale* and *D. hercoglossum* with a double pseudo-testcross strategy using SRAP and RAPD methods. In root plants, Chen et al., (2010) constructed a genetic map in an F_1population derived from an interspecific cross in cassava by combining AFLP, SSR, SRAP and expressed sequence tag (EST)-SSR markers. In total, they assembled 355 markers into 18 linkage groups covering a genetic distance of 1,707.9 cM, which served as a foundation for QTL mapping in this species. In grass species, Xie et al., (2011) used SSR and SRAP markers to construct two genetic maps of male and female parental lines respectively in diploid orchardgrass (*Dactylis glomerata* L.) using a pseudo-test cross strategy. In total, they assembled 164 SSR markers and 108 SRAP markers on these two genetic maps. In a fruit tree, Luohanguo (*Siraitia grosvenorii* C. Jeffrey), Liu et al., (2011) used SRAP and ISSR markers to assemble a genetic map consisting of 170 SRAP markers and 29 ISSRs in 25 linkage groups. In a fiber crop, Chen et al., (2011) used SRAP, ISSR and RAPD markers to construct a genetic map in Kenaf (*Hibiscus cannabinus* L.) that is one of the most economically important fiber crops globally.

4. QTL MAPPING

A common application of genetic maps is QTL mapping of complex traits. Since QTL are often underpinned by multiple genes in a genome, it is difficult to tag QTL using procedures for tagging Mendelian loci. In general, if a complex trait is changed into a simple Mendelian trait and the underlying QTL is Mendelized, many strategies are available to map and clone the Mendelized genes. In fact, most cloned QTL have been accomplished through such a Mendelized strategy by developing near-isogenic lines (NILs). However, most QTL are not easily Mendelized, so it is necessary to construct genetic maps first and then perform QTL mapping.

In canola, we used SRAP and SSR to construct a genetic map in a doubled haploid (DH) line population that was developed from a synthetic yellow-seeded line and a conventional canola cultivar through microspore culture (Chen et al., 2009). Data for three complex traits including days to flowering, oil content and seed yield at three locations for three years were collected and used in QTL mapping. For oil content, 27 QTL on 14 linkage groups and for seed yield, 18 QTL on 11 linkage groups were identified while days to flowering was suggested to be controlled by a single genetic locus in this mapping population. In rapeseed, Chen et al., (2007) used 208 SSR and 189 SRAP markers to construct a genetic map for a DH line population and performed QTL mapping of yield-related traits in *B. napus*. They also developed a fixed immortalized population from randomly permutated intermating of these DH lines. They collected data for six yield-related traits, plant height, height of lowest primary effective branch, length of main inflorescence, silique length, number of primary branches and silique density. After QTL mapping in the DH line and immortalized populations, they identified 29 common QTL between the two populations, suggesting that there are some chromosomal regions containing QTL for multiple traits. In another QTL mapping report, Fu et al., (2007) constructed a genetic map in *B. napus* to map gene loci controlling the yellow seeded trait. They developed 420 SSR, RAPD and SRAP markers and assembled two genetic maps, of which one contained 26 linkage groups and another which had 20 linkage groups. After QTL analysis, they identified 19 QTL with one common in the two mapping populations. Further analysis allowed them to identify the collinear genomic region of chromosome 5 in Arabidopsis.

In cotton, Lin et al., (2005) developed a mapping population by crossing *G. hirsutum* and *G. barbadense* and performed QTL mapping with a genetic map constructed with SRAP, SSR and RAPD markers. Using 437 SRAP, 107 RAPD and 205 SSR markers, they constructed a genetic map with 566 markers assembled into 41 linkage groups, of which 28 were assigned to the corresponding known chromosomes. This genetic map was used to perform QTL mapping for fiber traits. In total, 13 QTL for fibre traits including two QTL for fibre strength, four QTL for fibre length and seven QTL for micronaire value were identified, of which six QTL were assigned into the A-subgenome, another six QTL into the D-subgenome while one QTL was not assigned. Three QTL for micronaire value were identified to cluster on linkage group 1, suggesting that the flanking molecular markers of these three QTL might be useful in marker-assisted selection for this trait.

Similarly, Zhang et al., (2009b) reported on QTL mapping in cotton using SRAP and other markers. They assembled a genetic map containing 509 SSR, 58 intron targeted intron/exon splice junction (IT-ISJ), 29 SRAP and 8 morphological loci in 60 linkage groups. Among these 60 linkage groups, 54 were assigned into 26 chromosomes. This genetic map was used to identify QTL for fiber quality traits in five environments. In total, thirteen QTL including four QTL for fiber length, two QTL for fiber strength, two QTL for fiber fineness, three QTL for fiber length uniformity, and two QTL for fiber elongation were identified. Eleven out of 13 QTL were assigned into the A-subgenome and other two QTL, into the D-subgenome.

In chrysanthemum (*Dendranthema morifolium*), Zhang et al., (2011b) performed QTL mapping of inflorescence-related traits using SRAP markers. They constructed two genetic maps in a F_1segregating population using a double pseudo-testcross mapping strategy. With 500 SRAP primer combinations, they produced 896 polymorphic loci and assembled 333 SRAP markers into 57 linkage groups on one genetic map, 342 SRAP markers into 55 linkage groups on the second genetic map. The results indicated that the distribution of these SRAP markers on these two genetic maps was quite uniform. Using these two genetic maps, they mapped 12 QTL for three inflorescence traits, of which each four QTL underpinned specified flower diameter, ray floret layer number, and ray floret length, respectively.

In radish, Xu et al., (2012) recently constructed a genetic map with 592 molecular markers including 287 SRAP, 135 RAPD, 78 SSR, 49 ISSR, 29 randomly amplified microsatellite polymorphism (RAMP), and 14 resistant gene analogs (RGA). They used this genetic map to analyze QTL that controlled root cadmium accumulation. They mapped four QTL on linkage groups 1. 4. 6 and 9. The QTL on linkage group 9 was a major one that accounted for 48.64% of phenotypic variance, suggesting that this QTL might be applied for marker assisted selection to improve radish root quality by reducing cadmium concentration.

5. GENE TAGGING AND CLONING

SRAP technology has several merits for gene tagging. Since SRAP detection uses unlimited primer combinations and there are multiple loci detected in a single SRAP PCR reaction, SRAP technology is advantageous over other molecular marker systems for gene tagging. After many genetic loci in a genome are screened quickly, closely linked SRAP markers to a trait of interest can be identified easily. We intensively used SRAP to perform gene tagging and cloning in Brassica species and worked on several economically important traits such as yellow-seeded canola and rapeseed, disease resistance and glucosinolates.

Yellow-seeded oilseeds in Brassica species are suggested to be related to high oil content so it is worthwhile to characterize the genes controlling seed coat color. Using SRAP technology, we cloned and characterized a gene controlling seed coat color and plant hairiness traits in *B. rapa* (Zhang et al., 2009a). We used 1,100 SRAP primer combinations to screen pooled DNA from yellow-seeded and black-seeded individuals based on the bulk segregant analysis (BSA) strategy and found 48 SRAP primer combinations that produced polymorphic loci in the pooled DNA samples from yellow-seeded and black-seeded individuals. Then we tested more pooled DNA and identified 13 SRAP markers that were linked to the gene of interest. Sequencing these SRAP markers allowed the identification of a chromosomal region that was further used to develop new SCAR markers. With new SCAR markers and chromosome walking, we eventually identified the candidate gene and characterized the gene by complementary transformation of the corresponding mutant in Arabidopsis. The functional copy of the candidate gene recovered the phenotype of the Arabidopsis mutant and the non-functional copy in *B. rapa* mutant did not so the candidate was confirmed to underpin the yellow-seeded mutation in *B. rapa*.

Similarly, we used SRAP to tag other genes controlling the seed-coat color trait in yellow sarson, another yellow-seeded *B. rapa* and also in yellow-seeded *B. napus* canola (Rahman et al., 2007, 2010). Yellow sarson is bright yellow-seeded and there is no color variation under different environments while all yellow-seeded canola accessions developed with yellow sarson are not pure yellow and the seed coat color varies due to differences in maturity and environments. Using SRAP screening, we identified several SRAP markers linked to one seed coat color gene. After extended flanking regions were sequenced, one closely-linked SRAP marker was successfully converted into SNP and SCAR markers. Meanwhile, we analyzed a yellow-seeded canola line that was developed with yellow sarson and found that three genes controlled the yellow-seeded color in this *B. napus* line (Rahman et al., 2010). We identified one SRAP marker that was linked closely to one seed color locus and confirmed that this locus was located on linkage group N9 of our ultradense genetic map of *B. napus*. The second locus was mapped on linkage group N13 of our ultradense genetic map. To identify SRAP markers for the third locus, we screened 768 SRAP primer combinations and eventually found one SRAP that was linked closely to the yellow seeded color locus. These SRAP, SNP and SCAR markers can be used in marker assisted selection of yellow-seeded trait in oilseed crops of *B. rapa* and *B. napus*.

SRAP technology is an effective molecular marker system to analyze qualitative and quantitative resistance to plant diseases. In general, qualitative and quantitative resistances are conferred by oligogenic or multigenic loci, respectively. In canola, blackleg is a major disease and qualitative resistance is available. We used the previously described ultradense genetic map to tag resistance genes to blackleg in *B. napus*. After screening 384 SRAP primer combinations, we identified two SRAP markers that were linked to a blackleg resistance gene. By compared the linked SRAP markers with the molecular markers on the ultradense *B. napus* genetic map that was constructed with another mapping population, we found that one SRAP marker corresponded to a SRAP marker on N10. Therefore we took the flanking SRAP markers of the mapped resistance locus on N10 and identified other SRAP markers on the genetic map that were also polymorphic in the mapping population of the blackleg resistance gene. Eventually, further analysis allowed us to identify two blackleg resistance genes in the region where one resistance gene was suggested by other researchers (Long et al., 2010).

In several reports, SRAP markers were used to map genes controlling resistance to plant diseases in several crop species. For instance, Yi et al., (2008) used SRAP, STS and SSR markers to tag a resistance gene (*Pm4b*) to powdery mildew in wheat. They tested 240 SRAP primer combinations and identified two SRAP markers linked to the *Pm4b* gene. Eventually, they mapped the *Pm4b* gene on chromosome 2AL that was flanked by SRAP, STS and SSR markers. In another study on gene mapping in wheat, Chen et al., (2012) used SSR, SRAP and TRAP markers to tag a wheat strip rust resistance gene. Using 400 SSR, 315 pairs of SRAP primers, and 40 pairs of TRAP primers to screen F_1, F_2 and BC_1 mapping populations, they constructed a fine map flanking the resistance gene locus on chromosome arm 2AS and suggested that the mapped resistance gene should be a novel one.

In rice, Zhao et al., (2010) searched for SSR markers linked to a dominant resistance gene (RSV1) to rice stripe virus and then used the SRAP method to find closely linked markers. They located RSV1 into a region flanked by SSR and SRAP markers. In maize, a new dominant resistance gene to maize head smut was tagged by SSR-BSA and SRAP-BSA methods (Li et al., 2012). Closely linked molecular markers were identified and used to transfer the resistance gene from the resistant source to elite lines via marker assisted selection to breed head smut resistant hybrid cultivars in maize.

In eggplant, Mutlu et al., (2008) tagged a Fusarium wilt resistance gene using SRAP, SRAP-RGA, RAPD, and SCAR markers. They used 2316 primer combinations to identify molecular markers linked to the resistance gene, of which two SRAP markers were closely linked to the resistance gene. The

SRAP markers were converted into SCAR markers and used in marker assisted selection of the Fusarium wilt resistance in eggplant.

Besides plant disease resistance, genes underpinning other traits have been tagged using SRAP technology, For instance, genes controlling two important traits, sex determination and tuberculate fruit in cucumber were tagged using SRAP technology (Li et al., 2008; Zhang et al., 2010a). In cucumber, there are three major gene loci, F/f, M/m, and A/a that determine various sex types. Li et al., (2008) analyzed M/m gene locus and identified 8 SRAP markers linked to this gene locus. Additionally, they used SRAP markers to perform chromosome walking and converted some SRAP markers to co-dominant SCAR markers through sequencing SRAP fragments. Eventually they identified very closely linked SRAP markers at a genetic distance of less than one cM. Similarly, Zhang et al., (2010a) performed gene tagging of cucumber tuberculate fruit. They found that the tuberculate fruit (Tu) was controlled by a dominant gene and used a BSA strategy to identify molecular markers linked to this dominant gene. After testing 736 SRAP primer combinations, they found 9 SRAP markers that were linked to the Tu gene and used SSR markers to anchor this gene on chromosome 5, further indicating that they would use the mapping results to clone the Tu gene later.

Male sterility is a commonly used method to produce hybrid seeds for exploiting heterosis in crops. Since genic male sterility is usually controlled by a few genes, SRAP technology is useful to tag the genes underpinning male sterility. For example, Zhang et al., (2011c) used SRAP and SSR markers to tag a dominant genic male sterile gene in *B. oleracea*. They performed BSA analysis with SRAP and SSR markers. By screening polymorphisms between fertile bulks and sterile bulks with 26,417 SRAP primer pairs, they identified 14 SRAP markers that were linked to the male sterility gene MS-cd1. After sequencing the SRAP fragments, three SRAP markers were converted into SCAR markers that were very closely linked to the MS-cd1 gene. Moreover, through comparative genomics with SRAP sequences, they identified a collinear region on chromosome A10 in *B. rapa* corresponding to a collinear genomic region of chromosome 5 in Arabidopsis, which could lead to cloning of this gene in the future.

SRAP technology has also been used to tag quantitative traits using the same approaches as described previously in qualitative traits. In alfalfa, Castonguay et al., (2010) used SRAP to identify polymorphic genetic loci that controlled superior tolerance to freezing. Through BSA analysis, they found four SRAP markers that were associated with freezing tolerance and the frequency of their occurrence reflected changes in response to selection. In another report, SRAP was used to tag a major QTL controlling cadmium accumulation in oat (Tanhuanpaa et al 2007). The concentration of toxic cadmium in oat grains is often over the accepted limit and must be reduced. SRAP, RAPD and retrotransposon-microsatellite amplified polymorphism (REMAP) markers were used to perform BSA analysis in an F2 population and four molecular markers were identified to be associated with cadmium concentration in oat grains. All these four markers were located on the same linkage group, suggesting that this mapped QTL had major effect on grain cadmium concentration in oat.

6. GENETIC DIVERSITY

Genetic diversity analysis is necessary in plant breeding, plant systematics and evolution, plant pathology. SRAP is an adequate molecular marker system for genetic diversity analysis in plants and fungi. Since SRAP has many features such as simplicity, reliability, flexibility, detection of multiple loci and cost-effectiveness, which allows beginners and experienced people to perform SRAP routinely with limited facilities or in well-established genomics labs. Since genome sequence information is not necessary for SRAP detection, SRAP can be used to perform genetic diversity analysis in a wide range of living organisms. We first used SRAP to analyze the genetic diversity of parental lines that were used to produce hybrid cultivars in *B. napus* (Riaz et al., 2001). As expected, we found that there was a positive correlation of genetic distance and hybrid performance. In celery, we used SRAP to tag a major resistance locus to celery mosaic virus (Ruiz et al., 2001).

In melons, Ferriol et al., (200) used SRAP and AFLP to analyze 69 accessions selected from morphotypes and unclassified types that belong to two subspecies, *Cucurbita pepo* ssp. *pepo* and ssp. *ovifera*. Among these accessions, some commercial cultivars and Spanish landraces represent diversified types in Europe. Their results showed that SRAP markers characterized well the morphological variability and the evolutionary history of these tested accessions better than AFLP markers. Molecular markers were used to identify new types for the development of new cultivars. The genetic diversity in the landraces of *C. pepo* spp. *ovifera* was detected with molecular markers, which is useful for preserving the diversity in this species.

In grasses, Budak et al., (2004a; 2004b; 2005) used SRAP to analyze genetic diversity and ploidy complexity in buffalograss. They found that SRAP markers were abundant and that they could distinguish genetic diversity among closely related cultivars. Their data showed that among several molecular markers (SSRs, ISSRs, SRAPs, and RAPDs), SRAP estimated the highest mean genetic dissimilarities in buffalograss. Additionally, they used SRAP and other markers to analyze ploidy complex and geographic origin of the *Buchloe dactyloides* genome and identified a significant correlation between the ploidy levels such as diploid, tetraploid, pentaploid, and hexaploid and the numbers of alleles detected using nuclear DNA markers. SRAP again was the best one among three molecular markers (ISSR, SSR, and SRAP, r = 0.39, 0.39, and 0.41).

Similarly, Gulsen et al., (2009) used SRAP, peroxidase gene polymorphism (POGP), ISSR and RAPD to study the relationship of ploidy levels, geographic locations and genetic diversity in bermudagrass. They found that there was a significant correlation between ploidy levels in diploids, triploids, tetraploids, pentaploids, and hexaploids and band frequencies of molecular markers ($r = 0.62$, $P < 0.001$), suggesting that ploidy levels resulted in genome variation and genetic diversity. Geographic locations of Cynodon accessions also contributed to genetic diversity based on molecular marker analysis. They suggested that combining several molecular markers would be more efficient to evaluate genetic diversity and genetic structure in bermudagrass and eventually broaden genetic basis for developing new cultivars.

In elephant grass, Xie et al., (2009) used SRAP markers to study the genetic diversity and relationships of commonly used cultivars in China. They generated 1,395 genetic loci with 62 SRAP primer combinations with an average of 22.5 genetic loci per primer combination. They found that SRAP loci were very polymorphic (72.8%) and used these SRAPs to estimate the genetic diversity within and between elephant grass cultivars. The results showed the genetic diversity within cultivars was less than that among tested cultivars and the relationship of those tested cultivars was also estimated.

In cereal crops, Zaefizadeh and Goleiv (2009) analyzed genetic diversity and relationships among durum wheat landraces by SRAP marker and phenotypic differences. They used 65 SRAP markers and 27 traits to perform cluster analysis of 40 subconvars of *Triticum durum* landraces from the region of North West Iran and Azerbaijan. Traits failed to detect any geographic association in durum landraces while 12 combinations of SRAP markers were distinguishable among these landraces, suggesting that SRAP technology is useful for genetic diversity and evolutionary relationship analysis, marker assisted selection and genetic map construction in durum wheat. Yang et al., (2010) used SRAP markers to analyze the genetic diversity of hulless barley cultivars from Sichuan, Gansu, Tibet, Qinghai and Yunnan provinces of the Qinghai-Tibet Plateau in China. With 20 SRAP primer combinations, they detected 153 polymorphic loci and used these SRAP markers to classify 68 hulless barley accessions into four major groups using a unweighted pair-grouping method with arithmetic averages (UPGMA) analysis. They concluded that SRAP was an effective method to perform genetic diversity in hulless barley and develop new cultivars.

In rice, Dai et al., (2012) developed indica- and japonica-specific markers using SRAP, TRAP, and SSR markers and performed genetic diversity analysis of Asian *Oryza sativa* varieties. In general, rice varieties are classified into *O. sativa* ssp. *japonica* kata and ssp. *indica* kata. In this report, they used 45 rice varieties in a cultivated and wild rice collection to study the genetic diversity in rice. By developing 90 indica- and japonica-specific genetic loci, they could easily distinguish typical indica and japonica subspecies and determined whether a domesticated rice variety came from the indica or japonica type.

In alfalfa, Vandemark et al.,(2006) used SRAP markers to analyze genetic relationships among historical sources of alfalfa germplasm in North American. Their results showed that SRAP detected highly polymorphic loci (>90%) in alfalfa, which distinguished nine original sources of Medicago germplasm based on genetic similarity calculated with SRAP markers. They suggested that SRAP technology is an adequate marker system for detecting polymorphisms in alfalfa.

In sesame, Zhang et al., (2010b) performed genetic diversity analysis using SRAP and SSR markers. They analyzed 404 landraces from a sesame collection in China. Using11 SRAP and 3 SSR markers, they produced 175 fragments, of which 126 were polymorphic with an average polymorphism rate of 72%. They calculated several parameters such as Jaccard's genetic similarity coefficients, Nei's gene diversity and Shannon's information index and constructed a dendrogram with all the 404 landraces. According to the dendrogram, landraces from different agro-ecological zones did not cluster together, suggesting that geographical locations did not represent the greater genetic variation among the sesame landraces. They concluded that SRAP and SSR markers would be useful to study sesame genetic diversity and understand the relationship of those indigenous landraces, which would guide the collection, protection and utilization of sesame landraces in breeding purposes.

In banana and plantain, Youssef et al., (2011) used SRAP and AFLP markers to analyze 40 Musa accessions including commercial cultivars and wild species. They developed 353 SRAP and 787 AFLP markers to perform cluster analysis using an unweighted pair-grouping method with arithmetic averages (UPGMA) and principal coordinate (PCO) analysis. They eventually assigned all the 40 accessions into corresponding *Eumusa*, *Australimusa*, *Callimusa* and *Rhodochlamys* sections and species. They found that SRAP and AFLP polymorphism amongst sections and species and the relationships within *Eumusa* species and subspecies were not consistent and suggested that SRAP produced threefold more specific and unique loci than AFLP. Therefore, the data showed that SRAP markers were effective to distinguish *M. acuminata, M.*

balbisiana and *M. schizocarpa* in the *Eumusa* section, and also triploid plantains and cooking bananas.

In grape, Guo et al., (2012) used SRAP markers to study genetic variability and relationships of cultivated wine-type *Vitis vinifera* and wild *Vitis* species. They selected 76 grape genotypes representing indigenous and new varieties and wild *Vitis* species from China and other countries. After testing 100 SRAP primer combinations, they selected 19 primer combinations based on primer performance to produce 228 genetic loci, of which 78.63% were polymorphic with an average polymorphism information content value of 0.76. The SRAP markers were used to perform cluster analysis to evaluate Nei and Li's similarity coefficients by unweighted pair-group method of arithmetic averages (UPGMA) analysis. Additionally, they performed principal coordinate analysis (PCoA) to plot all 76 grape genotypes which showed a similar cluster pattern to that in the dendrogram, representing their geographical origins and taxonomic classification of these grape varieties. All the results indicated that three main groups including table grape of *V. vinifera*, table grape of Euro-America hybrids and wine grape of *V. vinifera*, wild *Vitis* species were identified and also the table *V. vinifera* group was genetically different from the wine-type *V. vinifera* and wild *Vitis* species originated from America and China. So they suggested that SRAP markers are informative and grape germplasm in China contains abundant genetic diversity.

In medicinal plants, Ortega et al., (2007) analyzed genetic diversity of cultivated and non-cultivated mashua, *Tropaeolum tuberosum* that were grown in six communities in the Cusco region of Perú and selected from the germplasm collection at the International Potato Center (CIP) using SRAP markers. Mashua is used as a medicinal plant, possibly due to a high concentration of glucosinolates in mashua roots. DNA fingerprinting generated by SRAP markers showed that mashua is a genetically variable crop. The genetic analysis also showed that most non-cultivated accessions were likely feral races resulting from escape from cultivation rather than wild relatives. In another medicinal plant, Wang et al., (2012) analyzed the genetic diversity of 35 wild goat's rue accessions (*Galega officinalis* L.) collected from Russia and Europe countries using ISSR and SRAP markers. Although there was some discrepancy between ISSR and SRAP markers, the clustering patterns of genotypes were relatively consistent between these two kinds of molecular markers in this study. They indicated that both markers were useful for goat's rue germplasm characterization, improvement, and conservation.

In ornamental plants, Hao et al., (2008) used SRAP technology to perform genetic diversity analysis of 29 ornamental and medicinal Paeonia. Dendrogram and principle component analysis indicated that SRAP markers well characterized the genetic relationships of these 29 peony cultivars, which is useful to guide parent selection and molecular marker assisted selection in Paeonia breeding. In another ornamental plant, Feng et al., (2009b) performed genetic analysis of diversity and population structure of *Celosia argentea* and related species using SRAP markers. They included 16 populations of *C. argentea*. and 6 populations of *C. cristata*. from China. Using 10 SRAP primer combinations, they produced 507 scored bands, of which 274 were polymorphic. With UPGMA cluster analysis, they constructed a phylogenetic tree and calculated genetic distances of all 22 populations. The results showed that the genetic distances of all populations were coincident with their geographic origins. Additionally, they identified one SRAP marker separating accessions in *C. argentea* from those in *C. cristata* and suggested that the extensive genetic diversity in *C. argentea* populations would be very useful for breeding and conservation of *C. argentea* varieties in the future.

In chrysanthemum, Zhang et al., (2011a) did a genetic diversity study on two flowering traits of chrysanthemum, initial blooming time and the duration of flowering. They identified two pairs of major genes with high levels of inheritance. Using SRAP technology, they performed association mapping of these two traits and identified SRAP markers that were significantly associated with phenotypes, suggesting that SRAP markers might be useful in chrysanthemum breeding. In another report on ornamental plants, Soleimani et al., (2012) used wild, cultivated, and ornamental pomegranates (*Punica granatum* L.) in Iran to perform analysis of genetic diversity and population structure with SRAP molecular markers. They produced 133 SRAP markers with 13 SRAP primer combinations to evaluate the genetic diversity of 63 pomegranate genotypes from five different geographical regions of Iran. Their data showed that the average polymorphism information content value was 0.28 and the genetic distance was 0.10 to 0.37 with an average of 0.24 in all 63 genotypes. Cluster analysis allowed them to identify the relationship between ornamental and wild genotypes. They found that the genetic variation of genotypes from various regions was bigger than that of intra regions. They concluded that SRAP markers could be an effective marker system in the analysis of genetic diversity and population structure in pomegranate.

In woody plants, Li et al., (2010) did genetic diversity analysis of sea buckthorn which is grown as a nutritious berry crop. They produced 191 polymorphic loci using SRAP technology to perform cluster analysis of 77 accessions, of which 73 *Hippophae rhamnoides* were classified into 2 groups and 4 *H. salicifolia*, into 1 group. They associated SRAP markers with dried-shrink disease (DSD) resistance and suggested SRAP markers are useful for breeding new sea buckthorn lines with resistance to DSD. Feng et al., (2009a) reported on the genetic diversity analysis of *Pinus koraiensis* using SRAP markers. They obtained 24 to 33 loci per primer combination and used 143 SRAP markers to analyze 480 samples collected from 24 provinces in China. They found that there was no significant difference in genetic diversity among provinces. However, genetic variation of intra population accounted for 93.355% of the total variation.

In fungi, Sun et al., (2006) used SRAP markers to classify *Ganoderma lucidum* strains. They performed genetic diversity analysis with 31 accessions collected from several countries. Using 75 polymorphic loci, they classified all 31 accessions into five groups. The results showed that *G. lucidum* strains were significantly different from *G. sinense* and *G. lucidum* in China, also different from *G. lucidum* in Yugoslavia. They suggested that SRAP markers are useful in taxonomy and systematics of Ganoderma strains within basidiomycetes. In another fungus, Tang et al., (2010) analyzed Chinese *Auricularia auricula* strains using SRAP and ISSR markers. They found both SRAP and ISSR markers were abundant in *A. auricula* and could be used to effectively distinguish all tested strains. After phylogenetic analysis, they classified 34 *A. auricula* strains into four or five major groups using the UPGMA method. They suggested that genetic diversity information would be used in *A. auricula* breeding programs to develop new medicinal mushroom. Fu et al., (2010) performed genetic diversity analysis in 23 elite *Lentinula edodes* strains from China using RAPD, ISSR and SRAP markers. In total, they used 16 RAPD primers, 5 ISSR primers and 23 SRAP primer combinations to produce 138, 77 and 144 bands, respectively. After UPGMA clustering analysis, they classified all 23 *L. edodes* strains into three or four groups. However, all groups showed high levels of similarity, showing a low level of genetic diversity in all tested strains.

7. OTHER APPLICATIONS

SRAP amplification is actually a small portion of all possible sampling of a genome. So SRAP can be used to produce a reduced genome samples when multiple SRAP reactions are pooled. As described previously, pooled SRAP produces can be directly sequenced using next generation sequencing technologies. When replacing genomic DNA with cDNA samples, SRAP is adequate to perform gene expression profiling and also construct cDNA genetic maps.

More recently, Yu et al., (2012) used SRAP markers to distinguish fertile somatic hybrids of *G. hirsutum L.* and *G. trilobum* produced by protoplast fusion. They obtained fertile somatic hybrids by symmetric electrofusion of protoplasts of tetraploid upland cotton *G. hirsutum* and wild cotton *G. trilobum*. These hybrids were confirmed using morphological characteristics, flow cytometric analysis, and molecular markers including RAPD, SRAP and AFLP.

In aquaculture, Ding et al., (2010) used SRAP and SCAR markers to differentiate two cultured populations in grass carp (*Ctenopharyngodon idella*). Through cloning and sequencing SRAP fragments, they developed SCAR markers to characterize individuals from the cultured population and the wild population, showing different frequencies of SCAR alleles (87% in the cultured population and 6% in the wild), suggesting that this SCAR might serve as a specific molecular marker for the cultured population. They also identified eight SRAP fragments that shared high similarities to functional genes.

8. SUMMARY REMARKS

SRAP was first used to construct a genetic map and tag genes in *Brassica oleracea* in 2001 (Li and Quiros, 2001). This molecular marker technology is simply performed with one round PCR to amplify multiple or occasionally over a hundred loci in a genome. In its PCR reaction mixture, two random primers are included, which leads to maximum flexibility in primer designing and primer labelling. There is no limitation on primer combinations and one labelled primer may be combined with any number of unlabelled primers. Most SRAP products fall into a size range of 100 to 1000 base pairs, which can be separated in both polyacrylamide and agarose gels. In automatized detection, one SRAP primer is fluorescently labelled and SRAP products can be analyzed using advanced instruments such as an ABI genetic analyzer, which dramatically increases throughput of SRAP molecular marker detection.

There is a wide range of applications of SRAP technology such as genetic map construction, genetic diversity analysis, gene tagging and cloning. Since SRAP detects multiple loci in one reaction, it is feasible to construct ultradense genetic maps with over 10,000 SRAP molecular markers. SRAP has advantages over other molecular detection techniques in gene tagging and cloning and allows screening thousands of loci shortly to pinpoint the genetic position underlying the trait of interest. Sequencing SRAP products enhances the applications of SRAP technology. In well characterized genomes, SRAP sequences are used to identify the chromosomal region of mapped genes while in species without a known whole genome sequence, sequences of SRAP markers on a genetic map allow arranging sequence contigs and assembly of a whole genome sequence.

SRAP molecular technology is very useful in plant breeding. In QTL mapping, common QTL for the same trait of interest can be effectively identified. Since SRAP has a high throughput feature, multiple mapping populations can be analyzed effectively to construct several genetic maps. In addition, the same set of SRAP primers allows detection of the same genetic loci, which can used to align several genetic maps. SRAP is effective and efficient in marker assisted selection in plant breeding since thousands of samples can be analyzed inexpensively. SRAP technology has been commonly used in analysis of genetic diversity of many plant species. Currently, SRAP are used in most crops, tree species, ornamental and medicinal plants.

REFERENCES

1. H. R. C. Budak, I. Shearman, R. E. Parmaksiz, T. P. Gaussoin, D. Riosdan, 2004aMolecular characterization of Buffalograss germplasm using sequence-related amplified polymorphism markers. *Theor Appl Genet.* 108328334
2. H. Budak, R. C. Shearman, I. Parmaksiz, I. Dweikat, 2004bComparative analysis of seeded and vegetative biotype buffalograsses based on phylogenetic relationship using ISSRs, SSRs, RAPDs, and SRAPs. *Theor Appl Genet.* 1092808
3. H. Budak, R. C. Shearman, O. Gulsen, I. Dweikat, 2005Understanding ploidy complex and geographic origin of the *Buchloe dactyloides* genome using cytoplasmic and nuclear marker systems. *Theor Appl Genet.* 111154552
4. Y. Castonguay, J. Cloutier, A. Bertrand, R. Michaud, S. Laberge, 2010SRAP polymorphisms associated with superior freezing tolerance in alfalfa (*Medicago sativa* spp. *sativa*). *Theor Appl Genet.* 12016119
5. G. Chen, J. Geng, M. Rahman, X. Liu, J. Tu, T. Fu, G. Li, P. B. E. Mc Vetty, M. Tahir, 2009Identification of QTL for oil content, seed yield, and flowering time in oilseed rape (*Brassica napus*). *Euphytica* 142161174
6. M. Chen, C. Wei, J. Qi, X. Chen, J. Su, A. Li, A. Tao, W. Wu, 2011Genetic linkage map construction for kenaf using SRAP, ISSR and RAPD markers. *Plant Breed.* 130679687
7. S. Chen, G. Chen, H. Chen, Y. Wei, W. Li, Y. Liu, D. Liu, X. Lan, Y. Zheng, 2012Mapping stripe rust resistance gene YrSph derived from *Tritium sphacrococcum* Perc. with SSR, SRAP, and TRAP markers. *Euphytica* 1851926
8. W. Chen, Y. Zhang, X. Liu, B. Chen, J. Tu, T. Fu, 2007Detection of QTL for six yield-related traits in oilseed rape (*Brassica napus*) using DH and immortalized F(2) populations. *Theor Appl Genet.* 11584958
9. X. Chen, Z. Xia, Y. Fu, C. Lu, W. Wang, 2010Constructing a genetic linkage map using an F1 population of non-inbred parents in cassava (*Manihot esculenta* Crantz). *Plant Molecular Biology Reporter* 28676683
10. X. Dai, Y. Yang, L. Zhou, L. Ou, M. Liang, W. Li, G. Kang, B. Chen, 2012Analysis of indica- and japonica-specific markers of *Oryza sativa* and their applications. *Plant Systematics and Evolution* 298287296
11. W. Ding, Z. Cao, L. Cao, 2010Molecular analysis of grass carp (*Ctenopharyngodon idella*) by SRAP and SCAR molecular markers. *Aquaculture International* 18575587
12. F. Feng, M. Chen, D. Zhang, X. Sui, S. Han, 2009aApplication of SRAP in the genetic diversity of *Pinus koraiensis* of different provenances. *African Journal of Biotechnology* 810001008
13. N. Feng, Q. Xue, Q. Guo, R. Zhao, M. Guo, 2009bGenetic diversity and population structure of *Celosia argentea* and related species revealed by SRAP. *Biochem Genet.* 4752132

14. M. Ferriol, B. Pico, F. Nuez, 2003Genetic diversity of a germplasm collection of *Cucurbita pepo* using SRAP and AFLP markers. *Theor Appl Genet.* 107271282
15. F. Y. Fu, L. Z. Liu, Y. Chai, L. Chen, T. Yang, M. Jin, A. , X. Yan, Z. Zhang, J. Li, 2007localization of QTLs for seed color using recombinant inbred lines of *Brassica napus* in different environments. *Genome* 50840854
16. L. Fu, H. Zhang, X. Wu, H. Li, H. Wei, Q. Wu, L. Wang, 2010Evaluation of genetic diversity in *Lentinula edodes* strains using RAPD, ISSR and SRAP markers. *World Journal of Microbiology and Biotechnology* 26709716
17. M. Gao, G. Li, B. Yang, D. Qiu, M. Farnham, C. F. Quiros, 2007High-density *Brassica oleracea* map: Identification of useful new linkages. *Theor Appl Genet.* 115277287
18. O. Gulsen, S. Sever-Mutlu, N. Mutlu, M. Tuna, O. Karaguzel, R. C. Shearman, T. P. Riordan, T. M. Heng-Moss, 2009Polyploidy creates higher diversity among *Cynodon* accessions as assessed by molecular markers. *Theor Appl Genet.* 118130919
19. O. Gulsen, A. Uzun, I. Canan, U. Seday, E. Canihos, 2010A new citrus linkage map based on SRAP, SSR, ISSR, POGP, RGA and RAPD markers. *Euphytica* 173265277
20. D. Guo, J. Zhang, C. Liu, G. Zhang, M. Li, Q. Zhang, 2012Genetic variability and relationships between and within grape cultivated varieties and wild species based on SRAP markers. *Tree Genetics & Genomes* 8789800
21. Q. Hao, Z. Liu, Q. Shu, R. Zhang, J. D. Rick, L. Wang, 2008Studies on *Paeonia* cultivars and hybrids identification based on SRAP analysis.*Hereditas* 1453847
22. N. Jones, H. Ougham, H. Thomas, I. Pasakinskiene, 2009Markers and mapping revisited: finding your gene. *New Phytol.* 18393566
23. A. Levi, C. E. Thomas, T. Trebitsh, A. Salman, J. King, J. Karalius, M. Newman, O. U. K. Reddy, Y. Xu, X. Zhang, 2006An extended linkage map for watermelon based on SRAP, AFLP, SSR, ISSR, and RAPD markers. *J Am Soc Hortic Sci.* 131393402
24. G. Li, M. Gao, B. Yang, C. F. Quiros, 2003Gene to gene alignment between the Arabidopsis and *Brassica oleracea* genomes. *Theor Appl Genet.* 10716880
25. G. Li, C. F. Quiros, 2001Sequence-related amplified polymorphism (SRAP), a new marker system based on a simple PCR reaction: Its application to mapping and gene tagging in Brassica. *Theor Appl Genet.* 103455461
26. H. Li, C. J. Ruan, T. Silva, A. Jaime, B. Q. Liu, 2010Associations of SRAP markers with dried-shrink disease resistance in a germplasm collection of sea buckthorn (*Hippophae rhamnoides* L.). *Genome* 53447457
27. W. Li, X. Xu, G. Li, Q. Guo, S. Wu, Y. Jiang, H. Dong, M. Weng, D. Jin, J. Wu, Z. Ru, B. Wang, 2012Characterization and molecular mapping of RsrR, a resistant gene to maize head smut. *Euphytica* (DOIs10681-012-0747-4, in press).
28. W. Li, J. Zhang, Y. Mou, J. Geng, P. B. E. Mc Vetty, S. Hu, G. Li, 2011Integration of Solexa sequences on an ultradense genetic map in *Brassica rapa* L. *BMC Genomics* 12: 249.
29. Z. Li, J. Pan, Y. Guan, Q. Tao, H. He, L. Si, R. Cai, 2008Development and fine mapping of three co-dominant SCAR markers linked to the M/m gene in the cucumber plant (*Cucumis sativus* L.).*Theor Appl Genet. 117*125360
30. Z.. Lin, D. He, X. Zhang, Y. Nie, X. Guo, C. Feng, et al.2005Linkage map construction and mapping QTL for cotton fiber quality using SRAP, SSR and RAPD. *Plant Breed.* 124180187
31. L. Liu, X. , J. Wei, J. Qin, C. Mo, 2011The first genetic linkage map of Luohanguo (*Siraitia grosvenorii*) based on ISSR and SRAP markers. *Genome*541925
32. Y. Long, Z. Wang, Z. Sun, D. W. G. Fernando, P. B. E. Mc Vetty, G. Li, 2010Identification of two blackleg resistance genes and fine mapping of one of these two genes in a *Brassica napus* canola cultivar 'Surpass 400'. *Theor Appl Genet.* 12212231231
33. N. Mutlu, F. H. Boyaci, M. Göçmen, K. Abak, 2008Development of SRAP, SRAP-RGA, RAPD and SCAR markers linked with a Fusarium wilt resistance gene in eggplant. *Theor Appl Genet.* 117130312

34. O. Ortega, E. Duran, C. Arbizu, R. Ortega, W. Roca, D. Potter, C. F. Quiros, 2007Pattern of genetic diversity of cultivated and non-cultivated mashua, *Tropaeolum tuberosum*, in the Cusco region of Perú. *Genetic Resources and Crop Evolution* 54807821
35. M. Rahman, G. Li, D. Schroeder, P. B. E. Mc Vetty, 2010Inheritance of seed coat color genes in *Brassica napus* (L.) and tagging the genes using SRAP, SCAR and SNP molecular markers. *Molecular Breed.* 26439453
36. M. Rahman, Z. Sun, P. B. E. Mc Vetty, G. Li, 2007Development of SRAP, SNP and SCAR molecular markers for the major seed coat color gene in *Brassica rapa. Theor Appl Genet. 115*11017
37. A. Riaz, G. Li, Z. Quresh, M. Swati, C. F. Quiros, 2001Genetic diversity of oilseed *Brassica napus* inbred lines based on sequence-related amplified polymorphism and its relation to hybrid performance. *Plant Breed.* 12015
38. J. J. Ruiz, B. Pico, G. Li, V. D'Antonio, B. Falk, C. F. Quiros, 2001Identification of markers linked to a Celery Mosaic Virus resistance gene in celery. *J Amer Soc for Hortic Sci.* 1264432435
39. M. H. Soleimani, M. Talebi, B. E. Sayed-Tabatabaei, 2012Use of SRAP markers to assess genetic diversity and population structure of wild, cultivated, and ornamental pomegranates (Punica granatum L.) in different regions of Iran. *Plant Systematics and Evolution* 29811411149
40. S. Sun, W. Gao, S. Lin, J. Zhu, B. Xie, Z. Lin, 2006Analysis of genetic diversity in Ganoderma population with a novel molecular marker SRAP. *Applied Microbiology and Biotechnology 72*537543
41. Z. Sun, Z. Wang, J. Tu, J. Zhang, F. Yu, P. B. E. Mc Vetty, G. Li, 2007An ultradense genetic recombination map for *Brassica napus*, consisting of 13551 SRAP markers. *Theor Appl Genet.* 1148130517
42. L. Tang, Y. Xiao, L. Li, Q. Guo, Y. Bian, 2010Analysis of genetic diversity among Chinese *Auricularia auricula* cultivars using combined ISSR and SRAP markers. *Current Microbiology* 61132140
43. P. Tanhuanpää, R. Kalendar, A. H. Schulman, E. Kiviharju, 2007A major gene for grain cadmium accumulation in oat (*Avena sativa* L.). *Genome*5058894
44. G. J. Vandemark, J. J. Ariss, G. A. Bauchan, R. C. Larsen, T. J. Hughes, 2006Estimating genetic relationships among historical sources of alfalfa germplasm and selected cultivars with sequence related amplified polymorphisms. *Euphytica* 152916
45. Z. Wang, J. Wang, X. Wang, H. Gao, N. I. Dzyubenko, V. F. Chapurin, 2012Assessment of genetic diversity in *Galega officinalis* L. using ISSR and SRAP markers. *Genetic Resources and Crop Evolution 59*865873
46. W. Xie, X. Zhang, H. Cai, L. Huang, Y. Peng, X. , 2011Genetic maps of SSR and SRAP markers in diploid orchardgrass (*Dactylis glomerata* L.) using the pseudo-testcross strategy. *Genome* 54212221
47. X. Xie, F. Zhou, X. Zhang, J. Zhang, 2009Genetic variability and relationship between MT-1 elephant grass and closely related cultivars assessed by SRAP markers. *Journal of Genetics* 88281290
48. L. Xu, L. Wang, Y. Gong, W. Dai, Y. Wang, X. Zhu, T. Wen, L. Liu, 2012Genetic linkage map construction and QTL mapping of cadmium accumulation in radish (*Raphanus sativus* L.).*Theor Appl Genet.* 12565970
49. D. Xue, S. Feng, H. Zhao, H. Jiang, B. Shen, N. Shi, J. Lu, J. Liu, H. Wang, 2010The linkage maps of *Dendrobium* species based on RAPD and SRAP markers. *Journal of Genetics and Genomics* 37197204
50. P. Yang, X. Liu, X. Liu, W. Yang, Z. Feng, 2010Diversity analysis of the developed qingke (hulless barley) cultivars representing different growing regions of the Qinghai-Tibet Plateau in China using sequence related amplified polymorphism (SRAP) markers. *African Journal of Biotechnology*985308538
51. M. A. Yeboah, C. Xuehao, C. Feng, G. Liang, M. Gu, 2007A genetic linkage map of cucumber (*Cucumis sativus* L.) combining SRAP and ISSR markers. *African Journal of Biotechnology* 627842791

52. Y. Yi, H. Liu, X. Huang, L. An, F. Wang, X. Wang, 2008Development of molecular markers linked to the wheat powdery mildew resistance gene *Pm4b* and marker validation for molecular breeding. *Plant Breed.* 127116120
53. M. Youssef, A. C. James, R. Rivera-Madrid, R. Ortiz, Medrano. R. M. Escobedo-Gracia, 2011Musa genetic diversity revealed by SRAP and AFLP. *Molecular Biotechnology* 47189199
54. J. Yu, S. Yu, C. Lu, W. Wang, S. Fan, M. Song, Z. Lin, X. Zhang, J. Zhang, 2007High-density linkage map of cultivated allotetraploid cotton based on SSR, TRAP, SRAP and AFLP markers. *Journal of Integrative Plant Biology* 49716724
55. X. Yu, B. Chu, R. Liu, J. Sun, J. J. Brian, H. Wang, Z. Shuijin, Y. Sun, 2012Characteristics of fertile somatic hybrids of *G. hirsutum* L. and *G. trilobum* generated via protoplast fusion. *Theor Appl Genet.* 2012 Jul 10.
56. M. Zaefizadeh, R. Goleiv, 2009Diversity and relationships among durum wheat landraces (subconvars) by SRAP and phenotypic marker polymorphism. *Research Journal of Biological Sciences* 4960966
57. F. Zhang, S. Chen, F. Chen, W. Fang, Y. Deng, Q. Chang, P. Liu, 2011aGenetic analysis and associated SRAP markers for flowering traits of chrysanthemum (*Chrysanthemum morifolium*). *Euphytica* 1771524
58. F. Zhang, S. Chen, F. Chen, W. Fang, Y. Chen, F. Li, 2011bSRAP-based mapping and QTL detection for inflorescence-related traits in chrysanthemum (Dendranthema morifolium). *Molecular Breed.* 27: 11-23,
59. J. Zhang, Y. Lu, Y. Yuan, X. Zhang, J. Geng, Y. Chen, S. Cloutier, P. B. E. Mc Vetty, G. Li, 2009aMap-based cloning and characterization of a gene controlling hairiness and seed coat color traits in *Brassica rapa*. *Plant Molecular Biol.* 69553563
60. W. Zhang, H. He, Y. Guan, H. Du, L. Yuan, Z. Li, D. Yao, J. Pan, R. Cai, 2010aIdentification and mapping of molecular markers linked to the tuberculate fruit gene in the cucumber (*Cucumis sativus* L.). *Theor Appl Genet.* 120645654
61. W. Zhang, J. Pan, H. He, C. Zhang, Z. Li, J. Zhao, X. Yuan, L. Zhu, S. Huang, R. Cai, 2012Construction of a high density integrated genetic map for cucumber (*Cucumis sativus* L.) *Theor Appl Genet.* 124249259
62. Y. Zhang, X. Zhang, W. Hua, L. Wang, Z. , 2010bAnalysis of genetic diversity among indigenous landraces from sesame (*Sesamum indicum* L.) core collection in China as revealed by SRAP and SSR markers. *Genes & Genomics* 32207215
63. X. Zhang, J. Wu, H. Zhang, Y. , A. Guo, X. Wang, 2011cFine mapping of a male sterility gene MS-cd1 in *Brassica oleracea*. *Theor Appl Genet.* 1232318
64. Z. Zhang, M. Hu, J. Zhang, D. Liu, J. Zheng, K. Zhang, W. Wang, Q. Wan, 2009bConstruction of a comprehensive PCR-based marker linkage map and QTL mapping for fiber quality traits in upland cotton (*Gossypium hirsutum* L.) *Molecular Breed.* 244961
65. F. Zhao, Z. Cai, T. Hu, H. Yao, L. Wang, N. Dong, B. Wang, Z. Ru, W. . Zhai, 2010Genetic analysis and molecular mapping of a novel gene conferring resistance to rice stripe virus. *Plant Molecular Biology Reporter* 28512518

CHAPTER 11

Global Gene Expression Defines Faded Whorl Specification of Double Flower Domestication in *Camellia*

Xinlei Li[1,2], Jiyuan Li[1,2], Zhengqi Fan[1,2,] Zhongchi Liu [3], Takayuki Tanaka[4] & Hengfu Yin [1,2]

[1] *Research Institute of Subtropical Forestry, Chinese Academy of Forestry, Fuyang, Zhejiang, 311400, China.*

[2] *Key Laboratory of Forest genetics and breeding, Chinese Academy of Forestry, Fuyang, Zhejiang, 311400, China.*

[3] *Departments of Cell Biology and Molecular Genetics, University of Maryland, College Park, MD, USA.*

[4] *Department of Plant Science, School of Agriculture, Tokai University, Minami-aso-mura, Aso-gun, Kumamoto, 869-1404, Japan.*

ABSTRACT

Double flowers in cultivated camellias are divergent in floral patterns which present a rich resource for demonstrating molecular modifications influenced by the human demands. Despite the key principle of ABCE model in whorl specification, the underlying mechanism of fine-tuning double flower formation remains largely unclear. Here a comprehensive comparative transcriptomics interrogation of gene expression among floral organs of wild type and "formal double" and "anemone double" is presented. Through a combination of transcriptome, small RNA and "degradome" sequencing, we studied the regulatory gene expression network underlying the double flower formation. We obtained the differentially expressed genes between whorls in wild and cultivated *Camellia*. We showed that the formation of double flowers tends to demolish gene expression canalization of key functions; the faded whorl specification mechanism was fundamental under the diverse patterns of double flowers. Furthermore, we identified conserved miRNA-targets regulations in the control of double flowers, and we found that miR172-AP2, miR156-SPLs were critical regulatory nodes contributing to the diversity of double flower forms. This work highlights the hierarchical patterning of global gene expression in floral development, and supports the roles of "faded ABC model" mechanism and miRNA-targets regulations underlying the double flower domestication.

INTRODUCTION

Human selection of plant traits through breeding is the foundation of modern day agriculture that supports human subsistence. The process of plant domestication is rooted in the genetic modification of genes that regulate plant development and growth. Therefore, plant domestication is a special class of evolution influenced by human behavior[1].

The showy double flower in many ornaments is a typical example of "domestication syndrome" (traits that are different to their progenitors) and thought to be controlled by genetic selection of a small number of genes[2,3]. It is recognized that several groups of floral homeotic genes, mainly the ABC(E) genes, played central roles in directing floral forms both in floral organ identify and number[4,5]. The well-established ABCE model on flower development elegantly explains how a few key regulatory genes could control floral organ identity, which then informs (or directs) subsequent floral organ development characteristic of the specific floral organ[4,5]. Under the guidance of the ABCE model, double flower domestication in many cases was found to be associated with the alterations of the C function gene[3]. For example, in the cultivated rose, the incensement of petal numbers was correlated with the reduced domain of C class gene expression[6]. Although the ABCE model was shown to be largely conserved in many other higher plants, refinements were proposed by individual studies in plant species with noncanonical floral structures. For example, the 'inside-out' whorl specification in *Lacandonia* was matched with recombination of B and C function genes[7]; the petaloid bracts in dove tree were likely associated with ectopic expression of petal and stamen genes8; the expression analysis of C function orthologous gene in daffodil indicated that the corona was a structural innovation between petal and stamen9. These studies largely supported the ABCE model by showing how modified expression domains of A, B, C, E genes led to morphological innovations5. Nevertheless, to form a functional organ with defined shape, the subsequent morphogenesis of a floral organ is far more complicated. As examined in *Arabidopsis* and *Antirrhinum*, the petal development required organized cell division and growth in different zones controlled by gene and plant hormones10, 11. Hence, it has been difficult to investigate molecular mechanisms of floral organogenesis.

In recent years, global transcriptomic analyses of floral organs in the basal and eudicot flowering plants provided another means to understand the evolution of floral forms. In parallel to the molecular genetics studies in model plants, the feasibility of obtaining gene expression profiles at the genome scale was revolutionary in the field of floral evolution12, 13. It appeared that the ABCE model was established based on the study of highly derivative plant species; the 'shift of boundary' (floral structure evolution involved changes of spatial expression of ABCE genes) or 'faded whorl specification' (gradient expression of regulatory genes was rudimentary for establishing floral genetic programs) was proposed based on comparisons with the basal angiosperm flowers5. Genomics characterizations from the broader phylogenetic sampling revealed that the faded whorl specification might represent a 'default' state, and the diversity of floral forms required the canalization of gene expression patterns such as, the whorl-specific expression of floral regulatory genes12.

It is now clear that multiple pathways were involved in the regulation of floral forms. The microRNA was an important part of post-transcriptional regulation of gene expression and played critical roles in plant floral development14. For example, miR172 targeted A function gene *AP2* and regulated petal development; expression of miR172-resistant *AP2* by the *AP3* promoter caused a special double flower phenotype in *Arabidopsis*15. The miR169-NFYA regulation was required for restraining C class gene expression in *Antirrhinum* and *Petunia*16. Moreover, miRNA-target regulatory modules were found to be required for gamete cell differentiation in stamen (miR156-SPLs)17, determination of floral organ boundaries and numbers (miR164-NACs)18, and outgrowth of stamens and petals (miR319-TCPs)19. With the increasing power of high throughput sequencing, the sequencing of small RNA and Parallel Analysis of RNA Ends (PARE, degradome) offer effective and high throughput ways in identifying miRNAs and their targets in plant species even in the absence of a reference genome20, 21. Indeed, in *Camellia azalea*, the small RNA and transcriptome sequencing were shown to be useful in identifying key regulatory genes including conserved and novel miRNAs, and evolution of miRNA genes was predictable in the context plant evolution22,23. It is unclear if these miRNAs target genes are involved in the regulation of double flower development.

Cultivated *Camellia japonica* is a world-wide ornamental flower and has been domesticated over centuries. Its remarkable diversity of floral forms imparts a rich resource for understanding the genetic regulation of floral patterning and forms24. For instance, the double flowers in *Camellia* include 5 distinctive types: semi, formal, anemone, rose, and peony doubles, mainly distinguished by the number and arrangement of petals and stamens. The wild *Camellia* flower was defined as a single flower, with one row of overlapping petals (usually less than 8), and a columnar stamen cluster and one normal pistil in the center; the anemone form was characterized with several outer rows of large petals, with many petaloid stamens, and few degenerated stamens and pistils. The center of anemone double was clustered with many small petals, and the petaloid stamens and degenerated stamens and pistils were generally covered in the small petal cluster. The formal double completely lacked stamen and pistil organs with replacement of many rows of petals (Fig. 1A). The cherished varieties of 'Wabisuke' and 'Higo' camellias somehow represented mitigated and augmented stamen growth respectively25. Double flowering in camellia cultivars is also a phenomenon of interest especially for its proliferated and redundant floral structure. An unusual flowering phenomenon had been observed in an unnamed camellia cultivars characterized by a formal double flower of perfect symmetrical form26. Previous studies in *Camellia* characterized the expression of A-, and C- function genes and suggested that double flower domestication were involved in multiple pathways27. Interestingly, the expression of C function ortholog (*CjAG*) displayed opposite patterns in petals of formal and anemone doubles (expression of *CjAG* suppressed in formal double, but upregulated in anemone double)27. Despite the extensive knowledge on genetic regulation of floral development in the model plant species, it remains unclear how the floral patterns in cultivated *Camellia* were achieved.

In this work, we performed high-throughput transcriptome sequencing in wild, formal double and anemone double types of *Camellia* to capture the genome-wide gene expression patterns related to double flower formation. We identified differentially expressed genes among organ types in both wild and cultivated *Camellia* and analyzed their functional properties by comparing homological structures. Small RNA and degradome sequencing identified conserved miRNA-targets pairs and their potential contribution to different types of double flowers. We found that double flower domestication in camellia tends to deteriorate specified expression patterns

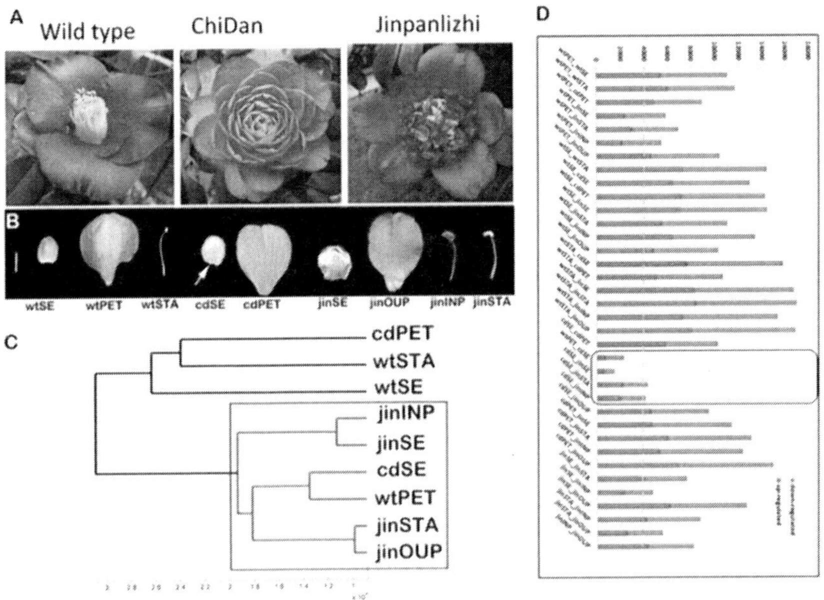

Figure 1. Comparative transcriptomics analysis in wild and doubled camellias. (**A**) Overviews of wild and double flower varieties. From left to right, wild, formal double "ChiDan", anemone double "Jinpanlizhi". (**B**) Floral tissues for the transcriptomics study. Arrow indicates the pigmentation in sepal of ChiDan. White bar, =1 cm. (**A,C**) dendrogram plot of clustering of samples by the mean expression levels. Distance less than 20,000 was highlighted by red lines (inside the blue rectangular). (**D**) The summary of numbers of DEGs among floral organs. Red, up-regulated genes; green, down-regulated genes. wtSE, wild type sepal; wtPET, wild type petal; wtSTA, wild type stamen; cdSE, ChiDan sepal; cdPET, ChiDan petal; jinSE, Jinpanlizhi sepal; jinOUP, Jinpanlizhi outer petal; jinINP, Jinpanlizhi inner petal; jinSTA, Jinpanlizhi stamen.

of regulatory genes. Further, the development of inner floral organs, a characteristic that distinguishes between double flower types, appear to have promiscuous organ identity influenced by positional effects of gene expression in the zones of floral bud. Moreover, some key miRNA-targets, such as miR172-AP2, miR156-SPLs, were identified as important molecular signatures for the classification of double flower types. These findings may highlight the orchestration of genes controlling floral organ identity, miRNA-targets underlying the floral patterning and their contributions to the double flower domestication.

RESULTS AND DISCUSSION

Transcriptome sequencing for gene discovery and gene expression profiling in wild and cultivated *C. japonica*

To generate a comprehensive genomic resource in *C. japonica* for gene discovery, RNAs from a mixed sample containing leaf, shoot, floral bud, sepal, petal, stamen and carpel were sequenced by Illumina Hiseq platform, which yielded over 45.7 million 2 × 125 pair-end reads. Raw reads were filtered and assembled using the Trinity software, and 104, 810 unigenes were obtained with N50 of 893 bp (Supplementary Table S1). The unigene sequences were assessed by alignments to various databases for annotation (details in Supplementary Table S2), and 37384 unigenes were annotated with at least one hit in all databases (Supplementary Table S2). Notably, 78.3% of annotated unigenes (29265) were over 299 bp indicating a competent transcriptome in *C. japonica* for gene identification. The assembly of 104,810 unigenes served as the reference for subsequent RNA-seq analysis.

Cultivated *Camellia* contains more than 5 types of double flowers with distinctive floral forms. To understand the global signatures of gene expression between different organ types, we constructed 27 libraries of 9 tissue types from wild (sepal, petal, stamen), formal double (sepal, petal) and anemone double flowers (sepal, outer petal, inner petal, petaloid stamen) with 3 biological replicates for next generation sequencing (Fig. 1A,B). Around 12.3 million reads (single end 50 bp in length) on average were generated for each library with a mean value of 92.8% for over Q30 bases (Supplementary Table S3). The above transcriptome assembly was served as a reference for gene expression quantification. The mapped reads were converted into RPKM for the expression level of unigene (see Material and Methods). To assess the expression data between replicates and tissues, the Pearson's correlation coefficient between every pair was calculated (Supplementary Figure S1). The correlation values among replicates were mostly over 0.9 (Supplementary Figure S1), which indicated that the results were reliable. The samples 'jinOUP-3′ and 'jinSTA-3′ were poorly correlated (Pearson's correlation <0.2) with other two biological replicates and therefore removed for further analysis. 'jinSE-1′ and 'cdSE-2′ displayed low correlation within biological replicates (Pearson's correlation <0.73) and were also removed from analysis.

Global analysis of gene expression between wild and cultivated *Camellias* identified DEGs among various floral organs

The primary objective of this study was to capture the global expression signatures between flower types by identifying significant differentially expressed genes (DEGs) between organs types and between different double flower types. To generate a top-down view of gene expression profiles, we performed clustering of global expression of mean values of replicates. The Euclidean distances between organ types were calculated and clustered to display the resemblance of organs at the level of gene expression. We showed that, in wild type *Camellia*, each of the floral organs, sepal, petal and stamen, exhibit distinct morphology and easily separable (Fig. 1B). While some of the sepals from formal double (ChiDan) exhibit some petal-like pigmentation spots (Fig. 1B). In contrast, large numbers of inner petal, and petaloid stamen arise and cluster together in the center of the anemone double flower (Fig. 1A) suggesting the continuous differentiation and proliferation of inner petal/stamen like organs. To determine the genes that were preferentially expressed in specific organs, we performed statistical analysis to identify significant DEGs between tissue types (with fold change > = 2, P-value <0.05). The numbers of up and down regulated unigenes were generally comparable between pairs of organs (Fig. 1D). It was not surprising that, between each pair of wild *Camellia* organs, there were over 10,000 DEGs considering distinctive floral organ identities (Fig. 1D). Nevertheless, the putative homologous organs displayed large numbers of DEGs between wild and different types of double flowers. For instance, wild and formal double sepal (wtSE_cdSE) had over 12,000 DEGs and wild and anemone double stamen (wtSTA_jinSTA) had the largest number of DEGs (Fig. 1D). Also, markedly less DEGs among cdSE_wtPET, cdSE_jinSE, cdSE_jinINP and jinSTA_cdSE were revealed (Fig. 1D) underlying complex regulatory networks might be responsible during the domestication of *Camellia* double flowers, and development of these organs might require some concurrent processes. Both formal doubled and anemone doubled *Camellia* cultivars displayed highly proliferated growth of petal or petal-like organs, while the differentiation of extra organs varied remarkably in terms of shape and cell types (Fig. 1A). These results suggested that the double flower domestication may have a role in shaping the global gene expression landscapes during the whorl specification and outgrowth.

Common and specific sets of DEGs in structural homologous organs of wild and double flowers

To investigate further characteristics of DEGs, we examined the DEGs between homologous organs in wild and cultivated *Camellia*. In sepals, the DEGs between wild-formal double and wild-anemone double were considerably overlapped (more than 80% of total DEGs, Fig. 2A); majority of DEGs were found in both double flower varieties compared to wild *Camellia*, and almost all DEGs had the same expression pattern of up- or down- regulation (Fig. 2B). These data suggested that domestication of formal and ammonia double might result in the same modifications of sepal development despite the ChiDan sepals had some petal-like pigmentations. The petals in double flower cultivars were diverse in shape and cell types-formal double displayed a gradient change of petal size, and outer petal and inner petal in anemone double were distinctive in petal shape with the filament-like structures in inner petals28. The analysis of DEGs between petals showed a different and more complexed pattern than the analysis of sepals. By comparing DEGs from formal double and anemone double to wild, the large proportion of DEGs was specific to the double flower type, and small number of DEGs was shared (Fig. 2C). When comparing stamen-like organ and inner petal from anemone double to wild stamen, we found majority of DEGs had consistent expression patterns (both up- or down-regulated in double flower than wild *Camellia*) despite the morphological distinctions of inner organs, while considerably more DEGs were found to be not overlapped between inner petal and petaloid stamens (Fig. 2D,E). This result suggested the inner petal and stamen of anemone double flower might share some conserved gene expression patterns during development which were different from normal stamen development, but may more related to petal development. The comparisons of up- and down- regulated DEGs showed that about half of up- regulated genes of inner petal and petaloid stamen were not shared, and a relatively large number of down-regulated DEGs was shared (Fig. 2E). Similar to the scenario in sepal, most of genes had consistent regulation patterns, and few genes showed opposite up- or down repression trends in different types of double flowers (Fig. 2E). Significant overlap of DEGs between sepals and inner organs from anemone double versus wild stamen (Fig. 2A,D) implied derivative origins of those double flower organs during domestication.

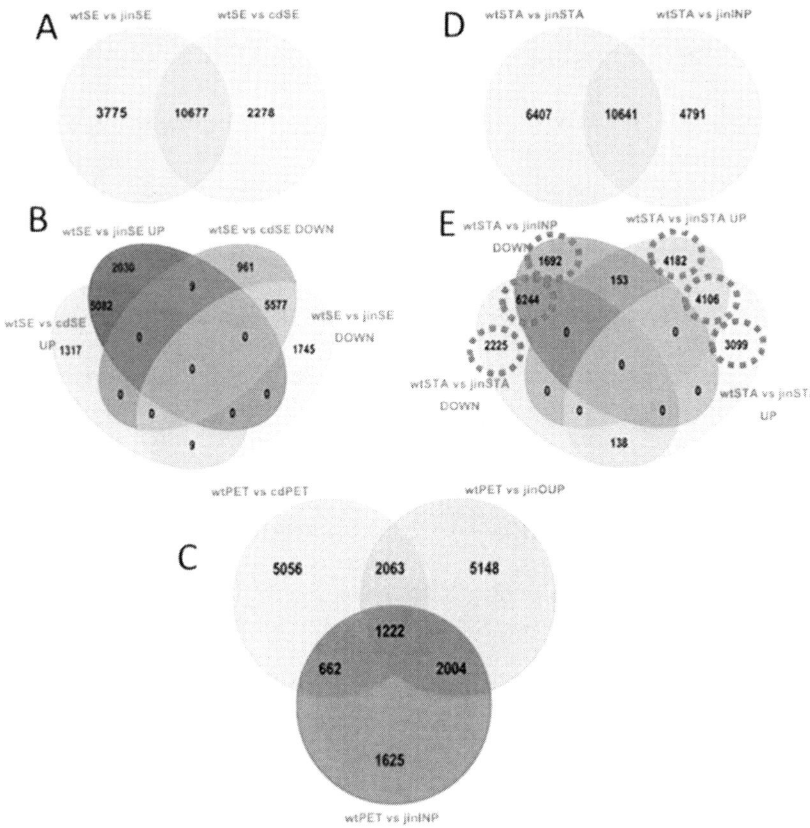

Figure 2. The venn diagram plots of distribution of DEGs. (**A**) Common and special DEGs between sepals of wild and double flower cultivars. (**B**) 4-way venn diagram of up and down regulated DEGs between sepals of wild and double flower cultivars. (**C**) Common and special DEGs between petals of wild and double flower cultivars. (**D**) Common and special DEGs between stamen like organs of wild and double flower cultivars. (**E**) 4-way venn diagram of up and down regulated DEGs between stamen and inner organs of wild and double flower cultivars. The sample labels were described in Fig. 1, and UP indicated up-regulated genes, DOWN indicated down-regulated genes in each comparsion.

Enrichment of biological processes entails the domestication paths relevant to doubled floral types

To gain insights from DEGs, the Gene Ontology (GO) enrichment analysis was performed to identify over-representative GO terms in different categories of DEGs. Significant enriched GO terms were analyzed between pairs of organ types (sepal, petal, stamen) in wild *Camellia*, and some organ-wise significant biological processes, such as photosynthesis, stress responses, hormone signaling, highlighted the activities related to cell growth and differentiation during floral organ development (Supplementary Dataset 1). Meanwhile, the significant enriched biological process could reveal the cumulative gene expression that was required for floral organ development. In order to capture a holistic view in floral development of double flowers, we performed GO enrichment analysis in 10 pairs of organ types (Supplementary Dataset 1; Fig. 3A) based on the whorl identities in wild and doubled cultivars. Significant biological processes were identified ($ks < 0.05$, FDR corrected) and combined as a matrix keeping only the significant GO terms for comparative analysis to test the resemblance of homolog between assayed pairs of organ types. The topology of sample pairs was calculated by the furthest neighbor distance and clustered to reveal the resemblance of organ pairs (Fig. 3B). We found 4 subgroups were formed, and the organ types in wild *Camellia* were classified into independent subgroups, coincided with the structural homology (Fig. 3B). The formal double simply consisted of sepal and petal, and the GO signature was closely correlated with sepal and petal pairs in wild and anemone double (Fig. 3B). In anemone double, a diverse pattern was revealed in which adjacent whorls appeared to be more closely related (Fig. 3B). Whereas the 'identity' of inner organs of anemone double seemed to be not strictly constrained, for example, significant processes in 7 (jinOUP-jinSTA) and 8 (jinINP-jinSTA) were clustered together (Fig. 3B). To understand how gene functions were related to the hierarchical relationships, we screened the most significant GO terms in all combinations through ($ks <= 0.00001$), and 44 highly enriched biological processes were identified (Fig. 3C). We found that, the sepal/petal in formal double displayed much more processes comparing to wild sepal/petal which were not found in wild and anemone doubled *Camellia* (Fig. 3C), and some processes were identified in wild petal/stamen and sepal/stamen. Considering the complete lack of stamen and pistil development in formal double, this result suggested gene expression were deployed across whorl specification. Nevertheless, a small number of GO terms were revealed between adjacent whorls in anemone double flower (Fig. 3C). In the anemone double, sepal/outer petal, outer petal/inner petal, inner petal/petaloid stamen displayed 7, 2, 12 GO terms respectively (Fig. 3C). These data uncovered a spread-out domain of gene expression which supported the faded whorl specification origin of higher plants, and domestication of double flowers tends to be subversive to the canalization of gene expression.

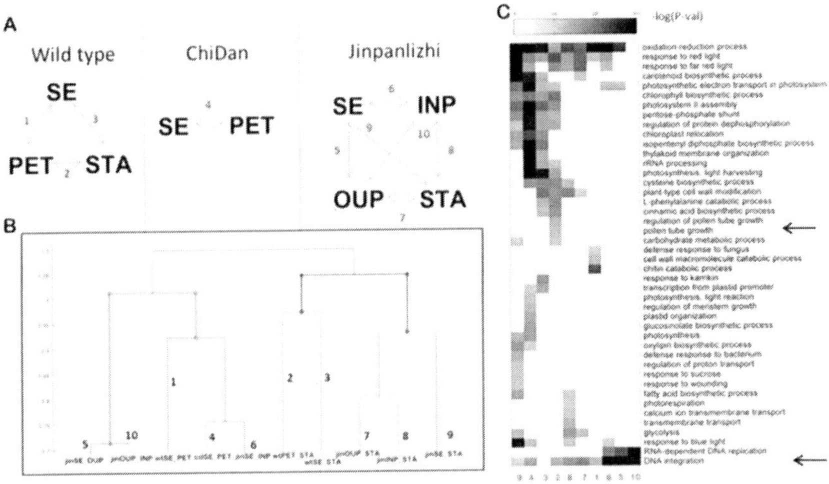

Figure 3. Functional characterizations of DEGs among homologous structural organs. (**A**) The overview of comparisons among organ types in wild and double flower cultivars. (**B**) The computed resemblances based on the presence and significance of all enriched GO terms. (**C**) the highly enriched ($KS < 0.00001$) GO terms and their distributions among comparisons. Arrows indicate GO terms related to pollen tube growth and DNA integration respectively. The numbers indicate the pairs for comparison described in (**A**) SE, sepal; PET, petal; STA, stamen; OUP, outer petal; INP, inner petal.

ABCE genes in the formation of double flowers

We showed the domestication of different types of double flowers potentially deconstructed the constrains of gene expression required for the whorl development. To gain insights of transcriptional activities in double flower development, we focused on several gene families containing important floral regulators. The MADS-box transcription factors included some best characterized floral homeotic genes of ABCE model29. We performed sequence and phylogenetic analyses to identify homologous genes in *C. japonica* (Fig. 4A; See 'Materials and methods' for details). We found 28 homologs including A, B, C, E function genes and flowering timing regulators (Fig. 4A).

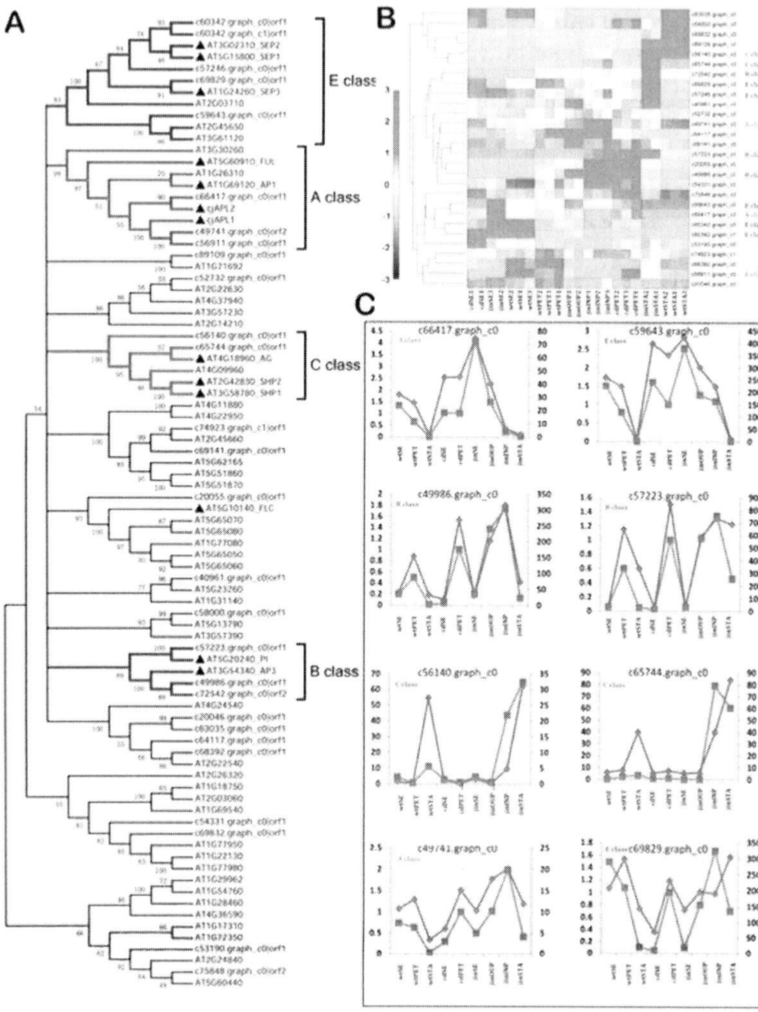

Figure 4. Identification and expression analyses of MADS-box genes. (**A**) The phylogenetic tree of MADS-box genes in Arabidopsis and *C. japonica*, and subgroups of functional characters were identified by annotation from Arabidopsis. Colored branch indicated a subgroup of MADS genes based on the ABCE genes in Arabidopsis. (**A**,**B**) heatmap plot of expression patterns of MADS-box genes from *C. japonica*. The genes that were classified in ABCE categories as shown in (**A**) were labeled. Expression levels of samples were normalized by z-score and plotted. (**C**) Comparisons of expression profiles of key MADS-box genes by RNAseq and qPCR analyses. In each panel, the blue line indicated values of qPCR experiments referring to the left y-axis, and the orange line indicated RPKM values of RNA-seq referring to the right y-axis.

Previous studies showed A function genes were up-regulated in different double flower cultivars and C function gene (*CjAG*) had different expression patterns in formal and anemone double flowers27,28. We found the transcriptomics results were in good agreement with previous work: A function genes (c66417. graph.c0, Fig. 4C) were up-regulated in double flower organs consistent with more petals and a lack of carpels and reduced stamens in these double flowers (Fig. 4B); AG-like (including *Plena*-like) genes were not detectable in formal double consistent with a lack of stamens or carpels in formal double. However C class gene was upregulated in inner organs of anemone double (Fig. 4B,C); this was surprising as many inner organs displayed petal-like structures27. We selected 31 genes and validated the expression profiles through real-time PCR experiment (Primers in Supplementary Table S5). The expression patterns of candidates were compared to the RNA-seq data. We found 21 of those profiles were significantly correlated in Q-PCR and RNA-seq, and 7 correlations were not significant (Supplementary Figure S2).

Expression of unigenes that were homologous to ABCE genes was characterized to understand the roles in double flower domestication. The A-class unigenes (c66417. graph_c0) displayed expected expression pattern in wild *Camellia*-highly expressed in sepal and petal, undetectable in stamen (Fig. 4C). In the inner petal of anemone double, unigene c59643. graph_c0 (a class E gene), not c66417. graph_c0 (a class A gene close to *CjAPL2*), was expressed abundantly (Fig. 4C). The inner petal displayed the characteristics of petal identity, but placed in the central zone (whorl 3) of floral bud27, and the expression pattern suggested that the E class gene might be essential to inner organ differentiation. The B- class genes had similar expression levels in petals of wild, formal double, and outer petals of anemone double, but its high expression in inner petal of anemone was also notable (Fig. 4C). The expression of ABCE homologs was predictable in wild *Camellia* suggesting the well-defined floral organ differentiation required conserved functions of those genes. However, in domesticated cultivars, the formal double tends to disrupt the deployment of homeotic regulators to achieve the petal conversion; the inner whorls of anemone double displayed obscure expression domains crossing the borders of floral organ types.

Conserved miRNA-target regulations with functional diversifications in *Camellia* and double flowers

To investigate the molecular mechanism underlying the altered domain allocation, we asked whether the micro-RNA (miRNA) and target circuit may contribute to the double flower. To generate the sequence resource for miRNA and target identification, we performed small RNA sequencing in wild shoot and floral bud and compared the miRNA expression between wild type petal and inner petal and petaloid stamen of anemone double flower. 18.5~21.5 million short reads per library were obtained and short reads between 18–30 bp in length were subjected to bioinformatics pipeline for miRNA isolation using transcriptome as the reference (Supplementary Table S4). In total 127 miRNAs genes were identified with mature and precursor sequences (Supplementary Dataset 3). To further understand the roles of miRNAs, we also sequenced 2 'degradome' libraries of wild type shoot buds (spring) and floral buds (fully differentiated before opening, average 1.5 cm in length) to identify the recognition targets of miRNAs. Over 80 million reads per the degradome library were generated for target identification (Supplementary Table S4). The slice sites of miRNAs were searched against the sequencing end of degradome libraries by CleaveLand software version 4.0[20,30], and a total of 771 potential targeted genes with annotation information were predicted (alignment score >5) (Supplementary Dataset 4). 87 and 79 highly accurate recognition sites (P-value < 0.05) were revealed in shoot and floral bud, and 80 targets were identified when combining sequencing datasets from 2 libraries for target identification (Fig. 5A, Supplementary Dataset 4).

Among these highly confident targets, we found many evolutionarily conserved miRNA-target pairs in *Camellia* (Supplementary Dataset 4). Some transcription factors including *Auxin Response Factors, MYB, Homeodomain-containing genes, AP2, SPL, TCP* which had important functions in plant development were identified (Supplementary Dataset 4); meanwhile some targets like galactinol synthase (c49174. graph_c0), anthocyanidin glucosyltransferase (c67697. graph_c0) were potentially newly evolved in the regulation of metabolism in *Camellia* (Supplementary Dataset 4). Hence, these results suggested that the present resource of miRNA and their targets were comprehensive for further functional evaluation in double flower development.

It's known that miR172-AP2 and miR169-NFYA regulations were critical for restricting the C-class function during floral patterning, and other regulations such as miR164-CUC, and miR156-SPL were involved in the determination of floral organ number and shape in parallel to ABC functions[14,16,31]. We showed that the recognition of predictive target sites was verified in degradome sequencing (Fig. 5B). In *Arabidopsis*, SPL was found to be required for anther sporogenic tissues differentiation and expressed abundantly in the anther tissues[17,32,33]. In *Camellia*, miR156 was expressed abundantly in stamen where *SPL1* and *SPL2* were expressed at a low level[22]. In this study, SPLs were expressed just in stamen filaments but not anther tissues (Fig. 6) suggesting the silencing by miR156. It indicated that suppression of SPLs by miR156 was a key step of inner petal and stamen-like organ formation. The expression level of SPLs in petal and stamen organs may present a distinctive molecular event for determination of double flower type.

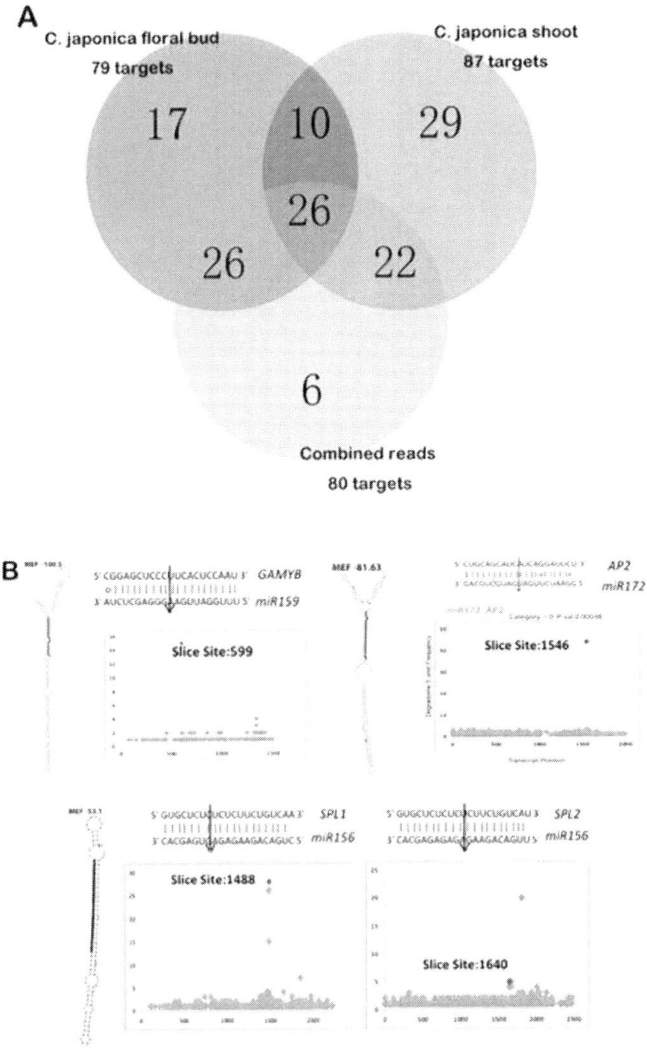

Figure 5. Analysis of miRNA-Targets from small RNA and degradome sequencing. (**A**) The venn plot of highly confident miRNA targets from degradome sequencing. The degradome libraries of shoot and floral bud were analyzed independently and also merged together to identify targets. (**B**), in each panel, left displayed the secondary structure of miRNA precursors and blue color highlighted the mature miRNA sequences. Right, the T-plot of key miRNA targeted slicing sites validated in the degradome sequencing. The slicing sites were indicated by arrows.

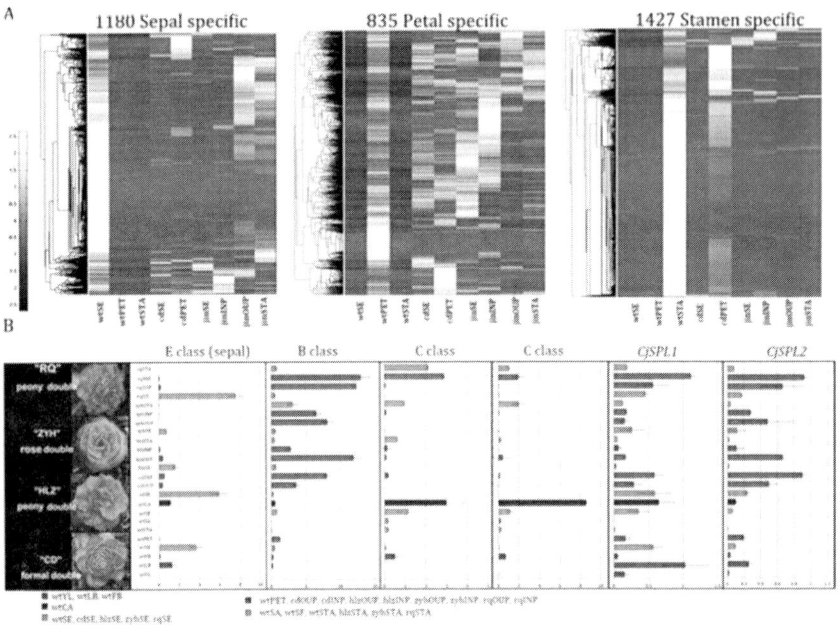

Figure 6. Whorl-specific genes and expression analysis of key regulators in various double flowers in *Camellia*. **(A)** Whorl specific genes displayed disrupted expression patterns in doubled flower culvars. The genes with preference in one organ of wild *Camellia* sepal, petal and stamen were identified through expression analysis. The mean expression of samples was obtained and normalized by the z score method. Normalized expression of clusters was plotted in the heatmap figure. (**B**) Expression analysis of floral regulators in different double flower cultivars. "RQ", Rongqiu, peony double; "ZYH", ZhuangYuanHong, rose double; "HLZ", HongLuZhen, peony double with modified arrangements of inner petal and stamen whorls; "CD", ChiDan, formal double. Organ types were grouped by the positional homology and indicated in different colors. wtYL, wild type young leaf; wtLB, wild type leaf bud; wtFB, wild type floral bud; wtSE, wild type sepal; wtPET, wild type petal; wtSTA, wild type stamen; wtSA, wild type anther; wtSF, wild type filaments; wtCA, wild type carpel; cdSE, ChiDan sepal; cdOUP, ChiDan petals in outer whorls; cdINP, ChiDan petals in the inner whorls; hlzSE, HongLuZhen sepal; hlzOUP, HongLuZhen outer petal; hlzINP, HongLuZhen inner petal; hlzSTA, HongLuZhen stamen; zyhSE, ZhuangYuanHong sepal; zyhOUP, ZhuangYuanHong outer petal; zyhINP, ZhuangYuanHong inner petal; zyhSTA, ZhuangYuanHong stamen; rqSE, RongQiu sepal; rqOUP, RongQiu outer petal; rqINP, RongQiu inner petal; rqSTA, RongQiu stamen.

Collapsed expression of whorl-specific genes in different types of double flowers

We showed that distinctive gene expression is related to morphological alterations in double flowers, and post-transcriptional regulation mediated by miRNA played important roles in fine-tuning floral organ development. To further characterize the well-defined floral organs in wild *Camellia*, we identified some whorl-specific genes and analyzed the expression among the floral organs in double flowers. Using the RNA-seq abundance data of wild *Camellia*, we identified the whorl specific unigenes with more than ten-fold level in one organ than the rest expression level in other organs; we found 1180 unigenes were sepal-specific, 835 petal-specific, and 1427 stamen-specific; the whorl-specific expression tends to be diverted in double flower cultivars; hierarchical expression of whorl-specific genes revealed diverted patterns in domesticated cultivars despite highly preferential expression in specific organ of wild *Camellia* (Fig. 6A). Meanwhile, some abovementioned ABCE genes were found in the whorl specific categories, such as B and C class genes. Interestingly, one E class gene (c60342. graph_c1) was identified as a sepal-specific gene. To understand how the expression of whorl-specific genes and ABCE genes were altered in double flowers, we performed qPCR analysis in wild and 4 types of double flower cultivars including 2 peony double, 1 rose double and 1 formal double (Fig. 6B). We showed that the sepal-specific E class gene was indeed highly abundant in sepals, in wild as well as cultivated double flowers. However in peony and formal double, there is a slight induction of expression of this E gene in petals (Fig. 6A,B). The B class gene was up-regulated in all assayed double flower petals (Fig. 6B) confirming the previous study in which multiple B class copies had differential expression levels contributing to petal development28. The reduced expression of C class genes was expected in formal double27, however, low expression was detected in the inner whorls of petals–usually with deformed shapes in rose and formal doubles (Fig. 6B); in another peony double cultivar "RongQiu" (cultivar with more and enlarged inner petals in floral center), the expression of C class was evidently abundant which was similar to anemone double (Fig. 6B). Taken these results together we speculated that the domestication in *Camellia* has resulted in fundamental deconstruction of the gene expression deployment which consisted of not only ABCE function genes but also the overlapping spatial domains of present unknown factors. We showed the post-transcriptional regulation of SPL genes were specific to stamen development; specifically, the miR156 targeted silencing was required in the anther tissue formation. We found *CjSPL1/2* were expressed in whole stamens and filaments, but not detectable in anther (Fig. 6B). Detailed analysis in peony and rose double revealed that the expression levels of *CjSPL1/2* were significantly lower in inner stamenoid than inner petals (Fig. 6B). Since the reduction of SPLs expression was mediated by miR156, the expression levels SPLs could be a molecular character of types of double flowers.

CONCLUSIONS

This work analyzed the genome-wide gene expression profiles in wild and cultivated double *Camellia*, and captured the dynamics of differential expressed genes with functional characteristics. We showed that the double flower domestication deconstructed the designated expression patterns of key components such as ABCE genes and other floral regulators, and revealed the diversity of double flower types relied on the faded whorl specification mechanisms. Finally, through small RNA and degradome sequencing, we identified the conserved miRNA-target regulations in *Camellia*; detailed expression comparisons indicated that key regulatory pairs, such as miR172-AP2 and miR156-SPLs, had divergent functions in *Camellia* floral development, and were central nodes of directing diverse inner organ growth in different double flowers.

METHODS

Plant materials and growth conditions

Camellia plants were grown in the greenhouse of Research Institute of Subtropical Forestry (Fuyang, Zhejiang, China) under natural light. For RNA extraction, different floral organs from wild and cultivated camellias (floral bud around 1.5–2 cm, before opening) were collected and frozen immediately in liquid nitrogen and stored in −80 °C freezers before use. Three biological replicates of all samples were collected from at least three individuals during the flowering seasons of year 2014 and 2015.

RNA extraction and quality assessment

Total RNAs were extracted using the PLANTpure kit (Cat. RN33, Aidlab, Beijing, China) and stored in freezer before use. For QPCR analysis, cDNA was synthesized by a PrimeScriptII 1st Strand cDNA Synthesis Kit (Cat. 6210 A, Takara, Dalian, China) supplied with DNase treatment columns to avoid DNA contamination. For next-generation sequencing experiments, the RNA quality and quantity was determined using a Nanodrop 1000 spectrophotometer (Thermo Fisher Scientific, Wilmington, DE) and a

Bioanalyzer RNA nano chip (Agilent Technologies, Singapore). The quality and quantity of RNAs were assessed by OD 260/280 and 260/230 and RNA integration number (RIN) to meet the standards for library construction of transcriptome, small RNA and degradome sequencing as described21,34.

Transcriptome, small RNA and degradome sequencing and data processing

Approximately, 5 ug of total RNA for each tissue sample was used for the construction of libraries using mRNA-Seq Sample Prep kit (Illumina Inc., San Diego, CA) according to the manufacturer's protocol. Equal quantities of libraries (approximately 5 ng per sample) with different indices were mixed and stored in −80 °C freezer before sequencing. Sequencing was performed in a v3 flowcell on an Illumina HiSeq 2500 sequencer by Biomarker Technologies (Beijing, China), generating 2 × 101 bp and 1 × 60 bp reads. About 120 million high quality RNA-Seq reads (with quality score >20 for each base) were pooled from Illumina sequencing of each of the 9 samples (three biological replicates of 3 stages) and were then assembled into contigs using Trinity35. All sequencing data were deposited in NCBI Short Read Archive under BioProject ID PRJNA331772. We quantified transcript levels in reads per kilobase of exon mode per million mapped reads (RPKM)36. Small RNA libraries were constructed as previously described and sequenced by a Hiseq 2500 machine generating 1 × 51 bp reads. The 'degradome' sequencing libraries were constructed as described and sequenced by a Hiseq 2500 as described before (Biomarker technologies, Beijing, China). All sequencing reads were trimmed to yield clean reads for downstream analysis, and detailed parameters for trimming were as previously described in the37. All sequencing data were deposited in NCBI Short Read Archive under BioProject ID PRJNA331772.

Identification of miRNA and targets

For each small RNA sequencing library, the clean reads range from 18–30 bp were obtained and then filtered through RNA databases before miRNA identification22. The transcriptome assemble from *C. japonica* was used as reference for precursors identification. The unannotated reads were undergone the miRDP package pipeline to search for miRNAs38. The core miRDeep output figures were generated for each miRNA and accessed. The precursors and mature sequences were aligned in PNRD database for annotation of miRNAs39. To quantify the abundance of miRNA, the TPM value was defined as 'counts of reads mapped to miRNA × 1,000 000'/'reads mapped to the reference genome'40.

The sequencing reads from degradom library were obtained for target site identification. The CleaveLand4.0 pipeline was used for target scanning30. The reads were mapped to *C. japonica* transcriptome and the alignment scores and p-value were calculated according the signatures (abundances of potential slicing end based on reads distributions). The t-plots were generated to visualize the miRNA directed slicing to targets. The alignment score over 5 was used to select potential miRNA targets; and p-value less than 0.05 was used to identify highly confident targets. The 2 degradome libraries were run through the target identification process separately, and the merged data combing the 2 libraries went through for the target identification as well.

DEG identification and clustering and GO enrichment analysis

The RPKM values were used to calculate the significance of gene expression changes. The differential expression analysis was carried out by using the Bioconductor package package EdgeR41. The DEGs were identified by the thresholds of log2 (RPKM) ratio >= 1, P-value < 0.05, false discovery rate <0.001. The ks statistic test (Kolmogorov-Smirnov test) was performed to identify the enrichment significance of GO term in DEG datasets42; p-value correction was performed using the Benjamini-Hochberg (BH) method. The GO terms with p-values less than 0.05 were considered significantly enriched. The enriched GO terms from each comparison were consolidated into a matrix, and the distance between comparisons was calculated by the furthest neighbor distance and clustered. The Hierarchical clustering of gene expression was performed by clustergram function in Matlab Bioinformatics toolbox with minor changes of appearance.

Phylogenetic analysis

The MADS-box genes were searched by using *Arabidopsis* genes sequences. The Matlab script of reciprocal blast against two protein databases (RBH) was used to find orthologs of MADS-box sequences in *C. japonica* and *Arabidopsis*. The phylogenetic tree was built by MEGA5 using the Neighbor Joining Tree method with minor modifications43.

Quantitative PCR analysis

To validate the gene expression of RNA-seq, Real time quantitative PCR was performed on a QuantStudio 7 Flex Real-Time PCR System (Applied Biosystems, USA). The regular ΔΔct method was used by comparing target gene and the internal reference gene (Glyceraldehyde 3-phosphate dehydrogenase, GAPDH)44. All amplifications were performed 3 times as technical replicates, and 3 biological replicates were used for each experiment. The SYBR Premix Ex Taq II kits (Cat. RR820a, Takara, Dalian, China) were used to perform the real-time PCR analyses. The relative expression values were calculated by using the default software of the QuantStudio 7 system. The primers used in the study were listed in Supplementary Table S5.

AVAILABILITY OF DATA AND MATERIALS

All relevant data are published within the paper and its supporting additional files.

ADDITIONAL INFORMATION

Publisher's note: Springer Nature remains neutral with regard to jurisdictional claims in published maps and institutional affiliations.

ACKNOWLEDGEMENTS

This work was supported by National Science Foundation of China (NSFC) Grant 31470697, Nonprofit Research Project (CAFYBB2016SZ001) of Chinese Academy of Forestry, and Forestry Industry Research Special Funds for Public Welfare Projects (201504707). We also acknowledge the supports from the Zhejiang province science and technology plan (2013C32075). We thank Drs. N. Ye and T. Yin from Nanjing Forestry University (Nanjing, China) for the help of bioinformatics analysis and sharing computing resources. We are grateful to SL. Xu and WX. Liu for technical assistances in RNA preparation and the gene expression analysis.

REFERENCES

1. Beirnaert, A. & Vanderweyen, R. Contribution à l'étude génétique et biométrique des variétiés d'Elaeis guineensis Jacq. In Publ. Inst. Nat. Etude Agron. Congo Belge. Ser. Sci. Vol. 27 1–101 (1941).
2. Corley, R. H. V. & Tinker, P. B. Selection and breeding. In The Oil Palm 133–187 (Blackwell 2003).
3. Hardon, J. J., Corley, R. H. V. & Lee, C. H. Breeding and selecting the oil palm 63–81 (Academic Press, London 1987).

4. Singh, R. et al. The oil palm SHELL gene controls oil yield and encodes a homologue of Seedstick. Nature 500, 340–344 (2013).
5. Ordway, J. et al. Expression of sep-like genes for identifying and controlling palm plant shell phenotypes. Vol. WO2015010131A3 (Malaysia 2015).
6. Ooi, L. C. -L. et al. Non-tenera Contamination and the Economic Impact of SHELL Genetic Testing in the Malaysian Independent Oil Palm Industry. Frontiers in Plant Science 7 (2016).
7. Kwong, Q. B. et al. Development and Validation of a High-Density SNP Genotyping Array for African Oil Palm. Molecular Plant 9, 1132–1141 (2016).
8. Rosenquist, E. A. An overview of breeding technology and selection in Elaeis guineensis. In International Oil Palm Development Conference - Agriculture 5–25 (Palm Oil Research Institute Malaysia, Kuala Lumpur 1990).
9. Tang, T. L. The possible use of pisifera in plantation. Malay Agric J 48, 57–68 (1971).
10. Rao, V., Law, I. H., Shaharudin, Z. & Chia, C. C. Ekona and AVROS - a tale of two pisiferas. In 1999 PORIM Int. Palm Oil Congr. (PORIM, Kuala Lumpur 1999).
11. Sterling, F. & Alvorado, A. Ekona and Calabar as alternative sources of male parents in the commercial production of the oil palm seed. ASD Oil Palm Papers 11, 23–32 (1995).
12. Xue, Y. et al. Spread of an Inactive Form of Caspase-12 in Humans Is Due to Recent Positive Selection. American Journal of Human Genetics 78, 659–670 (2006).
13. Kelly, M. & Semsarian, C. Multiple Mutations in Genetic Cardiovascular Disease. Circulation: Cardiovascular Genetics 2, 182 (2009).
14. Ingles, J. et al. Compound and double mutations in patients with hypertrophic cardiomyopathy: implications for genetic testing and counselling. Journal of Medical Genetics 42, e59 (2005).
15. Zheng, X. et al. A high-performance computing toolset for relatedness and principal component analysis of SNP data. Bioinformatics 28, 3326–3328 (2012).

CHAPTER 12

Characterizing Haploinsufficiency of *SHELL* Gene to Improve Fruit form Prediction in Introgressive Hybrids of Oil Palm

Chee-Keng Teh[1], Siti Dalila Muaz[1], Praveena Tangaya[1], Po-Yee Fong[1], Ai-Ling Ong [1], Sean Mayes[2], Fook-Tim Chew [3], Harikrishna Kulaveerasingam[1] & David Appleton [1]

[1]*Biotechnology & Breeding Department, Sime Darby Plantation R&D Centre, Selangor, Malaysia.*

[2]*School of Biosciences, University of Nottingham, Nottingham, UK.*

[3] *Department of Biological Sciences, National University of Singapore, Singapore, Singapore.*

ABSTRACT

The fundamental trait in selective breeding of oil palm (*Eleais guineensis* Jacq.) is the shell thickness surrounding the kernel. The monogenic shell thickness is inversely correlated to mesocarp thickness, where the crude palm oil accumulates. Commercial thin-shelled *tenera* derived from thick-shelled *dura* × shell-less *pisifera* generally contain 30% higher oil per bunch. Two mutations, *sh*^{MPOB} (M1) and sh^{AVROS} (M2) in the *SHELL* gene – a type II MADS-box transcription factor mainly present in AVROS and Nigerian origins, were reported to be responsible for different fruit forms. In this study, we have tested 1,339 samples maintained in Sime Darby Plantation using both mutations. Five genotype-phenotype discrepancies and eight controls were then re-tested with all five reported mutations (sh^{AVROS}, sh^{MPOB}, sh^{MPOB2}, sh^{MPOB3} and sh^{MPOB4}) within the same gene. The integration of genotypic data, pedigree records and shell formation model further explained the haploinsufficiency effect on the *SHELL* gene with different number of functional copies. Some rare mutations were also identified, suggesting a need to further confirm the existence of cis-compound mutations in the gene. With this, the prediction accuracy of fruit forms can be further improved, especially in introgressive hybrids of oil palm. Understanding causative variant segregation is extremely important, even for monogenic traits such as shell thickness in oil palm.

INTRODUCTION

Elaeis, the only genus under the Arecaceae family, produces edible oil. *Elaeis guineensis* Jacq. of West African origin is the predominantly used species, planted due to oil yield superiority, whereas the planting of another related species, *E. oleifera* (HBK) Cortes is usually restricted to Latin America, where the species originated from. In oil palm, three naturally occurring fruit forms (*dura*, *tenera* and *pisifera*) can be recognized based on their shell thickness, which is controlled by a single gene with two co-dominant alleles[1]. The thinner shell usually confers additional mesocarp to the fruit, allowing more oil to accumulate. Hence, planters in Southeast Asia in the 1960s switched from planting Deli *dura* (thick-shelled) fruit to *tenera* (thin-shelled) fruit by making hybrids between Deli *dura* and AVROS *pisifera* (shell-less, typically female sterile), realizing a 30% increment in oil/hectare[2,3].

Two independent mutations in the *SHELL* gene – a type II MADS-box transcription factor which is homologous with *Arabidopsis* SEEDSTICK (STK) and rice OsMADS13 – were recently reported to be responsible for the fruit forms[4]. The two mutations, sh^{MPOB} (T → C, for leucine → proline amino acid change) and sh^{AVROS} (A → T, for lysine → asparagine amino acid change) were identified among the descendants of a Nigerian *tenera* accession and a Congo-derived AVROS *pisifera* accession, respectively[4]. These mutations result in amino acid changes within the MADS-box domain and disrupt heterodimerization (sh^{MPOB}) and DNA binding (sh^{AVROS})[4,5], respectively. The homozygous wild-type $sh^{DeliDura}$ nucleotides at both variant positions are carried by *dura* palms. Palms which are heterozygous for either of the two mutations ($sh^{DeliDura}/sh^{MPOB}$ or $sh^{DeliDura}/sh^{AVROS}$) produce *tenera* palms. *Pisifera* palms are homozygous for either mutation or heteroallelic for both mutations. In other words, the most obvious causative mechanism for the co-dominant fruit form trait is mainly due to haploinsufficiency, wherein only a single functional copy of *SHELL* gene in a diploid genome is insufficient to maintain the wild-type function i.e. *dura*.

From the mass screening of commercial *tenera* materials, another three non-synonymous variants (sh^{MPOB2}, sh^{MPOB3} and sh^{MPOB4}) within the same MADS-box coding regions controlling the fruit form phenotype in a similar manner were then identified6. An example of potential use would be for *pisifera* breeding, where both *pisifera* and *tenera* are assessed, but *dura* discarded. The markers developed could be used to remove the undesired *dura* seeds derived from *tenera* × *tenera* crosses at the pre-nursery stage. This will improve both genetic gain over generations and trial management through reduced land requirement. Also, the *SHELL* genetic testing could potentially increase the oil yield of planted areas by eliminating the lower-yielding, non-*tenera* contaminants prior to field planting6. The reported mutations in the *SHELL* gene are applicable in Nigerian and the commercial AVROS accessions, separately. However, some breeding institutes are aiming to broaden the genetic base of AVROS accessions through introgression, or have switched to other genetic resources for paternal lineages, such as Ekona, Yangambi, LaMe and *Serdang* fertile *pisifera* (SP). A better understanding of the heterozygous action of the loss-of-function mutations in *SHELL* gene is also essential to further improve fruit form predictability, especially for introgressive hybrids of oil palm.

RESULT AND DISCUSSION

In this study, we carried out *SHELL* genetic testing on Sime Darby Plantation's repertoire of genetic materials using the sh^{MPOB} and sh^{AVROS}tests. The objectives were to validate the *SHELL* markers on these materials and understand the effect of introgression from different *pisifera* sources on fruit form prediction. As the first phase of validation, three fruit forms (120 *dura*, 305 *tenera* and 162 *pisifera*) belonging to four paternal stocks were selected (Table 1). The genetic stratification of the 587 palms was determined through principal component analysis (PCA) based on our OP200K genotyping array7. Four main clusters i.e. I) Ekona, II) AVROS, III) AVROS x SP, IV) AVROS-SP × Ekona (Fig. 1A) were identified and these showed a good agreement with the known pedigree information (Fig. 1B). The AVROS *pisifera* is selected for superiority in growth uniformity, precocity and mesocarp oil content when combined with Deli *dura* origins to produce commercial *tenera* planting materials. However, the narrow genetic base of AVROS (derived from only one palm selected from the self-pollination of the SP540 palm which was again self-pollinated, although there is evidence for some illegitimacy from the final self-pollination) may hinder future breeding progress in oil palm8. Oil palm breeders, hence, are actively looking for other genetic resources to enrich the genetic diversity of the current AVROS accession (cluster II). One of these is SP, which was initially developed in the 1960's for shell-less fertile *pisifera* planting materials9, but seed germination and oil extraction from *pisifera* fruits still remains challenging2. By having AVROS × SP (cluster III), the breeders aimed to further reduce shell thickness in the commercial *tenera* without complete loss of the shell. Another examined option was the Ekona origin (cluster I) which originated from Cameroon. The trial result showed that Ekona *pisifera* gave smaller trunk height increments, as good or even better bunch production and oil/bunch than the AVROS *pisifera*10, 11. In Sime Darby, we have further introgressed cluster I into cluster III to generate AVROS-SP × Ekona hybrids (cluster IV) (Fig. 1B). This relationship was clearly reflected as cluster IV was located in between cluster I and cluster III (Fig. 1A).

Table 1: The first validation set consists of three fruit forms belong to four paternal stocks.

Accession	No. of families	D	T	P	Total
AVROS	25	73	197	84	354
AVROS × SP	15	12	70	58	140
AVROS-SP × Ekona	1	23	24	10	57
Ekona	1	12	14	10	36
Total	42	120	305	162	**587**

The fruit forms of individual palms were phenotyped based on visual inspection of shell thickness and the diagnostic presence of a fiber ring around the kernel in the *tenera* fruit. To validate the accuracy of *SHELL* markers, the fruit form data was compared to the genotype data. The two mutations were defined as M1 for sh^{MPOB} and M2 for sh^{AVROS}, respectively. In general, the prediction accuracy of the *SHELL* markers was successfully validated (100% accuracy), except in the AVROS-SP × Ekona accession (Fig. 2) (Supplementary Table 1). The AVROS accession and the Ekona accession carried sh^{AVROS} and sh^{MPOB} mutations, respectively, which may have arisen independently in these accessions derived from central

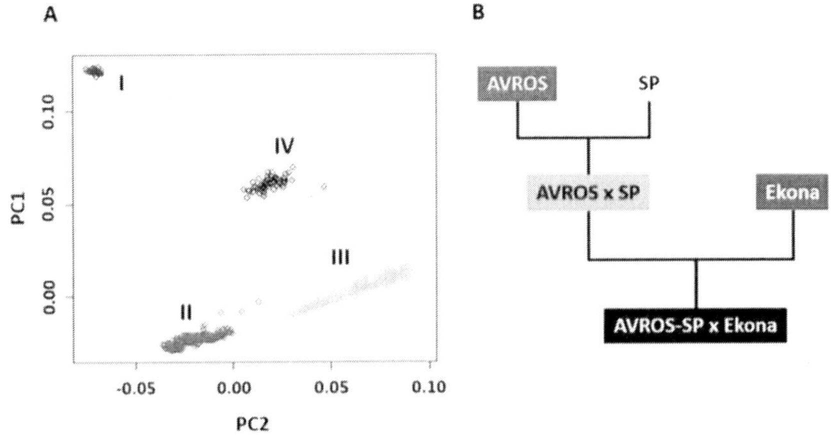

Figure 1. Genetic stratification of 587 oil palm samples, as the first validation set representing four paternal stocks. (**A**) Four clusters i.e. I) Ekona, II) AVROS, III) AVROS x SP, and IV) AVROS-SP × Ekona corresponding to their known origins were detected. PC1 and PC2 indicate the scores of principle components 1 and 2, respectively. (**B**) A pedigree of the four origins. SP – *Serdang* fertile *pisifera*.

and western Africa. The introgressive hybridization between AVROS × SP and Ekona origins introduced polymorphism for both M1 and M2 (Fig. 2) (Supplementary Table 1) into their progenies. Using single marker prediction, neither M1 nor M2 correctly predicted the segregating *dura*, *tenera* and *pisifera* fruit forms. We then expanded the *SHELL* testing using a second validation set with a total of 752 palms representing AVROS, MPOB, Congo, Tanzanian and four Introgressive hybrid origins (AVROS × Nigerian, Cameroon × Congo, Deli-Nigerian × URT-Calabar and URT × Calabar) (Table 2) (Supplementary Table 2). The same result was observed, where the M2 was carried by Congo and Tanzanian originated from central and eastern Africa. The fruit forms of the pure accessions (except AVROS accession with 72.00% of accuracy) thus can be predicted accurately using a single marker, but not in the introgressed hybrids (Table 2) (Supplementary Table 2).

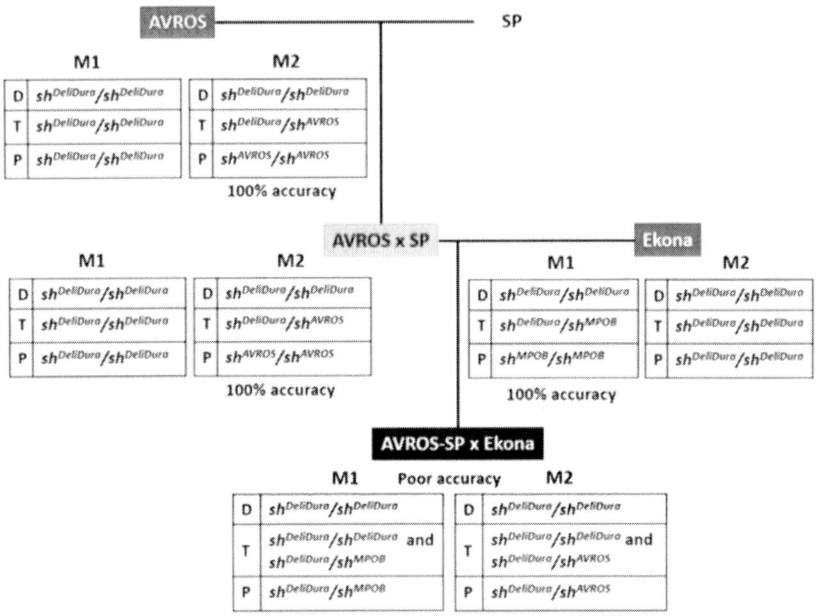

Figure 2. Fruit form prediction using *SHELL* markers in different accessions. Single marker prediction was highly accurate in the accessions tested, except in AVROS-SP (*Serdang* fertile *pisifera*) × Ekona.

Table 2: The prediction accuracy of fruit form based on haploinsufficiency model and genotypes of assayed samples.

Accession	N	Predicted fruit form			Observed fruit form			Observed genotype			Accuracy (%)
		D	T	P	D	T	P	D	T	P	
AVROS	11	3	5	3	2	4	5	(*T/T, A/A*)	(*T/T, T/A*)	(*T/T, T/T*)	72.00
*Nigerian × AVROS	281	69	206	6	69	206	6	(*T/T, A/A*)	(*C/T, A/A*) & (*T/T, T/A*)	(*C/T, T/A*), (*C/C, A/A*) & (*T/T, T/T*)	100.00
*Cameroon × Congo	57	13	29	15	12	29	16	(*T/T, A/A*)	(*C/T, A/A*) & (*T/T, T/A*)	(*C/T, T/A*), (*C/C, A/A*) & (*T/T, T/T*)	96.00
*Deli-Nigerian × URT- Calabar	95	11	84	0	11	84	0	(*T/T, A/A*)	(*C/T, A/A*) & (*T/T, T/A*)		100.00
*URT × Calabar	1	0	0	1	0	0	1			(*C/T, T/A*)	100.00
MPOB	136	136	0	0	136	0	0	(*T/T, A/A*)			100.00
Tanzanian	131	97	33	1	97	33	1	(*T/T, A/A*)	(*T/T, T/A*)	(*T/T, T/T*)	100.00
Congo	40	14	20	6	14	20	6	(*T/T, A/A*)	(*T/T, T/A*)	(*T/T, T/T*)	100.00

To interpret this further, we mapped all the genotype-phenotype relationships onto an AVROS-SP *tenera* × Ekona *tenera* hybrid population (Fig. 3). The AVROS-SP *tenera* containing a single heterozygous mutation at M2 ($sh^{DeliDura}/sh^{DeliDura}$; $sh^{DeliDura}/sh^{AVROS}$) was hybridized to the Ekona *tenera* that carried the other single heterozygous mutation at M1

($sh^{DeliDura}/sh^{MPOB}$; $sh^{DeliDura}/sh^{DeliDura}$), and the cross produced four genotypes in the hybrid population, according to Mendelian inheritance. In this population, we confirmed the presence of the haploinsufficiency effect on the *SHELL* gene. Not all loss-of-function mutations of an encoded protein are deleterious. Some heterozygous loss-of-function mutations, mainly due to haploinsufficiency, even give selective advantages12. The scenario was clearly demonstrated in thin-shelled *tenera*, which produce 30% more oil with full fertility compared to their *dura* and *pisifera* parents. The *tenera* palms in this study contained only a single functional copy of a gene per diploid genome causing an inability to produce enough functional protein as wild-type *dura* palms in the cells; this leads to a haploinsufficiency of the heterodimer which mediates the effects of the *SHELL* gene. Hence, we scored the wild-type $sh^{DeliDura}/sh^{DeliDura}$ which occurred at both M1 and M2 as two functional copies; one functional copy for a single mutation at either the M1 or M2 variant position per diploid; and nil functional copy for multiple mutations at both M1 and M2 positions. Next, we link the haploinsufficiency effect to the shell formation model through SHELL-SEP-like protein heterodimerization and DNA binding[4,5,6] (Fig. 3). The *dura* palms with two functional copies were not disrupted in shell formation at all, as expected. As for *tenera*, the two genotypes possible only contributed one functional copy of the SHELL protein, which explained the thinner shell compared to *dura*. One of the alleles led to an inactive protein heterodimer due to failure in either DNA binding for sh^{AVROS} carriers, or SHELL-SEP-like dimerization itself for sh^{MPOB} carriers. All the shell-less *pisifera* palms were heteroallelic for both mutations (nil functional copy). Therefore, it is important to determine the polymorphism at M1 and M2 first to decide which prediction method should be used. Here, we can observe the clear effect of the causal variants influencing the shell thickness when introgression between certain populations takes place. By shifting single marker analysis to functional copy counting, the prediction accuracies of fruit form in the four introgressive hybrids were significantly improved, ranging from 96.00 to 100.00% (Table 2). Nevertheless, we could not observe any association between both mutations and fertile *pisifera* (with the availability of the four fertile *pisifera*) in AVROS-SP accessions, indicating that the *SHELL* gene itself might not be responsible solely for female sterility in *pisifera* (Supplementary Table 1).

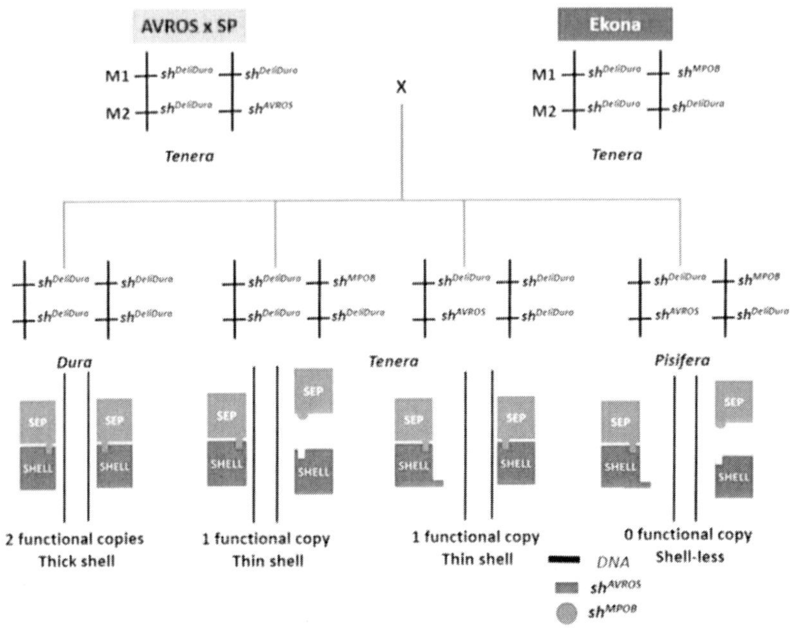

Figure 3. The mapping of genotype-phenotype relationships on an AVROS-SP *tenera* × Ekona *tenera* hybrid population and SHELL-SEP-like protein interaction model in three fruit forms influenced by functional copy (haploinsufficiency effect). The *pisifera* fruit form was due to trans-compound heterozygosity in *SHELL* gene.

The global accuracy in the second validation set was extremely high (99.34%), but unexpected low accuracy (72.00%) in one AVROS accession drew our attention (Table 2). The phenotype-genotype discrepancies might be due to the presence of other novel mutants on the *SHELL* gene, as reported6. Hence, a total of 13 palms (5 phenotype-genotype discrepancies and 8 controls) were tested using all five reported mutant *SHELL* alleles including sh^{AVROS}, sh^{MPOB}, sh^{MPOB2}, sh^{MPOB3} and sh^{MPOB4} within the MADS-box domain. The three AVROS discrepancies were successfully rectified based on sh^{MPOB2} and sh^{AVROS}mutations (Table 3), achieving 100.00% prediction accuracy in the accession. However, this did not apply to two Cameroon × Congo discrepancies, indicating probable occurrence of other novel mutations (Table 3). We thus expect another heterozygous mutation for V599 *pisifera* and V927 *tenera*. Interestingly, we also discovered a heteroallelic case for three mutations in V121 *pisifera* in Nigerian × AVROS accession. This implied that some accessions/origins can possess more than two mutations or a different cluster of mutations, although it can be rare. However, this will become prominent during introgression of wild germplasm. The compound mutation in the *SHELL* gene for fruit form phenotype in oil palm was firstly discussed in this study. Multiple mutations that occur in the same gene and in different genes, are defined as compound mutation and double mutation, respectively13, 14 (Fig. 4). All the reported mutations are located in the *SHELL* gene, therefore known as compound mutations which can be heterozygous or homozygous. The loss-of-function compound mutations can happen in cis (mutations on the same chromosome) (Fig. 4B) and trans forms (mutations on the different chromosome) (Fig. 4C) too. To date, the haploinsufficient *SHELL* gene for *pisifera* is mainly explained by trans-compound heterozygous mutations only, particularly in introgressive hybrids. Nevertheless, we should not ignore the possible existence of cis-compound heterozygous palms that are still able to produce one functional copy to express the *tenera* fruit form (Fig. 4B). Hence, these individual palms can be misinterpreted as *pisifera* if based on heteroallelic tests for both mutations only.

In conclusion, the haploinsufficiency effect on *SHELL* gene responsible for fruit form variation in oil palm was further characterized based on identity-by-descent. The reported sh^{AVROS}, sh^{MPOB}, sh^{MPOB2}, sh^{MPOB3} and sh^{MPOB4} within MADS-box domain should be adopted together to improve prediction accuracy, especially in the introgressive hybrid materials. Furthermore, the existence of cis-compound heterozygous palms should be confirmed through continuous screening of various germplasm. This study provides a good reference for other crop and animal breeding programs where the understanding of causative variant segregation is extremely important, even for monogenic traits such as oil palm shell thickness.

Table 3: The prediction accuracy of fruit form based on haploinsufficiency model and genotypes of five discrepancies and eight controls tested in the second validation set.

Sample ID	Accession	Sh^{MPOB2}	Sh^{MPOB4}	Sh^{MPOB}	Sh^{AVROS}	Sh^{MPOB3}	Predicted Fruit Form	Observed Fruit Form	Accuracy
V907	*Cameroon × Congo	*A/A*	*A/A*	*T/T*	*A/A*	*C/C*	D	D	Concordant
V670	*Congo	*A/A*	*A/A*	*T/T*	*A/A*	*C/C*	D	D	Concordant
V749	*MPOB	*A/A*	*A/A*	*T/T*	*A/A*	*C/C*	D	D	Concordant
V121	*Nigerian × AVROS	*A/A*	*T/A***	*C/T***	*T/A***	*C/C*	P	P	Concordant
V152	*Tanzanian	*A/A*	*A/A*	*T/T*	*T/T****	*C/C*	P	P	Concordant
V685	*Congo	*A/A*	*A/A*	*T/T*	*T/T****	*C/C*	P	P	Concordant
V646	*Congo	*A/A*	*A/A*	*T/T*	*T/A***	*C/C*	T	T	Concordant
V086	*Nigerian × AVROS	*A/A*	*A/A*	*C/T***	*A/A*	*C/C*	T	T	Concordant
C2	AVROS	*C/A***	*A/A*	*T/T*	*A/A*	*C/C*	T	T	Corrected
C1	AVROS	*C/A***	*A/A*	*T/T*	*T/A***	*C/C*	P	P	Corrected
C3	AVROS	*C/A***	*A/A*	*T/T*	*T/A***	*C/C*	P	P	Corrected
V927	Cameroon × Congo	*A/A*	*A/A*	*T/T*	*A/A*	*C/C*	D	T	Miss
V599	Cameroon × Congo	*A/A*	*A/A*	*C/T**	*A/A*	*C/C*	T	P	Miss

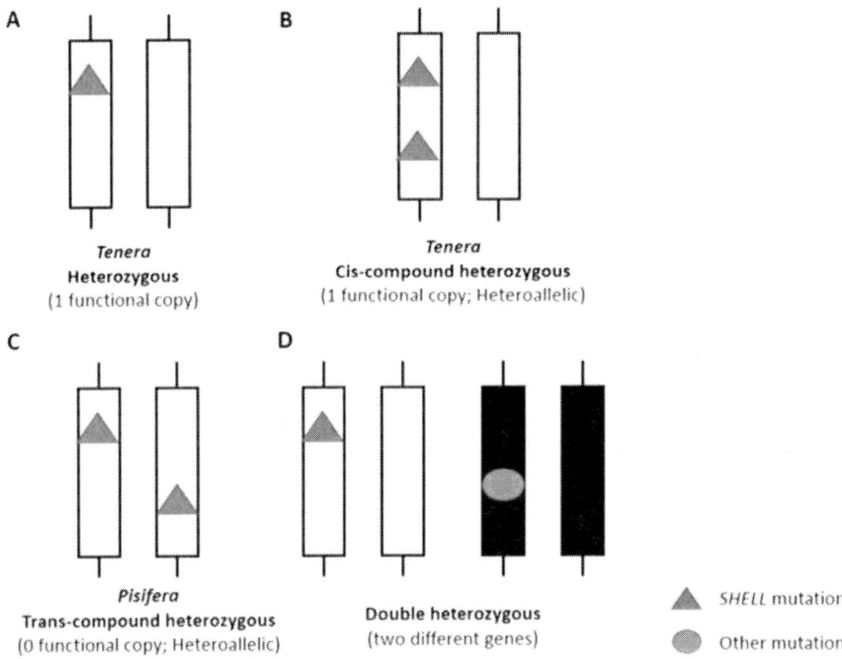

Figure 4. The zygosity of loss-of-function mutations in *SHELL* gene. (**A**) *Tenera* palm that carried a single heterozygous mutation, only produce 1 functional copy of the gene. (**B**) Cis-compound heterozygous *tenera* palms produce 1 functional copy, when both mutations occur on the same chromosome. (**C**) Trans-compound heterozygous *pisifera* palms produce 0 functional copy, when both mutations occur on different chromosome. (**D**) In double heterozygous individuals, the mutations occurred in different genes.

METHODS

Sampling and DNA Preparation

A total of 1,339 oil palm samples were selected and divided into two validation sets. The first validation set included 587 palms representing four different accessions, i.e. AVROS, AVROS × SP, AVROS-SP × Ekona and Ekona maintained at Sime Darby Plantation R&D Centre in Malaysia (Table 1). The validation set consisted of three different fruit forms (thick-shelled *dura*, thin-shelled *tenera* and shell-less *pisifera*). The pedigree was provided by the breeders as indicated in Fig. 1B. Subsequently, the study was extended to the second validation set derived from more diverse backgrounds, also with a larger overall group size (752 palms) (Table 2). The background included four pure accessions (AVROS, MPOB, Tanzanian and Congo) and four introgressed hybrid populations (Nigerian × AVROS, Cameroon × Congo, Deli-Nigerian × URT Calabar and URT × Calabar). Total genomic DNA was isolated from 0.1 g of young leaf tissue (frond 0) using the DNAeasy Plant Mini Kit (Qiagen).

Genetic stratification analysis

To understand the genetic structure, the first validation set was genotyped using genome-wide SNP markers on a OP200K Infinium array7 (Illumina). The genotypic data of 200,000 SNPs were filtered based on minor allele frequency (MAF) >0.01 and >90% call rate. Subsequently, the genetic stratification among 586 oil palm samples was analyzed in the R package SNPRelate15, using the default parameters.

SHELL genotyping

About 3 ng of genomic DNA was used as template for both specific allele tests. The probe design for sh^{MPOB}, sh^{AVROS}, sh^{MPOB2}, sh^{MPOB3} and sh^{MPOB4} mutations was done at their reported genomic positions: 3,078,161 bp, 3,078,154 bp, 3,078,180 bp, 3,078,125 bp and 3,078,178 bp on Chromosome 2, namely CM002082.14, 6 based on the requirement of Kompetitive Allele Specific PCR™ (KASP™) genotyping platform. Two fluorophores FAM and HEX were used to distinguish the KASP™genotyping data. In the clustering plot, the samples marked red were homozygous for alleles reported with HEX, whereas those marked blue were homozygous for FAM allele. The heterozygous samples appear as green.

Genotype-phenotype analysis

The fruit form census was performed based on visual examination of shell thickness and the presence or absence of a fiber ring in each cross-sectioned fruitlet. In addition, fertility status of AVROS-SP *pisifera* was also recorded based on the presence of embryo in the palm kernel. The observed fruit form and fertility data were then compared to the predicted data based on the reported sh^{MPOB} and sh^{AVROS}variants4. From there, accuracy (in %) of the single marker and haploinsufficiency models in each accession, except 15 samples with missing data were evaluated (Supplementary Table 1). The single marker method predicted the fruit forms based on either sh^{MPOB} or sh^{AVROS}variant. But, the haploinsufficiency model considered both variants simultaneously. The wild-type *sh*^{DeliDura} occurred at both M1 and M2 per chromosome, with 2 functional copies of the M1-M2 haplotype for *dura*; a single mutation at either M1 or M2 per chromosome as 1 functional copy for *tenera*; and compound mutations at M1 and M2 on both haplotypes as 0 functional copy for *pisifera*. The analysis was then extended to another three mutations, including *sh*^{MPOB2}, *sh*^{MPOB3} and *sh*^{MPOB4} tested on five discrepancies and eight controls based on the same haploinsufficiency model.

ADDITIONAL INFORMATION

Publisher's note: Springer Nature remains neutral with regard to jurisdictional claims in published maps and institutional affiliations.

ACKNOWLEDGEMENTS

We would like to acknowledge the contribution of our breeders from Oil Palm Breeding Unit. We also thank Molecular Breeding & Bioinformatics Unit, Sime Darby for sampling of oil palm materials and analytical supports. The study was conducted in Sime Darby Plantation R&D Centre, which is fully supported by a grant from Sime Darby Plantation Division, Malaysia.

REFERENCES

1. Beirnaert, A. & Vanderweyen, R. Contribution à l'étude génétique et biométrique des variétiés d'Elaeis guineensis Jacq. In Publ. Inst. Nat. Etude Agron. Congo Belge. Ser. Sci. Vol. 27 1–101 (1941).
2. Corley, R. H. V. & Tinker, P. B. Selection and breeding. In The Oil Palm 133–187 (Blackwell 2003).
3. Hardon, J. J., Corley, R. H. V. & Lee, C. H. Breeding and selecting the oil palm 63–81 (Academic Press, London 1987).
4. Singh, R. et al. The oil palm SHELL gene controls oil yield and encodes a homologue of Seedstick. Nature 500, 340–344 (2013).
5. Ordway, J. et al. Expression of sep-like genes for identifying and controlling palm plant shell phenotypes. Vol. WO2015010131A3 (Malaysia 2015).
6. Ooi, L. C. -L. et al. Non-tenera Contamination and the Economic Impact of SHELL Genetic Testing in the Malaysian Independent Oil Palm Industry. Frontiers in Plant Science 7 (2016).
7. Kwong, Q. B. et al. Development and Validation of a High-Density SNP Genotyping Array for African Oil Palm. Molecular Plant 9, 1132–1141 (2016).
8. Rosenquist, E. A. An overview of breeding technology and selection in Elaeis guineensis. In International Oil Palm Development Conference - Agriculture 5–25 (Palm Oil Research Institute Malaysia, Kuala Lumpur 1990).

9. Tang, T. L. The possible use of pisifera in plantation. Malay Agric J 48, 57–68 (1971).
10. Rao, V., Law, I. H., Shaharudin, Z. & Chia, C. C. Ekona and AVROS - a tale of two pisiferas. In 1999 PORIM Int. Palm Oil Congr. (PORIM, Kuala Lumpur 1999).
11. Sterling, F. & Alvorado, A. Ekona and Calabar as alternative sources of male parents in the commercial production of the oil palm seed. ASD Oil Palm Papers 11, 23–32 (1995).
12. Xue, Y. et al. Spread of an Inactive Form of Caspase-12 in Humans Is Due to Recent Positive Selection. American Journal of Human Genetics 78, 659–670 (2006).
13. Kelly, M. & Semsarian, C. Multiple Mutations in Genetic Cardiovascular Disease. Circulation: Cardiovascular Genetics 2, 182 (2009).
14. Ingles, J. et al. Compound and double mutations in patients with hypertrophic cardiomyopathy: implications for genetic testing and counselling. Journal of Medical Genetics 42, e59 (2005).
15. Zheng, X. et al. A high-performance computing toolset for relatedness and principal component analysis of SNP data. Bioinformatics 28, 3326–3328 (2012).

CHAPTER 13

Genetic Dissection of Blackleg Resistance Loci in Rapeseed (*Brassica napus* L.)

Harsh Raman[1], Rosy Raman[1] and Nick Larkan[2]

[1] *Graham Centre for Agricultural Innovation (an alliance between NSW Department of Primary Industries and Charles Sturt University), Wagga Wagga Agricultural Institute, Wagga Wagga, Australia*

[2] *Saskatoon Research Centre, Agriculture and Agri-Food Canada, Saskatoon, Canada*

1. INTRODUCTION

Blackleg disease caused by the heterothallic ascomycete fungus *Leptosphaeria maculans* (Desm.) Ces. et de Not. (anamorph: *Phoma lingam* Tode ex Fr.), is the major disease of Brassica crops such as turnip rape *(Brassica rapa* L. syn. *B. campestris;* 2n = 2x = 20, genome AA), cabbage (*B. oleracea* L.; 2n = 2x = 18, genome CC), rapeseed (syn. canola or oilseed rape *B. napus* L.; 2n = 4x = 38, genome AACC), and *B. juncea* L. (Indian or brown mustard; 2n = 4x = 36, genome AABB) grown in temperate regions of the world. It was recorded for the first time on stems of red cabbage [1]. *B. napus* originated as a result of natural interspecific hybridization and genome doubling between the monogenomic diploid species, *B. rapa* and *B. oleracea,* in southern Europe approximately 10,000–100,000 years ago [2, 3]. However, it was selected and grown as an oilseed crop only 300-500 years ago [4, 5]. *B. napus* originally evolved as a spring or semi-winter type under the Mediterranean climates, and spread rapidly from southern to northern Europe after the development of winter *B. napus* varieties [6].

Both spring and winter types are affected by blackleg disease, particularly in Australia, Europe and North America. Currently *B. napus* is the world's third most important oilseed crop, grown on an area of over 23 million hectares and produce almost 53.3 million tonnes annually [7]. Increase in *B. napus* production has been attributed to the development and release of high yielding superior varieties including hybrids having traits such as high oil content, improved protein quality and herbicide resistance for better crop management.

Among the bacterial, fungal, viral and phytoplasmic-like diseases, blackleg is the most important global disease of *B. napus* crops and causes annual yield losses of more than $900 million in Europe, North America and Australia [8-10]. *L. maculans* has an ability to kill plants even at the seedling stage, infecting cotyledons, leaves, stems, roots and pods. Under epiphytotic conditions, this disease can cause yield losses of up to 90 per cent [11 - 13]. Therefore, control of blackleg disease has been one of the major objectives of many *B. napus* breeding programs.

2. SYMPTOMS

Blackleg disease causes two distinct symptoms; leaf lesions and stem canker. Outbreak of the fungus is characterised by dirty-whitish spots on leaves with small dark fruiting bodies (pycnidia). Black lesions are generally also seen on the leaves and deep brown lesions with a dark margin can be seen on the base of stem [11]. In severe epidemic conditions fungus girdles the stem at the crown, leading to lodging of the plant and possible severance of the stem. Typical lesions of blackleg can also occur on pods. Pod infection may leads to premature pod shatter and seed infection.

3. BIOLOGY OF THE PATHOGEN AND EPIDEMIOLOGY OF THE *L. MACULANS*

The pathogen can infect several crucifers, including cruciferous weeds. Up to 28 crucifer species have been reported as hosts [14]. During infection, the pathogen grows systemically down towards the tap root of the plant, producing severe disease symptoms at the adult plant stage characterised by stem cankers. *L. maculans* reproduces both asexually and sexually on host species and can complete several disease cycles during a single growing season. The fungus survives as mycelium, pycnidia and pseudothecia on crop residues, mainly on stubble [15, 16] subsisting from one season to the next. Sexual mating occurs on crop residues, resulting in the production of ascospores which can travel up to 8 km [17]. High humidity and moderate temperatures during vegetative growth promote disease development [18].

In Australia and most parts of Europe, *L. maculans* infection generally occurs during the seedling stage from infected seed and wind-dispersed ascospores (sexual spores), released from pseudothecia. In western Canada and Poland, asexual pycnidospores are the primary source of inoculum [19], dispersed largely by rain-splash. Under high humidity conditions, ascospores and pycnidiospores adhere to cotyledons or young leaves and germinate to produce hyphae which penetrate through stomata and wounds [9, 20, 21] and grow into sub-stomatal cavities without forming appressoria [22]. After entering into substomatal cavities, the fungus grows between the epidermis and palisade layer and then into intercellular spaces in the mesophyll of lamina. The fungus then reaches the vascular strands and grows within the plant asymptomatically, until eventually invading and killing cells of the stem cortex and causing the stem canker symptom [22-24]. Variability for virulence in *L. maculans* for the first time was reported in 1927 [25]. Australian populations of *L. maculans* have a high level of genetic variability as compared to European and North American isolates [26], along with a high diversity of avirulence genes [27]. Molecular analyses of populations of *L. maculans* have shown high gene flow within and between populations. Isolates of *L. maculans* are usually classified either on the basis of their aggressiveness or into pathogenicity groups [28].

4. MANAGEMENT OF THE *L. MACULANS*

Various practices such as crop rotation, stubble management, time of sowing, seed dressing and foliar application of fungicide, and deployment of genetic resistance have been employed to control this disease and subsequently reduce yield losses [9, 29]. Deployment of host resistance has been used as the most cost-effective and environmentally sound measure for disease control in various crops including in rapeseed. This strategy has been extensively used to manage blackleg disease especially in Australia, Canada, France, and Germany.

5. EVALUATION OF GERMPLASM FOR *L. MACULANS* RESISTANCE

An efficient and reliable method for phenotyping resistance to *L. maculans* is required for germplasm evaluation and predictive breeding including molecular mapping and gene cloning research. Various criteria are used to assess disease severity, such as severity of cotyledon or stem canker lesions, which rely principally on scales or estimates of the percent of diseased leaf tissue at either seedling (intact and detached leaf) or at adult plant stages. Symptom expression can vary with the environmental conditions, test locations (glasshouse, environment chamber and field conditions), and the method of inoculations (cotyledon, leaf and stem).

Resistance of *B. napus* germplasm to *L. maculans* is tested on the basis of disease reaction under glasshouse and/or field conditions. Cotyledon inoculations, performed under controlled conditions in either a growth chamber or glasshouse, allow for large scale and efficient screening of germplasm. Various environmental conditions such as temperature, light intensity and humidity can be reliably controlled, expediting the development of suitable resistant cultivars [30] as selections can be performed at early stages of plant development. This method also overcomes some of the uncertainties inherent in field testing with its dependence upon growing environment and further reduce the genotype by environment (G x E) interactions. Growth conditions are typically maintained with at 18°C to and 22°C. For uniform infection, a spore suspension is used to inoculate wounded cotyledons of 7 to 15 day-old seedlings [31-33]. Alternatively, seedlings can be sprayed with a spore suspension at up to the third leaf stage and kept at 100 % humidity for 48-72 hr. Spore suspensions of *L. maculans* are generally raised from single-spore isolate cultures grown on different media such as V8-agar, malt-agar and rapeseed leaf extract-agar [21, 22, 34]. Published studies used spore concentrations in the range from 4×10^6 to 1×10^8 spores per ml [31 - 33].

Doubled haploid (DH) populations were screened for resistance to *L. maculans* in the glasshouse at three plant growth stages: cotyledon, true leaf and adult plant, as well as under field conditions and reported a high correlation ($r \geq 0.82$) for disease severity between glasshouse and field grown lines [33]. Similar observations were also made by McNabb et al [35]. High correlation coefficient values suggest that the resistance to *L. maculans* can be evaluated at all three stages [33]. However among three stages, cotyledon stage was the most promising as inoculum-droplets can be kept at the inoculation site as compared to true leaves.

Assessment of adult plants for resistance to *L. maculans* populations under field conditions is considered very important for the selection of resistant germplasm by the rapeseed breeders. Inoculum is provided by either spreading infected stubble in a disease nursery or spraying plants with fungal spore suspension. Two measures; disease severity and disease incidence are commonly used for evaluating resistance to *L. maculans*. However, disease severity is much more difficult to estimate than disease incidence, due to the G x E interactions and unreliable and inconsistent estimation of canker lesions, even within the same genotype, particularly when infection is not uniform. The use of increased sample size (25 to 50 plants/genotype) and reliable and congenial growing conditions for the disease development will allow better estimation of canker lesions.

Assessment of blackleg resistance under field conditions is usually performed by exposing the plants to a mixed population of *L. maculans* races, which can make the detection of race-specific *R*-genes difficult. No relationship between the degree of cotyledon-lesion development at the seedling stage and crown canker development in mature plants was observed in the intercross population derived from Maluka/Niklas [36]. This study concluded the limited value of the cotyledon test in screening for adult plant blackleg resistance. Similarly a lack of correlation between cotyledon (seedling) resistance and stem (adult plant) resistance in *B. napus* and B genome sources has also been reported [37]. Recently, a poor correlation between seedling and field reactions was reported in the DH from Skipton/Ag-Spectrum which could have been due to the prevalence of different pathotypes under field conditions as contrary to cotyledon test, where often a specific isolate is used for phenotyping [32]. In order to mimic field conditions and increase reliability of disease development, an ascospore shower test [38] has been used for germplasm evaluation and varietal release in Australia. In this test, stubble with mature pseudothecia is sprayed with distilled water until run-off, producing 'ascospore shower'. The infected plants can then be assessed for resistance at both the cotyledon and adult plant stages. This method has shown a high correlation with canker lesions scored under field conditions [39].

6. NATURAL GENETIC VARIATION FOR RESISTANCE TO *L. MACULANS*

The introgression of blackleg resistance (*R*) genes into *B. napus* germplasm for blackleg disease management is one of the major objectives of breeding programs aiming to release cultivars in disease-prone areas. Genetic variation for resistance to *L. maculans* exists within *B. napus* germplasm [39, 40, 41]. Some other Brassica species such as *B. rapa*, *B. juncea*, *B. nigra* (black mustard; 2n = 2x = 16, genome BB) and *B. carinata* (Abyssinian or Ethiopian mustard; 2n = 34, genome BBCC), as well as other crucifers such as *Sinapis arvensis* have been reported to carry resistance [42-53]. Some of these sources were utilised in transferring resistance into *B. napus* breeding lines and cultivars. A continuous variation for blackleg resistance in a world-wide collection of *B. rapa* genotypes was reported [54]. None of genotypes were completely susceptible or completely resistant to either *L. maculans* pathotypes used. However, some *B. rapa* accessions that were either highly resistant or completely susceptible were identified (Raman et al., unpublished) in a set of differential cultivars currently being used in Australia [39].

It has been reported that all B genome Brassica species; *B. nigra, B. carinata* and *B. juncea* carry complete resistance to *L. maculans* which remains effective throughout the life of the plant [40], however susceptible *B. juncea* cultivars have also been identified [55] demonstrating that complete resistance is not a feature of all B genomes. Some B genome resistance genes have been introgressed into *B. napus* lines. [47, 56-59]. Earlier studies have shown that C genome species of the *Brassica* are susceptible to blackleg [50, 53, 60]. However, a recent study [61] evaluated three accessions of *B. oleracea* var. *virids*, collected from the USDA germplasm collection and found that the accession NSL6146 was moderately resistant to *L. maculans*.

Genetic resources for adult plant resistance are very limited and most of them are derived from the French cultivar Jet Neuf [62]. Efforts are currently being made to identify both qualitative and quantitative resistance in the Australian Brassica Germplasm Improvement Programs.

7. INHERITANCE OF RESISTANCE TO *L. MACULANS*

Genetic inheritance studies revealed that resistance to *L. maculans* is complex. Resistance is either described as qualitative (also referred as monogenic/seedling/race-specific resistance/vertical resistance) or quantitative (also referred as polygenic/adult plant/race non-specific resistance/horizontal resistance) in *Brassica*.

7.1. Qualitative Resistance

Monogenic inheritance was reported in several spring and winter cultivars of *B. napus* such as Cresor, Maluka, Dunkeld, Maluka, Skipton, and Major [32, 63-67]. Eighteen major genes for resistance to *L. maculans*; *Rlm1* to *Rlm11*, *RlmS*, *LepR1* to *LepR4*, *BLMR1* and *BLMR2*, have been identified in Brassica species; *B. rapa*, *B. napus*, *B. juncea* and *B. nigra* [31, 32, 40, 45, 68-73]. Six of them, *Rlm1*, *Rlm2*, *Rlm3*, *Rlm4*, *Rlm7* and *Rlm9* were identified in *B. napus*, all of them except *Rlm2* were clustered genetically on chromosome A07 [74]. *Rlm2* was mapped on chromosome A10 [45]. The *Rlm5* and *Rlm6* were identified in *B. juncea*; *Rlm8* and *Rlm11* in *B. rapa*, and *Rlm10* was identified in *B. nigra*. Four resistance genes; *LepR1*, *LepR2*, *LepR3*, and *LepR4* were introgressed into *B. napus* from *B. rapa* subsp. *sylvestris* (Table 1).

TABLE 1. Molecular mapping of qualitative genes for resistance to *Leptosphaeria maculans* in *Brassica*.* BC: Backcross population, DH: Doubled haploid population. # loci not mapped with molecular markers to date.

Species	Locus	*Population	Phenotyping stage	Marker type	Mapping strategy	chromosome	Linked markers/interval	Reference
B. napus	*Rlm1*	Maxol/S006 (140 DH)	Cotyledon inoculation	RAPD	Bulked segregant analysis	A7	T04 680 (14cM)	74
		Quinta/Score (110 F_2)	Cotyledon inoculation	RAPD	Bulked segregant analysis	A7	C02 1375/O15.1360 (19cM)	57, 74
		Maxol/Westar-10 (96 DH)	Cotyledon inoculation and stem canker	SSR, DArT	Whole genome mapping	A7	Xna12a-02a/Xra2-a05b	82
		Columbus/Westar-10	Cotyledon inoculation and stem canker	SSR	A7 chromosome specific mapping	A7	Xol12-e03a/Xna12-a02a	82
B. napus	*Rlm2*	Glacier/Score (110 F_2)	Cotyledon inoculation	RAPD	Bulked segregant analysis	A7	M08.1200, M08.600, P02.700	57, 74
		Glacier/Yudal (BC189)	Cotyledon inoculation	RAPD	Bulked segregant analysis	A7	M03 1200, M08.600, P02.700	74
		Darmor/Samourai (133 DH)	Cotyledon, Field	RAPD	Bulked segregant analysis	A7	M03 1200 (10cM),	74
B. napus	*Rlm3*	Maxol/S006 (140DH)	Cotyledon inoculation	RAPD	Bulked segregant analysis	A7	Q12.750 (7cM)	74
B. napus	*Rlm4*	Quinta/Score (110F2)	Cotyledon inoculation	RAPD	Bulked segregant analysis	A7	C02 1375 (3.6 cM)/ O15.1360 (~33 cM)	74
		Skipton/Ag-Spectrum	Cotyledon and Stem canker	SSR	Whole genome mapping		BRMS075 (~0.7 cM)	3[illegible]
B. juncea	*Rlm6*	Recombinant lines (*B. napus-B. juncea*)	Cotyledon and field test	RAPD/ RFLP	Bulked segregant analysis	B8	OPG02.800, OPT01	47
B. napus	*Rlm7*	2311.1/Darmor (221 F_2)	Cotyledon	RAPD	Bulked segregant analysis	A7	T12.650 (4cM)	74, 85
B. napus	*Rlm9*	Darmor-bzh/Yudal (132 DH)	Cotyledon	RAPD	Bulked segregant analysis	A7	T12.650/C02 1375	74
B. nigra	*Rlm10*	Addition lines (Darmor/Junius)	Cotyledon test	Isozyme RAPD	Whole genome mapping	B4	OPA11.1200, OPC19.3300	83, 84
B. rapa ssp. *sylvestris*	*LepR1*	6270/Springfield (DHP95)	Cotyledon inoculation and field resistance	RFLP	Whole genome analysis	A2 (N2)	pR4b, pO85h, pW180b, pN181a, pW207a	31
B. rapa ssp. *sylvestris*	*LepR2*	6279/3027 (DHP96)	Cotyledon inoculation and field resistance	RFLP	Whole genome analysis	A10 (N10)	pN21b, pR34b, pN53b	31
B. napus	*LepR3*	Surpass400/Westar (N-o-1)-BC	Cotyledon inoculation	SSR	A1 and A10 chromosome specific mapping		sR12281a (2.2 cM) sN2428Rb (0.7 cM)	69

		Topas (DH16516)/ Surpass400	Cotyledon inoculaton	SSR, SCAR	A10 chromosome specific mapping	A10 (N10)	Ind10-12	79
B. rapa ssp. *sylvestris*	*LepR4*	16S/PAS12//16S (BC_1S_1)	Cotyledon inoculation Disease nursery (field)	SSR	A genome specific marker analysis	A6	sN2189b (8.8cM) sR9571a (8.3 cM)	77
B. napus	*BLMR1*	Surpass400/Westar (1513 F_3BC_1)	Cotyledon inoculation	SRAP, SNP	Selective genotyping	A10 (N10)	80E24a (0.1 cM)	70
B. napus	*BLMR2*	Surpass400/Westar (1513 F_3BC_1)	Cotyledon inoculation	SRAP, SNP	Selective genotyping	A10 (N10)	R278 (1.2 cM)	70
B. napus	*LmFr1*	Cresor (resistant)/ Westar (susceptible)	Field/artificial inoculation	RFLP	Whole genome analysis	Linkage group 6 (A7)	cDNA011/cDNA110	64
B. napus	*LEM1*	Major (resistant)/Stellar (susceptible)	Cotyledon Stem inoculation	RFLP	Whole genome analysis	A7	TG5D9b/WG5A1A	65
B. napus	*cLmR1*	Shiralee/90-3046 (153 DH lines)	Cotyledon incoulation	RFLP, RAPD	Bulked segregant analysis	A7	RAPD654 (~4.8cM)	66
B. napus	*cLmR1*	Shiralee/PSA12 (BC1 lines)	Cotyledon incoulation	RFLP, EST, SCAR	Bulked segregant analysis	A7	est126M9a/est149M9d	140
B. napus	*cLmR1*	DH12075/PSA12 (BC1 lines)	Cotyledon incoulation	RFLP, EST, SCAR	Bulked segregant analysis	A7	est126M9a/est149M9d	140
B. napus	*cLmR1*	Maluka/90-3046 (34 DH lines)	Cotyledon incoulation	RFLP, RAPD	Bulked segregant analysis	A7	RAPD654 (~4.8cM)	66
B. napus	*cRLM (cRLMm)*	Maluka/Westar	Cotyledon, Adult	RFLP, AFLP, RAPD	Bulked segregant analysis	A7	22-25	67
B. napus	*cRLM (cRLMrb)*	RB87-62/Westar	Cotyledon, adult plant	RFLP, AFLP, RAPD	Bulked segregant analysis	A7	22-25	67
B. napus	*cRLM (cRLMc)*	Cresor/Westar	Cotyledon, adult plant	RFLP, AFLP, RAPD	Bulked segregant analysis	A7	22-25	67
B. napus	*Rlm.wwai-A1*	Skipton/Ag-Spectrum (DH)	Cotyledon incoulation	SSR	Whole genome analysis	A1	Xpbcessrna16-Xbrms017b	32
B. napus	*QRlm.wwai-A10a*	Skipton/Ag-Spectrum (DH)	Cotyledon inoculation	SSR	Whole genome analysis	A10	Xcb10079d-Xcb10079c	32
B. napus	*Rpg$_3$Dun*	Westar/Dunkeld (F_2)	Cotyledon inoculation	SRAP	Bulked segregant analysis	A7	$NA_{12}A_{10}$-200/$NA_{12}A_{12}$-190, $BG_{30}SA_{12}$-480/$BG_{30}SA_{12}$-475/ BN204	81
B. juncea	*#Rlm5*	150-2-1, 151-2-1, Aurea, Picra	Cotyledon inoculation	-	-	-	-	71
B. repa	*#Rlm8*	156-2-1	Cotyledon inoculation	-	-	-	-	71

B. rapa	*#Rlm11*	02-159-4-1	Cotyledon inoculation	-	-	-	-	72
B. napus	*#Rlm5*	Surpass400	Cotyledon inoculation	-	-	-	-	73
B. juncea	*LMJR1*	AC Vulcan/UM3132 (F_2)	Cotyledon test	RFLP, SSR	Whole genome mapping	J13 (B3)	PN199RV (22.1 cM), sBb31143F (8.7 cM)	43
B. juncea	*LMJR2*	AC Vulcan/UM3132 (F2)	Cotyledon test	RFLP, SSR	Whole genome mapping	J18 (B8)	PN120cRI, sB1534	43
B. juncea	*rjlm2*	B genome introgression lines	cotyledon	RAPD, RGA & SCAR	B genome-specific	not defined	B5-1520, C5-1000, RGALm	80

Recently, two genes *BLMR1* and *BLMR2* in Surpass 400; an Australian cultivar developed from an interspecific cross between wild *B. rapa* subsp. *sylvestris* (resistant) from Sicily and *B. oleracea* subsp. *alboglabra* were identified [70, 76]. However, *LepR1* to *LepR4* genes are thought not be related with *Rlm* genes on the basis of their map locations, except for *Rlm2* and *LepR3*, which are phenotypically different [31, 69, 77]. It appears that loci *LepR3*, *BLMR1* and *BLMR2* localised on chromosome A10 control resistance to *L. maculans* in Surpass 400. However, Van de Wouw et al. [73] demonstrated that two independently segregating *L. maculans* avirulence (Avr) genes, *AvrLm1* corresponding to *Rlm1* (on chromosome A7) and *AvrLmS*, are responsible for inducing resistance in this cultivar. Subsequently, Larkan *et al.* [78] investigated the interaction of *AvLm1* and *AvLmS* isolates with *B. napus* populations segregating for the resistance genes *Rlm1* (from the French cultivar Quinta) and *LepR3* (from Surpass 400). This study reported that (i) *AvrLm1* interacts in a gene-for-gene manner with both *Rlm1* and *LepR3*, (ii) *AvrLmS* is not responsible for triggering the *LepR3* mediated defence response, (iii) Surpass 400 does not contain *Rlm1*, and (iv) *Rlm1* and *LepR3* may be the same genes located in two distinct loci or may have evolved as two functional genes. Recently, *LepR3* has become the first functional *B. napus* resistance gene to be cloned and was shown to encode a receptor-like protein. Additionally, *LepR3*-transgenic *B. napus* and *AvrLm1*-transgenic *L. maculans* were used to demonstrate that *AvrLm1* conveys avirulence to *LepR3*. The shared genomic location of *LepR3* and *BLMR1* also suggested that these were the same gene [79]. Several other genes such as *LmR1, ClmR1,LmFr1, cRLMm, cRLMrb, aRLMrb,* and *LEM1* have also been identified using uncharacterised isolates, which are thought to be allelic to known *R*-genes [45, 68, 74]. Qualitative resistance conferred by single major genes is usually dominant and expressed at the seedling growth stage. Qualitative *R*-genes explain majority of phenotypic variation for blackleg resistance at adult plant stage [32, 74]. However, digenic mode of inheritance has also been reported in *B. napus and B. juncea* populations [40, 80].

7.2. Quantitative Resistance

Quantitative inheritance for field resistance has been reported in segregating populations derived from *B. napus*, *B. juncea* and their hybrid derivatives [30, 32, 65, 80, 86]. Some of the QTLs identified are given in Table 2. Quantitative genetic analysis revealed that significant non-additive genetic variance for all measures of disease severity indicated the presence of strong dominance/epistasis at loci controlling blackleg resistance [36]. In the literature, the term 'QTL' as a quantitative locus has been used even when a large percent of genotypic variation is explained by the major locus. In classical genetics, QTL refers to genes that have, low heritability, non-Mendelian and quantitative accumulative effects. The majority of genetic analyses have utilised doubled-haploid (DH) populations, which are not suitable to infer modes of inheritance. Advanced intercross populations are required to interpret such phenomena, as used in [74].

8. GENE-FOR GENE INTERACTIONS

Host resistance genes (*R*-genes) interact in paired combination with pathogen avirulence (*Avr*) genes to condition resistance [89]. Two types of interactions may occur; compatible and incompatible. Compatible interaction occurs when there is an absence of an effective host defence response, due to a lack of a resistance allele in host (*r*) or an allele for virulence (*avr*) at the corresponding pathogen locus. An incompatible interaction occurs when there is no disease development due to the presence of both an effective host resistance allele (*R*) with an allele for *Avr* at the corresponding pathogen locus [90]. Biochemically, gene-for-gene interactions have been interpreted as the interaction of a race-specific pathogen elicitors with either cultivar-specific plant receptors or alternatively with a cultivar-specific signal transduction compounds [91]. Differential interaction between specific *R*-genes in the host (Brassica) and corresponding *Avr* genes of the pathogen *(L. maculans)* was first studied at the seedling stage using a cotyledon inoculation test in *B. napus* [92] and subsequently verified [57]. Qualitative and quantitative resistance differ with respect to host-pathogen interaction, as the latter does not appear to (but not proven) follow the gene-for gene hypothesis, being more effective against diverse pathogen populations (non-race specific). While quantitative resistance normally provides partial resistance to the pathogen and it is less likely to be rapidly overcome by shifting pathogen populations.

TABLE 2. Significant QTLs associated with blackleg resistance (scored as Internal infection due to canker development at adult plant stage) identified from mapping populations, * QTL with consistent effect, # range of LOD and R^2 varied with method of regression analysis (simple and composite interval mapping).* refers to predicted markers from supplementary figures ESM7-10 shown in Kaur et al [81]

Mapping Population	Stubble, Location	Flanking markers	Chromo-some	LOD# score	%Genetic variance (R^2)#	Additive effect	Reference
Av-Sapphire/ Westar10	*B. napus*, Lake Bolac, Australia	E34M15_S190/ E35M53_S416	A1	2.5-5.6	14-16	Not known (-)	86*
		E34M15_S218/ E35M53_S350	A2	1.3-3.8	4-26	-	
	Dahlen, Australia	CB10443_W258_ S269	C1	0.8-2.9	3-8	-	
	Lake Bolac	E36M47_W197/ E34M62_W127	LG1	1.0-3.6	4-10	-	
Caiman$_3$/Westar	*B. napus*, Lake Bolac	BRMS056/ E34M50_W140	A1	2.9-3.5	20-22.7	-	86*
	Dahlan	E35M53_C455/ E34M15_W271	A10	1.4-3.0	5-34	-	
		Not shown	C5	4.4-5.6	19-23	-	
Camberra$_4$/Westar	Lake Bolac	E36M55_C306/ E33M57_C306	A5	0.5-2.6	1.5-33	-	86*
	Dahlen	Na12D10_w203	A1/C1	2.9-5.1	17-18	-	
	Lake Bolac	E36M62_W414/0 1ju1fE07_cl_3b	A10	0.3-2.7	2-31	-	
		Not shown	C7-2	2.7-3.7	13-24	-	
		E33M59_W107/ E33M53_C75	LG2	2.1-2.8	14-28	-	
Darmor/Samourai	*B. napus*, Le Rheu	*BN483*	A1	2.3	6.7	Samourai	88
		BN239	A2	2.3-3.02	8.1-14.6	Darmor	
		BN182.1	A6	1.9-2.8	6.2-10.0	Samourai	
		At17	A10	2.7	11.0	Samourai	
		BN204	C2	2.0-2.2	8.0-8.4	Darmor	
		BN167	C4	2.4-3.2	6.7-12.2	Darmor	
		Vers6.9	C8	1.9	-	Samourai	

Darmor/Yudal	*B. napus*, Le Rheu, France	*OPE02.1200*	A2	2.4-5.5	3.8-8.5	Darmor	87 Delourme et al, 2008; comm. pers.)
		OPW08.1620	A4	3.3	4.8	Darmor	
		OPW05.750-Bzh	A6	5-12.2	7.2-20	Darmor	
		Bras023	A7	4.5	6.9	Darmor	
		CB10026b	A8	7.2	13.0	Darmor	
		OPW15.1470	A9	3.3	4.8	Darmor	
		Fad8	C2	5.5-6.6	8.3-13.3	Yudal	
		OPD08.1310	C4	4.7-9.5	6.7-15.2	Darmor	
		OPH06.CD1	C8	4.2	6.2	Darmor	
Rainbow/Av-Sapphire	Lake Bolac	E33M57_R105	A9-2	3.7	13	-	
Skipton/Ag-Spectrum	*B. napus* Mixed stubble,	*Xbras123/Xem1-bg11-237*	A2	7.0	11.5	Ag-Spectrum	32
	Wagga, Australia	*Xbrms319-Xbrms176*	A9	2.9	5.0	Skipton	
		Xcb10172-BnFLC10	A10b	2.2	6.2	Skipton	
		Xbrms287a-Xcb10034	C1	4.2	11.5	Ag-Spectrum	
		Xol10-c10/Xna12-c03	C2a	6.8	16.6	Skipton	
		Xpbcessrna13/Xol13-d02a	C3	4.2	24.5	Skipton	
		Xem1-bg23-89/Xol12-e03	C6	6.1	14.5	Ag-Spectrum	
	B. napus, ATR Beacon stubble, Wagga, Australia	*Xol12-f11/Xpbcessrbr21*	A1a	6.1	26.1	Ag-Spectrum	

At least ten *Avr* genes have been identified in *L. maculans*, many of which map to two gene clusters; *AvrLm1-AvrLm2-AvrLm6* and *AvrLm3-AvrLm4-AvrLm7-AvrLepR1* ([71, 72, 86, 87]. Four of the *Avr* genes; *AvrLm1, AvrLm6, AvrLm4-7* and *AvrLm11* have been cloned. It has shown that although *AvrLm1* and *AvrLm6* are physically clustered together in the *L. maculans* genome, they are not allelic forms of a single gene [85, 96]. However, *AvrLm4* and *AvrLm7* are allelic variants of a single *Avr* gene that corresponds to the two resistance genes; *Rlm4* and *Rlm7* [71, 85]. It has also been demonstrated that *AvrLm1* interacts with two distinct resistance loci; *Rlm1* and *LepR3*, though these loci are located on different chromosomes (A7 and A10, respectively) [78]. The cloning and characterisation of additional Brassica *R*-genes and *L. maculans* Avr genes will lead to a better understanding of how these functional redundancies developed. In the recent years, understanding of *L. maculans/ Brassica* interactions has increased our ability to deploy appropriate *R*-

genes in new cultivars and manage blackleg disease with the increased knowledge of the distribution of *Avr* alleles in *L. maculans* populations [27, 94, 98]. Currently, it seems that the genes involved in race-specific resistance and polygenic non-specific resistance are distinct. A better understanding of the mechanisms underlying quantitative resistance would help our understanding of the relationships between quantitative and major resistance genes [99].

9. ALIEN GENE INTROGRESSION FOR BLACKLEG RESISTANCE

Deployment of *R*-genes has been used as the most cost-effective and environmentally sound measure for disease control in various crops since a century ago when first *R*-genes were identified [100]. Conventional plant breeding methodologies have played an important role in gene introgression for disease resistance, especially in easily-crossed genetic backgrounds. As a result several cultivars rated for resistance to *L. maculans* now dominate commercial cultivation worldwide. There has been a continuous threat of 'breakdown' of resistance, especially when a resistant cultivar is grown extensively on large acreages over long period of time. For example, 'breakdown' of resistance in cultivar Surpass 400 occurred within three years of its release [101, 102] due to the evolution and spread of more virulent strain of *L. maculans*. 'Breakdown' of resistance implies that the resistance has not changed rather the pathogen population has shifted/been selected for virulence. The effectiveness of *Rlm1* in France was also greatly reduced from 1997 to 2000 following wide deployment of *Rlm1* varieties, effectively selecting for enrichment of the virulent *avrLm1* allele in *L. maculans* populations [34]. Interestingly, a similar enrichment for the virulent *avrLm1* allele was documented after the 'breakdown' of *LepR3* resistance in Australia [103]. Due to the threat of current resistance being rendered ineffective by shifting *L. maculans* populations, new effective sources of resistance are constantly in demand. In order to enlarge genetic variation for resistance to *L. maculans*, interspecific and intergeneric donor sources have been utilised. This has been achieved by conventional sexual crossing [44, 52, 75, 104] or via laboratory tools such as somatic hybridization [105], and embryo culture. Roy [52] crossed *B. juncea* and *B. napus* to introgress genes for blackleg resistance but none of the interspecific hybrids achieved the same level of *B. juncea* resistance as the donor parent. Wide hybrids (interspecific, intergeneric or intertribal) have also been produced either by sexual crossing followed by embryo culture or by somatic hybridisation as a result of protoplast fusion to transfer genes for blackleg resistance [106, 107].

Previous studies have reported hybrids between *B. napus* and *Arabidopsis thaliana*, belonging to different tribes; the Brassiceae and Sisymbrieae, respectively [108]. These hybrids were further utilized for identifying genetic regions associated with blackleg resistance [49]. Two regions localised on chromosome 3 of *A. thaliana* were shown to be linked with resistance to *L. maculans*.

Crouch *et al.* [75] transferred genes for resistance to *L. maculans* derived from *B. rapa* subsp. *sylvestris* into *B. napus*, using a resynthesised amphidiploid, as a result of hybridisation between *B. rapa* subsp. *sylvestris* and *B. oleracea* subsp. *alboglabra*. As a result, several cultivars derived from the re-synthesized *B. napus* lines were released for commercial cultivation in Australia such as Surpass 400, Surpass 404CL, Surpass 501TT, Surpass 603CL, Hyola 43, and Hyola 60. The *R*-genes *LepR1*, *LepR2* and *LepR4* have also been introgressed into *B. napus* via conventional interspecific crosses [75, 109]. Introgression of genes for resistance to *L. maculans* from *Sinapis arvensis*, *Coincya momensis* and *B. juncea* into *B. napus* was attempted [110]. Hybrid derivatives of *B. napus* and *S. arevensis*, and *B. napus* and *C. momensis* showed a high levels of resistance at the seedling (cotyledon) and/or adult plant stages. The offspring from asymmetric hybrids between *B. napus* and *B. nigra*, *B. juncea* and *B. carinata* were analysed for the presence of B genome markers and resistance to *L. maculans* [111]. This study revealed that resistance is conserved in one triplicate region in the B genome. Often, the majority of wide-hybrid derivatives exhibit unwanted traits and low frequencies of recombination between the different species which complicate the development of *B. napus* cultivars resistant to *L. maculans* by traditional breeding [43, 47]. Linkage drag due to suboptimal/undesired genes can be eliminated using the application of high density genome-wide molecular markers such as SNPs [112]. However, Rouxel and Balesdent [93] cautioned that before important breeding efforts are devoted to introgression of resistance genes from distant species into Brassica, there is a need thoroughly to evaluate their genetic control, putative redundancy and potential durability in the field.

Using transgenic technology, *R*-genes from other organisms can also be transferred irrespective of natural barriers to crossing. However, it is possible that transferred genes may not always contribute novel resistance specificities to the transgenic crop. Although several approaches have been used to induce host resistance in plants [113, 114] no major breakthrough has been made for an efficient management of blackleg disease. For example, Hennin *et al.* [115] demonstrated the expression of *Cf9* gene, which confers *Avr9*-dependant resistance to *Cladosporium fulvum* in tomato, along with co-expression of *Avr9* produced increased resistance to *L. maculans* in transgenic *B. napus* plants. Manipulation of plant defense responses is resource-expensive [116] and may be deleterious to the plant. Plants need to be selected for both appropriate expression of beneficial defense responses and avoidance of unnecessary ones [117], making artificially-induced constitutive expression of these responses an impractical solution to engineering resistance.

10. DURABILITY OF RESISTANCE TO *L. MACULANS*

Durable disease resistance can be achieved by utilisation of one or more single dominant *R*-genes [118]. However, the effectiveness of the specific *R*-genes depends on the *L. maculans* population structure, i.e. on the frequency of the corresponding *Avr* allele, which is known to differ according to regions/countries [27, 94] and the rapid evolution of virulent pathotypes. For example, the mean number of virulence alleles per isolates was reported to be higher in Australia (5.11 virulence alleles) than in Europe (4.33) and Canada (3.46) [27]. It has been suggested that there is a fitness cost associated with pathogen evolution from avirulence to virulence to overcome host resistance [38, 119].

Previous research has shown that different qualitative gene sources for resistance vary in providing effective durable resistance over period of time. For example, Light *et al.* [120] reported that the adult plant survival of French winter lines such as Doublol (*Rlm1*), Capitol (*Rlm1, Rlm3*), Columbus*1 (*Rlm1, Rlm3*), Carolus (*Rlm1, Rlm2, Rlm3*) and Rlm_EX (*Rlm7*) was higher than the Australian cultivar, AV-Sapphire and concluded that French winter canola cultivars have effective resistance under Australian conditions.

Single resistance genes do not always provide a durable resistance as has been shown in a field experiment using the *Jlm1/Rlm6* gene introgressed into *B. napus* from *B. juncea* [121]. Several incidences on the breakdown/ineffectiveness of race-specific resistance genes in Surpass 400 (*(LepR3, RlmS)*), in Vivol and Capitol (*Rlm1),* and *Rlm6* genes in *Brassica* have been reported in literature particularly when they were grown extensively [34, 94, 122]. As a consequence, breeders have to develop new cultivars and replace 'old' cultivars in order to change pathogen specificity of *R-gene* even without the knowledge of comprehensive distribution of *Avr* genes. The latter is now feasible and being used in order to monitor the pathogen population [123]. In order to avoid selection pressure against a particular *Avr* gene in the pathogen population, pyramiding of several host *R*-genes and deployment of quantitative resistance is being practiced in several crops such as in wheat, and barley. However, this strategy has not resulted in greater durability of resistance [124, 125]. In contrast, a recent study [121] demonstrated that a major *R*-gene (*Rlm6*) is more durable when expressed in a genetic background that also has quantitative resistance, indicating the need to identify and combine both qualitative and quantitative loci for blackleg resistance. Although the proposed strategy may be useful for blackleg disease management in areas where 'less' disease pressure and low variability with *L. maculans* populations exists, in Australia polygenic resistance derived from the French cultivar Jet Neuf [87], was reported to become less effective over time [37]. Additionally, several Australian cultivars which are reported to harbour both qualitative and quantitative loci for blackleg resistance are susceptible to natural populations of *L. maculans* Delourme et al [99]. It is difficult to know whether this evolution results from a change in virulence, or in aggressiveness in the pathogen populations since these polygenic-resistance cultivars may also carry specific *R*-genes [99]. In order to keep the frequency of isolates virulent towards any race–specific gene under a 'threshold' level, an integrated approach based upon best farm practices such as crop rotation, stubble management, application of fungicides and deployment of resistance genes including rotation of race-specific genes [126] needs to be implemented for sustainable canola production, especially in areas where *L. maculans* populations are highly diverse and rapidly evolving.

11. MOLECULAR DISSECTION OF QUALITATIVE AND QUANTITATIVE RESISTANCE LOCI

Molecular markers have been applied to identify loci associated with resistance to *L. maculans,* which relies on the availability of sequence variation among parental genotypes of mapping populations and diversity panels. Several genotyping methods based upon DNA hybridisation such as Restriction Fragment Length Polymorphism (RFLP) and Diversity Arrays Technology (DArT); PCR-based techniques such as Randomly Amplified Polymorphic DNA (RAPD), Simple Sequence Repeats (SSR) and Amplified Fragment Length Polymorphism (AFLP); Sequence-related amplified polymorphism (SRAP); and sequence-based analysis such as Single Nucleotide Polymorphism (SNP); Restriction site-associated DNA, (RAD) and Genotyping by Sequencing (GBS) have been developed for molecular analyses [81, 127 - 136]. RFLP, RAPD, SSR, SRAP and AFLP markers have all been used to map loci for resistance to *L. maculans* (Table 1). New marker technologies such as DArT, 60K SNP Infinium array, RAD and GBS are currently being developed and applied for mapping of blackleg resistance loci. These high-throughput approaches are expected to complement or replace low-throughput marker assays that were used previously to facilitate genetic and physical map-based cloning of resistance genes.

Loci for resistance to *L. maculans* have been mapped using linkage/QTL mapping and association mapping approaches [32, 74, 82, 86, 137] using structured (F_2, doubled-haploid (DH) and backcross) and unstructured (diversity sets/breeding lines) populations (Table 1). Bulked Segregant Analysis approach, used for the first time [138], is particularly useful when a limited number of traits are to be mapped and resources (money and time) required for extensive genotyping are limited [81]. Whole-genome analysis has been used to locate both qualitative and quantitative loci associated with resistance to *L. maculans* [32, 86]. Generally, it requires the framework linkage map of all 19 chromosomes (linkage groups) for linkage (QTL) analysis.

11.1. Qualitative Resistance

The majority of genes for resistance to *L. maculans* have been genetically mapped with molecular markers (Table 1) on chromosomes A1, A2, A6, A7, A10, B3, B4 and B8 in *Brassica species: B. rapa, B. napus, B juncea* and *B. nigra* [31, 32, 45, 68-70, 99]. None of the race-specific genes have been mapped on the C genome yet. Previous linkage mapping studies revealed that at least five resistance genes (*Rlm1, Rlm3, Rlm4, Rlm7* and *Rlm9)* are localised in a cluster within a 35 cM genomic region on chromosome A7 [32, 45, 64-68, 74, 82]. This genomic region showed extensive inter- and intra-genomic duplications, as well as intra-chromosomal tandem duplications [140]. Whether some of these *R*-genes are allelic remains unknown. For example, it was concluded that at least four resistance genes *Rlm3, Rlm4, Rlm7,* and *Rlm9* could correspond to a cluster of tightly linked genes, to a unique gene with different alleles, or to a combination of these two hypotheses. However, *Rlm1* has been shown to be linked with *Rlm3* but is not allelic [74].

A major gene named *LmFr1* controlling adult plant resistance to blackleg was tagged in the DH population from French cultivar Cresor (resistant to *L. maculans*) and Westar (susceptible to *L. maculans*) with RFLP markers [64]. Similar study [65] mapped loci for blackleg resistance in a DH population from Major/Stellar and found that genetic control of resistance vary with inoculation techniques. A major gene designated as *LEM1* was mapped to linkage group 6 based on qualitative/quantitative scores of the interaction phenotype on inoculated cotyledons with a single ascospore-derived PG2 isolate, PHW1245. However, four other putative QTL for resistance were also identified on linkage groups LG8, LG17 and pair 4. This study further showed that none of the QTL that were associated with resistance at the seedling and stem stage had a significant effect in conferring resistance in the field. This may be attributed due to use of different pathogen population (PHW 1245 in cotyledon and stem experiments and natural *L. maculans* population in field experiment). The *Rpg3Dun* gene was mapped in an F_2 population from Westar/Dunkeld and identified a suite of SCAR markers that showed cosegregation with resistance to *L. maculans* [81]. Recently, the whole genome average interval mapping approach was applied to localise both qualitative and quantitative trait loci controlling blackleg resistance [32] in a DH population derived from the Australian *B. napus* vernalisation responsive cultivars, Skipton and Ag-Spectrum. Marker regression analyses revealed that at least fourteen genomic regions were associated with blackleg resistance, explaining 19.5% to 88.9% of genotypic variation. A major qualitative locus, designated *RlmSkipton (Rlm4)*, was mapped on chromosome A7, within 0.8 cM of the SSR marker BRMS075 (Table 1).

Genomic regions of chromosome A10 harbours *Rlm2*, which has been shown to be the most common *R*-gene in winter *B. napus* varieties, such as Samourai, Eurol, Bristol, Symbol, Andol, Kintol, Akamar, Colvert, Synergy, and Tapidor, [41]. Chromosome A10 also harbours *LepR2*, *LepR3* and *BLMR2* genes derived from *B. rapa* subsp. *sylvestris* sources [31, 69, 70]. *LepR1* and *LepR2* were mapped on chromosomes A2 and A10, respectively with RFLP markers [31]. Genetic analysis revealed that both genes confer resistance independently and therefore are additive. *LepR1* was a dominant nuclear gene while *LepR2* was an incompletely dominant gene. This study further showed that *LepR1* generally conferred a higher level of resistance than *LepR2*. Both genes exhibited race-specific interactions with pathogen isolates.

The blackleg resistance gene *Rlm6* has been identified on B genome chromosome 8 [47]. *Rlm6* has been successfully introgressed to *B. napus* (AACC) from *B. juncea* (AABB) [47, 141] and provides excellent resistance to *L. maculans* isolates [58], though this gene has not yet been deployed in commercial cultivars [47, 58].

11.2. Quantitative Resistance

The genetic basis of quantitative resistance has been investigated only in limited *B. napus* cultivars such as in Darmor; a derivative of Jet Neuf [119, 137, 142]. However, a number of DH populations have recently been utilised for identification of loci for quantitative resistance under field conditions [32, 86, 143], and are currently being validated (Raman *et al.*, unpublished, Larkan *et al.*, unpublished). Thirteen quantitative trait loci (QTL) on 10 linkage groups associated with quantitative field resistance to *L. maculans* were identified in a DH population from Darmor-*bzh*/Yudal

[87]. Their detection was dependent upon phenotypic method used; seven QTL for mean disease index and six QTL for per cent survival (percentage of lost plants due to canker) and were also dependant on growing environment (year of evaluation). However, only four of the QTL were stable across experiments. These QTL accounted from 23% to 57% of the genotypic variation (Table 2). The unexplained variation was described due to non-detected additive QTL, G x E interaction and incomplete map coverage. This study further showed that resistance to *L. maculans* is influenced with growth habit. For example, one QTL, located close to a dwarf gene (*bzh*), was detected with a very strong effect, masking the detection of other QTL. This study further showed that these dwarfing genes also affect other traits such as earliness, and glucoinsolate content.

In order to validate the stability of QTL for field resistance to *L. maculans*, QTL were mapped and characterised in $F_{2:3}$ population from Darmor (resistant)/Samourai (susceptible) revealing only four QTL on LG3, LG11 and DY5 and DS6 that were consistent in Darmor/Yudal and Darmor/Samourai populations [143]. This study found that the genetic background and inoculum pressure are the major factors of the QTL instability and therefore suggested that QTL mapping must be carried out separately for each population. The genomic regions carrying the most consistent resistance QTL in Darmor do not correspond to the two regions on N7 (A7) and N10 (A10) identified as carrying race specific resistance genes to *L. maculans* [74]. The position of *Rlm2* on N10 (A10) corresponds to a QTL identified for adult plant resistance in the Darmor/Samourai DH population [88]. The cultivar Samourai carries both the resistance allele at this QTL and *Rlm2*. Since it has been reported that no French isolates of *L. maculans* carry *AvrLm2* [34], two hypotheses were proposed to explain this co-location; either the *Rlm2* gene has a residual effect at the adult plant stage, similar to that suggested in other pathosystems, or genes linked to *Rlm2* are responsible for part of variation for resistance at this QTL [99].

QTL for blackleg resistance were identified in four mapping populations derived from the crosses Caiman/$Westar_{10}$, Camberra/$Westar_{10}$, ^{AV}Sapphire/$Westar_{10}$ and Rainbow/^{AV}Sapphire [86]. Multiple QTLs were identified accounting for 13–33% of phenotypic variance. A recent study [32] identified seven significant QTL associated with blackleg resistance, scored on the basis of internal disease score, on chromosomes A2, A9, A10, C1, C2, C3 and C6 in a DH population derived from Skipton/Ag-Spectrum. The genotypic variation explained by the individual QTL ranged from 5% to 24.5%. Both parents contributed the alleles for blackleg resistance. This study showed poor correlation between canker lesion scores over the two years (2009, 2010). Some of the genomic regions for blackleg resistance may be the same as reported previously that have been identified using both classical QTL and association mapping approaches

[31, 69, 87, 137, 144, 145]. The conservation of QTL between Australian and French studies is interesting and suggests the non-specificity of these QTL, irrespective of the environment, genetic background and G x E interactions [32]. However, it is possible that some of the original donor gene sources in French and Australian parental lines used for mapping resistance genes may be the same.

The majority of mapping populations used to map blackleg resistance genes in *B. napus* so far have been comparatively small (Table 1). The development of a high density map utilising larger populations, comprising several hundred to thousands lines, will allow for the precise mapping of resistance loci. Stability of QTL resistance needs to be tested in different environments. Although QTL mapping studies provide comprehensive information on the nature of inheritance, location, magnitude and allelic effects of QTL, much of the information tends to be 'population' specific. In biparental (structured) populations, generally two alleles at each locus are sampled and therefore trait-marker association may not be highly relevant to diverse genetic backgrounds. The validation of trait-marker association is necessary before their use for routine marker-assisted breeding (MAS). Association mapping can be utilised for investigating linkage disequilibrium close to loci of interest in a diverse germplasm [145-149] and therefore offers an alternative to linkage and QTL mapping. This approach has been applied in determining and confirming the markers located within the QTL associated with resistance to *L. maculans* previously identified in Darmor and established their usefulness in MAS [137]. A diverse set of an oilseed rape collection, comprised of 128 lines showing a large spectrum of responses to infection by *L. maculans*, was characterised using 72 SSR and other markers. At least 61 marker alleles were found to be associated with resistance to stem canker. Some of these markers were associated with previously identified QTL, which confirms their usefulness in MAS. Markers located in regions not harbouring previously identified QTL were also associated with resistance, suggesting that new QTL or allelic variants are present in the collection [137]. Genome-wide association based on 1513 markers enabled identification and validation of genomic loci associated with blackleg resistance. This study detected significant marker - race-specific blackleg resistance associations ($P<0.01$) at the seedling and adult plant stages. Loci for resistance were located on chromosomes A1, A2, A3, A5, A6, A7, A10, C1, and C2. Both studies suggested that association mapping is an efficient approach for identifying novel loci/alleles associated with blackleg resistance in diverse germplasm [137, 142]. Superior molecular marker allele(s) associated with resistance to *L. maculans* may be captured by canola breeding programs. Molecular markers associated with seedling and stem canker resistance will help identify accessions carrying desirable alleles and facilitate QTL introgression to develop elite germplasm having new gene/allele combinations for blackleg resistance [32].

12. HOST *R*-GENE CLONING AND CANDIDATE GENE ANALYSIS

At least 20 *R*-genes and several allele variants and haplotypes of cloned *R*-genes have been identified in plants [151-158]. Molecular analyses revealed that these genes belong to large multiple gene families, which encode nucleotide binding site- leucine–rich repeats (NBS-LRRs), serine-threonine-kinases, leucine zipper and protein kinase domains, and toll/interleukin-1 receptor domains [159-164]. These genes are often clustered in many plant species including crops such as rice, maize and soybean and transduce the hypersensitive response to defend against pathogen attack [164-167]. At least 30 CC-NBS-LRR and TIR-NBS-LRR non-redundant genes have been identified in *B. rapa* [167]. Two major gene clusters for resistance to *L. maculans* exist on chromosomes A7 [74] and A10 [31, 69, 70], along with other genes dispersed on different chromosomes. It is possible that some of these *R*-genes may represent to multiple copies of the same functional gene. A recent study has shown that at least eight functional copies of *FLOWERING TIME LOCUS C* (*FLC*) exist within *B. napus* [6] which may modulate flowering time and other functions in different cultivars [168].

In *B. napus*, only few studies aimed at characterizing the genes underlying the resistance to *L. maculans* have been attempted. The recent cloning of the first functional *B. napus* resistance gene *LepR3* revealed a receptor-like protein responsible for conferring resistance to *AvrLm1 L. maculans* isolates [79]. Resistance genes effective against *L. maculans* have also been cloned in *A. thaliana* [169-171], which encode Toll interleukin-1 receptor-nucleotide binding (TIR-NB) or TIR-NB-LRR class proteins. Based on the synteny between *B. napus* and *A. thaliana*, it was deduced that several *B. napus* resistance genes are localised in a region of A7 (N7) that corresponds to the chromosome segment on *Arabidopsis* chromosome 1 which harbours *RLM1*Col [139, 167]. However, a recent report detailing the gene responses to *L. maculans* infections suggests very different responses in *B. napus* and *A. thaliana* [172]. Both salicylic acid and ethylene signaling was triggered in *B. napus*, possibly due to the hemibiotrophic nature of the infection. This stands in contrast to the JA signaling observed in *A. thaliana*, suggesting *L. maculans* is acting as a necrotroph during infection of susceptible *A. thaliana* lines. Since many *R*-genes are conserved and share sequence similarity, degenerated primers based on conserved motifs of *R*-genes have also been used to localise potential resistance gene loci in Brassica species such as *B. oleracea* (on chromosomes C1 (O1), C4 (O4), C8

(08) and C9 (09) and *B. napus* on linkage groups LG1a, LG1b, LG2, LG5, LG8, LG12, LG13, LG14, LG15 and LG18 [173, 174]. However their association with loci controlling resistance to *L. maculans* have not yet been established/validated.

In order to clone genes controlling blackleg resistance in *B. napus* population, high resolution mapping of *LmR1* and *ClmR1* loci was performed using 2500 backcross lines from two crosses between PSA12 and Shiralee, and PSA12 and Cresor, respectively [140], and reported that both resistance loci are located in a highly duplicated genomic region on chromosome A7. This region contained several genes encoding protein kinases or LRR domains. It is reported that the SCAR marker (BN204) that showed cosegregation with *RpgDun* locus for resistance to *L. maculans* is derived from a region showing 92% amino acid identity with the defense-related gene serine threonine 20 (ste-20) protein kinase of *Arabidopsis thaliana* [81]. A proteomic approach has also been utilised to understand gene expression in response to *L. maculans* infection [176]. However, candidacy of any of these genes has not yet been reported.

Recently an alternative approach for identifying candidate *R*-genes has been employed based on genomics [177]. Next-generation massively parallel sequencing platforms such as the Roche 454 genome sequencer FLX instrument, the Illumina Genome Analyser (HiSeq), and the ABI SOLiD System have revolutionized genome sequencing by providing high throughput and cost-effective high coverage sequencing [179-182] and has enabled much quicker identification of candidate genes [178]. Molecular markers associated with *RlmSkipton* (*Rlm4*) locus in the DH population from Skipton/Ag-Spectrum were aligned with the complete genome sequence *B. rapa* as reported in [32]. Eighteen candidate genes, designated as *BLR1-18* with disease resistance characteristics, several of which were clustered around a region syntenic to *Rlm4*. Among candidates, *BLR2* and *BLR11* were the promising candidates for *Rlm4*-mediated resistance [178]. High resolution mapping and gene sequencing of different sources of *L. maculans* resistance will allow for a better understanding of the structural organisation and function of *R*-genes. Recently, the reference genome of *B. rapa* has been published [182] and genomes of *B. oleracea*, *B. nigra* and *B. napus* are expected to be published in coming years. Re-sequencing of whole genomes of known blackleg-resistant genotypes will allow identification of genetic variation between individuals, which can provide molecular genetic markers and insights into gene function [183]. Sequencing of different *R*-genes and understanding their function will also enable us to manipulate resistance to *L. maculans*, as genes with different specificities can be created.

13. PREDICTIVE BREEDING FOR RESISTANCE TO *L. MACULANS* USING MOLECULAR MARKERS

Success of new disease resistance genes relies heavily on the successful transfer of target genomic regions from donor sources and the development of rigorous selection methods. Molecular markers have been used to improve the effectiveness and efficiency of selection strategies in predictive breeding in several agricultural crops. However, the development of molecular markers in *B. napus* and their application in breeding is a challenging exercise due to the large genome size, amphidipliod (4X) nature, open-pollination and lower research funding as compared to other key crops such as wheat, barley, maize and soybean. The *B. napus* genome is highly complex and homologous recombination plays a major role in chromosome rearrangements such as duplications and reciprocal translocations. These arrangements further add to the complexity of molecular analysis and interpretation. *B. napus* chromosomes C6 and A7, which harbours *Rlm1*, *Rlm3*, *Rlm4*, *Rlm7* and *Rlm9* genes for resistance, produced a reciprocal translocation in some cultivars such as in Westar, Marnoo, Monty and Maluka [185, 186] which makes analysis of resistance genes difficult [142].

In most of the breeding programs, selection for blackleg is conducted once a year during the growing season, hampering selection efficiency. Several studies suggest a significant correlation between cotyledon test and canker lesion scores. Therefore, cotyledon tests can be used for selection for resistance to *L. maculans*. However, in many developed countries, it is costly and laborious to perform, particularly as compared to molecular marker analysis, when several tests need to be carried to screen large populations. Furthermore, analysis of different blackleg resistance genes in a canola breeding program using a differential set of *L. maculans* isolates at various stages of the breeding cycle is a very slow process [39]. Interpretation of *R*-gene content using a differential set of control *B. napus* varieties, especially of Australian origin, is a challenging exercise, as majority of cultivars used are heterozygous and/or heterogeneous [32, 41]. In addition, phenotypic tests are dependent upon the growing environment (microclimate conditions and other factors such as powdery and downy mildew), which can complicate scoring of inoculated seedlings. Molecular markers generally out-perform conventional seedling assays, in both efficiency and reliability. It is also possible to identify haplotypes using molecular markers and then validate trait-marker associations, in conjunction with comprehensive phenotyping and conventional allelism tests.

The published literature suggests that little effort has been made to evaluate the allelic relationship among the known genes from different sources, to test stability of majority of QTL or qualitative genes identified over diverse growing environments, or to test their usefulness in achieving long term durable control of the disease. Table 1 also suggests that majority of markers are not very closely linked (<1cM) with resistance loci. Diagnostic or perfect markers for resistance genes are required for routine MAS and will assist allele enrichment strategies in breeding programs, although this is not always possible, even if the complete gene is cloned and characterised for its functionality [187]. The linkage between molecular markers and *Xbn204* flanking the *RlmSkipton* locus was verified in an F_2population derived from Skipton/Ag-Spectrum [32]. The results showed that SSR markers linked to *RlmSkipton* are suitable for enrichment of favourable alleles for blackleg resistance in breeding programs. A separate study [82] validated the map location of *Rlm1* in the DH population derived from Maxol/Westar with SSR and DArT markers. Previously, *Rlm1* and *Rlm3* genes were mapped on chromosomes A7 in the Maxol (resistant to blackleg)/S006 (susceptible to blackleg) utilising RAPD markers and with single spore isolates with known *Avr* genotypes in the *B. napus* European cultivars, Columbus and Maxol [41, 71, 74]. RAPD markers are not amenable for high throughput marker analysis, as they are assayed on low-throughput agarose or polyacrylamide gel systems. Validation of a large array of genes for blackleg resistance in diverse segregating populations representing *B. napus* germplasm is a challenging exercise. However, an association mapping approach can be employed to test trait-marker associations in a large set of germplasm as demonstrated recently [137, 142].

14. CONCLUSIONS

It is now clear that major resistance genes will be overcome in time, as has been seen in many crop plants. Therefore, there is constant need to identify new sources of both qualitative and quantitative resistance loci and to properly utilise the resources available to us so that resistance can be deployed long term. Recent advances in molecular marker systems, such as the development of highly-parallel systems for genotyping and sequencing, have created new opportunities and strategies to select for qualitative and quantitative traits, including resistance to *L. maculans*. Strategies for deploying resistance in breeding programs will vary with individual breeding programs; monitoring introgression of specific loci, using whole-genome marker scans (genomic selection) or identifying individual plants that may offer the greatest opportunity for genetic gain. This is now becoming reality as several genome-wide signals associated with blackleg resistance have been identified (but need to be validated) and alleles at these loci can be selected efficiently and at a cheaper rate with new marker technologies. Development and validation of tightly-linked molecular markers amenable to high throughput marker screening with both qualitative and quantitative resistance and cost effective systems will enable the increased adoption in *B. napus* breeding programs. In addition to genetic resistance, deployment of agronomic practices such as use of rotation and stubble management will remain key management tools for reducing pathogen inoculum for subsequent crops.

15. ACKNOWLEDGEMENTS

Authors are thankful to Dr Regine Delourme, INRA, Le Rheu, Cedex France for providing critical comments and QTL information for quantitative resistance in the Darmor/Yudal population.

REFERENCES

1. Tode HI. *Fungi Mecklenburgenses Selecti*. Fasciculus. 1791;II(51, Plate XVI, Fig 126).
2. Olsson G. Species Crosses within the Genus *Brassica*. I: Artificial *Brassica napus*. Hereditas. 1960;46:351-96.
3. U N. Genomic Analysis in Brassica with Special Reference to the Experimental For-mation of *B. napus* and Peculiar Mode of Fertilisation. Jpn J Bot. 1935;7:389-452.
4. Gómez-Campo C, Prakash S. Origin and Domestication. In: C.Gómez-Campo (Ed.), Biology of Brassica Coenospecies. Elsevier, Netherlands, 33-58. 1999.
5. Baranyk P, Fábry A. History of the Rapeseed (*Brassica Napus* L.) Growing and Breed-ing from Middle Age Europe to Canberra. Proceedings 10th International Rapeseed Congress, Canberra, Australia (http://regionalorgau/au/gcirc/4/374htm). 1999.
6. Zou X, Suppanz I, Raman H, Hou J, Wang J, Long Y, et al. Comparative Analysis of *FLC* Homologues in Brassicaceae Provides Insight into Their Role in the Evolution of Oilseed Rape. PLoS One 2012 ; 7(9): e45751.
7. FAO: Food and Agriculture Organisation (2012) http://faostat.fao.org/.
8. Fitt B, Brun H, Barbetti M, Rimmer S. World-Wide Importance of Phoma Stem Can-ker (*Leptosphaeria maculans* and *L. biglobosa*) on Oilseed Rape (*Brassica napus*). Europe- an Journal of Plant Pathology. 2006;114(1):3-15.
9. West JS, Kharbanda PD, Barbetti MJ, Fitt BDL. Epidemiology and Management of *Leptosphaeria maculans* (Phoma Stem Canker) on Oilseed Rape in Australia, Canada and Europe. Plant Pathol. 2001;50(1):10-27.
10. Howlett BJ. Current Knowledge of the Interaction between *Brassica napus* and *Leptosphaeria maculans*. Canadian Journal of Plant Pathology. 2004;26(3):245-52.
11. Marcroft S, Bluett C. Blackleg of Canola. Agriculture Notes, State of Victoria, Department of Primary Industries, May 2008, Ag1352, ISSN 1329-8062. 2008.
12. Henderson MP. The Black-Leg Disease of Cabbage Caused by *Phoma lingam* (Tode) Desmaz. Phytopathology. 1918;8:379-431.

13. Sosnowski MR, Scott ES, Ramsey MD. Infection of Australian canola cultivars (*Brassica napus*) by *Leptosphaeria maculans* is influenced by cultivar and environmental conditions. Australasian Plant Pathology. 2004;33(3):401-11
14. Petrie GA. Variability in *Leptosphaeria maculans* (Desm.) Ces et de Not., the Cause of Blackleg of Rape. PhD Thesis, University of Saskatchwen, Canada. 1969.
15. Hall R. Epidemiology of Blackleg of Oilseed Rape. Canadian Journal of Plant Pathology 1992;14:46-55.
16. Alabouvette C, Brunin B. Recherches Sur La Maladie Du Colza Due a *Leptospharia maculans* (Desm.) Ces. Et De Not. 1. Role Des Restes De Culture Dans La Conservation Et La Dissemination Du Parasite. Ann Phytopathol 2(3): 463-475. 1970.
17. Bokor A, Barbetti MJ, Brown AGP, MacNish GC, Poole ML, Wood P. Blackleg - Major Hazard to the Rapeseed Industry. Journal of Agriculture, Western Australia. 1975;16:7-10.
18. Ghanbarnia K, Fernando WGD, Crow G. Developing Rainfall- and Temperature-Based Models to Describe Infection of Canola under Field Conditions Caused by Pycnidiospores of *Leptosphaeria maculans*. Phytopathology 2009;99(7):879-86.
19. Ghanbarnia K, Fernando WGD, Crow G. Comparison of Disease Severity and Incidence at Different Growth Stages of Naturally Infected Canola Plants under Field Conditions by Pycnidiospores of *Phoma lingam* as a Main Source of Inoculum. Canadian Journal of Plant Pathology. 2011;33(3):355-63.
20. Chen CY, Howlett BJ. Rapid Necrosis of Guard Cells Is Associated with the Arrest of Fungal Growth in Leaves of Indian Mustard (*Brassica juncea*) Inoculated with Avirulent Isolates of *Leptosphaeria maculans*. Physiological and Molecular Plant Pathology. 1996;48(2):73-81.
21. Hua L, Sivasithamparam K, Barbetti MJ, Kuo J. Germination and Invasion by Ascospores and Pycnidiospores of *Leptosphaeria maculans* on Spring-Type *Brassica napus* Canola Varieties with Varying Susceptibility to Blackleg. Journal of General Plant Pathology 2004;70(5):261.
22. Hammond KE, Lewis BG, Musa TM. A Systemic Pathway in the Infection of Oilseeed Rape Plants by *Leptosphaeria maculans*. Plant Pathology. 1985;34:557-65.
23. Travadon R, Marquer B, Ribule A, Sache I, Masson JP, Brun H, et al. Systemic Growth of *Leptosphaeria maculans* from Cotyledons to Hypocotyls in Oilseed Rape: Influence of Number of Infection Sites, Competitive Growth and Host Polygenic Resistance. Plant Pathology. 2009;58:461-9.
24. Sprague SJ, Watt M, Kirkegaard JA, Howlett BJ. Pathways of Infection of *Brassica napus* Roots by *Leptosphaeria maculans*. New Phytologyst. 2007;176: 211-222.
25. Cunningham GH. Dry-Rot of Swedes and Turnips: Its Causes and Control. Wellington, New Zealand: New Zealand Department of Agriculture: Bulletin 133. 1927.
26. Kutcher HR, van den Berg CGJ, Rimmer SR. Variation in Pathogenicity of *Leptosphaeria maculans* on *Brassica Spp.* Based on Cotyledon and Stem Reactions. Can J Plant Pathol 1993;15:253-8.
27. Balesdent MH, Barbetti MJ, Li H, Sivasithamparam K, Gout L, Rouxel T. Analysis of *Leptosphaeria maculans* Race Structure in a Worldwide Collection of Isolates. Phytopa- thology. 2005 Sep;95(9):1061-71.
28. Koch E, Song K, Osborn TC, Williams PH. Relationship between Pathogenicity Based on Restriction Fragment Length Polymorphism in *Leptosphaeria maculans*. Molecular Plant-Microbe Interactions. 1991;4:341-9.
29. Aubertot J, West J, Bousset-Vaslin L, Salam M, Barbetti M, Diggle A. Improved Resistance Management for Durable Disease Control: A Case Study of Phoma Stem Canker of Oilseed Rape (*Brassica napus*). European Journal of Plant Pathology. 2006;114(1):91-106.
30. Cargeeg LA, Thurling N. Contribution of Host-Pathogen Interactions to the Expression of the Blackleg Disease of Spring Rape (*Brassica napus* L.) Caused by *Leptosphae- ria maculans* (Desm.) Ces. et de Not.. Euphytica. 1980;29:465-76.

31. Yu F, Lydiate DJ, Rimmer SR. Identification of Two Novel Genes for Blackleg Resistance in *Brassica napus.* Theoretical and Applied Genetics. 2005;110:969-79.
32. Raman R, Taylor B, Marcroft S, Stiller J, Eckermann P, Coombes N, et al. Molecular Mapping of Qualitative and Quantitative Loci for Resistance to *Leptosphaeria maculans*; Causing Blackleg Disease in Canola (*Brassica napus* L.). Theoretical and Applied Genetics. 2012;125(2):405-18.
33. Bansal VK, Kharbanda PD, Stringam GR, Thiagarajah MR, Tewari J. A Comparison of Greenhouse and Field Screening Methods for Blackleg Resistance in Doubled Haploid Lines of *Brassica napus.* Plant Disease 1994;78:276-81.
34. Rouxel T, Penaud A, Pinochet X, Brun H, Gout L, Delourme R, et al. A 10-Year Survey of Populations of *Leptosphaeria maculans* in France Indicates a Rapid Adaptation Towards the *Rlm1* Resistance Gene of Oilseed Rape. European Journal of Plant Pathology. 2003;109(8):871-81.
35. McNabb WM, Van Den Berg CGJ, Rimmer SR. Comparison of Inoculation Methods for Selection of Plant Resistant to *Leptosphaeria Maculans* in *Brassica napus.* Can J Plant Sci 1993;73:1199-207.
36. Pang ECK, Halloran GM. Genetics of Virulence in *Leptosphaeria maculans* (Desm.) Ces. et De Not., the Cause of Blackleg in Rapeseed (*Brassica napus* L.). Theoretical and Applied Genetics. 1996;93(3):301-6.
37. Salisbury P, Ballinger DJ, Wratten N, Plummer K, Howlett BJ. Blackleg Disease on Oilseed Brassica in Australia: A Review. Australian Journal of Experimental Agricultural. 1995;35:665-72.
38. Huang YJ, Li ZQ, Evans N, Rouxel T, Fitt BDL, Balesdent MH. Fitness Cost Associat- ed with Loss of the *Avrlm4* Avirulence Function in *Leptosphaeria maculans* (Phoma Stem Canker of Oilseed Rape). European Journal of Plant Pathology. 2006;114(1): 77-89.
39. Marcroft SJ, Elliott VL, Cozijnsen AJ, Salisbury PA, Howlett BJ, Van de Wouw AP. Identifying Resistance Genes to *Leptosphaeria maculans* in Australian *Brassica napus* Cultivars Based on Reactions to Isolates with Known Avirulence Genotypes. Crop and Pasture Science. 2012;63(4):338-50.
40. Rimmer SR, van den Berg CGJ. Resistance to Oilseed Brassica Spp. To Blackleg Caused by *Leptosphaeria maculans.* Can J Plant Pathol. 1992;14: 56-66.
41. Rouxel T, Willner E, Coudard L, Balesdent M-H. Screening and Identification of Resistance to *Leptosphaeria maculans* (Stem Canker) in *Brassica napus* Accessions. Euphy- tica. 2003;133(2):219-31.
42. Marcroft SJ, Purwantara A, Salisbury P, Potter TD, Wratter N, Khangura R, et al. Reaction of a Range of *Brassica Species* under Australian Conditions to the Fungus, *Leptosphaeria maculans*, the Causal Agent of Blackleg. Australian Journal of Experimental Agriculture. 2002;42(5):587-94.
43. Christianson JA, Rimmer SR, Good AG, Lydiate DJ. Mapping Genes for Resistance to *Leptosphaeria maculans* in *Brassica juncea.* Genome. 2006;49(1):30-41.
44. Roy NN. Interspecific Transfer of *Brassica juncea*-Type High Blackleg Resistance to *Brassica napus.* Euphytica. 1984;33: 295-303.
45. Delourme R, Chevre AM, Brun H, Rouxel T, Balesdent MH, Dias JS, et al. Major Gene and Polygenic Resistance to *Leptosphaeria maculans* in Oilseed Rape (*Brassica napus*). European Journal of Plant Pathology. 2006;114(1):41-52.
46. Pang ECK, Halloran GM. The Genetics of Blackleg [*Leptosphaeria maculans* (Desm.) Ces. et de Not.] Resistance in Rapeseed (*Brassica napus* L.). II. Seedling and Adult-Plant Resistance as Quantitative Traits. Theoretical and Applied Genetics. 1996;93:941-9.
47. Chèvre AM, Barret P, Eber F, Dupuy P, Brun H, Tanguy X, et al. Selection of Stable *Brassica napus-B. juncea* Recombinant Lines Resistant to Blackleg (*Leptosphaeria macu- lans*). 1. Identification of Molecular Markers, Chromosomal and Genomic Origin of the Introgression. Theoretical and Applied Genetics. 1997;95(7):1104-11.

48. Snowden RJ, Winter H, Diestel A, Sacristan MD. Development and Characterisation of *Brassica napus-Sinapus arvensis* Addition Lines Exhibiting Resistance to *Leptosphae- ria maculans*. Theoretical and Applied Genetics. 2000;101:1008-14.
49. Bohman S, Wang M, Dixelius C. *Arabidopsis thaliana*-Derived Resistance against *Leptosphaeria maculans* in a *Brassica napus* Genomic Background. Theoretical and Applied Genetics. 2002;105(4):498-504.
50. Sjödin C, Glimelius K. Screening for Resistance to Blackleg *Phoma lingam* (Tode Ex Fr.) Desm. Within Brassicaceae. Journal of Phytopathology. 1988;123(4):322-32.
51. Leflon M, Brun H, Eber F, Delourme R, Lucas MO, Vallee P, et al. Detection, Introgression and Localization of Genes Conferring Specific Resistance to *Leptosphaeria maculans* from *Brassica rapa* into *B. napus*. Theoretical and Applied Genetics. 2007 Nov;115(7):897-906.
52. Roy NN. A Study on Disease Variation in the Populations of an Interspecific Cross of *Brassica juncea* L. x *B. napus* L. Euphytica 27:145-149. 1978
53. Mithen RF, Lewis BG, Heaney RK, Fenwick GR. Resistance of Leaves of Brassica Species to *Leptosphaeria maculans*. Trans Br Mycol Soc 88:525-531. 1987.
54. Mitchell-Olds T, James RV, Palmer MJ, Williams PH. Genetics of *Brassica rapa* (Syn. *Campestris*). 2. Multiple Disease Resistance to Three Fungal Pathogens: *Peronospora parasitica*, *Albugo candida* and *Leptosphaeria maculans*. Heredity. 1995;75(4):362-9.
55. Keri M, van den Berg CJG, McVetty PBE, Rimmer SR. Inheritance of Resistance to *Leptosphaeria maculans* in *Brassica juncea*. Phytopathology. 1997;87:594-8.
56. Brun H, Levivier S, Somda I, Ruer D, Renard M, Chevre AM. A Field Method for Evaluating the Potential Durability of New Resistance Sources: Application to the *Leptosphaeria maculans-Brassica napus* Pathosystem. Phytopathology. 2000;90(9):961-6.
57. Ansan-Melayah D, Balesdent MH, Delourme R, Pilet ML, Tanguy X, Renard M, et al. Genes for Race-Specific Resistance against Blackleg Disease in *Brassica napus* L. Plant Breeding. 1998;117(4):373-8.
58. Plieske J, Struss D, Röbbelen G. Inheritance of Resistance Derived from the B-Genome of Brassica against *Phoma lingam* in Rapeseed and the Development of Molecu- lar Markers. Theoretical and Applied Genetics. 1998;97(5):929-36.
59. Barret P, Guérif J, Reynoird JP, Delourme R, Eber F, Renard M, et al. Selection of Stable *Brassica napus-Brassica juncea* Recombinant Lines Resistant to Blackleg (*Leptosphaeria maculans*). 2. A 'To and Fro' Strategy to Localise and Characterise Interspecific Introgressions on the *B. napus* Genome. Theoretical and Applied Genetics. 1998;96(8): 1097-103.
60. Monteiro AA, Williams PH. The Exploration of Genetic Resources of Portuguese Cabbage and Kale for Resistance to Several Brassica Diseases. Euphytica. 989; 41:215-225. 1.
61. Ananga AO, Cebert E, Soliman K, Kantety R, Pacumbaba RP, Konan K. Rapd Mark- ers Associated with Resistance to Blackleg Disease in Brassica Species. African Jour- nal of Biotechnology. 2006;5(22):2041-8.
62. Wittern I. Wittern I (1984) Untersuchungen Zur Erfassung Der Resistenz Von Winter- raps (*Brassica napus* L. Var. *oleifera* Metzger) Gegenüber *Phoma lingam* (Tode Ex Fr.) Desm. Und Zu Der Durch Den Erreger Verursachten Wurzelhals- Und Stengel- fäule. PhD Thesis, Universität Göttingen, Göttingen, Germany 1984.
63. Stringam GR, Bansal VK, Thiagarajah MR, Tewari JP. Genetic Analysis of Blackleg (*Leptosphaeria maculans*) Resistance in *Brassica napus* L. Using the Doubled Haploid Method. XIII Eucarpia Congress, July 6-11, Angers, France. 1992.
64. Dion Y, Gugel RK, Rakow GFW, Séguin-Swartz G, Landry BS. RFLP Mapping of Resistance to the Blackleg Disease [Causal Agent, *Leptosphaeria maculans* (Desm.) Ces. Et De Not.] in Canola (*Brassica napus* L.). Theoretical and Applied Genetics. 1995;91(8): 1190-1194.

65. Ferreira ME, Rimmer SR, Williams PH, Osborn TC. Mapping Loci Controlling *Brassica napus* Resistance to *Leptosphaeria maculans* under Different Screening Conditions. Phytopathology. 1995; 85:213-217.
66. Mayerhofer R, Bansal VK, Thiagarajah MR, Stringam GR, Good AG. Molecular Mapping of Resistance to *Leptosphaeria maculans* in Australian Cultivars of *Brassica napus*. Genome. 1997;40:294-301.
67. Rimmer SR, Borhan MH, Zhu B, Somers D. Mapping Resistance Genes in *Brassica napus* to *Leptosphaeria maculans*. Proceeding 10th International Rapeseed Congress, Can- berra, Australia. 1999. http://www.regional.org.au/au/gcirc/3/47.htm
68. Rimmer SR. Resistance Genes to *Leptosphaeria maculans* in *Brassica napus*. Can J Plant Pathol-Rev Can Phytopathol. 2006;28:S288-S297.
69. Yu F, Lydiate DJ, Rimmer SR. Identification and Mapping of a Third Blackleg Resistance Locus in *Brassica napus* Derived from *B. rapa* Subsp. *sylvestris*. Genome. 2008;51(1):64-72.
70. Long Y, Wang Z, Sun Z, Fernando DW, McVetty PB, Li G. Identification of Two Blackleg Resistance Genes and Fine Mapping of One of These Two Genes in a *Brassica napus* Canola Cultivar 'Surpass 400'. Theoretical and Applied Genetics. 2011;122(6): 1223-31.
71. Balesdent MH, Attard A, Kuhn ML, Rouxel T. New Avirulence Genes in the Phytopathogenic Fungus *Leptosphaeria maculans*. Phytopathology. 2002;92(10):1122-33.
72. Balesdent MH, Fudal I, Ollivier B, Bally P, Grandaubert J, Eber F, et al. The dispensable chromosome of *Leptosphaeria maculans* shelters an effector gene conferring avirulence towards *Brassica rapa*. New Phytol. 2013;198(3):887-98.
73. Van de Wouw AP, Marcroft SJ, Barbetti MJ, Hua L, Salisbury PA, Gout L, et al. Dual Control of Avirulence in *Leptosphaeria maculans* Towards a *Brassica napus* Cultivar with '*Sylvestris*-Derived' Resistance Suggests Involvement of Two Resistance Genes. Plant Pathology. 2008;58(2):305-13
74. Delourme R, Pilet-Nayel ML, Archipiano M, Horvais R, Tanguy X, Rouxel T, et al. A Cluster of Major Specific Resistance Genes to *Leptosphaeria maculans* in *Brassica napus*. Phytopathology. 2004;94(6):578-83.
75. Crouch JH, Lewis BG, Mithen RF. The Effect of a Genome Substitution on the Resistance of *Brassica napus* to Infection by *Leptosphaeria maculans*. Plant Breeding.1994; 112(4):265-278.
76. Buzza G, Easton A. A New Source of Blackleg Resistance from *Brassica sylvestris*. In GCIRC Technical Meeting Poznan, Poland Bulletin No18. 2002.
77. Yu F, Lydiate DJ, Hahn K, Kuzmisz Kuzmicz S, Hammond C, Rimmer SR. Identification and Mapping of a Novel Blackleg Resistance Locus *LepR4* in the Progenies from *Brassica napus* x *B. rapa* Subsp. *sylvestris*. Proc. 12th International Rapeseed Conference, Wuhan, China, March 2007.
78. Larkan NJ, Kuzmicz S, Yu F, Lydiate D, Genetic Evidence for the Recognition of the *Leptosphaeria maculans* Avirulence Gene *AvrLm1* by Two *Brassica napus* Resistance Genes: *Rlm1* and *LepR3*. Proceedings 17th Crucifer-Genetics Workshop; 2010; Saska- toon, Canada.
79. Larkan N.J., D.J. Lydiate, I.A.P. Parkin, M.N. Nelson, D.J. Epp, W.A. Cowling, et al. The *Brassica napus* blackleg resistance gene *LepR3* encodes a receptor-like protein triggered by the *Leptosphaeria maculans* effector *AvrLM1* New Phytologist. 2013, 197:595-605
80. Saal B, Struss D, RGA- and RAPD-derived SCAR markers for a Brassica B-genome introgression conferring resistance to blackleg in oilseed rape. Theoretical and Applied Genetics, 2005; 111: 281-290
81. Dusabenyagasani M, Fernando WGD. Development of a Scar Marker to Track Canola Resistance against Blackleg Caused by *Leptosphaeria maculans* Pathogenicity Group 3. Plant Disease. 2008;92(6):903-8.

82. Raman R, Taylor B, Lindbeck K, Coombes N, Barbulescu D, Salisbury P, et al. Molecular mapping and validation of *Rlm1* genes for resistance to *Leptosphaeria maculans* in canola (*Brassica napus* L). Crop & Pasture Science. 2012;63:1007–1017.
83. Delourme R, Piel N, Horvais R, Pouilly N, Domin C, Vallée P, et al. Molecular and Phenotypic Characterization of near Isogenic Lines at QTL for Quantitative Resistance to *Leptosphaeria maculans* in Oilseed Rape (*Brassica napus* L.). Theoretical and Applied Genetics. 2008(117):1055-67
84. Eber F., Lourgant K., Brun H., Lode M., Huteau V., Coriton O., Alix K., Balesdent M., Chevre A.M. Analysis of *Brassica nigra* Chromosomes Allows Identification of a New Effective *Leptosphaeria maculans* resistance Gene Introgressed in *Brassica napus*. Proceeding of the 13th International rapeseed congress, Prague 5–9 June 2011.
85. Parlange F, Daverdin G, Fudal I, Kuhn M-L, Balesdent M-H, ne, et al. *Leptosphaeria maculans* Avirulence Gene *Avrlm4-7* Confers a Dual Recognition Specificity by the *Rlm4* and *Rlm7* Resistance Genes of Oilseed Rape, and Circumvents *Rlm4*-Mediated Recognition through a Single Amino Acid Change. Molecular Microbiology. 2009;71:851-63.
86. Kaur S, Cogan NOI, Ye G, Baillie RC, Hand ML, Ling AE, et al. Genetic Map Construction and QTL Mapping of Resistance to Blackleg (*Leptosphaeria maculans*) Disease in Australian Canola (*Brassica napus* L.) Cultivars. Theoretical and Applied Genetics. 2009 Dec;120(1):71-83.
87. Pilet ML, Delourme R, Foisset N, Renard M. Identification of Loci Contributing to Quantitative Field Resistance to Blackleg Disease, Causal Agent *Leptosphaeria maculans* (Desm.) Ces. Et De Not., in Winter Rapeseed (*Brassica napus* L.). Theoretical and Applied Genetics. 1998;96:23-30.
88. Delourme R, Piel N, Horvais R, Pouilly N, Domin C, Vallée P, et al. Molecular and Phenotypic Characterization of near Isogenic Lines at QTL for Quantitative Resistance to *Leptosphaeria maculans* in Oilseed Rape (*Brassica napus* L.). Theoretical and Ap- plied Genetics. 2008;117(7):1055-67.
89. Flor HH. The Current Status of the Gene-for-Gene Concept. Annu Rev Phytopathol. 1971;9:275-96.
90. Keen NT. Gene-for-Gene Complementarity in Plant-Pathogen Interactions. Annu Rev Genet. 1990;24:447-63
91. Flor HH. Inheritance of Pathogenicity in *Melampsora lini*. Phytopathology 1942;32:653-69.
92. Williams PH, Delwiche PA, editors. Screening for Resistance to Blackleg of Crucifers in the Seedling Stage. Proceedings Eucarpia Conference, Breeding of Cruciferous Crops, Wageningen, Netherlands: 164-170; 1979.
93. Rouxel T, Balesdent MH. The Stem Canker (Blackleg) Fungus, *Leptosphaeria maculans*, Enters the Genomic Era. Mol Plant Pathol. 2005;6(3):225-41.
94. Balesdent MH, Louvard K, Pinochet X, Rouxel T. A Large-Scale Survey of Races of *Leptosphaeria maculans* Occurring on Oilseed Rape in France. European Journal of Plant Pathology. 2006;114(1):53-65.
95. Ghanbarnia K, Lydiate D, Rimmer SR, Li G, Kutcher HR, Larkan N, et al. Genetic mapping of the *Leptosphaeria maculans* avirulence gene corresponding to the *LepR1* resistance gene of *Brassica napus*. Theoretical and Applied Genetics. 2012;124(3):505-13.
96. Gout L, Fudal I, Kuhn ML, Blaise F, Eckert M, Cattolico L, et al. Lost in the Middle of Nowhere: The *Avrlm1* Avirulence Gene of the Dothideomycete *Leptosphaeria maculans*. Mol Microbiol 2006;60:67-80.
97. Fudal I, Ross S, Gout L, Blaise F, Kuhn ML, Eckert MR, et al. Heterochromatin-Like Regions as Ecological Niches for Avirulence Genes in the *Leptosphaeria maculans* Genome: Map-Based Cloning of *Avrlm6*. Mol Plant Microbe Interact 2007;20:459-70.
98. Stachowiak A, Olechnowicz J, Jedryczka M, Rouxel T, Balesdent MH, Happstadius I, et al. Frequency of Avirulence Alleles in Field Populations of *Leptosphaeria maculans* in Europe. European Journal of Plant Pathology. 2006;114(1):67-75.

99. Delourme R, Barbetti MJ, Snowdon R, Zhao J, Maria J. Manzanares-Dauleux. Genetics and Genomics of Disease Resistance Edwards D. Batley J, Parkin IAP, Kole C (Eds), USA: Science Publishers, CRC Press; 2011.
100. Jackson AO, Taylor CB. Plant-Microbe Interaction:Life and Death at the Interface. The Plant Cell. 1996; 8:1651-68.
101. Sprague S, Balesdent M-H, Brun H, Hayden H, Marcroft S, Pinochet X, et al. Major Gene Resistance in *Brassica napus* (Oilseed Rape) Is Overcome by Changes in Virulence of Populations of *Leptosphaeria maculans* in France and Australia. European Journal of Plant Pathology. 2006;114(1):33-40.
102. Li H, Sivasithamparam K, Barbetti MJ. Breakdown of a *Brassica rapa* Subsp. *sylvestris* Single Dominant Blackleg Resistance Gene in *B. napus* Rapeseed by *Leptosphaeria maculans* Field Isolates in Australia. Plant Disease. 2003;87(6):752.
103. Van de Wouw AP, Cozijnsen AJ, Hane JK, Brunner PC, McDonald BA, Oliver RP, et al. Evolution of linked avirulence effectors in *Leptosphaeria maculans* is affected by genomic environment and exposure to resistance genes in host plants.. PLoS Pathogens 2010;6: e1001180.
104. Howell PM, Sharpe AG, Lydiate DJ. Homoeologous Loci Control the Accumulation of Seed Glucosinolates in Oilseed Rape (*Brassica napus*). Genome. 2003;46:454-60.
105. Sjödin C, Glimelius K. Transfer of Resistance against *Phoma lingam* to *Brassica napus* L. By Asymmetric Somatic Hybridisation Combined with Toxin Selection. Theoretical and Applied Genetics. 1989;78:513-20.
106. Waara S, Glimelius K. The Potential of Somatic Hybridization in Crop Breeding Euphytica. 1995;85:217-33.
107. Gerdemann-Kncörck M, Sacristan MD, Braatz C, Schieder O. Utilization of Asymmetric Somatic Hybridization for the Transfer of Disease Resistance from *Brassica ni- gra* to *Brassica napus*. Plant Breeding. 1994;113(2):106-13.
108. Bauer-Weston B, Keller WA, Webb J. Production and Characterization of Asymmetric Somatic Hybrids between *Arabidopsis thaliana* and *Brassica napus*. Theoretical and Applied Genetics 1993;86:150-158
109. Yu F, Lydiate DJ, Gugel RK, Sharpe AG, Rimmer SR. Introgression of *Brassica rapa* Subsp. *sylvestris* Blackleg Resistance into *B. napus*. Molecular Breeding 2012;30:1495-1506.
110. Winter H, Gaertig S, Diestel A, Sacristan MD. Blackleg Resistance of Different Origin Transferred into *Brassica napus*. Proceeding 10th International Rapeseed Congress, Canberra, Australia. 1999. http://www.regional.org.au/au/gcirc/4/593.htm
111. Dixelius C, Wahlberg S. Resistance to *Leptosphaeria maculans* Is Conserved in a Specific Region of the Brassica B Genome. Theoretical and Applied Genetics. 1999;99(1): 368-72.
112. Michael TP, Alba R. The Tomato Genome Fleshed Out. Nature Biotechnology. 2012;30(8): 765-7.
113. Strittmatter G, Janssens J, Opsomer C, Botterman J. Inhibition of Fungal Disease Development in Plants by Engineering Controlled Cell Death. Nature Biotechnology. 1995;13(10):1085-9.
114. Keller H, Pamboukdjian N, Ponchet M, Poupet A, Delon R, Verrier J-L, et al. Pathogen-Induced Elicitin Production in Transgenic Tobacco Generates a Hypersensitive Response and Nonspecific Disease Resistance. The Plant Cell. 1999;11:223-35.
115. Hennin C, Hofte M, Diederichsen E. Functional Expression of *Cf9* and *Avr9* Genes in *Brassica napus* Induces Enhanced Resistance to *Leptosphaeria maculans*. Molecular Plant-Microbe Interactions. 2001;14(9):1075-85.
116. Bolton MD. Primary Metabolism and Plant Defense-Fuel for the Fire. Mol Plant-Microbe Interact. 2009;22:487-97.
117. Katagiri F, Tsuda K. Understanding the Plant Immune System. Molecular Plant-Microbe Interactions. 2010;23(12): 1531-1536.
118. Johnson R, Law CN. Cytogenetic Studies of the Resistance of the Wheat Variety Bersée to *Puccinia Striformis*. Cereal Rusts Bulletin 1973;1:38-43.

119. Vera Cruz C, Bai J, On˜ a I, Leung H, Nelson R, Mew T-W, et al. Predicting Durability of a Disease Resistance Gene Based on an Assessment of the Fitness Loss and Epidemiological Consequences of Avirulence Gene Mutation. Proceedings of the National Academy of Sciences USA 2000;97: 13500-13505.

120. Light KA, Gororo NN, Salisbury PA. Usefulness of Winter Canola (*Brassica napus*) Race-Specific Resistance Genes against Blackleg (Causal Agent *Leptosphaeria maculans*) in Southern Australian Growing Conditions. Crop and Pasture Science. 2011;62 162-8.

121. Brun H, Chevre AM, Fitt BDL, Powers S, Besnard AL, Ermel M, et al. Quantitative Resistance Increases the Durability of Qualitative Resistance to *Leptosphaeria maculans* in *Brassica napus*. New Phytol. 2010;185(1):285-99.

122. Sprague SJ, Marcroft SJ, Hayden HL, Howlett BJ. Major Gene Resistance to Blackleg in *Brassica napus* Overcome within Three Years of Commercial Production in South-eastern Australia. Plant Disease. 2006;90(2):190-8.

123. Van de Wouw AP, Stonard JF, Howlett BJ, West JS, Fitt BDL, Atkins SD. Determining Frequencies of Avirulent Alleles in Airborne *Leptosphaeria maculans* Inoculum Using Quantitative PCR. Plant Pathol. 2010;59(5):809-18.

124. Kolmer JA, Dyck PL, Roelfs AP. An Appraisal of Stem and Leaf Rust Resistance in North American Hard Red Spring Wheats and the Probability of Multiple Mutations to Virulence in Populations of Cereal Rust Fungi. Phytopathology 1991;81:237-9.

125. Crute R, Pink D. Genetics and Utilization of Pathogen Resistance in Plants. Plant Cell. 1996;8:1747-55.

126. Marcroft S, van De Wouw A, Salisbury P, Potter T, Howlett BJ. Rotation of Canola (*Brassica napus*) Cultivars with Differential Complements of Blackleg Resistance Genes Decreases Disease Severity. Plant Pathol. 2012;

127. Raman H, Raman R, Kilian A, Detering F, Long Y, Edwards D, et al. A Consensus Map of Rapeseed (*Brassica napus* L) Based on Diversity Array Technology Markers: Applications in Genetic Dissection of Qualitative and Quantitative Traits. BMC Genomics 2013; 14:277.

128. Li G, Quiros CF. Sequence-Related Amplified Polymorphism (SRAP), a New Marker System Based on a Simple PCR Reaction: Its Application to Mapping and Gene Tagging in Brassica. Theoretical and Applied Genetics. 2001;103(2):455-61.

129. Sun Z, Wang Z, Tu J, Zhang J, Yu F, McVetty P, et al. An Ultradense Genetic Recombination Map for *Brassica napus* Consisting of 13551 SRAP Markers. Theoretical and Applied Genetics. 2007;114(8):1305-17.

130. Durstewitz G, Polley A, Plieske J, Luerssen H, Graner EM, Wieseke R, et al. SNP Discovery by Amplicon Sequencing and Multiplex SNP Genotyping in the Allopolyploid Species *Brassica napus*. Genome. 2010;53(11):948-56.

131. Trick M, Long Y, Meng J, Bancroft I. Single Nucleotide Polymorphism (SNP) Discovery in the Polyploid *Brassica napus* Using Solexa Transcriptome Sequencing. Plant Bi- otechnology Journal. 2009;7:334-46.

132. Suwabe K, Iketani H, Nunome T, Kage T, Hirai M. Isolation and Characterization of Microsatellites in *Brassica rapa* L. Theoretical and Applied Genetics 2002;104:1092-8.

133. Suwabe K, Tsukazaki H, Iketani H, Hatakeyama K, Kondo M, Fujimura M, et al. Simple Sequence Repeat-Based Comparative Genomics between *Brassica rapa* and *Arabi- dopsis thaliana*: The Genetic Origin of Clubroot Resistance. Genetics. 2006;173(1):309-19.

134. Lombard V, Delourme R. A Consensus Linkage Map for Rapeseed (*Brassica napus* L.): Construction and Integration of Three Individual Maps from DH Populations. Theoretical and Applied Genetics. 2001;103(4):491-507.

135. Baird NA, Etter PD, Atwood TS, Currey MC, Shiver AL, Lewis ZA, et al. Rapid Snp Discovery and Genetic Mapping Using Sequenced RAD Markers. PLoS ONE. 2008;3(10):e3376.

136. Miller MR, Dunham JP, Amores A, Cresko WA, Johnson GA. Rapid and Cost-Effec- tive Polymorphism Identification and Genotyping Using Restriction Site Associated DNA (RAD) Markers. Genome Research. 2007;17:240-8.

137. Jestin C, Lodé M, Vallée P, Domin C, Falentin C, Horvais R, et al. Association Mapping of Quantitative Resistance for *Leptosphaeria maculans* in Oilseed Rape (*Brassica napus* L.). Mol Breed. 2011;27:271-87.

138. Michelmore RW, Paran I, Kesseli RV. Identification of Markers Linked to Disease-Resistance Genes by Bulked Segregant Analysis: A Rapid Method to Detect Markers in Specific Genomic Regions by Using Segregating Populations (Random Amplified Polymorphic DNA/Restriction Fragment Length Polymorphism). Proc Natl Acad Sci USA. 1991;88:9828-32,.

139. Raman H, Milgate A. Molecular Breeding for Resistance to Septoria Tritici Blotch. Cereal Res Commun. 2012; 40:451-466

140. Mayerhofer R, Wilde K, Mayerhofer M, Lydiate D, Bansal VK, Good AG, et al. Complexities of Chromosome Landing in a Highly Duplicated Genome: Toward Map-Based Cloning of a Gene Controlling Blackleg Resistance in *Brassica napus*. Genetics. 2005;171(4):1977-88.

141. Chèvre AM, Brun H, Eber F, Letanneur JC, Vallee P, Ermel M, Glais I, Li H, Sivasithamparam K, Barbetti MJ. Stabilization of Resistance to *Leptosphaeria maculans* in *Brassica napus - B. juncea* Recombinant Lines and Its Introgression into Spring-Type *Brassica napus*. Plant Disease. 2008;92(8):1208-14.

142. Raman H, Raman R, Taylor B, Lindbeck K, Coombes N, Eckermann P, et al. Blackleg Resistance in Rapeseed: Phenotypic Screen, Molecular Markers, and Genome-Wide Linkage and Association Mapping. Proceedings of the 17th Australian Research Assembly on Brassicas, 15-17 August, 2011, Wagga Wagga, pp 61-64.

143. Pilet ML, Duplan G, Archipiano M, Barret P, Baron C, Horvais R, et al. Stability of QTL for Field Resistance to Blackleg across Two Genetic Backgrounds in Oilseed Rape. Crop Sci 2001;41:197-205.

144. Piquemal J, Cinquin E, Couton F, Rondeau C, Seignoret E, Doucet I, et al. Construction of an Oilseed Rape (*Brassica napus* L.) Genetic Map with SSR Markers. Theoretical and Applied Genetics. 2005;111(8):1514-23.

145. Breseghello F, Sorrells ME. Association Analysis as a Strategy for Improvement of Quantitative Traits in Plants. Crop Sci. 2006;46:1323-30.

146. Breseghello F, Sorrells ME. Association Mapping of Kernel Size and Milling Quality in Wheat (*Triticum aestivum* L.) Cultivars. Genetics. 2006;172(2):1165-77.

147. Buckler ES, Holland JB, Bradbury PJ, Acharya CB, Brown PJ, Browne C, et al. The Genetic Architecture of Maize Flowering Time. Science. 2009;325(5941):714-8.

148. Flint-Garcia SA, Thornsberry JM, Buckler ES. Structure of Linkage Disequilibrium in Plants. Annu Rev Plant Biol 2003 54:357-74

149. Hirschhorn JN, Daly MJ. Genome Wide Association Studies for Common Diseases and Complex Traits. Nat Rev Genet. 2005;6:95-108.

150. Raman H, Stodart B, Ryan P, Delhaize E, Emberi L, Raman R, et al. Genome Wide Association Analyses of Common Wheat (*Triticum aestivum* L) Germplasm Identifies Multiple Loci for Aluminium Resistance. Genome. 2010;53(11):957-66.

151. Cloutier S, McCallum BD, Loutre C, Banks TW, Wicker T, Feuillet C, et al. Leaf Rust Resistance Gene *Lr1*, Isolated from Bread Wheat (*Triticum aestivum* L.) Is a Member of the Large *Psr567* Gene Family. Plant Mol Biol. 2007;65:93-106.

152. Feuillet C, Travella S, Stein N, Albar L, Nublat A, Keller B. Map-Based Isolation of the Leaf Rust Disease Resistance Gene *Lr10* from the Hexaploid Wheat (*Triticum aestivum* L.) Genome. Proc Natl Acad Sci U S A. 2003;100(25):15253-8.

153. Krattinger SG, Lagudah ES, Spielmeyer W, Singh RP, Huerta-Espino J, McFadden H, et al. A Putative ABC Transporter Confers Durable Resistance to Multiple Fungal Pathogens in Wheat. Science. 2009;323:1360-3.

154. Michelmore RW, Meyers BC. Clusters of Resistance Genes in Plants Evolve by Divergent Selection and a Birth-and-Death Process. Genome Res 1998;8:1113-30.

155. Srichumpa P, Brunner, S., Keller, B. & Yahiaoui, N. Allelic Series of Four Powdery Mildew Resistance Genes at the *Pm3* Locus in Hexaploid Bread Wheat. Plant Physiology 139, 885-895. 2005.

156. Martin G, Brommonschenkel S, Chunwongse J, Frary A, Ganal M, Spivey R, et al. Map-Based Cloning of a Protein Kinase Gene Conferring Disease Resistance in Tomato. Science. 1993;262(5138):1432-6.

157. Fu D, Uauy C, Distelfeld A, Blechl A, Epstein L, Chen X, et al. A Kinase-Start Gene Confers Temperature-Dependent Resistance to Wheat Stripe Rust. Science. 2009;323(5919):1357-60.

158. Baumgarten A, Cannon S, Spangler R, May G. Genome-Level Evolution of Resistance Genes in *Arabidopsis thaliana*. Genetics. 2003;165 309-19.

159. Persson M, Staal J, Oid S, Dixelius C. Layers of Defense Responses to *Leptosphaeria Maculans* Below the *Rlm1* - and Camelexin - Dependent Resistances. New Phytologist. 182, 470-482. 2009.

160. Dangl JL, Jones JDG. Plant Pathogens and Integrated Defence Responses to Infection. Nature 2001;411:826-33.

161. Dixon MS, Hatzixanthis K, Jones DA, Harrison K, Jones JDG. The Tomato *Cf-5* Disease Resistance Gene and Six Homologs Show Pronounced Allelic Variation in Leucine-Rich Repeat Copy Number. Plant Cell 10: 1915-1926. 1998.

162. Dixon MS, Jones D, Keddie JS, Thomas CM, Harrison K, Jones JD. The Tomato *Cf-2* Disease Resistance Locus Comprises Two Functional Genes Encoding Leucine-Rich Repeat Proteins. Cell 84: 451-459. 1996.

163. Song WY, Pi LY, Wang GL, Gardner J, Holsten T, Ronald PC. Evolution of the Rice *Xa21* Disease Resistance Gene Family. Plant Cell 9: 1279–1287. 1997.

164. Ellis J, Jones D. Structure and Function of Proteins Controlling Strain-Specific Pathogen Resistance in Plants. Current Opinion in Plant Biology. 1998;1:288-93.

165. Meyer JDF, Silva DCG, Yang C, Pedley KF, Zhang C, Mortel Mvd, et al. Identification and Analyses of Candidate Genes for *Rpp4*-Mediated Resistance to Asian Soybean Rust in Soybean. Plant Physiology. 2009;150:1 295-307

166. Hulbert SH, Bennetzen JL. Recombination at the *Rp1* Locus of Maize. Mol Gen Genet 226: 377-382. 1991.

167. Mun JH, Yu HJ, Park S, Park BS. Genome-Wide Identification of NBS-Encoding Resistance Genes in *Brassica rapa*. Molecular and General Genetics. 2009;282:617-31.

168. Deng W, Ying H, Helliwell CA, Taylor JM, Peacock WJ, Dennis ES. Flowering Locus C (*FLC*) Regulates Development Pathways Throughout the Life Cycle of *Arabidopsis*. Proceedings of the National Academy of Sciences. 2011;108 (16):6680-5

169. Staal J, Kaliff M, Bohman S, Dixelius C. Transgressive Segregation Reveals Two Arabidopsis TIR-NB-LRR Resistance Genes Effective against *Leptosphaeria maculans*, Causal Agent of Blackleg Disease. Plant J. 2006;46(2):218-30.

170. Staal J, Dixelius C. *Rlm3*, a Potential Adaptor between Specific *TIR*-NB-LRR Receptors and DZC Proteins. Communicative & Integrative Biology 2008;1(1):59-61.

171. Staal J, Kaliff M, Dewaele E, Persson M, Dixelius C. *Rlm3*, a TIR Domain Encoding Gene Involved in Broad-Range Immunity of Arabidopsis to Necrotrophic Fungal Pathogens. Plant Journal 2008;55(2):188-200.

172. Šašek V, Nováková M, Jindřichová B, Bóka K, Valentová O, BurketováLenka. Recognition of Avirulence gene *AvrLm1* from hemibiotrophic ascomycete *Leptosphaeria maculans* triggers salicylic acid and ethylene signaling in *Brassica napus*. Molecular Plant-Microbe Interactions 25:9, 1238-1250. 2012.

173. Vincente JG, King GJ. Characterisation of Disease Resistance Gene-Like Sequences in *Brassica oleracea* L. Theoretical and Applied Genetics. 2001;102:555-63.

174. Fourmann M, Charlot F, Froger N, Delourme R, Brunel D. Expression, Mapping, and Genetic Variability of *Brassica napus* Disease Resistance Gene Analogues. Genome. 2001;44(6):1083.

175. Sharma N, N. Hotte, M.H. Rahman, M. Mahammadi, M.K. Deyholos, N.N.V. Kav. Towards Identifying Brassica Proteins Involved in Mediating Resistance

to *Leptosphaeria maculans*: A Proteomics-Based Approach. Proteomics. 2008;8:3516-35.

176. Marra R, Li H, Barbetti MJ, Sivasithamparam K, Vinale F, Cavallo P, et al. Proteomic Analysis of the Interaction between *Brassica napus* Cv. Surpass 400 and Virulent or Avirulent Isolates of *Leptosphaeria maculans*. Journal of Plant Pathology. 2010;92:89-101.
177. Tollenaere R, Hayward A, Dalton-Morgan J, Campbell E, Lee JRM, Lorenc M, et al. Identification and Characterization of Candidate *Rlm4* Blackleg Resistance Genes in *Brassica napus* Using Next-Generation Sequencing. Plant Biotechnology Journal. 2012;10(6):709-15.
178. Wheeler DA, Srinivasan M, Egholm M, Shen Y, Chen L, McGuire A, et al. The Complete Genome of an Individual by Massively Parallel DNA Sequencing. Nature. 2008;452(7189):872-6.
179. Metzker ML. Sequencing Technologies- the Next Generation. Nat Rev Genet. 2010;11(1):31-46.
180. Ju YS, Kim J-I, Kim S, Hong D, Park H, Shin J-Y, et al. Extensive Genomic and Transcriptional Diversity Identified through Massively Parallel DNA and Rna Sequencing of Eighteen Korean Individuals. Nat Genet. 2011;43(8):745-52.
181. Consortium TPGS. Genome Sequence and Analysis of the Tuber Crop Potato. Nature. 2011;475(7355):189-95.
182. Wang X, Wang H, Wang J, Sun R, Jian Wu, Shengyi Liu, et al. The Genome of the Mesopolyploid Crop Species *Brassica rapa*. Nature Genetics. 2011;43:1035-9.
183. Imelfort M, Edwards D. Next Generation Sequencing of Plant Genomes. Briefings in Bioinformatics. 2009;10:609-18.
184. Osborn TC, Butrulle DV, Sharpe AG, Pickering KJ, Parkin IAP, Parker JS, et al. Detection and Effects of a Homeologous Reciprocal Transposition in *Brassica napus*. Genetics. 2003;165(3):1569-77.
185. Kelly AL. The Genetic Basis of Petal Number and Pod Orientation in Oilseed Rape (*Brassica napus*). PhD Thesis, University of New Castle, UK. 1996.
186. Wang J, Kaur S, Cogan N, Dobrowolski M, Salisbury P, Burton W, et al. Assessment of Genetic Diversity in Australian Canola (*Brassica napus* L.) Cultivars Using SSR Markers. Crop & Pasture Sci. 2009;60:1193-201.
187. Raman H, Ryan PR, Raman R, Stodart BJ, Zhang K, Martin P, et al. Analysis of *TALMT1* Traces the Transmission of Aluminum Resistance in Cultivated Common Wheat (*Triticum aestivum* L.). Theoretical and Applied Genetics. 2008;116:343–54.

CHAPTER 14

Targeted Recombination between Homologous Chromosomes for Precise Breeding in Tomato

Shdema Filler Hayut[1] , Cathy Melamed Bessudo[1] & Avraham A. Levy[1]

[1]*Department of Plant and Environmental Sciences, Weizmann Institute of Science, Rehovot 76100, Israel*

ABSTRACT

Homologous recombination (HR) between parental chromosomes occurs stochastically. Here, we report on targeted recombination between homologous chromosomes upon somatic induction of DNA double-strand breaks (DSBs) via CRISPR-Cas9. We demonstrate this via a visual and molecular assay whereby DSB induction between two alleles carrying different mutations in the *PHYTOENE SYNTHASE* (*PSY1*) gene results in yellow fruits with wild type red sectors forming via HR-mediated DSB repair. We also show that in heterozygote plants containing one *psy1* allele immune and one sensitive to CRISPR, repair of the broken allele using the unbroken allele sequence template is a common outcome. In another assay, we show evidence of a somatically induced DSB in a cross between a *psy1* edible tomato mutant and wild type *Solanum pimpinellifolium*, targeting only the *S. pimpinellifolium* allele. This enables characterization of germinally transmitted targeted somatic HR events, demonstrating that somatically induced DSBs can be exploited for precise breeding of crops.

INTRODUCTION

DNA double-strand breaks (DSBs) are one of the powerful forces that shape plant genomes. These DSBs may occur throughout the plant life cycle, in somatic or meiotic cells, spontaneously during the movement of replication forks or developmentally controlled as in the early stages of first meiosis[1,2,3]. They also may be induced through ionizing radiation, genotoxic drugs or through the activation of endonucleases[4]. Unrepaired DNA DSB may cause extreme types of damage including chromosome loss, leading to gamete sterility or cell death. Repair of DSBs may also be associated with insertion/deletion (indels) mutations. DSBs repair mechanisms are therefore essential for the maintenance of genome integrity. Understanding these mechanisms is critical for our ability to precisely engineer genomes, for example, for targeted mutagenesis, gene targeting or for other types of targeted chromosomes reshuffling[5].

DNA DSB repair mechanisms have been widely studied in many organisms, including plants[6,7,8,9,10]. Studies in plants have characterized the genes involved in DSB repair via Non-Homologous-End-Joining (NHEJ) or Homologous Recombination (HR) and tested the outcome of DSB repair in both somatic and meiotic tissues (See refs 9, 10, 11, 12 for review). NHEJ has been characterized in a broad range of species and tissues (mostly somatic), using multiple DSB inducing agents including site specific meganucleases[13], transposon excision[14] and custom-designed nucleases, such as zinc-finger nucleases[15], transcription activator-like effector nucleases (TALENs)[16] and Clustered Regulatory Interspaced Short Palindromic Repeat associated protein Cas9 (CRISPR-Cas)[17,18,19,20]. The emerging picture from these works suggests that NHEJ is a prominent repair pathway in somatic cells. This error-prone mechanism involves indels ranging from a few base pairs (bp) to several Kbps[21] at the DSB site and is often associated with microhomologies[22,23]. In addition, CRISPR-Cas-based systems prove to be highly efficient in a broad range of plant species[24] including tomato[25,26].

Several studies that addressed the mechanism of DSB repair via HR in somatic tissues were done in *Arabidopsis*, using transgenic assays that tested repair mechanisms such as intrachromosomal recombination, unequal crossover and so on. In all cases, DSB induction enhanced HR-repair rates[27,28]. Recombination rates from the unequal crossover assay were much lower than for intrachromosomal recombination[29]. Somatic DSB repair by an homologous chromosome, using an allelic sequence, was also studied in transgenic tobacco plants, using transposable element-induced breaks: HR repair occurred upon excision of the transposon; but was not detected with an immobile element[30]. DSB induction could also trigger HR-mediated repair using an ectopic genomic sequence template, albeit at very low frequencies14,31.

DSB induction of HR between endogenous (non-transgenic) recombination partners was shown in maize upon excision of the *Activator* (*Ac*)32, or the *Mutator* elements33. In both cases recombination occurred in *cis*, between repeats flanking the transposon in somatic tissues. By contrast, germinal *Ac* activity did not stimulate the rate of meiotic recombination between homologous chromosomes at the maize *bronze* locus34. This surprising result, which goes against a large body of studies in plants (see above) and other organisms35,36, might be due to a lack of coordination between Ac excision and meiotic recombination, a preference of meiotic HR for Spo11-induced breaks or another unknown reason. The ability to induce HR between homologous chromosomes at a specific genomic location would provide geneticists and breeders with a powerful tool for the targeted induction of crossover or gene conversion.

In this study, we tested whether targeted recombination between homologous chromosomes can be achieved in plants. For this purpose, we combined a series of new technologies, such as CRISPR-Cas, whole genome sequences, High-Throughput sequencing and a variety of mutants in tomato fruit colour, enabling both identification and quantification of resulting recombination products. We provide evidence for the occurrence of induced allele-dependent DSB repair in tomato, including the characterization of germinally transmitted HR events.

RESULTS

Tomato fruit colour assay for the analysis of DNA DSB repair

To estimate the rate of somatic NHEJ versus HR based DSB repair at an endogenous plant locus, we designed a series of fruit colour assays. These assays enabled us to measure only those events of NHEJ that are error-prone as we cannot detect events of precise end-joining which might occur21,37. Similarly, we were only able to detect the HR events between polymorphic homologous chromosomes, while recombination between sister chromatids, which is also known to take place29,38 could not be detected.

For the first fruit colour assay, we used two mutant lines of tomato, each with a different mutation in the *Phytoene synthase 1* (*PSY1*) gene. The *yellow flesh* e^{3756} (also known as *yellow flesh, locus* r^{3756}) allele is an EMS mutant

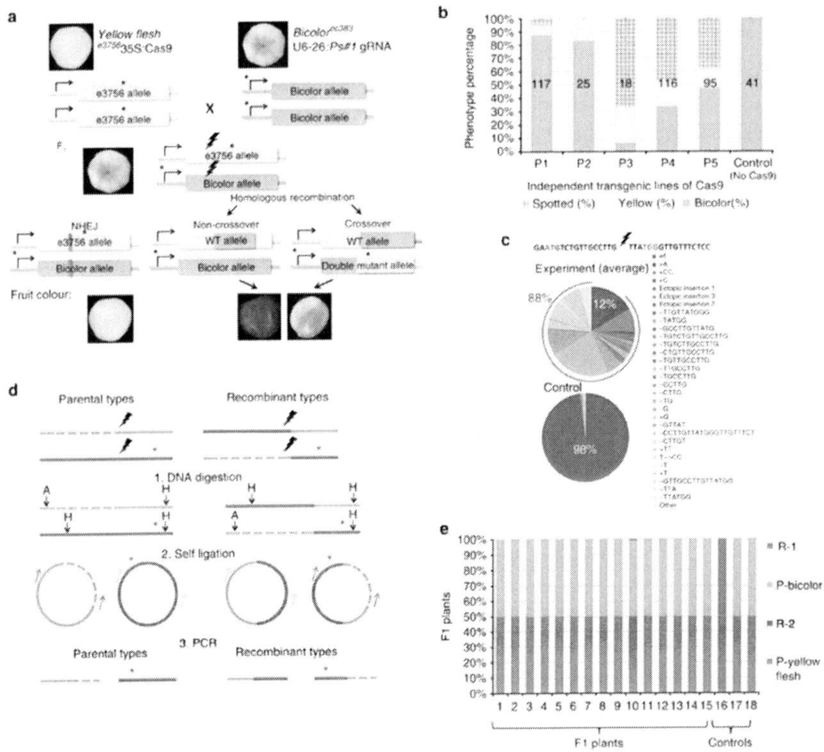

Figure 1: Tomato fruit colour DSB repair assay. (**a**) Crossing *yellow flesh* e^{3756} 35S:Cas9 and *bicolor*^{cc383} u6-26:Ps#1-sgRNA gives F_1 plants with a pale Bicolour fruit phenotype. F_1 plants expressing both Cas9 and gRNA were selected. The gRNA was designed for DSB induction (black lightning) in both alleles between the *yellow flesh* e^{3756} and *bicolor*^{cc383} mutations (*). In case of error-prone NHEJ repair (blue line) of *bicolor*^{cc383}, fruit colour is yellow. In cases of non-crossover or crossover, fruit colour is expected to be red or bicolour or yellow with red spots in case of late event. Note that whole red fruits were not obtained. Rather, fruits with red spots in a yellow or bicolour background were found and are shown together with additional products of HR-induced repair in Supplementary Fig. 1. (**b**) Fruit phenotype distribution in F_1 plants and control: Bicolour fruits are shown as orange boxes; Yellow fruits as yellow; Fruits with red sectors (putative somatic HR) are shown as red-stripped boxes. Each column represents a fruit population derived from cross of independent transgenic lines of Cas9 and a given u6-26:Ps#1-sgRNA line. The number of fruits analysed is shown on the column in black. (**c**) Sequences of the NHEJ DSB repair footprints and their relative frequency are shown. The CRISPR-Cas target sequence from the *PSY1* is shown on the top. The *PSY1* start codon is shown in red and the PAM in blue. The top pie represents an average of illumina Hiseq reads from 22

different F_1 plants of the cross of *yellow flesh* e^{3756} 35S:Cas9 × *bicolor*^{cc383} u6-26:Ps#1-sgRNA. The low pie represents an average of ilummina Hiseq reads from two plants of control F_1 population (*yellow flesh* e^{3756} × *bicolor*^{cc383} F_1 plants with no CRISPR-Cas). (**d**) Inverse PCR scheme for identification of recombinant DNA fragments (details in Supplementary Fig. 4). (1) DNA from separate leaves was digested with ApaI(A) and HindIII(H) and then blunted. (2) Each sample was self-ligated, and (3) amplified by two different primer sets (green and yellow). Blue- *Bicolor*; red- *Yellow flesh*; Dashed blue line- *Bicolor* deletion, *- *Yellow flesh* mutation. (**e**) Ratio of parental (P) versus recombinant (R) types (obtained from panel C) in individual plants. Plants 1–15- F_1 plants of the cross of *yellow flesh* e^{3756} 35S:Cas9 × *bicolor*^{cc383} u6-26:Ps#1-sgRNA; Plant 16- synthetic crossover control; Plants17–18-Yellow flesh × Bicolor (Cas9-) F_1 plants.

with a premature stop codon in *PSY1* leading to a yellow fruit phenotype39. The *bicolor*^{cc383} allele (also known as *yellow flesh Bicolor, locus* r^{Bi}) is a mutant with a 3.7 Kb deletion in the promoter of *PSY1* leading to a yellow–red moulted fruit phenotype (Fig. 1a). In order to monitor the CRISPR-Cas-induced mutations throughout plant development, starting from fertilization, we produced transgenic *yellow flesh* e^{3756} lines, expressing 35S:Cas9 and transgenic *bicolor*^{cc383} lines expressing a *PSY1* single guide RNA (u6-26:Ps#1-sgRNA). This u6-26:Ps#1-sgRNA was designed to induce a DNA DSB between the *bicolor*^{cc383} and *yellow flesh* e^{3756} mutations, on both alleles (Fig. 1a). It is located 1,086 bp downstream from the deletion in *bicolor*^{cc383} and 556 bp upstream from the *yellow flesh* e^{3756} mutation. A cross between *yellow flesh* e^{3756} 35S:Cas9 and *bicolor*^{cc383} u6-26:Ps#1-sgRNA is expected to yield F_1 plants with the dominant *bicolor*^{cc383} fruit phenotype. The same is expected for control plants that do not express either 35S:Cas9 or u6-26:Ps#1-sgRNA. Deviations from this phenotype in plants expressing both Cas9 and Ps#1-sgRNA are expected due to the induction of a DSB on one or both alleles followed by DNA repair. A NHEJ repair of the *bicolor*^{cc383} allele, or of both alleles, should yield a yellow fruit phenotype (sectors or whole fruit). The outcome of DSB repair by HR based mechanisms (inter-homologues crossover or non-crossover events), could be a red fruit in case of an HR event that occurred early in development, or fruits with red spots or sectors in a yellow or bicolour background in case of late events (Fig. 1a, Supplementary Fig. 1).

Upon DSB induction, a population of 50 *yellow flesh* e^{3756} 35S:Cas9 × *bicolor*^{cc383} u6-26:Ps#1-sgRNA F_1 plants, gave bicolour, yellow and yellow with red spots fruits. The distribution of fruit phenotypes varied when we used *yellow flesh* e^{3756} 35S:Cas9 transgenic lines originating from independent transformation events (as shown in Fig. 1b). As expected, in the absence of DSB induction, the control population of six *yellow flesh* e^{3756} × *bicolor*^{cc383} F_1 plants, showed only bicolour fruits (Fig. 1b). One of the advantages of our fruit colour assay is its ability to predict the inheritance of repair products in the next generation. Indeed, we expected that most of the F_2 seeds extracted from totally yellow fruits would give rise to a germinally transmitted mutation. To confirm that yellow fruits are indicative of NHEJ germinal events, we grew F_2 plants derived from yellow fruits. Using allele-specific PCR amplification of the *yellow flesh* e^{3756} and *bicolor*^{cc383} alleles and sequencing of the PCR products, we showed that in all cases tested seeds from yellow fruits yielded seeds carrying a germinally transmitted mutation at the DSB site of the *bicolor*^{cc383} allele (Supplementary Fig. 2). Some of the progeny from yellow fruits also showed mutation in the *yellow flesh* e^{3756} allele (Supplementary Fig. 2). Although we found many yellow fruits with small red sectors (Fig. 1b), we did not detect any fully red fruit among the F_1 plants. In addition, we grew a population of 400 F_2 plants derived from fruits with red spots, suggestive of somatic HR, but we did not detect any fully red fruit that would indicate germinally transmitted HR events.

CRISPR DNA DSB repair via both somatic NHEJ and HR

In order to identify, characterize and quantify somatic NHEJ events in F_1, we used 22 F_1 plants of *yellow flesh* e^{3756} 35S:Cas9 × *bicolor*^{cc383} u6-26:Ps#1-sgRNA and 2 F_1 plants of *yellow flesh* e^{3756} × *bicolor*^{cc383} as control. We collected four leaves from different branches of the plants. Then, we extracted their DNA, amplified the region flanking the induced DSB of both alleles by PCR and sequenced the resulting products using high-throughput sequencing Illumina HiSeq 2500 platform. Out of 250,000–850,000 reads per plant (PCR sample), an average of 88% of the reads of *yellow flesh* e^{3756} 35S:Cas9 × *bicolor*^{cc383}u6-26:Ps#1-sgRNA plants contained a mutation at the CRISPR DSB site, while only 2% of the illumina reads of *yellow flesh* e^{3756} × *bicolor*^{cc383}plants deviated from the WT sequence, presumably due to PCR and sequencing errors (Fig. 1c). The high rate of CRISPR-Cas DSB induction in the system, lead to a broad spectrum of mutations as a result of different NHEJ repair events. In addition, we found that some NHEJ signatures, such as the 4 bp CTTG deletion, were preferred over others at this locus (Fig. 1c, Supplementary Fig. 3).

For the measurement of HR repair, we designed an inverse PCR method that allowed the sequencing of the two allele-specific mutations which are 1.6 Kb apart (*yellow flesh* e^{3756} and *bicolor*^{cc383}), enabled to distinguish parental from recombinant molecules (Fig. 1d, Supplementary Fig. 4) and minimized the formation of false positive PCR products. The same DNA samples used for the somatic NHEJ sequencing (Fig. 1c) were used for the inverse PCR. In addition, we cloned two synthetic positive controls (recombinant-like clones) that were also treated by the same inverse PCR method. The products of the inverse PCR from each reaction (as shown in Fig. 1d) were sequenced by Illumina HiSeq 2500 paired-end sequencing. In this assay we got 5,000–50,000 reads per plant. The negative controls of *yellow flesh* e^{3756} × *bicolor*^{cc383} only showed the parental alleles in the absence of DSB induction, while the positive synthetic control showed the recombinant alleles (Fig. 1e). Most F_1 plants of *yellow flesh* e^{3756} 35S:Cas9 × *bicolor*^{cc383}u6-26:Ps#1-sgRNA showed only the parental alleles but some of them showed one of the recombinant alleles, suggesting somatic HR based repair.

Allele-specific DSB induction

The above cross between *yellow flesh* e^{3756} × *bicolor*^{cc383} did not provide enough SNPs to analyse in detail the HR repair products. Moreover, it did not enable to perform an allele-specific break, which is needed to perform a precise experiment where it is possible to distinguish between the broken chromosome and the repair template. Therefore, we designed a new DSB repair assay that provides several SNPs around the break site as well as in distal regions, through the use of *Solanum pimpinellifolium*^{LA1578}, a wild tomato accession with small red fruits and a sequenced genome showing the presence and location of multiple SNPs compared to *Solanum lycopersicum*, the edible tomato. In order to ensure allele-specific break, we mutated an allele in the *S. lycopersicum* M82 cultivar background, that is immune to u6-26:Ps#2-sgRNA. For that purpose, we transformed the red fruits cv. M82 with 35S:Cas9 and u6-26:Ps#2-sgRNA. Then, we selected for yellow fruits in T_0 and grew their T_1 seeds. From this T_1 population, we isolated a homozygote plant with an adenine insertion (+A) at the CRISPR-Cas9 DSB site and crossed it with the wild tomato accession. In this assay, the *S. pimpinellifolium*^{LA1578} is the only target for DNA DSB due to the +A insertion, in the M82 *psy1* allele, that disrupts the protospacer adjacent motif (PAM) and prevents Cas9 cleavage. The +A mutation of M82 allele is recessive and therefore F_1 plants are expected to have small red fruits. DSB repair in *PSY1* by NHEJ, or HR (crossover or non-crossover) leads to yellow fruits or red fruits with yellow sectors, depending on the developmental fruit stage when the repair occurred. NHEJ repair events are expected to leave small indels at the DSB site, while crossover and non-crossover events can be identified by the difference in SNPs patterns on both sides of the DNA DSB (Fig. 2a).

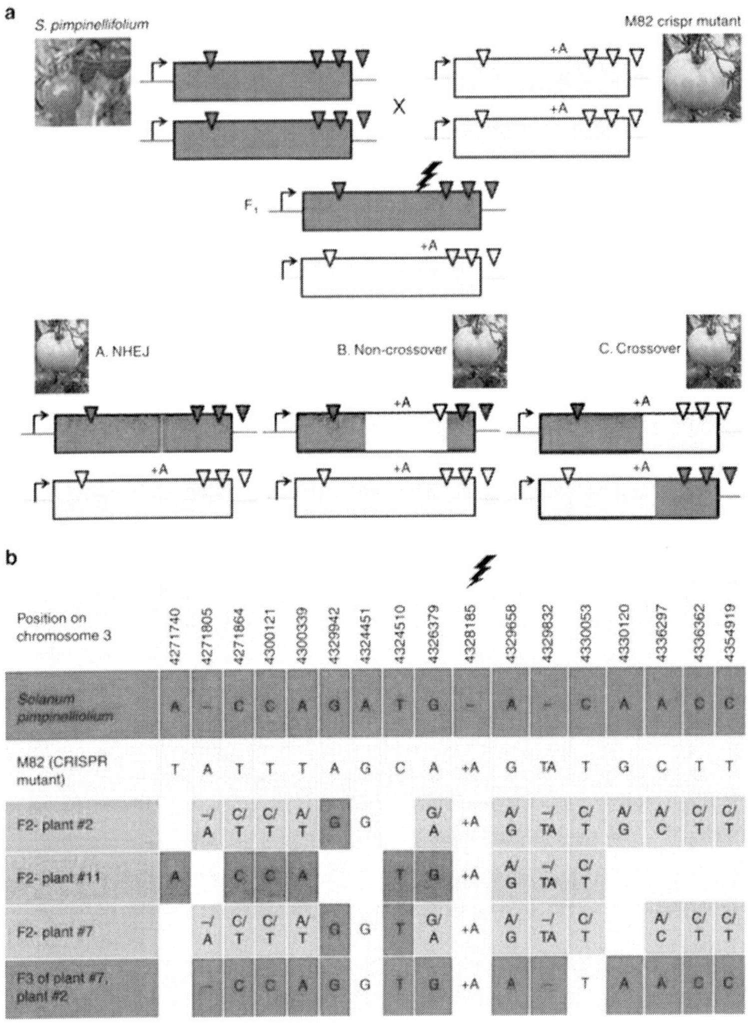

Figure 2: Tomato SNPs assay for crossover and non-crossover events. (a) An homozygote M82 CRISPR mutant (+A, +A) expressing 35S:Cas9 and u6-26:Ps#2-sgRNA was crossed with *S. pimpinellifolium*^{LA1578}. The F_1 is expected to give red fruits without DNA DSB and yellow fruit in case that the break was repaired by NHEJ, non-crossover or crossover. The SNPs pattern is allowing differentiating between repair mechanisms. Triangles are for SNPs; lightning mark the DSB site; blue line is for NHEJ indels. (**b**) Analysis of DNA DSB flanking markers in F_2 and F_3 plants. Red- homozygote for *S. pimpinellifolium*^{LA1578} SNPs; yellow- homozygote for M82 SNPs (including the +A CRISPR-Cas9 mutant); orange- heterozygote; empty cells are for missing data; lightning- DSB site.

For the analysis of somatic DSB repair, we PCR-amplified and sequenced by Illumina HiSeq 2500 paired-end sequencing the DSB DNA area from leaves of both parents (M82 35S:Cas9, U6-26:gRNA, +A homozygote and *S. pimpinellifolium*^{LA1578}) and from five F_1 plants (Supplementary Fig. 5). This sequencing yield was 600,000–900,000 reads per plant. We show that the M82 *psy1*$^{+A}$ allele was immune to DSB induction, with virtually no DSB footprints in the M82 (M82 35S:Cas9 and u6-26:Ps#2-sgRNA, +A homozygote) parent, supporting the designed allele-specificity of the gRNA (Supplementary Fig. 5). In addition, at least 50% of the reads gave the +A insertion while the *S. pimpinellifolium* allele was mutated (red colour in the pie chart of Supplementary Fig. 5). We found that only 7–18% of the F_1 plants reads were WT and the rest gave various indel patterns (Supplementary Fig. 5). To estimate the rate of germinal events, we documented the fruit colours on different branches and sequenced the fruit pericarp tissue by Illumina (Supplementary Fig. 6). With this assay, the fully yellow fruits might contain seeds that are germinal events of repair via NHEJ or HR (Fig. 2a). Moreover, crossover or non-crossover events should give +A, +A homozygote plant as the repair for template is the M82 *psy1*$^{+A}$ allele (Fig. 2a). In one of the F_1 Plants, we found yellow fruits that showed high +A, +A content by illumina and Sanger sequencing (Supplementary Figs 6 and 7). We grew the F_2 progeny of this F_1 plant and sequenced them by the Sanger method. The sequencing revealed F_2plants with SNPs patterns corresponding to germinal HR events (Fig. 2b). Plants #2 and #7 look like clear cases of non-crossover, both with conversion tracts of at least 5 Kb. These plants originate from different F_1 fruits on the same F_1 plant and therefore might correspond to the same recombination event. Plant #11 looks like a case of crossover (Fig. 2b), however, the analysis of flanking markers (Indels and SNPs), more than 20 kb away from both sides of the DSB site in plant #11 could not be performed due to plant death, therefore, we refer to this case as a putative crossover. To identify homozygote gene conversion products, and to better characterize the borders of the conversion tract, we sequenced F_3 plants from the progeny of plant F_2 #7. One of the F_3progeny of F_2 #7 (Shown at the bottom of Fig. 2b) is an homozygote product of gene conversion repair with a confirmed conversion tract of 5–6 kb length. Interestingly, the conversion tract seems interrupted with SNPs from *S. pimpinellifolium* within M82 (see discussion for possible mechanisms).

Quantification of the rate of allele-dependent repair

Thanks to the system we developed of allele-specific DSB induction, we were able to test allele-dependent repair, a signature of HR between homologues. Induction of DNA DSB on the *S. pimpinellifolium*^{LA1578}allele showed the +A signature, similar to the M82 *psy1*$^{+A}$ allele, at the broken site in many of the fruits and leaves we sequenced (Supplementary Figs 6 and 7). This excess in +A repair might be due to preferred NHEJ repair pattern or to allele-dependent repair mediated by HR. To distinguish between these two possibilities, we grew several plants of the M82 cultivar, all of them offspring of the same 35S:Cas9 u6-26:Ps#2-sgRNA. In this population, 22 plants were initially homozygote for the M82-WT allele of *PSY1*, while 14 plants were initially heterozygote M82-WT *PSY1*/ M82 $psy1^{+A}$. We grew the plants to the age of 4 weeks and collected four leaves from each of them. Then, we amplified DNA around the DSB by PCR and sequenced the PCR products with the Illumina HiSeq 2500 platform. For each plant, we calculated the percentage of each indel out of the total number of reads. In case that the +A mutation occurs independently in each chromosome, there should be twice more reads with new +A mutations in the WT (which has two potential targets) than in the heterozygote where only one target is available (Fig. 3a). It is possible though that in the WT a first mutation, repaired by NHEJ would generate the +A footprint. Following this event, a second mutation at the homologous target could be repaired by HR using the newly generated +A allele in the homologue. We chose to ignore this possibility, and to assume that all repair events in the WT occur by NHEJ in order to establish the most stringent threshold as control in our estimation of inter-homologue HR repair. To measure the expected allele-independent +A NHEJ footprint per chromosome, we used the 22 plants of the WT homozygote and calculated the per cent of +A reads divided by two to obtain the value of the occurrence of the +A mutation per allele. The following equation was used: $\text{Expected}=(\%(+\text{A reads})_{T=4\text{ weeks (wt, wt)}})/2$. The occurrence of a new +A mutation in the WT allele, when the second allele contains the +A mutation (in M82-WT *PSY1*/M82 $psy1^{+A}$ heterozygote plants) is calculated by taking the percentage of +A reads in the M82-WT *PSY1*/M82 $psy1^{+A}$ plants and deducing 50% (the initial per cent of reads originating from the M82 $psy1^{+A}$ allele). We used the following equation for the observed rate of +A mutation in heterozygote M82-WT *PSY1*/M82 $psy1^{+A}$ plants: $\text{Observed}=\%(+\text{A reads})_{T=4\text{ weeks, (wt, +A)}}-50\%$. When we compared the expected to the observed +A footprint, we found a significantly higher than expected rate of novel +A mutations in the heterozygote population (P=0.009, Wilcoxon rank sum test, $n_{\text{homozygote}}=22$, $n_{\text{hetrozygote}}=14$). Considering that the two populations are isogenic, this suggests that the repair at the site of DSB is dependent on the sequence of its homologous allele (Fig. 3b). The average allele-independent frequency of a +A footprint was 4% per allele in the M82-WT (under the assumption that all repair in the WT occurred by NHEJ), while the average +A footprint frequency in the M82-WT *PSY1* allele in M82-WT *PSY1*/M82 $psy1^{+A}$ heterozygote was 18% (Fig. 3b). This suggests that 18%–4%=~14% of the DSB repair events are allele-dependent and the rest occurs via NHEJ in an allele-independent manner. Note that probably, some of the repair events in the WT occurred via inter-homologue HR, therefore the rate of allele-dependent repair would be even higher than the 14% estimate.

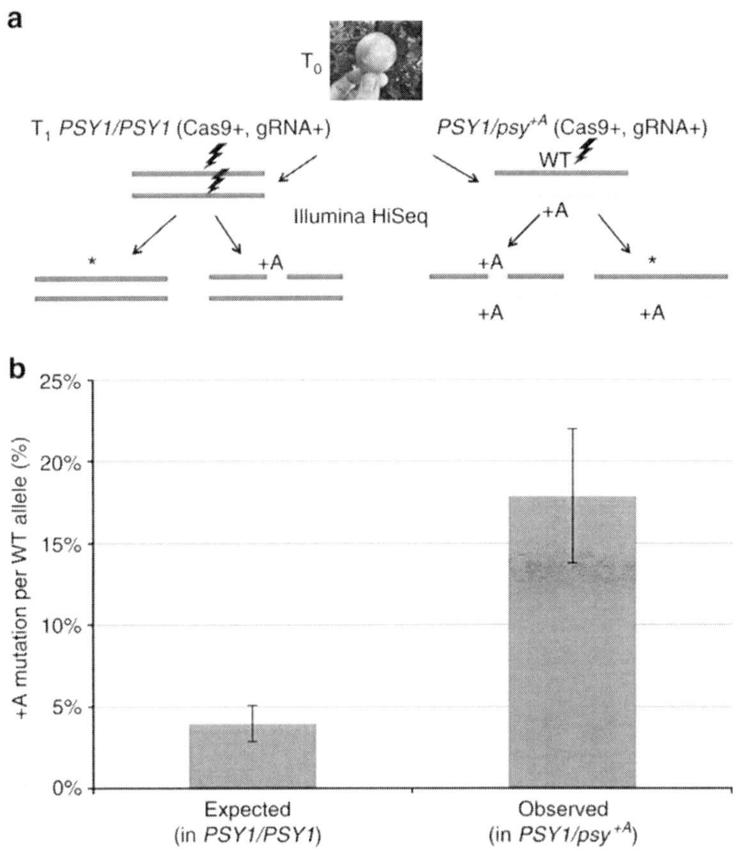

Figure 3. Quantification of allele-dependent repair. (**a**) Two plant populations were grown, both in the M82 background: one homozygote for *PSY1/PSY1* and the other heterozygote for the *PSY1/psy^{+A}* genotype. Progeny of these plants could give a +A SNP at the site of the break (lightning) or any other mutation (*). DNA was extracted from leaves of 4-week-old plants of both populations and sequenced by illumina. In the *PSY1/PSY1* plants, both alleles can be targeted, while in the *PSY1/psy^{+A}* plants, only the WT *PSY1* allele is targeted. (**b**) The per cent of +A mutation per WT allele in *PSY1/PSY1* plants served as expected value for allele-independent +A mutation. It was calculated by the following equations: Expected=$(\%(+A\ reads)_{T=4\ weeks\ (wt,\ wt)})/2$. To estimate the observed occurrence of the +A mutation when the second allele has a +A mutation (in M82-WT *PSY1*/ M82 *psy1^{+A}* heterozygote plants) as shown in panel A, we used the equation: Observed=$\%(+A\ reads)_{T=4weeks,\ (wt,\ +A)}-50\%$. The bars correspond to the s.e. for 22 *PSY1/PSY1* plants and 14 *PSY1/psy^{+A}* plants. The difference between the means was significant (P=0.009, Wilcoxon rank sum test, $n_{homozygote}=22$, $n_{hetrozygote}=14$).

Earlier studies on somatic DSB-induced HR repair were done mostly with transgenic assays for intrachromosomal recombination[27,28] or inter chromatids unequal crossover[29] or inter-homologues recombination[40]. The assays developed here were carried in an endogenous genomic context where the repair template origin could be tracked on the homologous chromosome. We present a series of evidence showing that targeted DSBs can be repaired via somatic HR using an homologous chromosome as the template. In addition, we demonstrated that some of these repair events can be transmitted germinally to the next generations. In one set of crosses, we showed that we can recover the WT allele through intragenic recombination between two defective *psy1* parental alleles (*bicolor*^{cc383} and *yellow flesh e*^{3756}), an event seen as red spots (Fig. 1a) and characterized through sequence analysis (Fig. 1e). In this cross, we did not recover fruits that were fully red, that would correspond to early germinal events. This might be due to the genomic context of the large deletion in the bicolor allele, or alternatively the 'cured' recombinant WT allele underwent a second round of NHEJ during development (the target site was not destroyed during HR), that would generate a loss of function (yellow) allele via NHEJ. Considering the high efficiency of NHEJ, this is a plausible scenario. In addition, in an assay of allele-specific DSB induction in a *S. pimpinellifolium* × *S. lycopersicum* F_1 hybrid, we found three cases of HR-dependent repair that were germinally transmitted to the F_2 and F_3 generations. Two cases corresponded to non-crossover events with conversion tracts of 5–6 Kb (Fig. 2b). The third case (F_2 plant #11) is a germinal HR event that might be either a crossover event or a non-crossover event—this could not be demonstrated due to the plant death. Finally, trying to quantify the ratio of HR versus NHEJ, we designed a sgRNA for allele-specific DSB induction in the *S. lycopersicum* background. This experimental set-up enabled to measure an excess of repair footprints originating from the homologous allele compared to expectation, suggesting that out of all the detectable DSB repair events at least 14% are allele-dependent and the rest is non-homologous.

It is interesting to compare HR-mediated repair in somatic versus meiotic cells. Overall little is known on inter-homologues recombination in somatic tissues probably owing to the low frequency of such events, to the lack of phenotypic markers and to the difficulty to retrieve germinal events. We could not detect any red sector in the absence of DSBs, and the presence of intragenic recombinant molecules was null or negligible. This is consistent with earlier studies in tobacco showing that the occurrence of somatic HR is very low in the absence of DSB induction for both reciprocal and non-reciprocal HR events[41,42]. The low rates of somatic HR between homologous chromosomes might be indicative of bottlenecks such as absence of the HR machinery found in meiosis that controls homologues pairing, synaptonemal complex formation and so on. Our results, showing a relatively high rate of HR-repair based on both case-studies and on quantitative assessments, indicate that DSBs are a major bottleneck, inducing somatic HR from 0 (in the absence of breaks) to ~14% per allele (allele-dependent repair measured in Fig. 3b) and that DSB-induced HR between homologues can occur in the absence of the meiotic HR machinery.

The rate of HR DSBs repair that we report here (of 14% per allele or more) is apparently higher than that reported during meiosis. Indeed, only a small fraction of meiotic breaks (~3–5%) evolves into crossovers[12,43] and a similar fraction is repaired as non-crossover[43,44]. However, meiotic non-crossover rates might be underestimated because such events may go unnoticed if they happen between sister chromatids or between homologous chromosomes with a short conversion tract that does not include SNPs as suggested by Drouaud *et al.*[44]. Moreover, the somatic HR DSB repair we report here could have happened during several cellular divisions during development (starting from the zygote) until error-prone repair destroys the target, while in meiotic cells, there is only one cell division when Spo11 induces hundreds of breaks in the genome. Therefore, it is premature to make meaningful statements on the difference in HR rates between somatic and meiotic recombination. Likewise, we presented several evidence on the occurrence of DSB-mediated HR repair, however in most assays we could not distinguish between crossover versus non-crossover repair mechanisms. The analysis of three germinal events in a polymorphic background enabled to perform this distinction but the sample (of two conversions and one putative crossover) is too small to draw conclusions.

We found a rather long (~5 kb) conversion tract where homozygous *S. pimpinellifolium* SNPs at positions 4324510, 4326379, 4329658 and 4329832 appeared within the M82 homozygous patch at the boundaries of the tract (Fig. 2b, F3 plant). It might be these SNPs were part of a large heteroduplex region in Holliday junctions and that mismatch repair at these positions generated a discontinuous conversion tract[45]. Alternatively, it might be that such chimeric tract was formed via the Synthesis-dependent-Strand-Annealing pathway and that multiple template switches happened during DNA synthesis. Such examples of multiple template switches were shown in plants[21,30] and in other species[46]. Earlier reports on meiotic conversion tracts showed mean values of 552 bp ref. 44. This length difference might reflect a difference between species (tomato versus *Arabidopsis*) or between meiotic and somatic cells or it might be due to the experimental set-up (Spo11 versus Cas9 DSB induction).

Finally, we showed that it is possible to induce germinally transmitted targeted HR between endogenous homologous chromosomes in somatic cells. It remains to be shown if and how efficiently targeted HR can occur during meiosis.

Our results suggest that custom-designed nucleases, such as CRISPR-Cas, can be used for precise reshuffling of chromosomal segments between homologous chromosomes in somatic cells. For example, it might be possible to transfer a disease resistance gene from a wild relative to the crop, without a long process of backcrossing which not only takes several generations in order to achieve isogenic lines, but also drags large segments of undesirable DNA flanking the desirable gene. In other words, we showed that somatic HR can be used for allelic replacement. Another potential application of DSB-induced somatic HR is 'targeted crossover', that is, the reciprocal exchange of large chromosomal segments at a precise site. Crossing over between homologous chromosomes during meiosis is the engine of breeding in sexually reproducing crops, but it takes place at uncontrolled and unpredictable sites in the genome. Obtaining rare recombination events—such as shown here for intra-allelic crossover—is a major hindrance in most breeding programs. We propose to exploit targeted somatic HR to segregate between undesirable genetic linkages. Somatic crossover does occur in plants[27,28,29,30,31,32], and can even reach high levels in some mutants[47] suggesting that the inter-homologue crossover machinery is available in somatic tissues and that targeted crossover is feasible. Interestingly, even-though the meiotic crossover machinery has been optimized during evolution, the targeted induction of a given DSB during meiosis would have to compete with the hundreds of naturally occurring other breaks as a substrate for crossover and counter to intuition, might turn out to be less efficient than somatic HR for targeted crossover induction.

METHODS

Plant material

All tomato plants were grown in greenhouse conditions with temperature ranging between 18 and 25 °C. The tomato (*S. lycopersicum*) mutant line of *yellow flesh* e^{3756}, *Bicolor*^{cc383}, M82 and *S. pimpinellifolium*^{LA1578} were kindly provided by the labs of Prof. Joseph Hirschberg and Prof. Daniel Zamir at the Hebrew university of Jerusalem (http://zamir.sgn.cornell.edu/mutants/)[39].

Plasmids and plant transformation

We used the 35S:Cas9 and u6-26:sgRNA constructs[48] and the u6-26 promoter[49,50] which was synthesized by IDT (Integretad DNA Technologies) based on *Arabidopsis thaliana* sequence. The primers used for construction of ps#1-sgRNA and ps#2f targets are specified in Supplementary Table 1. All tomato lines were transformed by *Agrobacterium tumefaciens* GV3101 using cotyledon transformation51.

Inverse PCR for HR detection

DNA samples for the Inverse PCR assay were extracted using a DNA purification kit (MACHEREY-NAGEL). For each plant 300 ng of DNA from leaves sample or control plasmid were treated separately: first, they were incubated over night with 10 × FD buffer, ApaLI (ThermoFisher scientific) and HindIII-HF (New England BioLabs). After 20 min of 80 °C inactivation, 150 ng of the digested fragments were blunted with T4 DNA polymerase (New England BioLabs) for 2 hs at room temperature. The T4 DNA polymerase was inactivated at 75 °C for 10 min and the linear DNA was self-ligated with Quick T4 DNA ligase (New England BioLabs) for 30 min at room temperature. Control plasmids were diluted 1:10,000 with DDW and mixed together for mimicking 'heterozygosity'. Then all samples were amplified by 18 cycles of PCR with Phusion High-Fidelity DNA polymerase (New England BioLabs) (for primers see primers list). The primers for this assay were designed for allele-specific amplification. Samples were pooled and sequenced by high-throughput sequencing.

For the cloning of the synthetic crossover-control plasmids, two PCR fragments were amplified from *yellow flesh* e^{3756} and *Bicolor*^{cc383} DNA samples using Phusion. High-Fidelity DNA polymerase (New England BioLabs) (for primers sequence see primers list) and then cloned using the GoldenBraid cloning system52 (https://gbcloning.upv.es/). First, each of the four amplicons was cloned into the pUPD plasmid. Then pUPD2 plasmid with 'ps' segment from *yellow flesh* e^{3756} was cloned with pUPD2 plasmid with 'y1' segment from *Bicolor*^{cc383} into pDGB3_alpha1 plasmid. In parallel, pUPD2 plasmid with 'ps' segment from *Bicolor*^{cc383}was cloned with pUPD2 plasmid with 'y1' segment from *yellow flesh* e^{3756}into pDGB3_alpha1 plasmid. These two 'synthetic allele' plasmids were pooled together and were subjected to inverse PCR and sequencing. For the specific sequence of synthetic plasmids see R1 and R2 in Supplementary Note 1.

DNA amplification and sequencing

DNA samples for high-throughput sequencing were amplified using Phusion. High-Fidelity DNA polymerase (New England BioLabs) and 18 PCR cycles (for specific primers of each experiment see primers list). Libraries were prepared as Blecher-Gonen *et al.*53. High-throughput Sequencing was performed at the G-INCPM unit at the Weizmann Institute of Science with the Illumina HiSeq 2500 platform for 2 × 125 paired-end reads.

DNA samples for Sanger sequencing were amplified using REDTaq (SIGMA-ALDRICH) with 35 PCR cycles (for primers see Supplementary Table 1). Then cleaned with Exonuclease I and Shrimp Alkaline Phosphatase (rSAP) (New England BioLabs). Sequencing was performed at the Biological services unit at the Weizmann Institute of Science with Applied Biosystems 3730 DNA Analyzer.

Data availability

Sequencing data associated with this study has been deposited in the NCBI Sequence Read Archive under BioProject accession PRJNA380032. The authors declare that all data supporting the findings of this study are available within the manuscript and its supplementary files or are available from the corresponding author upon request.

ADDITIONAL INFORMATION

How to cite this article: Hayut, S. F. *et al.* Targeted recombination between homologous chromosomes for precise breeding in tomato. *Nat. Commun.* **8,** 15605

Publisher's note: Springer Nature remains neutral with regard to jurisdictional claims in published maps and institutional affiliations.

REFERENCES

1. Baudat, F., Imai, Y. & de Massy, B. Meiotic recombination in mammals: localization and regulation. Nat. Rev. Genet. 14, 794–806 (2013).
2. Li, J., Hsia, A.-P. & Schnable, P. S. Recent advances in plant recombination. Curr. Opin. Plant Biol. 10, 131–135 (2007).
3. Lieberman-Lazarovich, M. & Levy, A. A. Homologous recombination in plants: an antireview. Methods Mol. Biol. 701, 51–65 (2011).
4. Mehta, A. & Haber, J. E. Sources of DNA double-strand breaks and models of recombinational DNA repair. Cold Spring Harb. Perspect. Biol. 6, a016428 (2014).
5. Puchta, H. & Fauser, F. in Advances in New Technology for Targeted Modification of Plant Genomes 1–20Springer, New York NY, USA (2015).
6. Amiard, S., Gallego, M. E. & White, C. I. Signaling of double strand breaks and deprotected telomeres in Arabidopsis. Front. Plant Sci. 4, 405 (2013).
7. Bleuyard, J.-Y., Gallego, M. E. & White, C. I. Recent advances in understanding of the DNA double-strand break repair machinery of plants. DNA Repair 5, 1–12 (2006).
8. Decottignies, A. Alternative end-joining mechanisms: a historical perspective. Front. Genet. 4, 48 (2013).
9. Gorbunova, V. & Levy, A. How plants make ends meet: DNA double-strand break repair. Trends Plant Sci. 4, 263–269 (1999).

10. Puchta, H. The repair of double-strand breaks in plants: mechanisms and consequences for genome evolution. J. Exp. Bot. 56, 1–14 (2005).
11. Edlinger, B. & Schlögelhofer, P. Have a break: determinants of meiotic DNA double strand break (DSB) formation and processing in plants. J. Exp. Bot. 62, 1545–1563 (2011).
12. Mercier, R., Mézard, C., Jenczewski, E., Macaisne, N. & Grelon, M. The molecular biology of meiosis in plants. Annu. Rev. Plant Biol. 66, 297–327 (2015).
13. Puchta, H., Dujon, B. & Hohn, B. Two different but related mechanisms are used in plants for the repair of genomic double-strand breaks by homologous recombination. Proc. Natl Acad. Sci. USA 93, 5055–5060 (1996).
14. Shalev, G. & Levy, A. A. The maize transposable element Ac induces recombination between the donor site and an homologous ectopic sequence. Genetics 146, 1143–1151 (1997).
15. Marton, I. et al. Nontransgenic genome modification in plant cells. Plant Physiol. 154, 1079–1087 (2010).
16. Cermak, T. et al. Efficient design and assembly of custom TALEN and other TAL effector-based constructs for DNA targeting. Nucleic Acids Res. 39, e82 (2011).
17. Shan, Q., Wang, Y., Li, J. & Gao, C. Genome editing in rice and wheat using the CRISPR/Cas system. Nat. Protoc. 9, 2395–2410 (2014).
18. Fauser, F., Schiml, S. & Puchta, H. Both CRISPR/Cas-based nucleases and nickases can be used efficiently for genome engineering in Arabidopsis thaliana. Plant J. 79, 348–359 (2014).
19. Schiml, S., Fauser, F. & Puchta, H. The CRISPR/Cas system can be used as nuclease for in planta gene targeting and as paired nickases for directed mutagenesis in Arabidopsis resulting in heritable progeny. Plant J. 80, 1139–1150 (2014).
20. Butler, N. M., Baltes, N. J., Voytas, D. F. & Douches, D. S. Geminivirus-mediated genome editing in potato (Solanum tuberosum L.) using sequence-specific nucleases. Front. Plant Sci. 7, 1045 (2016).
21. Gorbunova, V. & Levy, A. A. Non-homologous DNA end joining in plant cells is associated with deletions and filler DNA insertions. Nucleic Acids Res. 25, 4650–4657 (1997).
22. Lloyd, A. H. et al. Single molecule PCR reveals similar patterns of non-homologous DSB repair in tobacco and arabidopsis. PLoS ONE 7, e32255 (2012).
23. Charbonnel, C., Allain, E., Gallego, M. E. & White, C. I. Kinetic analysis of DNA double-strand break repair pathways in Arabidopsis. DNA Repair 10, 611–619 (2011).
24. Ma, X., Zhu, Q., Chen, Y. & Liu, Y.-G. CRISPR/Cas9 platforms for genome editing in plants: developments and applications. Mol. Plant 9, 961–974 (2016).
25. Brooks, C., Nekrasov, V., Lippman, Z. B. & Van Eck, J. Efficient gene editing in tomato in the first generation using the clustered regularly interspaced short palindromic repeats/CRISPR-associated9 system. Plant Physiol. 166, 1292–1297 (2014).
26. Čermák, T. et al. High-frequency, precise modification of the tomato genome. Genome Biol. 16, 232 (2015).
27. Puchta, H., Swoboda, P. & Hohn, B. Induction of intrachromosomal homologous recombination in whole plants. Plant J. 7, 203–210 (1995).
28. Orel, N., Kyryk, A. & Puchta, H. Different pathways of homologous recombination are used for the repair of double-strand breaks within tandemly arranged sequences in the plant genome. Plant J. 35, 604–612 (2003).
29. Molinier, J., Ries, G., Bonhoeffer, S. & Hohn, B. Interchromatid and interhomolog recombination in Arabidopsis thaliana. Plant Cell 16, 342–352 (2004).
30. Rubin, E. & Levy, A. A. Abortive gap repair: underlying mechanism for Ds element formation. Mol. Cell. Biol. 17, 6294–6302 (1997).
31. Puchta, H. et al. Double-strand break-induced recombination between ectopic homologous sequences in somatic plant cells. Genetics 152, 1173–1181 (1999).
32. Athma, P. & Peterson, T. Ac induces homologous recombination at the maize P locus. Genetics 128, 163–173 (1991).

33. Lowe, B., Mathern, J. & Hake, S. Active mutator elements suppress the knotted phenotype and increase recombination at the Kn1-O tandem duplication. Genetics 132, 813–822 (1992).
34. Dooner, H. K. & Martínez-Férez, I. M. Recombination occurs uniformly within the bronze gene, a meiotic recombination hotspot in the maize genome. Plant Cell 9, 1633–1646 (1997).
35. Rosu, S. et al. Robust crossover assurance and regulated interhomolog access maintain meiotic crossover number. Science 334, 1286–1289 (2011).
36. Preston, C. R. & Engels, W. R. P-element-induced male recombination and gene conversion in Drosophila. Genetics 144, 1611–1622 (1996).
37. Bétermier, M. et al. Is Non-homologous end-joining really an inherently error-prone process? PLoS Genet. 10, e1004086 (2014).
38. Vu, G. T. H. et al. Repair of site-specific dna double-strand breaks in barley occurs via diverse pathways primarily involving the sister chromatid. Plant Cell 26, 2156–2167 (2014).
39. Kachanovsky, D. E., Filler, S., Isaacson, T. & Hirschberg, J. Epistasis in tomato color mutations involves regulation of phytoene synthase 1 expression by cis-carotenoids. Proc. Natl Acad. Sci. USA 109, 19021–19026 (2012).
40. Gisler, B., Salomon, S. & Puchta, H. The role of double-strand break-induced allelic homologous recombination in somatic plant cells. Plant J. 32, 277–284 (2002).
41. Dulieu, H. L. Somatic variations on a yellow mutant in Nicotiana tabacum L. (a1+/a1a2+/a2) I. Non-reciprocal genetic events occurring in leaf cells. Mutat. Res. Mol. Mech. Mutagen. 25, 289–304 (1974).
42. Carlson, P. S. Mitotic crossing-over in a higher plant. Genet. Res. 24, 109–112 (2016).
43. Mercier, R. et al. Two meiotic crossover classes cohabit in arabidopsis: one is dependent on MER3,whereas the other one is not. Curr. Biol. 15, 692–701 (2005).
44. Drouaud, J. et al. Contrasted patterns of crossover and non-crossover at Arabidopsis thaliana meiotic recombination hotspots. PLoS Genet. 9, e1003922 (2013).
45. Sweetser, D. B., Hough, H., Whelden, J. F., Arbuckle, M. & Nickoloff, J. A. Fine-resolution mapping of spontaneous and double-strand break-induced gene conversion tracts in saccharomyces cerevisiae reveals reversible mitotic conversion polarity. Mol. Cell Biol. 14, 3863–3875 (1994).
46. Tsaponina, O. & Haber, J. E. Frequent Interchromosomal template switches during gene conversion in S. cerevisiae. Mol. Cell 55, 615–625 (2014).
47. Gorbunova, V. et al. A new hyperrecombinogenic mutant of Nicotiana tabacum. Plant J. 24, 601–611 (2000).
48. Johnson, R. A., Gurevich, V., Filler, S., Samach, A. & Levy, A. A. Comparative assessments of CRISPR-Cas nucleases' cleavage efficiency in planta. Plant Mol. Biol. 87, 143–156 (2015).
49. Waibel, F. & Filipowicz, W. U6 snRNA genes of arabidopsis are transcribed by RNA polymerase III but contain the same two upstream promoter elements as RNA polymerase II-transcribed U-snRNA genes. Nucleic Acids Res. 18, 3451–3458 (1990).
50. Li, X., Jiang, D., Yong, K. & Zhang, D. Varied Transcriptional Efficiencies of Multiple Arabidopsis U6 Small Nuclear RNA Genes. J. Integr. Plant Biol. 49, 222–229 (2007).
51. McCormick, S. Plant Tissue Culture Manual 311–319Springer (1991).
52. Sarrion-Perdigones, A. et al. Goldenbraid: an iterative cloning system for standardized assembly of reusable genetic modules. PLoS ONE 6, e21622 (2011).
53. Blecher-Gonen, R. et al. High-throughput chromatin immunoprecipitation for genome-wide mapping of in vivo protein-DNA interactions and epigenomic states. Nat. Protoc. 8, 539–554 (2013).

Index